Essentials in Soil Science

Essentials in Soil Science

Edited by **Henry Wang**

Callisto Reference

New York

Published by Callisto Reference,
106 Park Avenue, Suite 200,
New York, NY 10016, USA
www.callistoreference.com

Essentials in Soil Science
Edited by Henry Wang

International Standard Book Number: 978-1-63239-326-5 (Hardback)

Printed in China.

Contents

Preface

I am honored to present to you this unique book which encompasses the most up-to-date data in the field. I was extremely pleased to get this opportunity of editing the work of experts from across the globe. I have also written papers in this field and researched the various aspects revolving around the progress of the discipline. I have tried to unify my knowledge along with that of stalwarts from every corner of the world, to produce a text which not only benefits the readers but also facilitates the growth of the field.

The dependence of humans on soil and curiosity regarding it is leading to the investigation of alterations within soil processes. The diversity and dynamics of soil are creating scope for novel discoveries and insights, which enable us to comprehend the variation in soil processes. As a result, this allows us to take the required steps for the protection of soil and subsequently promoting soil health. The aim of this book is to give an updated account on the present state of knowledge about recent evaluations and practices in soil science. Moreover, it provides a descriptive assessment about the impact of residue/waste application on soil properties and the mechanism of plant development and adaptation. Interesting examples of simulation employing distinct models dealing with ecosystem respiration, carbon sequestration, soil landscape etc. have been illustrated. This book also contains information regarding analysis of areal data and geostatistics by employing various evaluation techniques. Current progresses in analytical methods used for obtaining answers to the diverse physical mechanisms, biological and chemical processes in soil have also been elucidated in this book.

Finally, I would like to thank all the contributing authors for their valuable time and contributions. This book would not have been possible without their efforts. I would also like to thank my friends and family for their constant support.

Editor

1

Phosphorus: Chemism and Interactions

E. Saljnikov and D. Cakmak
Institute of Soil Science
Serbia

1. Introduction

Phosphorus (P) is a limiting nutrient for terrestrial biological productivity. The availability of "new" P in ecosystems is restricted by the rate of release of this element during soil weathering. Soil P exists in inorganic and organic forms. Inorganic P forms are associated with amorphous and crystalline sesquioxides, and calcareous compounds. Organic P forms include the relatively labile phospholipids and fulvic acids and the more resistant humic acids. The intergrades and dynamic transformations between the forms occur continuously to maintain the equilibrium conditions (Hedley et al., 1982). Its low concentration and solubility (< 0.01 mg P kg^{-1}) in soils, however, make it a critical nutrient limiting plant growth.

In natural soil ecosystems the main source of inorganic phosphorus is rocks where the primary minerals are of the greatest importance, where in turn calcium phosphates are the most important (e.g. apatite) (Fig.1). In weathered soils, leaching of Ca ion results in formation of Al-phosphate (e.g. berilinite) and Fe-phosphate (e.g. stregnite); the complete list were given by Lindsay, (1979) and Lindsay, et al., (1989). By the definition, these minerals are characterized with three-dimensional atomic structure. As far as phosphorus concentration in the soils is concerned, it can be very low from 50 mg kg^{-1} and high up to 3500 mg kg^{-1} (Foth & Ellis, 1997; Frossard et al., 1995). Application of phosphorus from mineral phosphate results mainly in formation of amorphous compounds with soluble Al, Ca and Fe where the phosphorus is adsorbed on the surfaces of clay minerals, Fe and Al oxyhydroxides or carbonates and physically occluded by secondary minerals.

In natural ecosystems, P availability is controlled by sorption, desorption, and precipitation of P released during weathering and dissolution of rocks and minerals of low solubility (Sharpley, 2000). Due to high fixation and immobilization of phosphorus in the soil, the agriculturists apply high amounts of p-fertilizer, what results in greater input of P into soil that plant uptake. Application of phosphates can maintain or improve crop yields, but it can also cause changes in the chemical and physical properties of the soil, both directly and indirectly (Hera & Mihaila, 1981; Acton & Gregorich, 1995; Aref & Wander, 1998; Belay et al., 2002). The cumulative accumulation of available P in agricultural soils may partially saturate the capacity of a soil for P sorption, with resulting increase of P leaching into the subsoil layers (Ruban, 1999), or may sometimes reach depth more than 90 cm (Chang et al., 1991), suggesting that erosion, rather than leaching, would cause a threat to water bodies (Zhou and Zhu, 2003). Such a process of leaching is especially effective in soils of Stagnosol type with clear E horizon due to their lower adsorption capacity, with relatively shallow ground waters (Tyler, 2004; Väänänen et al., 2008). Fertilization with mineral P in the inorganic pools explains 96 % of the variation in the level of available phosphorus (Beck & Sanchez, 1994).

Great number of researchers studied many aspects of fate and behavior of applied P. However, the chemical processes and following plant availability of soil P remains a big challenge for scientists since it offer wide spectrum of uncertainties, and contradictions. This Chapter is devoted to explanation of mechanisms and distribution of different forms of phosphorus, its transformation and dynamics in the soil based upon the 40-years of experience in phosphate field application.

2. Materials and methods

2.1 Site description

The investigation was conducted at the Varna experimental station, 44°41'38'' and 19°39'10'' (near Belgrade, Serbia), where a wide range of different fertilization treatments has been undertaken since 1968. The soil type is Stagnosol (WRB, 2006), a loam textured Pseudogley developed on Pliocene loam and clay materials under aquic conditions at 109 m above sea level. Average annual precipitation of the site is 705 mm, and the average temperature is 12°C. The mineralogical composition of the studied soil was as follows: illite (50–70%), vermiculite (10–30%), and other clay minerals (kaolinite, chlorite, feldspar, quartz and amphibolites) (Aleksandrovic et al., 1965). The cultivated crops were winter wheat (*Triticum aestivum* L.) and corn (*Zea maize* L.), with crop residues removed. The soil cultivation was performed by a standard plowing to 25 cm depth.

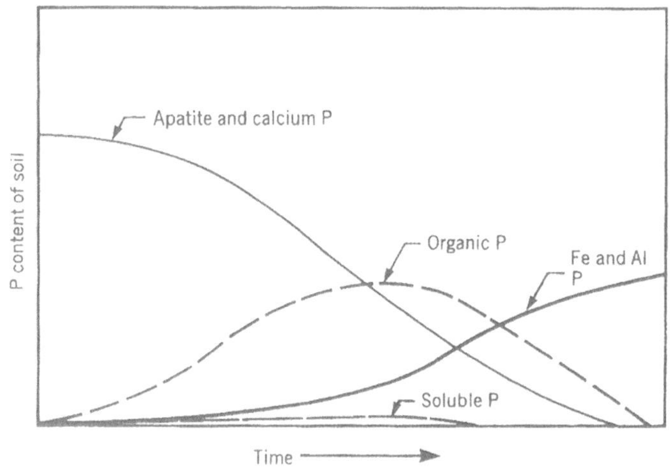

Fig. 1. Relative distribution of the major forms of soil P vs time of the soil development (Foth & Ellis, 1997)

Three rates (26, 39, and 52 kg P ha−1) of monoammonium phosphate (MAP) fertilizer (NH4H2PO4) were applied in combination with a consistent rate of N (urea) (60 kg ha−1) and K (KCl) (50 kg ha−1) for 40 yr of the experiment. The fertilized treatments were compared to the control, with no fertilizer applied. The experiment was arranged as a randomized block design, with each treatment randomized in three blocks for a total of 12 plots. Each plot was 5 by 11 m. Composite samples of five soil subsamples were taken from each plot in the three field replications from two depths: surface (0–30 cm) and subsurface (30–60 cm) layers in spring 2008.

2.2 Methods

Soil pH was determined with a glass electrode pH meter in a 1:2.5 water solution. Soil total C and N were measured with an elemental CNS analyzer, Vario model EL III (ELEMENTAR Analysasysteme GmbH, Hanau, Germany; Nelson & Sommers, 1996) Available P and K were determined by the Al-method of Egner–Riehm (Enger & Riehm, 1958), where 0.1 M ammonium lactate (pH = 3.7) was used as an extract. After the extraction, P was determined by spectrophotometry after color development with ammonium molybdate and SnCl2 (Enger & Riehm, 1958). Soil Ca and Mg were extracted by ammonium acetate and determined with a SensAA Dual atomic adsorption spectrophotometer (GBC Scientific Equipment Pty Ltd, Victoria, Australia; Wright & Stuczynski, 1996). Determination of CEC was performed by the steam distillation method after the treatment with 1 M ammonium acetate (Sumner & Miller, 1996). Exchangeable Al was determined by the titration method by Sokolov: the extraction with 1 M KCl (1:2.5) followed shaking for 1 h and titration with 0.01 M NaOH (Jakovljević et al., 1985).

Trace elements were determined with an ICAP 6300 ICP optical emission spectrometer (Thermo Electron Corporation, Cambridge, UK), after the soils were digested with concentrated HNO3 for extraction of hot acid-extractable forms, and by diethylenetriaminepentaacetic acid (DTPA) for extractable elements (Soltanpour et al., 1996). The F content was determined by ion-selective electrode, after the soil had been fused with NaOH for total F and after extraction with water for available F (Frankenberger et al., 1996). Soil granulometric composition was performed using the pipette method (Day, 1965). All chemical analyses were performed in two analytical replications. The Merck standards were used for the determinations on ICP and SensAA Dual. Before the determination of samples, three blank samples were read, which allowed correcting the results. For the verification of the results, a referent soil sample was determined for all the studied elements (NCS ZC 73005 soil, CNAC for Iron and Steel, Beijing).

Statistical analyses were performed with the SPSS version 16 software. The effects of treatments on all the variables were tested by ANOVA. Statistical differences between the treatments were determined using the *t* test (95%) Pearson for Fisher's LSD. The significance of their correlations was analyzed via the Pearson correlation matrix (SPSS, 2007).

3. Sequential analyses

The ways phosphates bound to soil particle are the parts of a puzzle whose solution can give many answers concerning their availability to plants and the possible leaching down the soil profile. The best way to obtain the answers is isolation of separate fractions of phosphorus in soil using series of solvents of different strength, i.e. sequential analysis.

3.1 Sequential extraction procedure

One of the most common phase divisions for sequential extraction was used in the experiment:

Exchangeable or *sorptive* (*adsorptive and ion exchange*) *phase.* This phase is used to estimate the maximum quantity of sorbed ions that geological material can release, without visible decomposition of some mineral phases. Neutral solutions of salts (NH_4OAc, $MgCl_2$, $CaCl_2$, $BaCl_2$, KNO_3, etc.) are usually used for this extraction phase. Their concentrations (and ion forces) must be high enough to initiate the most complete ion exchange and desorption from all substrates.

"Easily reducible" phase. Weak reduction means (for example, hydroxylamine) are used for selective reduction (solvent) of manganese oxyhydrates, but they are also used for the most mobile fraction of amorphous iron oxides. All the present microelements co-precipitated in these oxides to be detected in the solution.

"Moderate reducible" phase. For the amorphous iron oxides and the more crystalline manganese oxides, some stronger reduction means are used - oxalic acid, sodium dithionite and similar methods.

Organic-sulfide phase. Distinguishing of organic and sulfide metal fractions in a geological material is one of the disadvantages of sequential extraction. This problem is still unresolved. Pure nitric acid or its combinations with other acids is very effective, but it leads to a noticeable decomposition of silicate material. The use of hydrogen peroxide is acceptable at higher temperatures and low pH (about 2).

Residual phase. This is the least interesting phase of the ecochemical aspect as it includes silicate and oxide materials as well as incorporated metal ions, i.e. in natural conditions this fraction cannot be mobilized from geological material. Concentrated mineral acids and their mixtures are usually used for decomposition of this crystal matrix (Petrovic, et al., 2009) (Tab. 1).

Procedure	Step 1†	Step 2	Step 3	Step 4	Step 5
Petrovic et al. (2009)	1 M CH_3COONH_4 Exchangeable	0.1 M $NH_2OH \cdot HCl$ Bound to carbonates and easily reducible	0.2 M $(NH_4)_2C_2O_4$ and 0.2 M $H_2C_2O_4$ Moderately reducible	30 % H_2O_2 + 3.2 M CH_3COONH_4 Organic-sulphide	6M HCl Residual

Table 1. Reagent in the sequential extraction procedure used to study substrates metals (Fe, Al, Mn and Ca) and P

3.2 Chang and Jackson sequential analysis

3.2.1 Fraction of soil P extracted by 1M solution of ammonium-chloride (water-soluble P)

This fraction of phosphorus is closely linked with the dynamics of P bounding in soil. Such bounding of phosphorous ions can be characterized as an initial reaction. And it represents a non-specific adsorption and ligand exchange on mineral edges or by amorphous oxides and carbonates. This fraction is bound to Mn isolated in step 2 from Table 1 (Mn II) (r=0.994, **), which indicated its sorption on hydrated oxides of manganese. Due to specific further bounding of phosphorus, this fraction is very low in quantity (less than 1 % of the total mineral phosphorus) in acidic soils such as Stagnosol. However, due to application of mineral fertilizers and accumulation of phosphorus (Jaakola, et al. 1997) the processes of saturation of free spaces for adsorption of P in the soil (Vu et al, 2010) result in its significant increase, by about 6 times compared to the control plots in the studied experiment. Considering the low movement of phosphate ions along adepth the soil profile, which is slower than in the processes of bounding them into less soluble forms, such increase of P ions concentration is expressed distinct in the surface soil layer 0-30 cm.

Passage of this form of phosphorus into bounded-to-aluminum phosphorus is a process characteristic for acidic soils. The reverse process is also possible (correlation coefficients 0.974 and 0.780 for 0-30 and 30-60 cm, respectively). A very strong correlation between water-soluble P and available P (0.945 and 0.715 for 0-30 and 30-60 cm, respectively) proved that this form of phosphorus was available for plants.

3.2.2 Fraction of soil P extracted by 0.5M NH_4F solution (Al bound P)

Such an isolated fraction of phosphorus is a characteristic for monodent and bident bounds (Tisdale et al., 1993). Consequently, these compounds are very labile and are described as pseudo sorption (Van der Zee et al., 1987; Van der Zee et al., 1988). In acid mineral soils, such as Podzols, P is mostly retained by Al and Fe oxides by the ligand exchange mechanism where the OH^- or H_2O groups of sesquioxides surfaces were are displaced by dihydrogenphosphate anions (Simard et al., 1995). In certain soils, this bound is not strictly confined to Al but can bound bind to Si, as well (Manojlovic, et al. 2007). However, in the studied Stagnosol, the strong correlation of Al-P with Al extracted in the step 2 (Al II) could be attributed to carbonates and alumosilicates (r=0.998**). Somewhat increased content of this fraction versus to the available phosphorus indicates that not only modettant bounds are involved (Fig. 3).

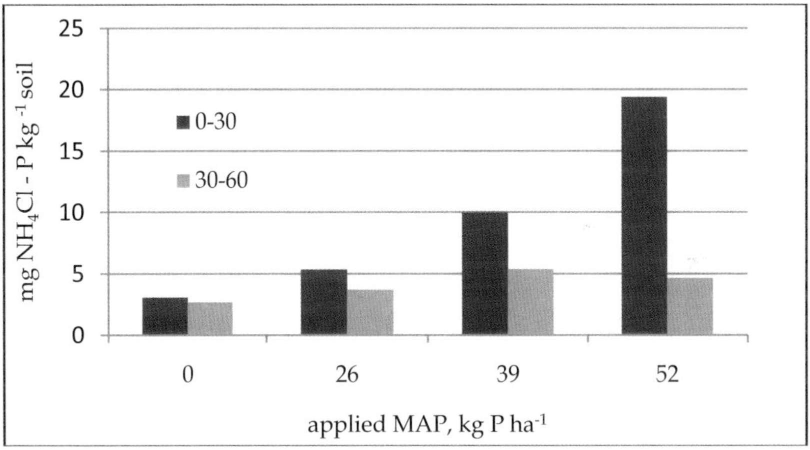

Fig. 2. Content of ammonium-chloride extractable P upon 40-years of phosphate application (water-soluble P)

It is obvious that this fraction of soil phosphorus is the most important for plants since there is a high correlation between the Al bound P and the available forms of phosphorus extracted by the Al-method (r=987**). Also, application of mineral phosphorus influences this fraction of soil P the most with the recorded increase of its content from 16.08% in the control to 34.51% in the treatment with 52 kg P ha^{-1}. This fraction of P is responsible for migration of phosphorus along the soil depth, which is confirmed by a significant correlation between the values at two depths (0.876**), as well as for the replenishment of the pool of other fractions of soil P.

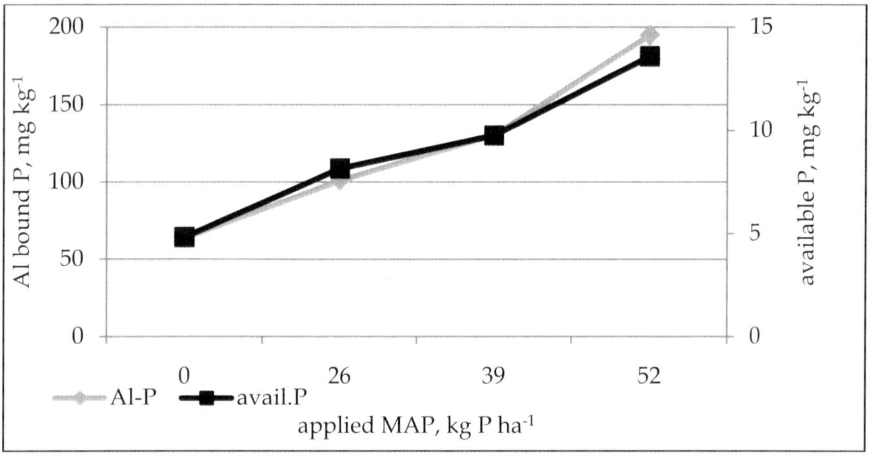

Fig. 3. Content of Al bound P and available P upon 40-years of application different amounts of phosphate fertilizer

3.2.3 Fraction of soil P extracted by M NaOH solution (Fe bound P)

Fraction of soil P isolated by such a strong reagent may have a high content in soil and mainly is greater than that bound to Al (Manojlovic et al., 2007, Mustapha et al., 2007), ranging between few hundred mg per kilogram. From the chemistry viewpoint, such bounded P is characteristic for the slow-flowing processes involving formation of covalent Fe-P or Al-P bonds on Fe and Al oxide surfaces (Willett et al. 1988) which can be an additional source of available P (Beck & Sanchez, 1994). However, the strength of this bound is quite high. Therefore, its availability is limited. That determines the absence of correlation between the mentioned fraction and the available P. However, in the layers of soils such as Stagnosol, due to constant wetting and alteration of oxidative - reductive conditions, the content of this form of P can be as low as less than 1 mg per kilogram due to passage into other forms (reducible and occluded). Its movement along the soil depth is also limited as indicated by the absence of correlation between the values at different depths. Sequential analysis didn't show marked correlation with the fraction of Fe, but the correlation with DTPA-extractable Fe was recorded (0.665*)

3.2.4 Fraction of soil P extracted by M Na dithionite, Na citrate solution (Reducible P)

In contrary to the previous types of bounding of P in soil, this fraction is characterized by the bounds within the particle. Such bounding results in the process of occlusion where the phosphate is adsorbed to the surface of Al hydroxide and is bound by poorly crystalline Fe oxides from that occluded in the crystalline Fe oxides (Delgado & Scalenghe, 2008). In this structure, the phosphate binds the Al- with Fe^{3+}hydroxide so the surface of Al phosphate particle is enveloped by a Fe^{3+}hydroxide skin. Such adsorbed phosphates are only indirectly available to plants. Thus, in the conditions determined by reduction processes the reduction of iron Fe^{2+} and the breakage of the earlier formed bounds take place, which makes this form of P available for plants. Although this fraction is small compared to other fractions of P in soils, under the oxidized conditions Fe-P represents the dominant fraction (Manojlovic et al, 2007, Mustapha et al., 2007). But under the conditions of soil undergoing alterations of wet and dry regimes with high content of available Fe (Cakmak, D. et al., 2010) the reducible-P can be of significant concentration up to 30% from the total mineral P. The high correlation found between the reducible and Al bound P (r=0.97**) indicates the indirect availability of this form of P under the alteration reduced conditions in Stagnosol.

3.2.5 Fraction of soil P extracted by M NaOH solution (Occluded P)

Chang & Jackson (1957) noticed that during the sequential extraction some soils, rich in Fe oxides, contain significant amounts of Fe-phosphate occluded within the oxide, which cannot be extracted by sodium dithionite and sodium citrate. This occluded phosphate can be extracted by repeated alkali solution. The P tied in this manner might be increased in quantity by constant addition of mineral phosphate fertilizer where its total content ranges between few milligrams to tens of milligram per kg soil; i.e. in small amounts from 1% to about 10% from the total mineral P (Manojlovic et al, 2007, Mustapha et al., 2007).

Under the alteration of reduced and oxidized conditions that predominate in Stagnosol, this form is chemically tied to the reducible form of P, especially, in the upper soil layer (r=0.890**). Also, within such soil particle, Al-phosphate can be present, which can be available under certain conditions within Fe-oxide (Fig.4). Its migration along the soil depth is limited and is of very low mechanic intensity. Absence of correlation with DTPA-extractable Fe indicates the un-availability of Fe in such compounds.

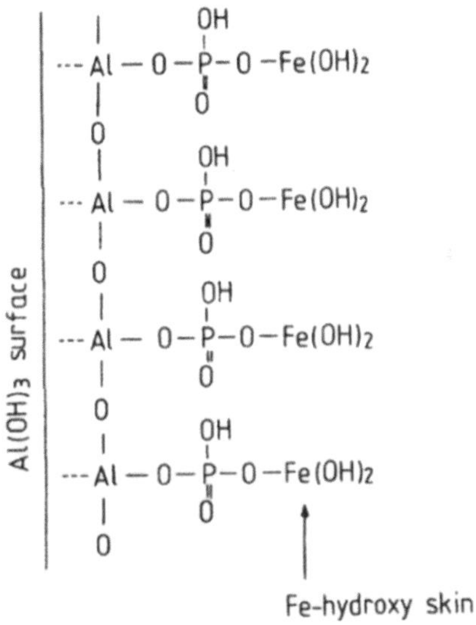

Fig. 4. Fe-hydroxy skin covering the phosphate adsorbed to Al oxide/hydroxide (Mengel & Kirkby, 2001)

3.2.6 Fraction of soil P extracted by M H_2SO_4 solution (Ca bound P)

In neutral to calcareous soil the concentration of phosphate in soil solution is governed mainly by the formation and dissolution of calcium phosphates. This in turn depends on soil pH and Ca^{2+} concentration in soil solution. The lower are Ca/P ratios in the Ca phosphates -

the higher is their solubility in water. However, in acidic soils in spite of significant amount of this fraction (up to 40% from total mineral forms) the Ca-P was widely dispersed in soil minerals and it was weakly changeable. This is supported by the absence of significant correlation between exchangeable Ca and Ca- bound P. Therefore, in such soils fertilization does not result in significant changes in the content of Ca-bound P (Hartikainen, 1989). However, relative increase of this fraction is possible in the subsurface soil layer due to leaching and accumulation of Ca ions, under acidic conditions, in deeper layers where it is transformed into non-labile phosphate fractions. This process of phosphate ageing is especially rapid in acid soils with a high adsorption capacity. The start of this process can also be detected by the negative significant correlation between water-soluble P and DTPA-extractable Ca (-0.590*).

4. BCR analysis

BCR method according to SMT standard protocol was applied for determination of P in soil (Ruban et al., 1999). BCR is a non-specific extraction procedure for determination of phosphorus in freshwater sediments, developed in the frame of the European Program, Standards, measurements and Testing (SMT) is used for certification campaign for a reference material. The SMT protocol was extended to soil material because bioavaialable forms of phosphorus are important not only for analysis of sediments but also of soils. The detailed description of the SMT protocol is given at Ruban et al., (2001). The Certified Reference material CRM 684 (River Sediment Extractable Phosphorus, from Po River, Italy) was analyzed to verify the results of analyses.

4.1 BCR procedure

Among numerous extraction schemes used, the procedures widely adopted are those developed by Williams (1976), Hieltjes & Lijklem (1980), Rutenberg (1992) and Galterman (1993). Together with the cited procedures, in the literature can be found other sequential schemes (Delaney, et al., 1997; Kleberg et al., 2000). Due to the large number of the existing procedures for extraction of phosphorus and due to the impossibility of comparison of the results from different source samples obtained via different laboratory procedures, the Program of Standardization of extraction scheme was initiated (Ruban et al, 1999).

To overcome the incompatibility of results a Program of the European Commission (SMT: Standards, Measurements and Testing, earlier BCR) initiated the project for selection of sequential extraction procedures for determination of forms of phosphorus in lake sediments. This project targeted the homogenization of extraction schemes to investigate a selected scheme in inter-laboratory investigations that includes expert European laboratories and to certificate traces in the referent material of the sediment. Four methods (Tab. 2) were chosen for testing. These methods were applied in the inter-laboratory investigations and served as the base for development of homogenized procedure for phosphorus extraction from lake sediments. For the determination of phosphorus, all the laboratories used spectrophotometry (Murphy & Riley, 1962). Along with this method, some laboratories used ion-chromatography such as ICP-AES. However, using the laters proved to be unsuitable, since ICP-AES allows determination of total phosphorus, while the method of ion-chromatography determines only orthophosphate.

Procedure	Advantages	Disadvantages
Williams	Simple and practical	Partial resorption of phosphorus extracted with NaOH onto CaCO3
Hieltjes - Lijklem	Simple and practical	Dissolution of small amounts of Fe-P and Al-P with NH_4Cl; hydrolysis of organic phosphor is unavoidable; no responce to bioavailability
Golterman	Extraction of specific compounds; allows extraction of organic P; supports information about bioavailability of fraction	Not practical; NTA i EDTA contaminate the determination of phosphorus; complicated preparation of the solution; in some sediments the extraction must be undertaken more times to obtain valid results
Rutenberg	Possibility of differentiation of different types of apatite; no distribution of phosphorus on the residual particle surfaces during extraction	Very long; extraction with butanol is very difficult

Table 2. Advantages and disadvantages of the methods for sequential extraction of soil P fractions (adapted from Ruban et al., 1999)

Based on the results of inter-laboratory investigations, the modified Williams scheme named SMT (1998) was proposed. The SMT scheme allows definition of the following forms of phosphorus: NaOH-extractable phosphorus (NaOH-P); phosphorus bound to oxides and hydroxides of Fe, Al and Mn; (Fe-Al-Mn-P) HCl-extractable phosphorus (HCl-P); phosphorus bound to Ca (Ca-P); organic phosphorus (Org-P), inorganic phosphorus (IP); concentrated HCl phosphorus, (conc. HCl-P); total phosphorus(TP) (Ruban et al., 1999).

After the compilation of the extraction scheme, certification of extractable phosphorus from the referent material CRM 684, sampled from Po River (Italy) near the Gorina city, was undertaked within the Project (SEPHOS - sequential extraction of phosphorus from fresh water sediment) (Ruban et al., 2001). After acceptance and certification of SMT procedure, valid comparison of worldwide results became possible, what is of great importance for understanding of biogeochemical cycle of phosphorus in actual systems.

In Tab. 3 there are phases of phosphorus and the reagents used for their extraction in both procedures: modified Chang & Jackson and BCR

Procedure	Step 1	Step 2	Step 3	Step 4	Step 5	Step 6
Manojlovic (2007) modified from Chang & Jackson (1957)	1M NH_4Cl Water soluble P	0.5 M NH_4F Al bound P	0.1 M NaOH Fe bound P	0.3 M Na dithionite, Na citrate Reducible P	0.1 M NaOH Occluded P	0.25 M H_2SO_4 Ca bound P
BCR (Ruban et al., 2001)	1M NaOH NaOH - P	3.5 M HCl HCl - P	1M NaOH Inorganic P	1M HCl + calcination Organic P	3.5 M HCl + calcination Conc HCl - P	

Table 3. Phases of soil phosphorus and the reagents used for their sequential extraction

4.1.1 Fraction of soil P extracted by 1M NaOH (Al-Fe-Mn bound P)

This fraction of phosphorus bound to oxides and hydroxides of aluminum, iron and manganese, so called oxide-hydroxide fractions was extracted in step 1 (Tab. 3). The associations of P and Fe are often found in sediments, where phosphorus is tied to complex compounds of iron through changes of ligands (Stumm & Morgan, 1981). In soils such as Stagnosol, this fraction of phosphorus is correlated with Ca from the second phase (Ca II, Table 1) (r=0.951, **) that indicates binding of phosphorus with carbonate fraction. Such bounds are quite labile, what is supported by a strong correlation between Al-Fe-Mn bound P with available P with corresponding coefficients 0.782 and 0.813 (for 0-30 and 30-60 cm, respectively). The correlation coefficients between HCl –P and available P were 0.939 and 0.902 (for 0-30 and 30-60 cm, respectively). The good correlation of the Al-Fe-Mn fraction with Al-P fraction from Chang and Jackson method (r=0.775, **) explains the leaching of Al-Fe-Mn bound P fractions from the surface to the subsurface soil layer (r=.901, **) (Fig. 5).

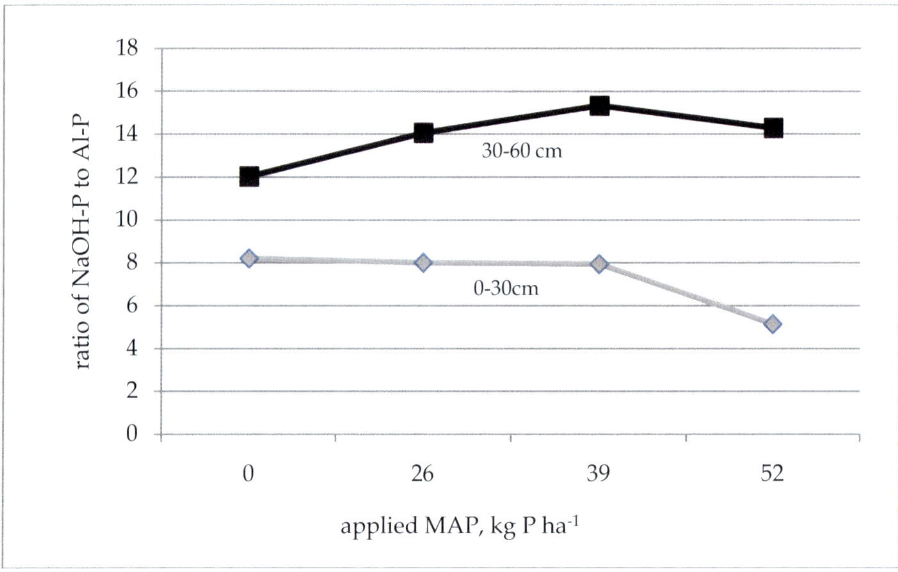

Fig. 5. Ratio of NaOH bound P (BCR) to Al bound P (Chang&Jackson) in the 0-30 cm and 30-60 cm soil intervals

4.1.2 Fraction of soil P extracted by 3.5 M HCl (Ca bound P)

This fraction generally represents the phosphorus in apatite (Williams et al., 1976; William et al., 1980) and phosphorus bound to Ca (Golterman, 1996, 1982) and was extracted in step 2 (Tab. 3). The adsorption of phosphorus in calcium carbonate is one of the mechanisms of formation of calcium phosphate in sediments. However, apart from the Fe bound phosphorus, formation of $CaPO_4$ is possible by sedimentation. The behavior and distribution of this fraction is similar to the fraction described above (HCL-P/CaII, r=0965,**). Its distribution along the soil profile is also analogous to the above described fraction of phosphorus (Fig. 6).

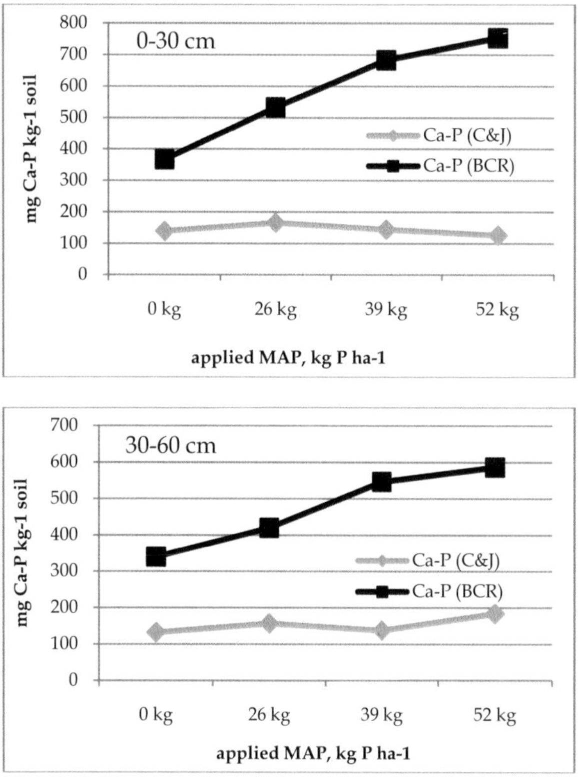

Fig. 6. Distribution of Ca bound P in the two applied methods: Chang&Jackson and BCR in the two soil intervals upon 40-years application of phosphate fertilizer

4.1.3 Fraction of soil P extracted by 1M NaOH (Inorganic P; IP)

This fraction is supposed to consist of the later two fractions (Al-Fe-Mn bound P and Ca-P) however, due to the different extracting reagent used for their separation, we have isolated this fraction by extraction in step 3 (Tab. 3). The fraction of inorganic phosphorus in our study was highly correlated with Ca II (Tab. 1), what is the consequence of the decrease of bioavailable phosphorus in the soil. Good correlation of IP with Ca and Al extracted in step 3 of the sequential analysis (Tab. 1) highlights the roles of Ca and Al in the fixation of phosphorus in soil. The decrease of the bioavailability of this fraction is supported by a moderately good positive correlation with the occluded phosphorus (r=.619 *).

4.1.4 Fraction of soil P extracted by 1M HCl+calcinations (Organic P; Org.P)

This fraction of phosphorus is an exact fraction with not precisely defined constitution that partially consists of phitite (De Groot & Golterman, 1993). The extraction of the organic P fraction was performed in step 4 (Table 3). Most of the organic soil phosphorus is present in the form of the inositol phosphate ester while the proportion of phospholipids and nucleic acids in soils is small due to the fact the two groups of phosphate esters are quickly dephosphorylated by microbial phosphatases (Flaig, 1966).

The moderately good correlation of the organic P with Fe extracted in step 4 (Table 1) indicates the importance of this phase in bonding of organic phosphorus. In contrary to other phosphorus fractions determined by BCR method the organic P fraction didn't show increases after 40-years of application of phosphate fertilizer on Stagnosol.
The absence of changes in the content of organic phosphorus after the 40-years of application of phosphate fertilizer (Fig.7) obviously was due to its low mobility; the microorganisms easier consumed the applied mineral phosphorus from fertilizer, which resulted in negligible changes of organic phosphorus.

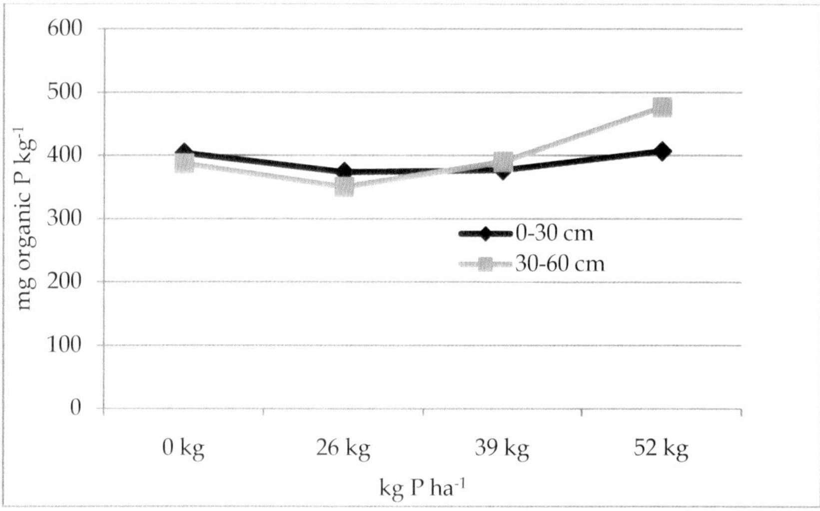

Fig. 7. Response of soil organic phosphorus after 40-years application of phosphate fertilizer

5. Microwave digestion method (Ethos Milestone) using HNO_3, HCl and HF (completely total P; TP)

Long-term application of mineral phosphate fertilizer on Stagnosol not only determined the changes in the content of total P extracted by these extraction solutions: known that more than 90% of phosphorus present in soil as insoluble and fixed. Because in soil the process of bounding of phosphate into unavailable forms takes place constantly, the content of total phosphorus can give information about which amount of added phosphate is tied, i.e. unused. Fertilization increases the amount of total forms of P both on surface and subsurface soil layers. But the ratio of the increase in the content of P between the sum of the entire mineral fraction extracted by Chang&Jackson method and the total P by microwave method is 1:6, which indicates the presence of very clear process of fixation and accumulation of P, i.e. formation of the secondary minerals (Fig. 8).
As mentioned earlier, for such type of soil, often due to mineral fertilization there are no increases in the amount of organic P (Sharpley & Smith 1983, Adeptu & Corey, 1976). The process of immobilization of P in deeper layers is closely linked to the mobility of Al-P, where according to the unpublished data, the correlation with the total P was $r=0.721^{**}$. Therefore, in such soils fertilization by phosphate demands a special caution, i.e. finding the exact ratio between the process of immobilization and the plant demands.

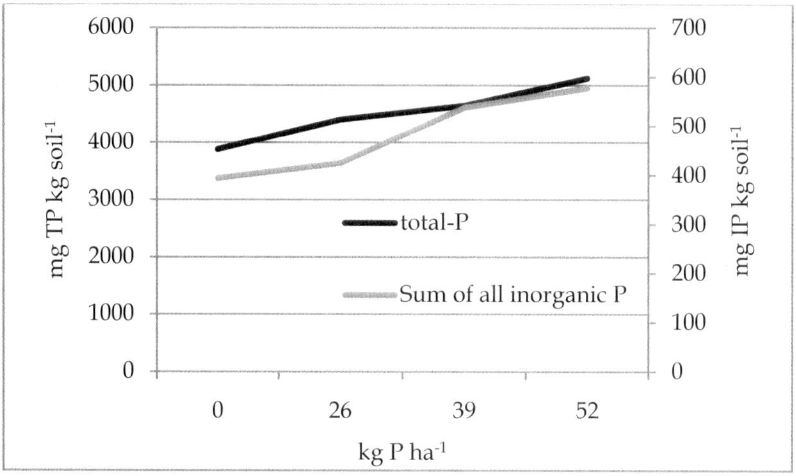

Fig. 8. Changes in the concentrations of total P extracted by microwave digestion and the sum of all inorganic P from Chang&Jackson method upon 40-years application of different rates of MAP

6. Soil agrochemical properties

The change in the basic soil properties and accumulation of micro- and trace elements, some with toxic species, upon the application of phosphate fertilizers are the important factors both for crop yield and for ecological concerns. They can maintain or improve crop yields, but they can also cause changes in the chemical and physical properties of the soil, both directly and indirectly (Hera & Mihaila, 1981; Acton & Gregorich, 1995; Aref & Wander, 1998; Belay et al., 2002). By affecting the basic soil properties (pH, organic C and N, cations, CEC, granulometic composition), phosphate fertilization may influence the solubility of certain elements, such as Al, F, Ca, and Mg (Lindsay, 1979; Kabata-Pendias & Pendias, 2001; Loganathan et al., 2006). On the other hand, mineral phosphate fertilizer could provide an abundance of available phosphorus in soil and increase the efficiency of metal-phosphate mineral formation (Ma et al., 1993; Berti & Cunningham, 1997; Hettiarachhchi et al., 1997; Cooper et al., 1998). Metal-phosphate minerals were shown to control metal solubility in soil suspension when available P was added (Santillian-Medrano & Jurinak, 1975) by inducing the formation of heavy metal phosphate precipitates (Cotter-Howells & Capron, 1996). Additionally, raw materials for P fertilizers contain certain amounts of trace elements and microelements, which may be incorporated into mineral fertilizer (Goodroad & Caldwell, 1979; Adriano, 2001; Kabata-Pendias & Pendias, 2001). The effect of fertilization on soil quality can be best evaluated through the use of long-term experiments (Mitchell et al., 1991; Nel et al., 1996). The general soil-chemical properties are given in Table 4.

Only 15% of the applied phosphorus is consumed by plants quickly after addition (Greenwood, 1981), the rest transforms into insoluble forms and non-labile fractions. Consumption of the applied P by plants becomes more and more difficult each year. However, constant application of fertilizer from year to year can replenish the capacity for the adsorption, and consequently the amount of the available phosphate (Stewart & Sharpley 1987; McCollum, 1991; Maroko, et al., 1999).

As would be expected, the content of P in the soil increased significantly ($p < 0,01$) depending on the amount of applied phosphate. So, in the treatment with 52 kg/Pha^{-1} an increase by about 3 times was detected in Stagnosol (Figure 9). In the deeper soil layer due to low mobility of phosphates this increase was only by 2 times as compared to the control treatment (Cakmak, D. et al., 2010).

6.1 Soil physical characteristics

Long-term application of phosphate, besides the direct influence on the content and forms of P in soil, also influences the other soil properties. It can affect the structure of clay minerals where fractions of P can replace Al ions from the tetrahedral structure thus destructing the structure and pulverizing clay minerals (Rajan, 1975). Long-term fertilization with MAP distninguished that phenomenon since the changes in soil texture were due to the amount of added P, i.e. in the experiment a significant increase of clay fraction was detected correspondingly to the rates of fertilizer, which is the result of fragmentation of clay particles (Fig. 9). In spite the mineral fertilization causes decreases in soil CEC (Belay et al., 2002) especially if the amount of organic matter did not change, in our case due to the increased content of clay the slightly significant increase of CEC in the treatments with 39 and 52 kg P ha^{-1} ($p < 0,01$) was recorded. The consequence is increases of the value of sum of base in the second depth under the higher rates of fertilizer ($p < 0.05$).

Treatment	Organic C	Total N	Avail. P†	Exch. Ca^{2+}†	Exch. Mg^{2+}	Mobile. Al	pH	CEC†	Sum of base
g P ha^{-1}	– – % – –		– – – mg 100 g^{-1} – – –					– – cmol kg^{-1} – –	
					0-30 cm				
0	0.99 ± 0.03	0.11 ± 0.001	4.8a‡ ± 0.6	237a ± 9	40.7 ± 4.2	3.7a ± 0.9	4.95a ± 0.0	16.4a ± 0.6	8.67a ± 0.2
26	1.06 ± 0.02	0.12 ± 0.003	8.2b ± 0.4	267b ± 2	43.8 ± 2.3	6.9b ± 1.5	4.80b ± 0.1	18ab ± 0.4	9.07b ± 0.2
39	1.03 ± 0.05	0.12 ± 0.004	9.6.c ± 0.3	251ba ± 7	45.9 ± 1.7	7.1b ± 1.0	4.84b ± 0.0	18.4b ± 0.2	9.34b ± 0.3
52	1.06 ± 0.04	0.12 ± 0.003	13.5d ± 2.8	241ba ± 6	43.8 ± 2.2	7.9c ± 0.9	4.76c ± 0.0	18.8b ± 0.4	8.94b ± 0.3
P	NS§	NS	***	*	NS	*	*	**	*
					30-60 cm				
0	0.81 ± 0.02	0.10 ± 0.000	3.5a ± 0.7	277 ± 3	48.1 ± 0.9	4.07 ± 0.6	5.03 ± 0.0	17.1 ± 0.5	9.87a ± 0.2
26	0.77 ± 0.05	0.09 ± 0.00	5.2b ± 1.6	262 ± 9	49.9 ± 4.4	5.51 ± 1.1	4.95 ± 0.1	16.9 ± 0.6	9.34a ± 0.5
39	0.83 ± 0.09	0.10 ± 0.01	6.1c ± 1.2	264 ± 16	48.2 ± 4.9	4.37 ± 0.1	5.02 ± 0.1	17.5 ± 0.4	10.44b ± 0.4
52	0.74 ± 0.03	0.09 ± 0.00	7.0d ± 1.5	272 ± 10	50.8 ± 1.8	4.68 ± 0.4	5.03 ± 0.0	17.9 ± 0.9	10.7b ± 0.6
P	NS	NS	***	NS	NS	NS	NS	NS	*

* Significant at $P < 0.05$.
** Significant at $P < 0.01$.
*** Significant at $P < 0.001$.
† Avail., available; Exch., exchangeable; CEC, cation exchange capacity.
‡ Within each depth increment, means with the same letter are not significantly different.
§ NS, not significant.

Table 4. Soil chemical characteristics of Stagnosol in 40-years phosphate fertilization experiment

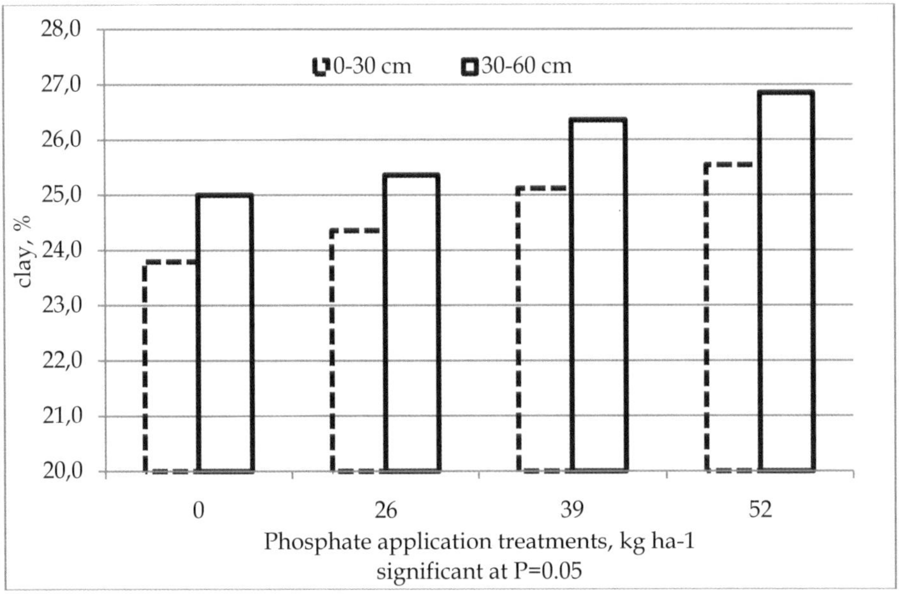

Fig. 9. Amount of clay fraction under long-term application of different rates of phosphate fertilizer in two depths on a Stagnosol soil.

6.2 Soil chemical characteristics

Acidity (pH). Mineral fertilizers can change the soil pH depending on the dominance of alkali or acidic components. Because in MAP the acidic components predominates, since in soil the process of nitrification leads to formation of nitrate ions, the long-term fertilization unavoidably results in acidification (Magdof et al., 1997; Belay et al., 2002; Saleque et al., 2004). Since the amounts of nitrogen components in the applied fertilizer were small and due to the negligible effect of phosphoric components, the decrease in soil pH after 40-years of application of MAP was slight but significant ($p < 0.05$ (Tab. 4). It should be mentioned that the other phosphoric fertilizers without nitrogen components did not have a negative impact on soil pH even after long application.

Available phosphorus (Al-method). Available phosphorus is a fraction of P that is considered available for plants. Chemical extraction is based on solvents, which more or less imitate adsorptive power of plant root. Phosphorus in the first four fractions obtained by the Chang&Jackson method by definition is more or less available to plants but this fact was not confirmed in the 40-year field experiment of application of MAP on Stagnosol. Absence of correlation between the available P and Fe-P is probably due to the low amount of Fe-phosphate found in this experiment (< 1 mg kg^{-1}). Based on the results of coefficient of correlation, the direct correlation between the available forms of P is recorded for water-soluble P and the P bound to Al at both depths (for 0-30 cm is 945** and 987** respectively; for 30-60 cm 715** 888** respectively). On the other hand, reducible phosphate is indirectly correlated with the available P via water-soluble and Al bound P. The bounds of occluded P in the first depth with water-soluble (r=.585*) and in the second depth with water-soluble and Al-P indicates on its possible availability under the extreme conditions of wetting and drying, characteristic for Stagnosol. Except the Ca-bound phosphorus, all the fractions of soil P showed increasing trends accordingly to the applied phosphate fertilizers where the Al-bound phosphorus showed the clearest increasing tendency (Fig. 10).

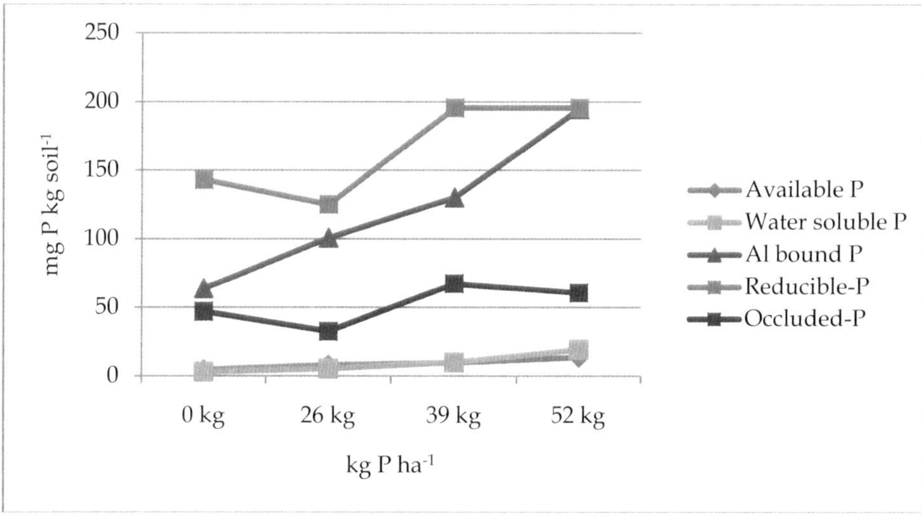

Fig. 10. Concentrations soil P extracted by different methods upon 40-years application of phosphate fertilizer

Mobile Al. One of the indirect consequences of phosphate fertilization that acidifies the soil is the increase in the amount of mobile Al. Soil acidity is caused by the formation of H^+ ions in the soil solution, which is neutralized by Al and Fe oxy/hydroxy complexes (Schwertmann et al., 1987) where the end product of neutralization are species of Al cations such as Al^{3+}, $Al(OH)^{2+}$, $Al(OH)_2^+$. Connection of these two processes is confirmed by a high correlation between soluble Al and decreases in soil pH (r=.897**). It is known that the solubility of Al progressively increases below 5.5 soil pH (H_2O) (Abrahmsen, 1984; McKenzie & Nyborg, 1984; Mrvic et al., 2007). In our experiments, in the 26 kgPha^{-1}, treatment in the subrface soil the amount of mobile Al is doubled versus to control (Tab. 4).

Exchangeable cataions and CEC. A recent research showed that the decrease of exchangeable Ca and Mg with P fertilization is caused by their replacement with H^+ ions and leaching to layers down the soil profile (Belay et al., 2002), immobilization by phosphates, and by their assimilation by plants. However, destruction of clay minerals as affected by phosphorus and increase in CEC reduces the leaching. In soils with pH above 5.0, increases in acidity results in destruction of minerals, i.e. exchangeable Ca^{2+} and in lesser degree Mg^{2+}, might be derived from structure of soil primary minerals thus increasing the Ca availability in soil solution (McLaughlin & Wimmer, 1999). This process of releasing Ca and Mg is much weaker than that of Al because the release of cations is proportional to charge, which results in higher amounts of cationic species in the soil solution. Under 40-years of phosphate application, the amount of exchangeable Ca is increased (Tab. 4), although the increase was not regular but significant in the surface soil due to the mentioned reverse processes.

7. Microelements

The content of microelements in the soils depends on their amounts in the soil-forming rocks and on the soil-forming processes. The role of microelements in the physiological and the biochemical processes is immeasurably great. Soil is a source of microelements for plants, animals and humans. Deficiency or excess amount of microelements in fodders and food products lead to the disturbance of exchange of substances and the appearance of diseases in plants, animals and people. Basic for the vital activity of the plants and other living organisms are manganese, copper, boron, zinc, molybdenum, nickel, cobalt, fluorine, vanadium, iodine. One of the anthropogenic sources of microelements in soil is agriculture, including the application of phosphate fertilizers.

Iron solubility is controlled both by pH and redox potential. Iron becomes more soluble at lower pH values and under reduction condition. Iron solubility is largely controlled by the solubility of the hydrous Fe^{3+}oxides. At higher pH levels the activity of Fe^{3+} in solution decreased for 1000 time for each pH unit rise (Lindsay, 1974). When soil is waterlogged, the reduction of Fe^{3+} into Fe^{2+} increases the solubility of Fe. On the other hand, application of phosphate and its immobilization by formation of compounds with Fe, results in decrease in the amount of DTPA-extractable Fe, what is supported by the absence of correlation between mobile Fe with reducible and occluded P. Application of phosphate during 40-years determined the antagonistic processes fully took place, i.e. a small increase in DTPA-extractable Fe in the treatments where the processes of acidification are active was detected ($p < 0.05$) (Tab. 7).

Zinc solubility is decreased with increasing of soil pH (Cakmak, I. et al., 1996) where phosphates can immobilize Zn as zinc phosphate $Zn_3(PO_4)_34H_2O$, although this theory is not fully proved (Jurinak & Inouye 1962). However, application of MAP has shown opposite results for the second depth (30-60 cm) (Tab. 6, 7). Consequently, in the subsurface soil the content of the total Zn increases ($p < 0.05$) versus control. This in turn probably resulted in leaching of Zn due to acidification of the surface soil. On the other hand, the decrease of the amount of DTPA-extractable Zn is evident ($p < 0.05$) what can explain the prevalence of the processes of immobilization of Zn by phosphorus.

Copper similarly to Zn responded to the changes in soil pH. However, its solubility is quite lowered due to very strong bounds of Cu with soil organic matter (Mc Bride 1989). Reduced conditions in Stagnosol ties Cu with iron forming cuprous ferrite, possibly controling Cu solubility, which depends on the solubility of Fe (Shuman, 1985), which was reflected in significant correlation between soluble Fe and Cu.

The amount of *fluorine* in phosphate fertilizers depends on geographic origin of the raw material and the degree of its treatment (Gooroad & Caldwell, 1979). In fertilizers such as MAP this amount is not negligible (Tab 5). There was no significant increase in the amount of total F (Tab. 6, 7); the values were near or within the world's average (Helmke, 2000; Glandey & Burns, 1985). Solubility of fluoride ions is controlled by the soil pH and the amount of soil Ca and P (Hurd-Karrer 1950). But it should be noted that P affects the availability of F only in case of very high amounts of P in soil. Therefore, in our experiments a statistically significant increase of soluble F was detected in the both soil depths in accordance with the rates of the applied fertilizer, which presumably were induced by the decrease of soil pH ($r=-0.92$ $p < 0.01$) (Pickering, 1985; Barrow & Ellis, 1986; Loganathan et al., 2006). The significant increase of the amount of soluble F in the subsurface soil might be due to its leaching from the upper soil layer.

7.1 Trace elements

The dominant sources of fertilizer contaminants are the raw materials used to manufacture phosphate fertilizer. The most common contaminant metals and metalloids found in phosphate rock are arsenic (As), cadmium (Cd), chromium (Cr), mercury (Hg), lead (Pb), selenium (Se) and fluoride. Therefore, phosphate fertilizer is considered as one of the most important sources of trace element contamination in agricultural soils. The content of some trace elements used in our 40-year experiment is given in Table 5.

Element	As	Hg	F	Cr	Cu	Zn	Cd	Ni	Pb	Fe
			---------- mg kg^{-1} ----------							
Concentration	4-15	0.003-0.005	19,600-26,700	15-315	10-20	10-38	3-16	4-39	10-20	4000-5000

Table 5. Content of trace elements in the applied monoammonium phosphate

7.2 Lead (Pb)

Amount of Pb in phosphate fertilizers ranges between 7-225 mg kg^{-1} (Kabata-Pendias, 2001). The amount of Pb in the applied phosphate ranged between 10 and 20 mg kg^{-1} (Stevanovic et al., 2009). It is known that lead-containing compounds are weakly soluble (Davies, 1995), but application of phosphates affects their solubility, resulting in their greater immobilization (McGowen et al., 2001; Dermatas et al., 2008).

Hettiarachchi & Pierzynski (1996) had more success with the solubility equilibrium approach. They studied the influence of P additions on Pb solubility in a Pb-contaminated soil. Prior to P additions, the soil was in equilibrium with cerrusite ($PbCO_3$), but after addition of a soluble P the equilibrium was shifted to the Pb controlled by hydroxypryromorphite [$Pb_5(PO_4)_3OH$], a less soluble Pb mineral phase (Sims & Pierzynski, 2005).

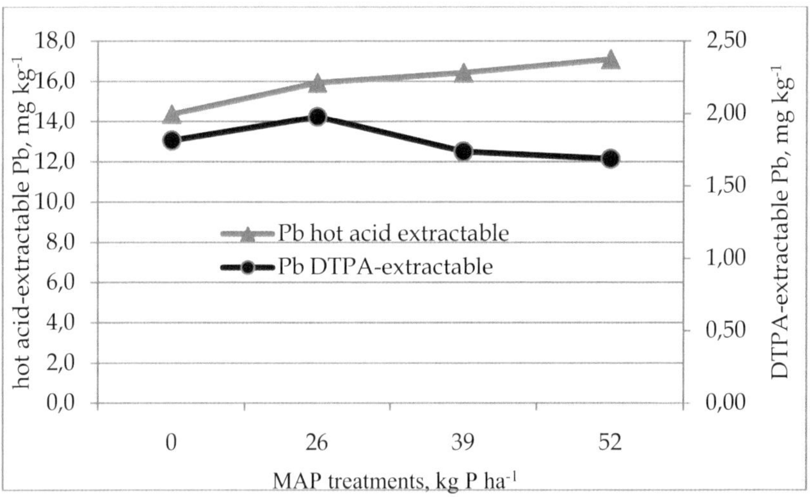

Fig. 11. Distribution of hot acid-extractable and diethylenetriaminepentaacetic acid (DTPA)-extractable Pb under different rates of phosphate fertilization on a Stagnosol soil

In our experiment, the amount of hot acid-extractable Pb in the 0- 30 cm soil interval increased significantly ($p < 0.05$) in accordance with the amount of added P fertilizers (Tab. 6). Significance was observed between the control and all the treatments. The content of hot acid-extractable Pb in the 0-30 cm interval of the control was less than that in the 30-60 cm interval, whereas in the fertilized treatments its content was higher in the upper than in the lower soil depth. Relatively small but significant increase of total P after the 40-years of phosphate application was due to the low content of P in the applied fertilizer, and to the low applied rates of fertilizer. Expectedly, based on the above-mentioned results, in contrast to the hot acid-extractable form, the amount of DTPA-extractable Pb decreased significantly ($p < 0.05$) in the 0-30 cm soil with the increase in fertilization rates (Fig. 11).

7.3 Cadmium (Cd)

Cadmium is in fertilized soils as divalent cation Cd_2^+ and the associated organic and inorganic complexes. As the cation, Cd solubility is highest in acidic soils, in soils with low CEC and soils with high anionic ligands (Hahne & Kroontje, 1973; Garcia-Miragaya & Page, 1976; Garcia-Miragaya & Page, 1978; Chubin & Street, 1981). Cadmium present in granular fertilizers rapidly dissolves and the solid-phase Cd in the fertilizer moves into the soil as free Cd_2^+. In contrast to the other trace metals such as Cu and Zn, organic ligands do not bind strongly free Cd_2^+ ions or the Cd complexes. In inorganic ligands the dominant Cd species present in soil solution tend to be more mobile than other trace metals, including Pb and Cu (Alloway, 1995).

Cadmium is a trace metal that is often bound to P fertilizers, especially to rock phosphates and it proved to be less mobile than the P carrier itself in the soil profile (Mulla et al., 1980; Adriano 2001; Kabata-Pendias & Pendias, 2001). However, the fertilizer we used (MAP) contains low concentrations of Cd (3–16 mg kg^{-1}) (Stevanovic et al., 2009). The low measured concentrations of this element and its solubility resulted in almost the same concentrations of Cd in both soil intervals. The minor changes in Cd observed in our studies were clearly due to the low levels of Cd in the source material (Tab. 6). About 71% of the accumulated Cd resided in the surface soil (0 to 15 cm) compared with only 45% for P (Adriano, 2001).

Treatment	Hot acid-extractable									
	Total F	Cu	Zn	Ni	Pb	Cd	Co	Cr	As	Hg
kg P ha^{-1}	---------- mg kg^{-1} ----------									
	0–30 cm									
0	334 ± 16.5†	12.7 ± 0.2	32.0 ± 0.4	26.3 ± 0.5	14.4a† ± 0.7	0.20a ± 0.00	11.5 ± 0.1	18.7a ± 0.6	4.5a ± 0.00	0.049a ± 0.001
26	337 ± 2.6	12.6 ± 0.1	32.3 ± 1.1	27.2 ± 0.3	15.9b ± 0.2	0.20a ± 0.00	12.5 ± 0.7	17.4ab ± 0.5	4.9ab ± 0.17	0.060b ± 0.004
39	339 ± 15.7	13.7 ± 0.8	32.0 ± 0.8	26.8 ± 1.2	16.4b ± 1.0	0.27b ± 0.02	11.3 ± 0.2	16.9ab ± 1.0	5.0b ± 0.20	0.061b ± 0.005
52	358 ± 8.5	12.6 ± 0.3	31.9 ± .0.3	26.2 ± 0.7	17.1c ± 0.1	0.27b ± 0.02	12.0 ± 0.4	15.9b ± 0.5	4.9a ± 0.13	0.092c ± 0.002
P	NS‡	NS	NS	NS	*	**	NS	*	*	***
	30–60 cm									
0	323a ± 10.1	12.5 ± 0.1	30.9 ± 0.1	27.0 ± 0.5	16.7a ± 0.3	0.20 ± 0.03	12.8 ± 0.8	18.6ab ± 0.8	4.7ab ± 0.2	0.04a ± 0.001
26	331a ± 6.4	13.3 ± 0.2	33.0 ± 0.8	28.0 ± 0.7	14.6b ± 0.6	0.20 ± 0.03	13.1 ± 0.4	19.9a ± 1.1	4.6a ± 0.1	0.06b ± 0.002
39	342a ± 3.1	12.8 ± 0.5	31.1 ± 0.6	27.1 ± 0.6	14.0b ± 0.8	0.20 ± 0.00	12.2 ± 0.8	16.8b ± 0.5	4.5a ± 0.1	0.06b ± 0.003
52	351b ± 11.2	13.3 ± 0.5	31.7 ± 0.1	27.3 ± 0.3	15.0ab ± 0.3	0.27 ± 0.02	12.0 ± 0.4	19.0a ± 0.6	5.0b ± 0.1	0.09c ± 0.002
P	*	NS	*	NS	**	NS	NS	*	*	***

Table 6. Concentrations of hot acid-extractable trace elements in Stagnosol of long-term phosphate fertilization experiment

Treatment	Water-extr. F	DTPA-extractable					
		Cu	Zn	Ni	Pb	Cd	Fe
kg P ha^{-1}				mg kg^{-1}			
				0–30 cm			
0	0.32a† ± 0.01	1.60a ± 0.04	0.77 ± 0.09	1.11 ± 0.08	1.81ab ± 0.09	0.07 ± 0.011	158a ± 3.4
26	0.69b ± 0.22	1.67ab ± 0.04	0.73 ± 0.03	1.31 ± 0.11	1.98b ± 0.04	0.07 ± 0.003	196b ± 8.4
39	0.76b ± 0.12	1.70b ± 0.01	0.74 ± 0.07	1.32 ± 0.12	1.74ab ± 0.08	0.06 ± 0.002	195b ± 9.4
52	0.92c ± 0.09	1.65a ± 0.02	0.88 ± 0.09	1.38 ± 0.07	1.65a ± 0.15	0.07 ± 0.008	194b ± 4.3
P	*	*	NS‡	NS	*	NS	**
				30–60 cm			
0	0.14a ± 0.03	1.57a ± 0.01	0.56 ± 0.02	1.22 ± 0.03	1.57 ± 0.02	0.04 ± 0.008	162 ± 6.2
26	0.31b ± 0.10	1.57a ± 0.02	0.47 ± 0.05	1.24 ± 0.07	1.64 ± 0.01	0.03 ± 0.002	173 ± 9.9
39	0.31b ± 0.05	1.57a ± 0.04	0.47 ± 0.06	1.29 ± 0.04	1.43 ± 0.10	0.04 ± 0.005	167 ± 5.4
52	0.38b ± 0.04	1.49b ± 0.02	0.41 ± 0.02	1.19 ± 0.06	1.45 ± 0.12	0.04 ± 0.004	166 ± 5.8
P	*	*	*	NS	NS	NS	NS

Table 7. Concentrations of water-extractable F and DTPA-extractable trace elements in Stagnosol of long-term phosphate fertilization experiment.

7.4 Arsenic (As)

Application of phosphates can enhance As mobility, phytoavailability, and phytotoxicity in lead arsenate contaminated soils amended with MAP (Peryea, 1991). The degree of arsenic toxicity depends on the phosphate concentration in the nutrient solution in certain soil-plant environments (Hurd-Karrer, 1939). The main predicators of As solubility are soil mineralogy, organic matter, soil pH and As oxidation state. Arsenic may form insoluble compounds with Fe and Al oxides, be absorbed by the organic matter or hydroxyl groups on clay minerals (Mitchell & Barr, 1995). Among various potassium and sodium salts, potassium phosphate was the most effective in extracting arsenic, attaining more than 40% extraction (Alam, et al., 2001). Phosphate could displace adsorbed or fixed As from sorbing complexes and thereby initially increase the amount of soluble As in soils. (Kabata-Pendias, & Pendias, 2001).

Because As is often a contaminant in phosphate fertilizers (Adriano, 2001; Kabata-Pendias & Pendias, 2001) produced from rock phosphates (O'Neill, 1995) we expected a higher concentration of As in the upper soil interval. However, because of their position in the periodic table of elements, where phosphorus, an essential element capable of being absorbed in relatively large amounts, and arsenic, a highly toxic element, chemically much like phosphorus, these elements often behave as antagonists. Anion antagonism between P and As (Parfitt, 1978), and their similar solubility under certain soil pH (Alloway, 1995), resulted in the movement of As down the soil profile (Tab. 6).

7.5 Chromium (Cr)

The content of Cr in phosphate fertilizers ranges between 66-245 mg kg^{-1} (Kabata-Pendias & Pendias, 2001). However, in the fertilizer used in our experiments the content of Cr was wider and ranged from 15-315 mg kg^{-1}. Chromium in soil could be found mainly in two forms: Cr^{3+} and Cr^{6+}. In the majority of soils, the relatively insoluble and less mobile Cr(III) form predominates and it geneally occurs as insoluble hydroxides and oxides (McGrath, 1995). Phosphorus states in soil and soil pH can determine the similarity of adsorption and solubility behavior between Cr^{3+} and Al^{3+}. Although the amount of Cr entering soil via the use of phosphate fertilizers is uncertain, it is likely to exist as Cr(III). Chromium (Cr III) in soil is not likely to be toxic (McGrath, 1995). Presence of Cr^{6+} in the soil surface horizons is due to its linkage to the oxidizing conditions and redox potentials. Once Cr is oxidized into the hexavalent form in the soil, it becomes more mobile and leachable (Adriano, 2001). Ions of $H_2PO_4^-$ and $HCrO_4^-$ are antagonists what is reflected in the increased mobility and solubility of Cr ion in soil solution. Data in the Table 4 confirms this hypothesis since in the control at both depths the concentration of Cr was nearly equal; however, at the surface depth the concentration of Cr is significantly decreasing correspondingly with the applied rate of phosphate; whilst at the subsurface depth the concentration of Cr significantly increases after the addition of phosphate (Adriano, 2001).

7.6 Nickel, Cobalt, Mercury (Ni , Co, Hg)

The content of Ni and Co did not show any differences in any of the treatments. The concentrations of these trace elements in the studied soil were within the range of average world concentrations. Most commercial fertilizers have Hg content below 50 ng per gram, but considerably higher amounts occur in phosphate fertilizers. Hg may be derived partly from the phosphate rocks and partly from the sulphuric acid used for dissolution of phosphates (Steinnes, 1995). Because of the initially low concentrations of hot-acid extractable Hg in the control treatment, the increased concentrations of Hg detected at both soil depths (Tab. 6) under the fertilized treatments are obviously due to the applied MAP. However, in spite of statistically significant increases in the content of Hg, its absolute amount is small and doesn't pose a risk for the environment even after 40-years application of MAP in Stagnosol.

8. Conclusions

Formation of the forms of soil phosphorus and their binding largely depends on soil pH. In neutral and alkali soils the main phosphate compounds are calcium phosphates, and their solubility depends mainly on the ratio Ca/PO_4. In such soils, the activity of Ca ions increased resulting in reduced solubility of phosphate while in neutral and acidic soil the adsorption and desorption of phosphate mainly occurs at Al and Fe oxide surfaces. The bounds Al-O-P forms much more labile forms than formations with double bounds of P.

Al bound P is the most labile form that supplies the plants with P-nutrient, and is the most responsible form for the movement of P along the soil profile and replenishment of other soil P-fractions. Both sequential analyses used for P fractionation proved that Al bound P fraction is closely related to the mobile Al of soil solution. Application of mineral phosphates results in the increase of Al bound P fraction thus increasing the amount of plant-available phosphorus in soil. In acidic soils, the application of MAP does't results in considerable changes in the amount of Ca-bound P and of organic P in soil. Also, the 40-year application of MAP destroys the structure of the clay minerals what in its turn increases the soil CEC. The long-term application of MAP didn't result in the accumulation of potentially toxic elements in their considerable concentrations, i.e. their concentrations were negligible in Stagnosol.

An important outcome is that soil P can exist in a series of "pools", which can be defined in terms of the extractability of P in different reagents. Also, the P in these pools can be related to the availability of P to plants, recognizing that there is a continuum of both extractability and availability. The most important concept is the reversible transfer of P among the most of soil phosphorus pools, what opens a possibility of the effective use of the applied phosphorus. These complex transformations of soil phosphorus grant wide opportunities for further researches.

9. References

Abrahamsen, G. & Miller, H.G. 1984. The effects of acid deposition on forest soil and vegetation. *Phil. Trans. R. Soc. B.*, Vol. 305, pp. (369–382). ISSN 1471-2970

Acton, D.F. & Gregorich, L.J. (Eds.). (1995). *The Health of Our Soils: Toward sustainable agriculture in Canada.* 1906/E. Centre for Land and Biological Resources Research, Research Branch, Agriculture & Agri-Food, ISBN 0-660-15947-3, Canada, Ottawa, ON.

Adeptu, J.A. & Corey, A.B., 1976: Organic phosphorus as a predictor of plant available phosphorus in soil of southern Nigeria. *Soil Sci.* Vol. 122, pp. (159-164). ISSN 0038-075x

Adriano, D.C. (2001). *Trace elements in terrestrial environments.* Sheridan Books, ISBN 0-387-98678-2, Ann Arbor, MI.

Alam, M. G. M., Tokunaga, S., & Maekawa, T. (2001). Extraction of arsenic in a synthetic arsenic-contaminated soil using phosphate, *Chemosphere,* Vol. 43, No.8 ,(June 2001), pp. (1035-1041)

Aleksandrovic, D., Pantovic, M., & Lotse, E., (1965). The minerological composition of some Pseudogley soils in Serbia, *Zemljiste i Biljka* Vol.14, pp. (333-351).

Alloway, B.J. (1995). *Heavy metals in soils.* Chapman and Hall, ISBN 0-7514-0198-6, London.

Aref, S., & Wander, M.M. (1998). Long-term trends of corn yield and soil organic matter in different crop sequences and soil fertility treatments on the Morrow Plots. *Adv. Agron.* Vol. 62, pp. (153–161). ISSN 0065-2113

Aydin, I., Aydin, F., Saydut, A. Hamamci, C. A., 2009. Sequential extraction to determine the distribution of phosphorus in the seawater and marine surface sediment. *J Hazard Mater.* Vol.168, No.2-3, (Sep.15), pp. (664-669), ISSN 0304-3894

Barrow, N.J., & Ellis, A.S., (1986). Testing a mechanistic model: III. The effects of pH on fluoride retention by soil. *Eur. J. Soil Sci.* Vol.37, pp. (287–293), ISSN 1351-0754

Beck, M. A., & Sanchez, P.A. (1994). Soil phosphorus fraction dynamics during 18 years of cultivation on Typic Paledult. *Soil. Sci. Soc. Am. J.*, Vol. 58, pp. (1424–1431), ISSN 0361-5995

Belay, A., Claassens, A.S., & Wehner, F.S. (2002). Effects of direct N and K and residual phosphorus fertilizers on soil chemical properties, microbiological components and maize yield under long-term crop rotation. *Biol. Fertil. Soils, Vol.* 35, pp. (420–427), ISSN 0178-2762.

Berti, W.R., & Cunningham, S.D. (1997). In-place inactivation of Pb in contaminated soils. *Environ. Sci. Technol,* Vol. 31, pp. (2673–2678), ISSN 0013-936X

Cakmak, D., Saljnikov, E., Mrvic, V., Jakovljevic, M., Marjanovic, Z., Sikiric, B., & Maksimovic, S. (2010). Soil properties and trace elements under 40-years of phosphate fertilization. *J. Environ. Quality,* Vol. 39, pp. (541-547), ISSN 0047-2425

Cakmak, I., Sari, N., Marschner, H., Ekiz, H., Kalayci, M., Yilmaz, A., & Braun, H.J. (1996). Phytosiderophore release in bread wheat and durum wheat genotypes differing in zinc efficiency. *Plant and Soil,* Vol.180, pp. (183-189), ISSN 0032-079X.

Chang, S.C., &. Jackson, M.L. (1957). Fractionation of soil phosphorus. *Soil Sci.* Vol. 84, pp. (133–144), ISSN 0038-075X.

Cooper, E.M.., Strawn, D.G., Sims, J.T., Sparks, D.L., &. Oneken, B.M. (1998). Effect of chemical stabilization by phosphate amendment on desorption of P and Pb from contaminated soil, In: *Agronomy abstracts*. pp. 343, ASA, ISSN-L: 00654671 0065 4671. Madison, WI;

Cotter-Howells, J., & Capron, S. (1996). Remediation of contaminated land by formation of heavy metal phosphates. *Appl. Geochem, Vol.* 11, pp. (335–342), ISSN 0883-2927.

Davies, B.E. 1995. Lead, In: *Heavy metals in soils*, B.J. Alloway (Ed.) pp. (206–220), Chapman and Hall, ISBN 0-7514-0198-6, London

Delgado, A., & Scalenghe, R. (2008). Aspects of phosphorus transfer from soils in Europe. *J. Plant Nutr. Soil Sci.* Vol. 171, pp. (552–575), ISSN 0718-9508.

De Groot, K.J. & Golterman, H.L. (1993). On the presence of organic phosphate in some Camargue sediments: evidence for the importance of phytate. *Hydrobiologia,* Vol. 252, pp. (105-116); ISSN 0018-8158

Dermatas, D., Chrysochoou, M., Grubb, D.G., & Xu, X. (2008). Phosphate treatment of firing range soils: Lead fixation or phosphorus release? *J. Environ. Qual.* Vol. 37, pp. (47–56), ISSN 0047-2425

Enger, H., & Riehm, H. (1958). Die Ammoniumlaktatessigsäure-Methode zur Bestimmung der leichtlöslichen Phosphorsäure in Karbonathaltigen Böden. (In German.) *Agrochimica* Vol. 3, No. 1, pp. (49–65), ISSN 0002-1857

Flaig, W. (1966). The Use of Isotopes in Soil Organic Matter Studies. Pergamon, ISBN 0-582-44245-1, New York, NY;

Frankenberger, W.T., Tabatabai, M.A., Adriano, D.C., & Doner, H.E. (1996). Bromine, chlorine, and fluorine, In: *Methods of soil analysis, Part 3. Chemical methods,* D.L. Sparks (Ed.), pp. 833–867. SSSA, ISBN 96-70096, Madison, WI;

Foth, H.D., & Ellis, B.G. (Jan. 31, 1997). *Soil fertility*, 2nd ed. CRC Press, Boca Raton, ISBN-10: 1566702437, FL.

Frossard, F., Brossard, M.., Hedley, M.J., & Metherell, A. (1955). Reaction controlling the cycling of P in soils.. In: *Phosphorus in the global environment,* H.Tiessen (ed.), pp. (107-137), JohnWiley&Sons, ISBN 0716750791, NY.

Garcia-Miragaya, J. & Page, A.L. (1976). Influence of ionic strength and inorganic complex formation on the sorption of trace amounts of Cd by montmorillonite . *Soil Sci. Sac.Amer. J.* Vol. 40, pp. (658-663), ISSN 0361-5995

Garcia-Miragaya, J. & Page, A.L. (1978). Sorption of trace quantities of cadmium by soils with different chemical and mineralogical composition *Water, Air & Soil Pollution* Vol. 9, No.3, pp. (289-299), ISSN 0049-6979

Gee, G.W., & Bauder, J.W. (1996). Particle fractionation and particle-size analysis. In: *Methods of soil analysis,* Part I, Series 5, C.A. Black (Ed.), pp. (545–567), ASA and SSSA, Madison, WI.

Gladney, E.S., & Burns, C.E. (1985). Compilation of elemental concentration data for samples SO-1 to SO-4. *Geostand. Newsl,* Vol. 9, pp. (38–43), ISSN 0150-5505

Golterman, H.L. (1982). Differential extraction of sediment phosphorus with NTA solutions. *Hydrobiologia,* Vol. 92, pp. (683-687), ISSN 0018-8158

Golterman, H.L. (1996). Fractionation of sediment phosphate with chelating compounds: a simplification, and comparison with other methods. *Hydrobiologia*, Vol. 335, pp. (87-95), ISSN 0018-8158

Goodroad, L.L., & Caldwell, A.C. (1979). Effects of phosphorus fertilizer and lime on the As, Cr, Pb, and V content of soils and plants. *J. Environ. Qual.* Vol. 8, pp. (493–496), ISSN 0047-2425

Greenwood, D.J. (1981). Fertilizer food production: world scene, *Fertilizer Research,* Vol. 2, pp. (31-51); ISSN 01671731.

Hartikainen, H. (1989). Effect of cumulative fertilizer dressing on the phosphorus status of mineral soils. I. Changes in inorganic phosphorus fractions. *Journal of Agricultural Science in Finland,* Vol. 61, pp. (55-59), ISSN 1795-1895

Hedley, M.J., Stewart, J.W.B., & Chauhan, B.S. (1982). Changes in inorganic and organic soil phosphorus fractions induced by cultivation and by laboratory incubations. *Soil.Sci.Am.J.* Vol. 46, pp. (970-976), ISSN 0361-5995

Helmke, P.A. (2000). The chemical composition of soil. In: *Handbook of soil science,* M.E. Sumner (Ed.), pp. B12–B17, CRC Press, Boca Raton, ISBN 0-8493-3136-6, FL.

Hera, C., & Mihaila, V. (1981). The changing of some agrochemical indices of the soil by application of the fertilizers. *Analele ICCPT* Vol. 47, pp. (319–327), ISSN 0253-1682

Hettiarachchi, G.M., & Pierzynski, G.M. (1989). The influence of phosphorus source and rate on soil solution cadmium and lead activities, In: *Agronomy abstracts.* pp. (333) ASA, CSSA and SSSA, ISSN 00654671, Madison , WI.;

Hettiarachhchi, G.M., Pierzynski, G.M., Zwonitzer, J., & Lambert, M. (1997). Phosphorus source and rate effects on cadmium, lead, and zinc bioavailability in metal-contaminated soil, *Extended Abstr., 4th. Int. Conf. on the Biogeochem. Trace Elements* (ICOBTE), ISBN 91-576-6626-1, Berkeley, CA., 23–26 June 1997.

Hieltjes, A.H.M., and Lijklem, L. (1980). Fractionation of inorganic phosphates in calcareous sediment. *J. Environ.*

Qual., Vol. 9, No. 3, pp. (405-407), ISSN 0047-2425

Hurd-Karrer, A. M. (1939). Antagonism of certain elements essential to plants toward chemically related toxic elements,. *Plant Physiology*, Vol.14, No. 1, pp. (9-29), ISSN 0-534-15162-0

Hurd-Karrer, A.M. (1950). Comparative fluorine uptake by plants in limed and unlimed soil. *Soil.Sci.* Vol. 70, pp. (153-159), ISSN 0038-075X.

Jaakola, A., Hartikainen, H., & Lemola, R. (1997). Effect of fertilization on soil phosphorus in long-term field experiment in southern Finland. *Agricultural and food science in Finland.* Vol. 614, pp. (313-322), ISSN 1239-0992.

Jakovljević, M., Pantović, M., & Blagojević, S. (1985). *Praktikum iz hemije zemljišta i voda.* (In Serbian.) Beograd. Poljoprivredni fakultet, Belgrade, Serbia.

Jurinak, J.J., & Inouye, T.S. (1962). Some aspects of zinc and cooper phosphate formation in aqueus systems. *Soil.Sci. Amer.Proc.* Vol. 26, pp. (144-147), 0038-075X

Kabata-Pendias, A., & Pendias, H. (2001). *Trace elements in soils and plants.* 3rd ed. CRC Press, Boca Raton, ISBN 0-8493-1575-1, FL;.

Linsdey, W.L. (1974). Role of chelation in micronutrient availability. In: *The plant root and its Environment,* ed. E.W. Carson, p.507-524, University Press of Virginia,

Lindsay, W.L. (1979). *Chemical equilibria in soils.* Wiley-Interscience, ISBN 10: 0471027049, New York.

Lindsay, W.L., Vlek, P.L.G., & Chien, S.H. (1989). Phosphate minerals, In: *Minerals in soil environments,* J.B. Dixon and S.B. Weed (ed.), SSS Book Ser. I., pp. (1089-1130), ASA, CSSA, and SSSA, ISBN 0891188444, Madison. WI.

Loganathan, P., Gray, C.W., Hedley, M.J., & Roberts, A.H.C. (2006). Total and soluble fluorine concentrations in relation to properties of soils in New Zealand. *Eur. J. Soil Sci.* Vol. 57, pp. (411–421), ISSN 1351-0754

Ma, Q.Y., Traina, S.J., & Logan, T.J. (1993). In situ lead immobilization by apatite. *Environ. Sci. Technol.* Vol. 27, pp. (1803–1810), ISSN 0013-936X

Magdoff, F., Lanyon, L., & Liebhardt, B. (1997). Nutrient cycling, transformation and flows: Implications for a more sustainable agriculture. *Adv. Agron.* Vol. 60, pp. (1–73), ISSN: 0065 -2113

Manojlovic, D., Todorovic, M., Jovicic, J., Krsmanovic, V.D., Pfendt, P.A., & Golubovic, P. (2007). Preservation of water quality in accumulation Lake Rovin: the estimate of the emission of phosphorus from inundation area. *Desalination,* Vol. 213, pp. (104-109), ISSN 00119164

Maroko, J.B., Buresh, R.J., & Smithson, P.C. (1999). Soil phosphorus fractions in unfertilized fallow–maize system on two tropical soils. *Soil Sci. Soc. Am. J.* Vol. 63, pp. (320–326), ISSN 0361-5995

McBride, M.B. (1989). Reaction controlling heavy metal solubility in soils. *Adv. Soil Sci.* Vol. 10, pp. (1-56), ISSN 0176-9340

McCollum, R.E. (1991). Build-up and decline in soil phosphorus: 30-yr trends on a Typic Umbraquult. *Agron. J.* Vol. 83, pp. (77–85), IOSSN 0002-1962

McGowen, S.L., Basta, N.T., & Brown, G.O. (2001). Use of diammonium phosphate to reduce heavy metal solubility and transport in smelter-contaminated soil. *J. Environ. Qual.* Vol. 30, pp. (493–500), ISSN 0047-2425

McGrath, S.P. (1995). Chromium and nickel, In: *Heavy metals in soils,* B.J. Allowey (Ed.), p. 152–174, Chapman and Hall, ISBN 0-7514-0198-6, London

McKenzie, R.C., & Nyborg, M. (1984). Influence of subsoil acidity on root development and crop growth in soils of Alberta and north-eastern British Columbia. *Can. J. Soil Sci.* Vol. 64, pp. (681–697), ISSN 0008-4271

McLaughlin, S.B., & Wimmer, R. (1999). Calcium physiology and terrestrial ecosystem processes. *New Phytol.* Vol. 142, pp. (373–417), ISSN 0028-646X

Mengel, K., & Kirkby, E.A. (2001). *Principles of plant nutrition.* 5th Ed. Kluwer Academic Publishers, ISBN 1-4020-0008-1, The Netherlands

Mitchell, C.C., Westerman, R.L., Brown, J.R., & Peck, T.R.. (1991). Overview of long-term agronomic research. *Agron. J.* Vol. 83, pp. (24–29), ISSN 0002-1962

Mitchell, P., & Barr, D. (1995). The nature and significance of public exposure to arsenic: A review of its relevance to South West England*; Environ. Geochem. Health,* Vol. 17, pp. (57-82), ISSN 0269-4042

Mrvic, V., Jakovljevic, M., Stevanovic, D., & Cakmak, D. (2007). The Forms of Aluminium in Serbia. *Plant Soil Environ.* Vol. 53, pp. (482–489), ISSN 1214-1178

Mulla, D.J., Page, A.L., & Ganje, T.J. (1980). Cadmium accumulation and bioavailability in soil from long-term phosphorus fertilization. *J. Environ. Qual.* Vol. 9, pp. (408–412), ISSN 0047-2425

Murphy. J., & Riley, P. (1962). A modified single solution method for determination of phosphate in natural waters. *Anal. Chim . Acta,* Vol. 27, pp. (31-36), ISSN 0003-2670

Mustapha, S., Yerima, S.I., Voncir, N., & Ahmed, B.I. (2007). Content and Distribution of Phosphorus Forms in Some Halpic Plinthaquults in Bauchi Local Goverment Area, Bauchi State, Nigeria. International *J. Soil Sci.* Vol. 2, No. 3, pp. (197-203), ISSN 1351-0754

Nel, P.C., Barnard, R.O., Steynberg, R.E., De Beer, J.M., & Groeneveld, H.T. (1996). Trends in maize grain yields in long-term fertilizer trial. *Field Crops Res.* Vol. 47, pp. (53–64), ISSN 1821-3944

Nelson, D.W., & Sommers, L.E. (1996). Total carbon, organic carbon, and organic matter, In: *Methods of soil analysis Part 3,* D.L. Sparks (Ed.), pp. (961–1010), SSSA, ISBN 96-700096, Madison, WI.

O'Neill, P. (1995). Arsenic. In: *Heavy metals in soils,* B.J. Alloway (ed.), Chapman and Hall, p. 105–119, ISBN 0-7514-0198-6, London

Parfitt, R.L. (1978). Anion adsorption by soils and soil materials. *Adv. Agron.* Vol. 30, pp. (1–50), ISSN 0065-2113

Peryea J. F. (1991). Phosphate-Induced Release of Arsenic from Soils Contaminated with Lead Arsenate, Issue 5, *Soil Sci. Soc. Am.J.*, Vol. 55, No. 5, pp. (1301-1306), ISSN 0361-5995

Petrovic Dj., Todorovic M., Manojlovic D., Krsmanovic V.D. (2009). Speciations of Trace Metals in the Accumulation Bogovina on the Crni Timok River. *Polish J. Environ. Stud.* Vol. 18, pp. (873-884), ISSN 1230-1485

Pickering, W. (1985). The mobility of soluble fluoride in soils. *Environ. Pollut,* Vol. 9, pp. (281–308), ISSN 0971-4871

Rajan, S.S.S. (1975). Phosphate adsorption and the displacement of structural silicon in allophane clay. *J. Soil Sci.* Vol. 26, pp. (250–256), ISSN 1351-0754

Ruban, V., Lopez-Sanchez, J.F., Pardo, P., Rauret, G., Muntau, H., & Quevauviller, Ph. (1999). Selection and evaluation of sequential extraction procedures for the determination of phosphorus forms in lake sediments. *J. Environ. Monit.* Vol. 1, pp. (51-56), ISSN 1464-0325

Ruban, V., Lopez-Sanchez, J.F., Pardo, P., Rauret, G., Muntau, H., & Quevauviller, Ph. (2001). Development of a harmonized phosphorus extraction procedure and certification of a sediment reference material. *Environ. Monit.* Vol. 3, pp. (121-125), ISSN 1464-0325

Ruttenberg, K. C. (1992). Development of a sequential extraction method for different forms of phosphorus in marine sediments. *Limnol. Oceanogr.* Vol. 37, pp. (1460-1482), ISSN 1541-5856

Saleque, M.A., Naher, U.A., Islam, A., Pathan, A.B.M.B.U., Hossain, A.T.M.S., & Meisner, A.C. (2004). Inorganic and organic phosphorus fertilizer effects on the phosphorus fractionation in wetland rice soils. *Soil Sci. Soc. Am. J.* Vol. 68, pp. (1635–1644), ISSN 0361-5995

Santillian-Medrano, J., & Jurinak, J.J. (1975). The chemistry of lead and cadmium in soil: Solid phase formation. *Soil Sci. Soc. Am. Proc.* Vol. 39, pp. (851–856), ISSN 0361-5995

Schwertmann, U., Süsser, P., & Nätcher, L.G. (1987). Proton buffer system in soils. *Pfanzenernähr. Bodenk.* Vol. 150, pp. (174-178), ISSN 0718-2791

Sharpley, A.N. & Smith, S.J. (1983). Distribution of phosphorus forms in virgin and fertilized soil and potential erosion losses. *Soil Sci. Soc. Am. J.* Vol. 47, pp. (37-47), ISSN 0361-5995.

Sharpley, A.N. (2000). Phosphorus availability. In: *Hanbook of soil science,* M.E.Sumner (ed.), pp. (D-18-D-38), CRC Press, Boca Raton, ISBN 10: 0849331366, FL.

Shuman, L.M. (1985). Fraction method for soil microelements. *Soil Sci.* Vol. 140, pp. (11-22), ISSN 0038-075X

Simard, R.R., Cluis, D., Ganbazo, G., & Beauchemin, S. (1995). Phosphorus status of forest and agricultural soils from a watershed of high animal density. *J. Environ. Qual. Vol.* 24, pp. (1010-1017), ISSN 0047-2425

Sims, J.T., & Pierzynski, G.M. (2005). Chemistry of Phosphorus in Soils, In: *Chemical Processes in Soil,* SSSA Book series 8, pp. (151-193), ISBN 0891188436, Madison, WI.

Soltanpour, P.N., Johnson, G.W., Workman, S.M.., Bentonjones, J.J., & Miller, R.O. (1996). Inductively coupled plasma emission spectrometry and inductively coupled plasma-mass spectrometry, In: *Methods of soil analysis. Part 3,* D.L. Sparks (ed.), pp. (91–139), SSSA, ISBN 96-700096, Madison, WI.

SPSS. (2007). *SYSTAT version 16 sofware: Statistics.* SPSS, Chicago, IL.

Steinnes, E. (1995). Mercury. In: *Heavy metals in soils,* B.J. Alloway (ed.), pp. (245–257), Chapman and Hall, ISBN 0-7514-0198-6, London

Stevanovic, D., Kresovic, M., Stojanovic, M., & Grubisic, M. (2009). Stanje proizvodnje i problemi primene mineralnih djubriva u Srbiji. *Proc. of the XXIII Conf. of Agronomist, Veterinarians, and Technologists.* Vol. 15. Serbia.

Stewart, J.W.B., &. Sharpley, V. (1987). Controls on dynamics of soil and fertilizer phosphorus and sulfur.. In: *Soil fertility and organic matter as critical components of production systems,* R.F. Follett, J.W.B. Stewart, and C.V. Cole (ed.)., pp. (101–121), SSSA Spec. Publ. 19. SSSA and ASA, ISBN 0 -7923-4833-8, Madison, WI.

Stumm, W. & Morgan, J. J. (1981). *Aquatic chemistry,* 2nd ed. Wiley, 780pp., ISBN-10: 0471511854, New York

Sumner, M.E., & Miller, W.P. (1996). Cation exchange capacity and exchange coefficients, In: *Methods of soil analysis. Part 3,* D.L. Sparks (ed.), pp. (1201–1229), SSSA, ISBN: 96-700096, Madison, WI.

Tisdale, S.L., Nelson, W.L., Beaton, J.D., & Havlin, J.L. (1993). *Soil Fertility and fertilizers.* 5th ed. Macmillian, ISBN 978-987-24977-1 -2 New York.

Tyler G. (2004). Vertical distribution of major, minor and rare elements in Haplic Podzol. *Geoderma.* Vol. 119, pp. (277-290), ISSN 0016-7061

Väänänen, R., Hristov, J., Taskanen, N., Hatikainen, H., Nieminen M., Ilvesniemi, H. (2008). Phosphorus sorption properties in podzolic forest soils and soil solution phosphorus concentration in undisturbed and disturbed soil profiles. *Boreal Envi. Research,* Vol. 13, pp. (553-567), ISSN 1239-6095

Van Der Zee S.E.A.T.M., Nederlof, M. M., Van Riemsdijk, W. H., & De Haan. F.A.M. (1987). Spatial Variability of Phosphate Adsorption Parameters, *J. Environ. Qual.* Vol. 17, pp. (682-688); ISSN 0047-2425

Van Der Zee, S.E.T.M., van Rimsdijk, W.H. (1988). Model for long-term phosphate reaction kinetics in soil *J. Envir. Qual.* Vol. 17, pp. (35-41), ISSN 0047-2425

Vu, D.T., Tang, C., Armstrong R.D. (2010). Transformation and availability of phosphorus in three contrasting soil types from native and farming systems: A study using fractionation and isotopic labeling techniques. *J. Soil Sediments.* Vol. 10, pp. (18-29), ISSN 1614-7480.

Willet, I.R., Chartres., C.J., & Nguyen, T.T. (1988). Migration of phosphate into aggregated particles of ferrihydrate. *J.Soil Sci.* Vol. 39, pp. (275-282), ISSN 0038-075X

Williams, J. D. H., Jaquet, J.M., & Thomas, R. L. (1976). Forms of phosphorus in the surficial sediments of Lake Erie. *J. Fish. Res. Bd. Can.*, Vol. 33, pp. (413–429), ISSN 0015-296X

WRB. (2006). World reference base for soil resources. Food and Agriculture Organization of the United Nations, ISBN 92-5- 105511-4, Rome

Wright, R.J., & Stuczynski, T. (1996). Atomic absorption and flame emission spectrometry. In: *Methods of soil analysis. Part 3,* D.L. Sparks (ed.), p. (65–90), ISBN 96-700096, SSSA, Madison, WI.

Zhou, Q, & Zhu, Y. (2003). Potential pollution and recommended critical levers of phosphorus in paddy soils of the southern Lake Tai area, China. *Geoderma* Vol. 115, pp. (45-54), ISSN 0016-7061

2

Classification and Management of Highly Weathered Soils in Malaysia for Production of Plantation Crops

J. Shamshuddin[1] and Noordin Wan Daud[2]
[1]*Department of Land Management, Faculty of Agriculture, Universiti Putra Malaysia, 43400 Serdang, Selangor,*
[2]*Department of Crop Science, Faculty of Agriculture, Universiti Putra Malaysia, 43400 Serdang, Selangor Malaysia*

1. Introduction

In Malaysia, Ultisols and Oxisols containing kaolinite, gibbsite, goethite and hematite in the clay fraction are very common especially in the upland areas, occupying about 72 % of the country's land area. The soils are highly weathered as they exist under tropical environment with high rainfall and temperature throughout the year, resulting in leaching of plant nutrients and accumulation of sesquioxides (Anda et al., 2008a). They are by nature devoid of basic cations (Ca and Mg) and available P (due to fixation by the oxides) and hence, their productivity is generally considered as low. The soils are mainly utilized for oil palm and rubber cultivation with great success due to excellent soil management practices. With the expertise available in the country, palm oil and rubber are produced in large amounts for the world market. However, cocoa growing on these soils produces low yield which are attributed to low pH and aluminum and/or manganese toxicity.

The problems of low productivity can be overcome by liming using ground magnesium limestone (Shamshuddin et al., 1991; Ismail et al., 1993; Shamshuddin and Ismail, 1995; Shamshuddin et al., 1998; Shamshuddin et al., 2009; Shamshuddin et al., 2010) or by applying basalt (Gillman et al., 2001; Anda et al., 2009; Shamshudin & Kapok, 2010). Basalt releases Ca, Mg, K, P and S on its dissolution into the soils (Gillman et al., 2002). Out of the six macronutrients needed by the growing crops in the field only N is not present in basalt.

Most of the Ultisols and Oxisols in the tropics is lacking in organic matter which can supply essential plant nutrients as well as improve soil structures. Normal organic matter applied for alleviating the infertility of Ultisols and Oxisols in Malaysia is compost (Anda et al., 2008b; Anda et al., 2010) and palm oil mill effluents.

This paper intends to classify the highly weathered soils in Malaysia and discusses the management of the soils for sustainable production of oil palm, rubber and cocoa. The information given in this paper is useful to students and researchers alike.

2. Highly weathered soils of Malysia

2.1 Classification

Due to high temperature in Malaysia silicate minerals become unstable due to the changes in the chemistry of their environment. The new chemical conditions are dominated by aqueous state. The consequence of such an environment is a tendency to hydrate the high-temperature silicate minerals. Major effect of hydration is a large portion of the minerals is dissolved integrally into the altering aqueous solution and it is transported as such into lakes and ocean. This process of chemical weathering is very intense in Malaysia due to the prevailing high rainfall and temperature. Residual products of rock and mineral weathering under this condition are quartz, secondary phyllosilicates and sesquioxides with or without muscovite, depending on the degree of weathering.

A chemical reaction denoting dissolution of feldspar (orthoclase) in water is as follows (Duff, 1993):

$$6H_2O + CO_2 + \underset{\text{(feldspar)}}{2KAlSi_3O_8} \rightarrow \underset{\text{(kaolinite)}}{A1_2Si_2O_8(OH)_4} + 4SiO(OH)_2 + K_2CO_3$$

In this reaction, K is lost via leaching into the groundwater. Further weathering of the clay mineral results in the formation of colloidal materials according to the following reaction (Duff, 1993):

$$H_2O + A1_2Si_2O_8(OH)_4 \rightarrow A1_2O_3.nH2O + SiO(OH)_2$$

Exposure of granite, shale, schist, sandstone, basalt, andesite and serpentinite in Malaysia over a long period of time to the forces of weathering results in the formation of highly weathered materials, dominated by kaolinite, halloysite, gibbsite, goethite and hematite in the clay fraction (Tessens & Shamshuddin, 1983). The overall products are collectively termed as soil materials, which are usually devoid of plant nutrients. The soils are classified as

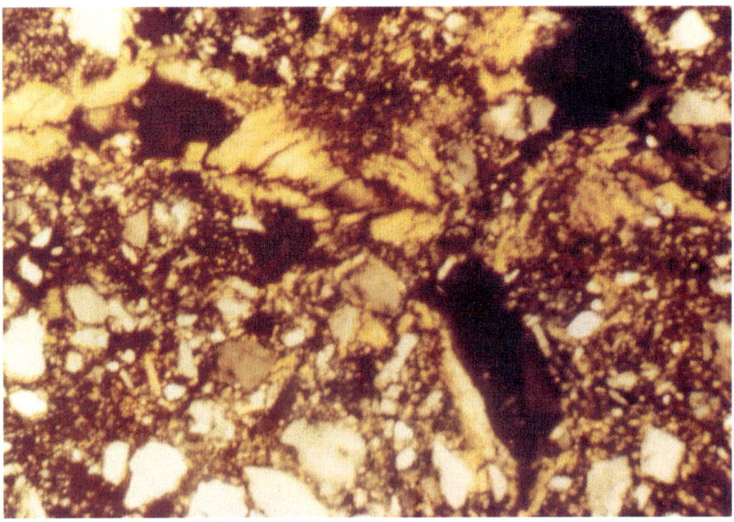

Fig. 1. Argillic horizon present in the B horizon of Ultisols (Courtesy of S Paramananthan)

either Ultisols or Oxisols according to Soil Taxonomy (Soil Survey Staff, 2010). Ultisol is defined by the presence of argillic horizon in the subsoil (Figure 1). Clays from the topsoil are moved and accumulated in the B horizon under intense leaching environment in the tropics.
By nature, Oxisols are considered as more weathered than Ultisols. They are dominated by kaolinite and sesquioxides (Anda et al., 2008a). The soils are defined by the presence of oxic horizon in the subsoil (Figure 2). The CEC of the soils is extremely low, with value < 16 $cmol_c/kg$ clay. As such, plant nutrients are mostly lost via leaching, further lowering the productivity of the soils.

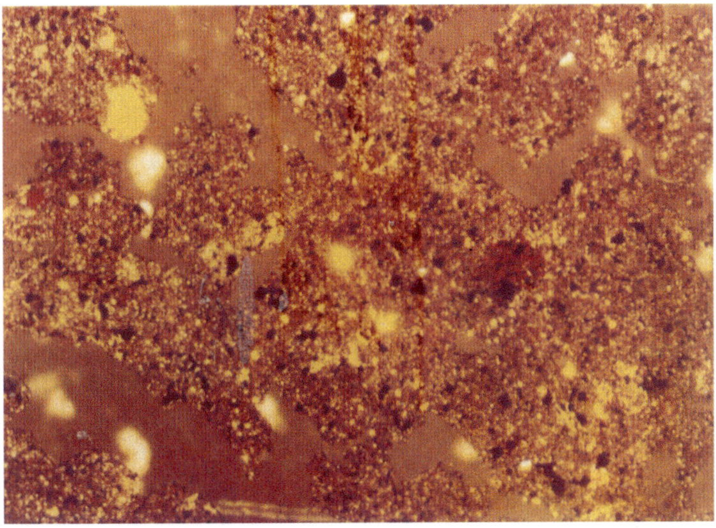

Fig. 2. Oxic horizon present in the B horizon of Oxisols (Courtesy of S Paramananthan)

2.2 Charge properties of Ultisols and Oxisols

It is known that soil materials are negatively- and/or positively-charged. A certain amount of negative charge in the soils is derived from within the phyllosilicates themselves via a process called isomorphic substitution. For instance, replacement of Si by Al in kaolinite present in the soils would result in excess of negative charges. Such charges are termed as negative permanent charges. Positive permanent charges are also existed in soils, being produced by isomorphic substitution of Fe by Ti in soils containing high amount of oxides of Fe (Tessens & Shamshuddin, 1983). This kind of isomorphic substitution is common in Oxisols having high amount of hematite and goethite, the active minerals in the soils.
Oxides of Fe and Al and the broken edges of phyllosilicates have another kind of charge known as variable charge. As the pH of the ambient solution changes the charge on the surfaces of these minerals also changes. When the pH is low, protons are chemisorbed onto the minerals to become net positively-charged. On the other hand, the minerals are net negatively-charged at high pH. The pH at which the net charge of the variable-charge mineral is zero is termed as pHo.
Each mineral has its own pHo value. The value for silicate is low, but for oxides it is high. Generally, soil is composed of many variable-charged minerals. The value reported for a particular soil is actually the resultant pHo value of the whole minerals in the soils. Silicate

is abundant in Ultisols, while oxides are abundant in Oxisols. Hence, the pHo value of Ultisols is lower than that of Oxisols (Tessens & Shamshuddin, 1983). This means that the mineralogy of the soil affects its pHo value.

We do not have soils completely composed of variable-charged minerals. Thus, overall charges in the soils have to be considered for the meaningful interpretation of the soil properties. The pH at which the net charge is zero, taken into account the whole soil materials, is termed as point of zero net charge (PZNC). Studies in Malaysia showed that PZNC is lower than pHo in the Ultisols and is higher than pHo in the Oxisols (Tessens & Shamshuddin, 1983). It is also observed that both pHo and PZNC values increase with increasing stage of soil weathering. As the soils weather more oxides are formed, leading to an increase in pHo and PZNC. The soils then become less productive and need special management practices for sustainable crop production.

Total charge in the soil can be subdivided into permanent and variable charge components, and can be represented by the following equation:

$$Qt = Qp + Qv$$

where

Q = charge

t = total

p = permanent v = variable

The amount of total charge in the soils can be manipulated by changing the Qv.

Qv is related to (pHo - pH)

Taking soil having net negative charge as an example, Qt can be increased by increasing the difference between pHo and pH. This can readily be done either by lowering pHo or increasing pH. The former can be implemented by incorporating basalt into the soils (Anda et al., 2009), while the latter is easily accomplished by liming (Shamshuddin et al., 1991) or applying basalt (Shamshuddin & Kapok, 2010). pHo can also be lowered, to a certain extent, by incorporating organic matter into the soil (Shamshuddin et al., 1987; Anda et al., 2008b).

A soil system is stable when the charge in it is low or at the minimal, suggesting that soil potential decreases as the charge decreases. Therefore, the potential in the soil is very low when pH is near its pHo. Under natural condition, soil pH tries to move to its pH_o in order to achieve maximal stability.

As soil pH increases with weathering, its pHo increases. It has been shown clearly that soil pH increases with increasing pH_o (Tessens & Shamshuddin, 1983). Primary minerals break up during the course of weathering and new minerals are formed, meaning that the silicates in the soils changes to oxides of Fe and Al, and consequently the pHo of the soils increases.

2.3 Soil pH

We have analyzed pH of hundreds of soil samples from all over Peninsular Malaysia and found that the values are mostly between 4 and 5 (Tessens & Shamshudin, 1983). The values are generally lower for the Ultisols than that of the Oxisols. The former ranges from 4.0 to 4.5, while the latter ranges from 4.5 to 5.0. The explanations for this phenomenon are as follows:

a. Soils will be at their greatest stability when the potential in them is zero, that is at the pH = pHo. The pHo of majority of the highly weathered soils in Malaysia is 4-5. Therefore, pH of the soils tries to approach 4-5 in order to remain stable; and

b. On weathering, Al^{3+} and Fe^{3+} in the phyllosilicates are released into the soil solution. Al^{3+} in the soil solution undergoes hydrolysis.

This allows us to determine the pK_a of Al which is 5. Likewise, Fe^{3+} hydrolysis releases proton into the soil solution. The pK_a value of Fe is 3.0. If these free reactions are allowed to take place without interruption, then the pH of the soil solution will go near their pK_a value in order to achieve equilibrium.

2.4 Effects of low pH on Al and Mn availability

As the soil pH of highly weathered Malaysian soils is low (< 5), Al on the exchange complex of the soils readily dissolves into the soil solution. In many cases, the Al in the soil solution is present at toxic level. The Al concentration in the soil solution increases as the pH lowers. Likewise, Mn may exist at toxic level at low pH. Exchangeable Al is lower in the Oxisols than in the Ultisols. It follows that there is less soil solution Al in Oxisols than in the Ultisols. This is consistent with the higher pH of the Oxisols as compared to that of the Ultisols. Low pH and high exchangeable Al have little effect on the growth of either oil palm or rubber, but they are expected to reduce the yield of cocoa significantly.

3. Oil palm cultivation

The most important agricultural crop in Malaysia right now is oil palm (Figure 3). The area covered by the so-called golden crop is estimated to be about 4.6 million ha, sporadically distributed throughout the length and breadth of the country. Currently, most of the oil palm is grown on upland areas where Ultisols and Oxisols occur. With good soil management practices oil palm grows very well, contributing to the wealth of the nation. At the current of production rate and high price in the marketplace some USD 16.7 billion is added into the economy annually, and so oil palm is indeed helping the economy of Malaysia going for a long run. In the case of Malaysia, the industry has to be protected at any cost in order to become a developed country comes the year 2020. By then Malaysia would have become a country with high income. If we keep doing what we do now and stick to the principle of sustainable crop production, it certainly will.

For oil palm cultivation, fertilizer input is very necessary where NPK fertilizers need to be applied regularly at the appropriate rates. This is because the Ultisols and Oxisols in the country are devoid of macronutrients, resulting from extreme leaching and weathering. As the soils are acidic in nature P-fertilizer recommended for application is phosphate rock, which is a slow release fertilizer of sort. As it dissolves slowly under low pH condition P is released and made available to the growing oil palm in the field. Phosphate is known to react with oxides/hydroxides of Fe in the soils, resulting in a slight increase in negative charge and pH (Tessens & Zaharah, 1983). This phenomenon has, to some extent, improved the productivity of the soils. However, we should not forget the fact that some of the applied P is fixed by the oxides of Fe, especially in Oxisols and consequently lost into the soils indefinitely or until some other reactions that dissolve the so-formed $FePO_4$. As such, phosphate fertilizer efficiency study is a popular topic of research in the tropical region, such as Malaysia and Brazil.

Oil palm is found to be acid tolerant, and hence, it can still grow even at the soil pH of 4.3 and at high Al concentration (Auxtero & Shamshuddin, 1991). Due to that Ultisols and Oxisols in Malaysia are considered suitable for oil palm cultivation. There is no need to apply lime onto the soils for oil palm cultivation as the area under oil palm is very large and therefore not economical. For sure, soil moisture has to be maintained at the optimal level for oil palm growth. We know for sure that oil palm can only be grown if the annual rainfall exceeds 1800 mm and it must be evenly distributed throughout the year. There should not be a dry period exceeding a few months. That it is so because oil palm is originated from swampy areas in African countries. The best growing areas in Peninsular Malaysia for cultivating oil palm is southern part (Johor) where rainfall is evenly distributed at 3000 mm/year. Kedah and Perlis (northern part) are probably a bit too dry for oil palm cultivation.

Under estate management, the yield of oil palm grown on Ultisols and Oxisols ranges from 20 to 30 tones fresh fruit bunches (ffb) per hectare per year. It will be higher on well managed soils planted with high yielding oil palm clone. According to a reliable source, this special clone can produce yield up to 40 t ffb/ha/year under special soil management practices. In the near future, we can expect more of this clone to be planted in Malaysia. On the average, the rate of oil extracted by the oil palm factory in Malaysia right now is 20 % of the ffb. So, we can get at least 4 tones of oil/ha/year. Malaysia is now aiming for a rate of 25 % extraction in the near future considering the advent of new extracting technology coming from R&D by the industry.

Fig. 3. A well managed oil palm estate in Malaysia

As seen in Figure 3 oil palm fronds are placed in between the planting rows. This practice helps maintain soil moisture in the oil palm estates. When the organic matter is decomposed (mineralized) plant nutrients (N, P, K) are released into the soils and can be taken up by the growing oil palm in the field.

After 25 years of production it is time to replant the oil palm. At this age the palm is no longer productive. Furthermore, it is too tall for harvesting the fresh fruit bunches using standard practice (knife stuck to a long wooden pole). Under the current practice burning of oil palm fronds and trunks are not allowed, not eco-friendly. Oil palm estates are now forced to practice zero-burning technology. In this technique of estate management, the fronds and trunks are cut and chopped off into small pieces and are buried into the ground. In this practice, we return the oil palm biomass to the soils for good. Again on mineralization, essential nutrients are added into the soils. It has been found that zero-burning technology is able to cut down the cost of production considerably where less fertilizer is applied to keep the oil palm growing. In the beginning a teething problem has arisen. The burying of plant biomass has encouraged the outbreaks of pests and diseases. However, these problems have since been taken care of by the estate management in the country.

Many areas under oil palm cultivation in Malaysia are undulating to steep land. These areas need to be protected lest soil erosion removes the fertile topsoil comes rainy season. In these areas contour terraces are constructed perpendicular to the slope of the land. The bare areas in between the terraces are planted with leguminous cover crops before oil palm canopy closes up. The cover crops fix some nitrogen from the air and added to the soils when they die. This is an excellent management practice for sustaining the productivity of Ultisols and Oxisols in Malaysia.

4. Rubber cultivation

Rubber has been grown commercially in Malaysia since 1903 when the first rubber estate was established in Melaka. Malaysia is the biggest consumer of pure latex and fifth in the consumption of natural rubber in the world. Asia is the biggest producer of natural rubber, led by Thailand, Indonesia and Vietnam. Surgical latex gloves and condoms are known for their high quality and accepted by the consumers. Currently, Malaysia is the biggest exporter of these products. The export earning from rubber in 2009 was worth USD 8.3 billion. Demand for natural rubber is expected to increase and it is projected that there will be a shortage of natural rubber in the near future. The price of SMR 20 was USD 0.77/kg in 2003 and USD 3.52/kg in 2010 (Malaysian Rubber Board, 2010). Rubber prices reached their highest in decades, fuelled by strong demand from China's auto industry.

Clones, in the case of rubber, refer to rubber plants produced from selection breeding process, which have been field-tested. There are several selection criteria in rubber breeding. These include vigor, resistance to diseases, resistance to wind damage, fewer and higher branches, bark thickness and yield. The clones are produced to improve the yield of rubber trees. In recent time, yield is not only latex but also timber from the trees, which has high demand in the marketplace. To satisfy for demand in latex and timber, the Malaysian Rubber Board has conducted research to produce rubber trees that can give high yield of latex and timber, resulting in the introduction of RRIM 2000 and RRIM 3000 series of clones. Clone RRIM 3000 is superior to that of RRIM 2000. Clone RRIM 3001, because of its vigorous growth and high yield, has been assigned a new name, i.e. Klon 1 Malaysia. Rubber plants exhibit a variety of responses to different water regimes. The ability of rubber plants to cope with water stress varies across and within clones (Shafar Jefri Mokhatar & Noordin Wan Daud, 2011).

For many years Malaysia was the major exporter of rubber. Now the country is number three in terms of latex production; Thailand and Indonesia have overtaken us of late. However, we still export rubber (scrap or solid rubber) to the world market, although we import some rubber (latex) in order to feed our own growing rubber industry. Due to the acute shortage of labor in Malaysia some rubber trees are left not tapped. The current area under rubber in Malaysia stands at 1.2 million ha, much less than what it used to be some 40 years ago. Most of our rubber are grown on Ultisols and Oxisols with little problem (Figure 4). Rubber seems to grow quite well on soils under well drained condition. Like oil palm it is acid tolerant and pH of 4-5 does not affect its growth. Under normal fertilizer input, rubber grows very well, giving high yield and contributing to the country's economic growth.

Like oil palm, the P-fertilizer for rubber cultivation is phosphate rock. This fertilizer has been applied in rubber estates for as far as we could remember. The benefit of phosphate rock has been clearly explained before. It works very well under acidic condition prevailing in the highly weathered soils of Malaysia. The preferred N-fertilizer in rubber estates is $(NH_4)_2SO_4$. The ammonium from this fertilizer undergoes nitrification which releases H^+ and hence, its long-term application would increase soil acidity slightly. Beside nitrogen, this fertilizer supplies S to the soils, which is another macronutrient. Rubber plant is said to be acid tolerant and so the acidity so produced by the application of the fertilizer would not affect its growth much. Unlike oil palm, Kedah and Perlis are quite suitable for rubber cultivation. If the rainfall is too high it will affect latex production because farmers are reluctant to tap rubber trees.

Fig. 4. A well managed rubber estate in Malaysia

Many rubber estates in the country are located on undulating or steep land, which are prone to soil erosion during rainy season. As such, contour terraces are needed to be constructed to prevent or reduce soil erosion, very much the case of oil palm estate. If this practice is carried out coupled with modern tapping technique there is no reason why Malaysia should not be a great rubber producer again, like it once was. Right now the price of rubber in the marketplace is very good and many farmers have since returned to the industry, for the good of the country.

Like oil palm estates, rubber estates are practicing zero-burning technology without fail. Malaysia is now in the forefront in promoting this eco-friendly technology for the good of us all. As has been said, practicing zero-burning in rubber estate management has reduced the cost of production and improved or protected the environment. This technology has won worldwide recognition and as such other countries are learning from us. Zero-burning technology is Malaysia's contribution to the world in reduction of global warming.

5. Cocoa cultivation

The fortune of cocoa industry in Malaysia is in the balance or in limbo. Way back in the 1970s, Malaysia was the 6th biggest cocoa bean producer in the world. The area under cocoa then was > 400,000 ha. Since then it is in the downward trend. Many cocoa plantations are either abandoned or replaced with oil palm. The cocoa area now is about 50,000 ha with production not enough to feed our own factories producing cocoa products. The government is now encouraging farmers to go back to planting cocoa by offering lucrative subsidy.

Some of the cocoa trees are planted on highly weathered soils with low fertility (Figure 5). Beside low price in the marketplace, outbreak of diseases has created havoc in cocoa industry. Furthermore, the Europeans are so used to the good taste of cocoa imported from Ivory Coast, Africa, which is grown on fertile soils with high pH and low Al saturation. Our soils, Ultisols and Oxisols, are acidic with pH ranging from 4 to 5, lower pH for the Ultisols. These soils contain toxic amount of Al and/or Mn, which also affect the taste of cocoa although it has been improved somewhat via innovative fermentation technology.

Current national average rate of cocoa production in Malaysia is low, only 0.8 t/ha/year. This is far too low compared to that of our counterpart in African countries. The target rate of production for country is 1.5 t/ha, which is still way below the potential cocoa yield of > 10 t/ha/year. In Sabah, a yield of > 2 t/ha/year can be obtained for cocoa planted on weathered soils with good soil management practice. At this rate of cocoa production, farmers can make money because the price of cocoa in the marketplace now is quite high.

For growing cocoa on the Ultisols and Oxisols, soil pH needs to be raised to above 5. At this pH, Al in the soil solution starts to precipitate as inert Al-hydroxides, rendering it unavailable to the growing cocoa in the field. Simultaneously, Mn in the soil solution is eliminated. Soil pH can be increased by liming (Shamshuddin et al., 1991; Shamshuddin et al., 2010). Likewise, soil pH can be increased by ground basalt application (Gillman et al., 2002; Shamshuddin & Kapok, 2010). Besides increasing pH, basalt supplies Ca, Mg, K, P and S into the soils for the crop's requirement. This agronomic practice helps reduce the cost of production as less fertilizer needs to be applied. The best technique is to apply basalt in combination with organic fertilizer. The organic fertilizer supplies nitrogen needed by cocoa for its healthy growth. Basalt is, however, taking a long time to disintegrate and dissolve completely. But it gives a long-term ameliorative benefit to the cocoa plant. As soil pH increases negative charges on the exchange complex of the variable charge minerals increases (Shamshuddin & Ismail, 1995). This helps retain basic cations in the topsoil and hence, soil productivity is further improved.

If the rainfall at an area is too much, the area is less suitable for cocoa cultivation. Under this condition, cocoa can be infested by diseases, resulting in the abortion of cherelles. This in the end would reduce the yield of cocoa. This being the case, it is not wise to grow cocoa in areas with heavy rainfall like Johor where the annual rainfall can exceed 3,000 mm. As it is

Sabah is the best growing area in Malaysia for cocoa production in terms of soils and climatic conditions.

Fig. 5. A well managed cocoa estate in Malaysia (Courtesy of Malaysian Cocoa Board)

6. Conclusion

Oil palm and rubber are suitable to be grown on Ultisols and Oxisols in Malaysia because the crops are acid and Al tolerant, but cocoa is not unless the soils are amended with lime or basalt. However, climate can somewhat affect their production except oil palm. With proper soil management practices, the yield of oil palm and rubber are high, contributing money that sustains the Malaysian economy. As the crops are planted either on undulating and/or sloping land, contour terraces are needed to be constructed to reduce or to prevent soil erosion that removes the fertile topsoil. Planting leguminous cover crops in between the rows of oil palm or rubber trees further helps improve soil conservation. On the other hand, cocoa requires lime or basalt application at the appropriate rate and time to increase soil pH, lower Al and increase basic cations. In so doing, cocoa can be grown sustainably on Ultisols and Oxisols in Malaysia.

7. Acknowledgements

The authors would like to acknowledge Universiti Putra Malaysia for technical and financial support.

8. References

Anda, M., Shamshuddin, J., Fauziah, C.I & Syed Omar, S.R. (2008a). Mineralogy and factors controlling charge development of three Oxisols developed from different parent materials. *Geoderma, 143,* 153-167.

Anda, M., Shamshuddin, J., Fauziah, C.I. & Syed Omar, S.R. (2009). Dissolution of ground basalt and its effects on Oxisol chemical properties and cocoa growth. *Soil Science, 174,* 264-271.

Anda, M., Shamshuddin, J., Fauziah, C.I. & Syed Omar, S.R. (2010). Increasing the organic matter content of an Oxisol using rice husk compost: changes in the composition and its chemistry. *Soil Science Society of America Journal, 74,* 1167-1180.

Anda, M., Syed Omar, S.R., Shamshuddin, J. & Fauziah, C.I. (2008b). Changes in properties of composting rice husk and their effects on soil and cocoa growth. *Communication in Soil Science and Plant Analysis, 39,* 2221-2249.

Auxtero, E.A. & Shamshuddin, J. (1991). Growth of oil palm (*Elaies guineensis*) seedlings on acid sulfate soils as affected by water regime and aluminum. *Plant and Soil, 137,* 243-257.

Duff, D. (1993). *Holmes' Principles of Physical Geology*. Chapman and Hall, London.

Gillman, G.P., Burkett, D.C. & Coventry, R.J. (2001). A laboratory study of basalt dust to highly weathered soils: effect on soil chemistry. *Australian Journal of Soil Science, 39,* 799-811.

Gillman, G.P., Burkett, D.C. & Coventry, R.J. (2002). Amending highly weathered soils with finely ground basalt rock. *Applied Geochemistry, 17,* 987-1001.

Ismail, H., Shamshuddin, J. & Syed Omar, S.R. (1993). Alleviation of soil acidity in a Malaysian Ultisol and Oxisol for corn growth. *Plant and Soil, 151,* 55-65.

Malaysian Rubber Board. (2010). Prices of SMR 20 by Malaysian Rubber Board. Http://www3.lgm.gov.my/mre/YearlyChart.aspx. Accecced on November 15, 2010.

Shafar Jefri Mokhatar & Noordin Wan Daud. (2011). Performance of *Hevea brasiliensis* on Haplic Ferralsols as affected by different water regimes. *American Journal of Applied Sciences,* 8(3), 206-211.

Shamshuddin, J. & Ismail, H. (1995). Reactions of ground magnesium limestone and gypsum in soils with variable-charge minerals. *Soil Science Society of America Journal, 59,*106-112

Shamshuddin, J. & Kapok, J.R. (2010). Effect of ground basalt application on the chemical properties of an Ultisol and Oxisols in Malaysia. *Pertanika Journal of Tropical Agricultural Science, 33,* 7-14.

Shamshuddin, J., Che Fauziah, I. & Sharifuddin, H.A.H. (1991). Effects of limestone and gypsum applications to a Malaysian Ultisol on soils solution composition and yields of maize and groundnut. *Plant and Soil, 134,* 45-52.

Shamshuddin, J., Fauziah, C.I. & Bell, L.C. (2009). Soil solution properties and yield of corn and groundnut grown on Ultisols as affected by dolomitic limestone and gypsum applications. *Malaysian Journal of Soil Science, 13,* 1-12.

Shamshuddin, J., Jamilah, I. & Mokhtaruddin, A.M. (1987). Chemical changes in soils affected by the application of palm oil mill effluent. *Pertanika, 10,* 41-47.

Shamshuddin, J., Sharifuddin, H.A.H. & Bell, L.C. (1998). Longevity of magnesium limestone applied to an Ultisol. *Communication in Soil Science and Plant Analysis, 29,* 1299-1313.

Soil Survey Staff. (2010). *Keys to Soil Taxonomy.* United States Department of Agriculture, Washington DC.

Tessens, E. & Shamshuddin, J. (1983). *Quantitative Relationship between Mineralogy and Properties of Tropical Soils.* UPM Press, Serdang, Malaysia.

Tessens, E. & Zaharah, A.B. (1983). The residual influence of P-fertilizer application on soil pH values. *Pedologie, 32,* 367-368.

3

Moisture and Nutrient Storage Capacity of Calcined Expanded Shale

John J. Sloan[1], Peter A.Y. Ampim[1], Raul I. Cabrera[1],
Wayne A. Mackay[2] and Steve W. George[3]
[1]*Texas AgriLife Research; Dallas, TX*
[2]*Mid-Florida Research & Education Center, Apopka, FL*
[3]*Texas AgriLife Extension Service; Dallas, TX*
USA

1. Introduction

Expanded shale (EXSH) is an important and increasingly popular soil conditioner with several horticultural applications, including its use as a soil amendment for clay textured soils (Sloan et al., 2002), as an ingredient in plant growing media (Sloan et al., 2010) or green roof substrates (Ampim et al., 2010). It is a lightweight material produced by firing mined lumps of shale at high temperatures in a rotary kiln in a process similar to that of clay ceramics. The resulting product can be screened to create various size fractions depending on the intended use. For example, Texas Industries (TXI) of North Texas, USA produces five size fractions of expanded shale that includes the following ranges, from smallest to largest, 0.07 to 0.60 mm, 0.60 to 2.0 mm, 2.0 to 4.8 mm, 4.8 to 6.4 mm, 6.4 to 9.5 mm, 9.5 to 12.7mm, 12.7 to 15.9 mm Expanded shale aggregates are suitable as components of planting media and soil amendments because, unlike most minerals, they are porous, stable, and resistant to decomposition (Ferguson, 2005). Expanded shale is believed to beneficially modify growth media properties by enhancing overall aeration, improving water and nutrient holding and release capacities, and promoting optimum plant growth (Blunt, 1988; Dunnett and Kingsbury, 2008).

Sloan et al. (2002) found that expanded shale consistently improved overall plant performance better than quartz sand, sphagnum peatmoss and cottonseed hull when they were used as amendments for poorly-drained Austin silty clay soils suggesting its superiority as a soil conditioner for the production of horticultural crops on soils with poor tillage characteristics. In a similar way, Nash et al. (1990) found that a potting medium comprising a mixture of peat moss and expanded shale increased the growth and quality of petunia and impatiens. Smalley et al. (1993) also found that amending soils with products containing expanded shale did not hamper plant performance. Though growth index and plant dry weight of Salvia (*Salvia splendens*) and Vinca (*Catharanthus roseus*) increased with increasing fertilizer levels for all their treatments, the greatest performance of these plants were observed for treatments amended with the product containing expanded shale, granite sand and composted poultry litter. In another application, Forbes et al. (2004 and 2005) discovered that expanded shale is a potential sorbent for phosphorus in subsurface flow wetlands because of its high hydraulic conductivity, large surface area and P sorption capacities. Forbes and et al. (2004) found that expanded shale retained 164 ± 110 g $P/m^2/yr$ in pilot-scale wetlands. Regardless of the generally good experimental reports for expanded shale and/or mixtures containing it, there is a need to understand the mechanisms controlling the basic water and nutrient retention properties of expanded shale and to ascertain its suitability for horticultural and soil amendment uses because expanded shale products can vary depending on geologic origin of the raw materials and differences in the production process (Ferguson, 2005).

In the United States of America, expanded shale is manufactured by multiple companies, most of which are members of the Expanded Shale, Clay, and Slate Institute (ESCSI). Expanded shale has a third to half the weight of regular rock or sand so it is easy to transport but has enough weight to avoid being carried away by water or wind. It is inert, inorganic and therefore not expected to degrade or react with agricultural or horticultural chemicals (TXI, 2009). Given the increasing interest in using EXSH as a soil amendment or as an ingredient in growing media, it is important to understand its interaction with water and nutrients. The first step towards realizing this goal is to thoroughly understand the basic properties of the material itself. The objective of this study therefore was to evaluate the dynamics of water and nutrient adsorption by EXSH.

2. Materials and methods

2.1 Basic chemical properties

2.1.1 pH and electrical conductivity

Untreated EXSH (10 g) was equilibrated with deionized water (10 mL) for 60 min on a reciprocal shaker and pH was measured in the supernatant with a hydrogen-specific combination electrode. Additionally, pH of the EXSH was measured using 0.01 M $CaCl_2$ as the equilibrating solution in order to minimize the possible effect of variable background salt concentrations due to soluble components in the EXSH (ASTM D 4972-89). Electrolytic conductivity (EC) of the EXSH was measured by equilibrating 20 g of EXSH with 20 mL of deionized water for 60 min on a reciprocal shaker. Then the supernatant was filtered through Whatman No. 2 filter paper and EC was measured with a dip-type electrode. The filtered supernatant was saved for determination of soluble constituents.

2.1.2 Soluble and labile elements

EXSH constituents that dissolve easily in water are immediately available to plants and microbes when applied to soil provided they are not immobilized in the soil. To determine the amounts of soluble elements in EXSH, the supernatants saved from the EC analysis were analyzed for Na, K, Ca, and Mg by flame atomic absorption spectrophotometry. In addition to soluble elements, labile elements in the EXSH were measured by extracting 20 g samples of EXSH with 20 mL of a buffered (pH 5) 1 mol L^{-1} acetic acid solution (Gibson and Farmer, 1986). Labile elements are not immediately soluble in water, but they are potentially available to plants and soil microorganisms when applied to soil.

2.1.3 Calcium carbonate equivalent

Calcium carbonate equivalent (CCE) is the measure of a material's ability to neutralize acidity relative to pure calcium carbonate. The comparison is useful for determining how a material will affect soil pH when mixed with soil. The CCE of EXSH was estimated by suspending 10 g of expanded shale in 50 mL deionized water and then titrating with 0.1 mole L^{-1} HCl to a pH 7 endpoint (modified from AOAC, 2005). The amount of pure $CaCO_3$ required to neutralize an equal amount of acidity was calculated and then CCE was determined by dividing the weight of calculated pure $CaCO_3$ by the actual weight of EXSH.

Cation Exchange Capacity (CEC): Methods that consist of extracting and summing exchangeable cations are not suitable for expanded shale because the manufacturing process produces significant amounts of Ca, Mg, K, and Na oxides that readily dissolve in water. On the other hand, expanded shale is essentially devoid of elements that were present in low concentrations in the original material and were subsequently volatilized during the heating process. One such element is nitrogen. Therefore, the ammonium saturation method was used to measure the CEC of the expanded shale (Chapman, 1965). For this method, the exchange complex of prewashed EXSH was saturated with NH_4^+, followed by replacement and extraction of NH_4^+ with Na^+. The extracted NH_4^+ was then quantified using a standard method consisting of alkaline steam distillation, boric acid capture, and titration with standard acid.

2.2 Water uptake

Water uptake experiments were designed to evaluate uptake and retention of water by individual EXSH aggregates rather than a bulk amount of the material (Fig.1). The volume of water measured in this manner includes only the internal pores of each EXSH aggregate. It should be noted that measuring the water holding capacity of a bulk volume of EXSH would give a higher value because it would also include water held not only in the internal pores, but also in the pore spaces created between adjacent aggregates.

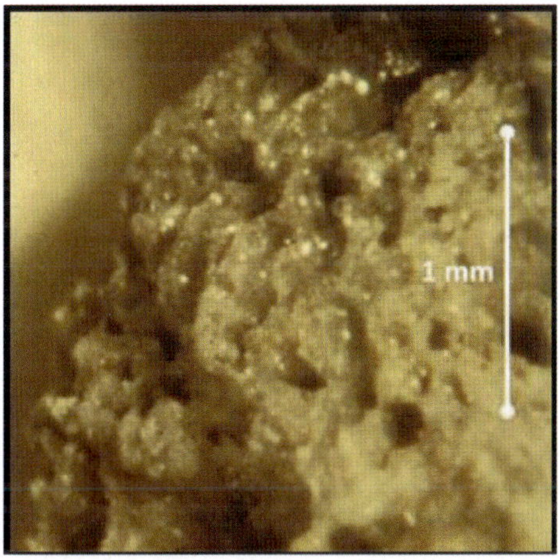

Fig. 1. Magnified view of a single expanded shale particle showing a highly porous exterior surface with range of pore sizes continuing into the interior of the particle

2.2.1 Maximum water holding content

To measure maximum water holding capacity, a quantity of EXSH was submerged in water for 72 h in order to completely saturate individual aggregates. At the end of that time, EXSH was removed from the water, weighed to determine saturated weight, and dried in an oven at 105°C for 48 hours or until a constant weight was obtained. The maximum water holding content was calculated as the difference between the saturated weight and the oven-dry weight divided by the oven-dry weight (w/w).

2.2.2 Water adsorption rate

A laboratory experiment was designed to quantify the rate at which water was absorbed by individual EXSH aggregates. Randomly selected individual EXSH aggregates were spread onto water-saturated filter paper. The filter paper was maintained in a saturated condition by the use of paper wicks connected to a water reservoir. This allowed unrestricted diffusion of water from the saturated filter paper into the EXSH aggregates through the point of direct contact. EXSH was left on the water-saturated filter paper for periods ranging from 2.5 minutes to 72 hours. After the specified time, EXSH was removed from the filter paper and oven-dried to determine water content.

2.3 Nutrient release from fertilizer-treated expanded shale

A bulk quantity of EXSH was soaked overnight in deionized water to remove soluble nutrients inherent to the material and then oven-dried for 24 h at 60°C. After drying, a quantity of this pre-washed EXSH was nutrient-loaded by soaking for 48 h in a soluble fertilizer solution containing 0.56 g L^{-1} NH_4-N, 1.0 g L^{-1} NO_3-N, 2.44 g L^{-1} urea-N, 1.76 g L^{-1} P, and 3.32 g L^{-1} K. Upon removal from the nutrient solution, the EXSH was rinsed with water to remove surficial deposits of nutrient solution and then allowed to air dry. Untreated and nutrient-loaded EXSH were mixed in ratios of 100:0, 75:25, 50:50, 25:75, and 0:100 (v/v) to create growing media with increasing proportions of nutrient-loaded EXSH. In order to assess the bioavailability of nutrients in the fertilizer-treated EXSH growing media, each blend was sequentially extracted six times with water and the water extracts were analyzed for selected fertilizer-added nutrients. More specifically, a 25 g quantity of each EXSH treatment was placed in a 125 mL flask, immersed in 40 mL of deionized water, and agitated on a reciprocal shaker for 30 min. Then the supernatant was filtered through Whatman No. 2 filter paper and analyzed for NH_4-N, NO_3-N and orthophosphate-P. This process was repeated five more times for a total of six extracts.

2.4 Bioavailability of adsorbed nutrients

Nutrient bioavailability in a growing medium is best demonstrated by a plant's ability to extract and utilize essential elements. A greenhouse experiment was initiated to test the bioavailability of fertilizer-applied nutrients in the EXSH growing blends described above. Four replicates of each expanded shale blend that contained 0, 25, 50, 75, or 100% nutrient-loaded EXSH were placed in 500 cm^3 greenhouse pots. The EXSH was covered with a thin layer of acid-washed sand so that Romaine lettuce seeds *(Lactuca sativa)* planted on the surface of the growing media would have an adequate germination bed. Approximately 20 seeds were sowed onto the sand layer followed by misting with deionized water twice a day to initiate seed germination. After emergence of lettuce seedlings, the number of plants was thinned to 5 plants per greenhouse pot. Plants were watered daily with deionized water, but no additional nutrients were supplied. After one month, above ground plant tissue was harvested and dried to determine yield. Plant tissue was analyzed for macro- and microelements.

3. Results and discussion

3.1 Basic chemical properties

3.1.1 Electrical conductivity (EC)

A 1:1 water extract of EXSH produced an EC of 1.6 dS m^{-1} (Table 1). Electrical conductivity (EC) is an indirect measurement of the soluble salt concentration of a material. Excessive salt is detrimental to plant growth. Several researchers have proposed different EC ranges that are limiting for plant growth. Mengel and Kirkby (1982) reported that EC values of less than 2 dS/m in soil will have negligible salinity effects on plant growth while Wright (1986) recommended an EC value of 2 dS/m for healthy and vigorous plant growth. In contrast, Gajdos (1997) indicated that EC values exceeding 1-3 dS/m could be detrimental to plants. In containers, however, Lemaire et al. (1985) found that only EC values greater than 3.5 dS/m were too high for vigorous plant growth. Based on published research findings (Lemaire et al., 1985; Mengel and Kirkby, 1982; Wright, 1986), the EC value measured for EXSH in this study will likely not create salinity problems when added to soil or potting media.

3.1.2 pH

Table 1 shows pH values for EXSH measured in water and in a 0.01 mole L^{-1} dilute $CaCl_2$ solution. The $CaCl_2$ solution was used to eliminate the effects of background salt concentrations in the EXSH. The dilute calcium chloride method for pH measurement is widely trusted to produce the most reliable results close to the actual pH in the root zone (Handrek and Black, 1994). As a result, the close agreement between the two values (pH 8.3 for water versus pH 8.5 for $CaCl_2$) indicates minimal effects due to soluble salts and suggests that both values accurately represent the pH of EXSH. The measured pH values for EXSH reflect the presence of calcium, magnesium, potassium, and sodium oxides (CaO, MgO, K_2O, and Na_2O). These oxides were formed as a result of the high temperatures used to manufacture the expanded shale. When EXSH is exposed to water, these oxides hydrolyze to form hydroxides, which can raise pH to as high as 13 in the equilibrium solution. The fact that the equilibrium pH of the EXSH tested in this study only increased to 8.3 suggests that the material contains only relatively small quantities of reactive oxides. Expanded shale materials with higher quantities of reactive oxides would have higher equilibrium pH values.

3.1.3 Calcium carbonate equivalent (CCE)

A material's capacity to neutralize acidity indicates its potential effect on pH when mixed with soil or other growing media materials, such as sphagnum peat moss, compost, sand, vermiculite, etc. Calcium carbonate equivalence can be an important value for alkaline materials, such as EXSH, that will be blended with other growing media ingredients because it predicts the how the EXSH will affect the final pH of the growing media. The calcium carbonate equivalent (CCE) shows the material's effectiveness relative to pure calcium carbonate, which is the standard material for neutralizing soil acidity. The EXSH analyzed in this study had a CCE of 0.2 to 0.3 percent. In other words, EXSH is less than 1% effective at neutralizing soil acidity as calcium carbonate limestone. This indicated that although measured pH values for EXSH were slightly above the broad range of 5.5 to 8 known to support plant growth (Bunt, 1988), it is unlikely that its addition to soil, peat moss, compost, or other growing media ingredients would result in long term increases in the pH of the growing medium. However, Sloan et al. (2010) reported that the pH of various organic growing media, including sphagnum peat moss, pine bark, compost, and biosolids, were slightly elevated when the proportion of EXSH exceeded 30% (v/v), especially when the initial pH was less than 5.0. It is likely that EXSH will have a positive impact on strongly acidic growing media, such as peat moss and pine bark due to it alkaline pH, but that it will have only a small impact on growing media with neutral to alkaline pH values, such as compost and biosolids.

Parameter	Measured Value	
	Mean	SD†
Electrical conductivity, dS m^{-1}	1.6	(± 0.2)
pH		
(1:1 - water)	8.25	(± 0.62)
(1:1 - 0.01 mol L^{-1} $CaCl_2$)	8.47	(± 0.21)
Calcium carbonate equivalent (CCE), %	0.2 - 0.3	
Water holding capacity, %	37.8	(± 1.35)
Cation exchange capacity (CEC), cmole kg^{-1}	2.75	(±0.15)
Extractable elements, mg/kg		
Water soluble		
Sodium (Na)	55.0	(± 6.5)
Potassium (K)	64.2	(± 6.0)
Magnesium (Mg)	29.0	(± 2.8)
Calcium (Ca)	88.3	(± 10.6)
Acid soluble§		
Sodium (Na)	87.3	(± 5.3)
Potassium (K)	126.8	(± 21.0)
Magnesium (Mg)	660	(± 40)
Calcium (Ca)	843	(± 59)

† Standard deviation of the mean. Shown in parentheses in table
§ 1 mol L^{-1} sodium acetate buffered to pH 5 with acetic acid.

Table 1. Basic chemical and physical properties of expanded shale (ExSh)

3.1.4 Soluble and labile elements

Expanded shale was extracted with water to measure the concentration of soluble elements and with acidified sodium acetate to measure potentially plant-available (labile) elements. Concentrations of water-soluble sodium (Na), potassium (K), magnesium (Mg), and calcium (Ca) in EXSH were small and similar in magnitude (Table 1). Measured concentrations for these elements are by far lower than guideline values established for cultivation which are < 857 mg/kg for Ca, <111.3 mg/kg for Na and <150 mg/kg for Mg (Fischer and Penningsfeld, 1979; Fisher and Meinken, 1995) and are consistent with the low EC values measured. When EXSH was extracted with acidified sodium acetate, only slightly higher amounts or Na and K were extracted, whereas the amounts of Mg and Ca were increased 20-fold and 10-fold, respectively. The data suggests that the heating process used to manufacture the expanded shale in EXSH produces small amounts of Na, K, Mg, and Ca salts that are easily soluble in water, plus larger amounts of Mg and Ca oxides that will dissolve in weak acid (pH 5). Potassium, Mg, and Ca are essential plant nutrients. Therefore addition of EXSH to soil will provide small amounts of immediately plant-available K, Mg, and Ca plus larger amounts of more slowly available Mg and Ca. However, the inherent concentrations of these elements in EXSH are too low to significantly affect plant nutrition - either beneficially or detrimentally when it is added to soil.

3.2 Water uptake

Two facets of water uptake by EXSH were investigated: 1) the maximum water-holding capacity and 2) the rate of water uptake. The maximum water holding capacity provides information on the total porosity in EXSH aggregates. After soaking in water for at least 72 hr, the EXSH tested in our study contained 37.8% water (Table 1). This value includes the amount of water in EXSH when 100% of the pores were saturated plus the small amount of water on the surfaces of its aggregates. The water-holding capacity of EXSH shows that it can be a significant source of water storage in soil. However, a typical silt loam soil will have a greater available water holding capacity than EXSH. Therefore, EXSH would be an appropriate soil amendment for soils that retain excess water and exhibit poor drainage characteristics. Sloan et al. (2002) reported that 3 to 6 mm diameter expanded shale, similar to that used in this study, effectively increased the root mass of pansy plants *(Viola x wittrockiana)* compared to un-amended clay soil during an excessively wet growing season.

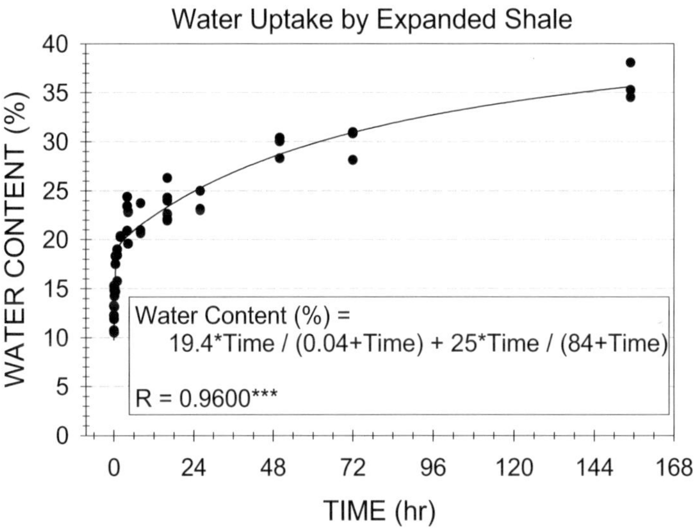

Fig. 2. Rate of water diffusion into expanded shale (EXSH) from a saturated porous medium. Data was fitted to a double rectangular hyperbolic function to create a predictive uptake equation. *** Significant at the 0.001 level of probability

The rate at which EXSH can absorb water is perhaps more important than its total water holding capacity. The water uptake rate provides information on the size of pores in the EXSH aggregates and can be related to the amount of water that will be available to plants. Figure 2 shows that EXSH absorbed 15% of its weight in water within 10 minutes and 20% within 2 hours. The water uptake rate slows considerably beyond 2 hours, but continued to increase to a maximum of 36% at 150 hours. The experiment was designed so that water could only enter the EXSH aggregates via capillary diffusion. When an EXSH aggregate was placed in contact with a saturated porous medium, water quickly diffused throughout the aggregate's surface layer until the exterior pores of the aggregate were saturated. A magnified view of an individual expanded shale aggregate shows the exterior pores to be relatively large and evenly distributed around the surface (Fig. 1). Water is held in these pores at relatively low tension and is therefore easily available for plant uptake, but it is also susceptible to rapid evaporation. This water accounts for 15 to 20% (w/w) of the total water holding capacity of EXSH. The remaining 16 to 21% of the total water is held in the much smaller interior pores of the EXSH aggregates. This water is held at relatively high tensions and is therefore mostly unavailable for plant use.

3.3 Nutrient release from fertilizer-treated EXSH

The porous nature of EXSH allows it to absorb water and therefore, any chemical constituents dissolved in water will also be absorbed by the EXSH aggregate. For this experiment, EXSH was loaded with nutrients by soaking it in a fertilizer solution for 48 hours, followed by rinsing in deionized water, and then air-drying. The reason for rinsing the fertilizer-impregnated EXSH was to ensure that nutrients released from the aggregates were from the internal pores and not residual surface deposits of fertilizer. Based on results from the water uptake experiment (Fig. 2), 5 to 10% of the pore volume was not saturated with fertilizer solution because the EXSH was soaked for only 48 hours whereas 150 hours was needed to attain complete saturation. Also, it is likely that 10 to 15% of the fertilizer was removed from the exterior EXSH during the rinsing process. These two factors probably reduced the total nutrient-supplying capacity of the fertilizer-treated EXSH by 15 to 25%.

A portion of each fertilizer-treated EXSH medium containing 0, 25, 50, 75, or 100% nutrient-loaded EXSH was sequentially extracted with water to determine the kinetics of nutrient release. Nutrients extracted with water are immediately available for plant uptake. Nitrate-nitrogen (NO_3-N), ammonium-nitrogen (NH_4-N), and phosphate-phosphorus (PO_4-P) were measured in all extracts. Although the fertilizer used to impregnate the EXSH with nutrients also contained urea-N, we did not measure that form of N in the water extracts. In general, the concentration of each constituent decreased with each sequential extract (Fig. 3). This demonstrates that a portion of soluble nutrients absorbed by EXSH can later redissolve and become available for plant uptake. If EXSH were present in a growing medium, individual aggregates would either absorb or release nutrients depending on the concentration gradient of the surrounding solution. In practical terms, if EXSH aggregates in the growing medium were surrounded by relatively pure water, then nutrients would diffuse from the interior pores towards the exterior pores due to a concentration gradient. In a similar but reverse process, nutrients from a fertilizer solution passed through the medium would possibly diffuse into the interior pores of the EXSH aggregates in response to the nutrient gradient. Phosphorus retention by expanded shale in constructed wetlands as reported by Forbes et al. (2004 & 2005) may have been related to this diffusion phenomenon.

Nitrate-N, NH_4-N and PO_4-P each have different binding mechanisms to solid particles in a growing medium. Nitrate-N has a negative charge and exists as a free (i.e., unbound) ion in typical soil conditions. It moves freely through the soil with soil water. Ammonium-N has a positive charge and is attracted to negatively charged exchange sites found in typical soils.

This allows it to be held more tightly in the soil and makes it less susceptible to loss through leaching. Orthophosphate-P is an oxyanion and tends to strongly bind to various soil components, primarily aluminum and iron minerals at low pH (<5.5), and calcium and magnesium minerals at high pH (>7.6).

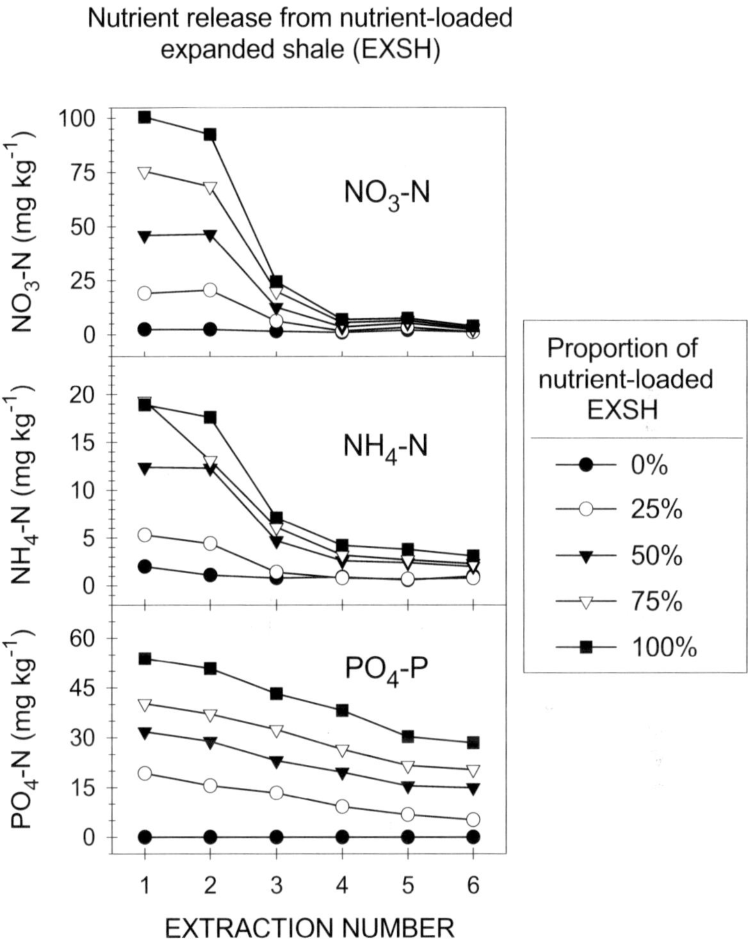

Fig. 3. Sequential extraction of NO_3-N, NH_4-N, and orthophosphate-P from expanded shale (EXSH) loaded with N, P, and K fertilizer

The different chemistries of NO_3-N, NH_4-N, and PO_4-P were apparent in the sequential extraction data presented in Fig. 3. Nitrate-N was easily extracted with water. This was best illustrated with the 100% fertilizer-treated EXSH treatment. The first extraction removed 100 mg NO_3-N per kg of EXSH. By the fourth extraction, the amount extracted had decreased to 7 mg kg^{-1}, but was still slightly greater than the concentration for the untreated 0% control (i.e., EXSH without fertilizer pre-treatment). The remaining extractions continued to remove small amounts of NO_3-N that were slightly greater than the control. The extraction of NO_3-N from fertilizer-treated EXSH was consistent with water uptake data that showed very rapid water uptake followed by a much slower uptake phase (Fig. 2). In the absence of any attractive force between NO_3-N and the EXSH aggregate, NO_3-N would diffuse out of the EXSH in a fashion similar to water. The rapid release of NO_3-N with the initial extractions corresponds to diffusion of NO_3-N from larger pores near the surface of the aggregate (Fig. 1). The much lower, yet significantly higher concentrations of NO_3-N released with the later extractions relative to the control suggest diffusion of NO_3-N from smaller pores deeper inside the EXSH aggregate.

Smaller amounts of NH_4-N were extracted from fertilizer-treated EXSH compared to NO_3-N because the fertilizer source used to pretreat the EXSH contained 1000 mg L^{-1} NO_3-N, but only 560 mg L^{-1} NH_4-N. However, the magnitudes of the difference between desorption of NO_3-N and NH_4-N could not be explained by fertilizer composition alone. It is likely that some of the NH_4-N was retained in the EXSH by cation exchange capacity. The CEC analysis indicated that EXSH had a small but significant CEC of 2.75 cmol kg^{-1} (Table 1), which was adequate to retain up to 385 mg kg^{-1} NH_4-N. Other than the quantity extracted, release of NH_4-N from fertilizer-treated EXSH followed a trend similar to that shown by NO_3-N suggesting that diffusion processes were a significant factor in controlling the release of NH_4-N from the porous EXSH.

Phosphorus release curves from fertilizer-treated EXSH were distinct from those for NO_3-N and NH_4-N (Fig. 3). The amount of PO_4-P released decreased linearly with each successive extraction. There was no rapid release with the initial extractions as seen for NO_3-N and NH_4-N. This suggests that other factors, in addition to diffusion, controlled the release of PO_4-P from EXSH. The presence of calcium- and magnesium- oxides in EXSH created conditions favorable for the adsorption and/or precipitation of PO_4-P. Experience with calcareous soils has shown that dicalcium phosphate ($CaHPO_4$) is commonly a first precipitation product following application of phosphate fertilizer (Sposito, 1989). EXSH contained a substantial amount of extractable Ca as well as pH>8.3 (Table 1) suggesting the possible precipitation of the relatively soluble dicalcium phosphate ($CaHPO_4$) mineral. Phosphate adsorption and precipitation are reversible processes, particularly in an EXSH medium, which is much less complex than a soil medium. Dissolution and release of adsorbed and precipitated phosphorus from EXSH would occur at a slower and more constant rate than would be expected from simple diffusion due to a concentration gradient. The practical implication of these results is that phosphorus-impregnated EXSH could function as a relatively effective slow-release phosphorus fertilizer. Forbes et al. (2004 & 2005) also found that expanded shale was effective at removing soluble P from effluent water in a constructed wetland.

3.4 Lettuce growth in fertilizer-treated EXSH

The true test of nutrient bioavailability is whether plants are able to extract and utilize nutrients from the growing medium. For this experiment, Romaine lettuce (*Lactuca sativa longifolia*) was grown in the same fertilizer-treated, EXSH media described in the preceding sequential extraction discussion. Briefly, the growing media consisted entirely of EXSH where 0, 25, 50, 75, or 100% (v/v) had been previously impregnated with fertilizer solution. Lettuce was grown for approximately 45 days with no additional fertilizer application and then harvested to determine shoot and root yields as well as nutrient content of the aboveground tissue (Fig. 4).

Fig. 4. Increasing the proportion of nutrient-loaded expanded shale (EXSH) in the growing media from 0 to 100% resulted in significant increases in the size of Romaine lettuce

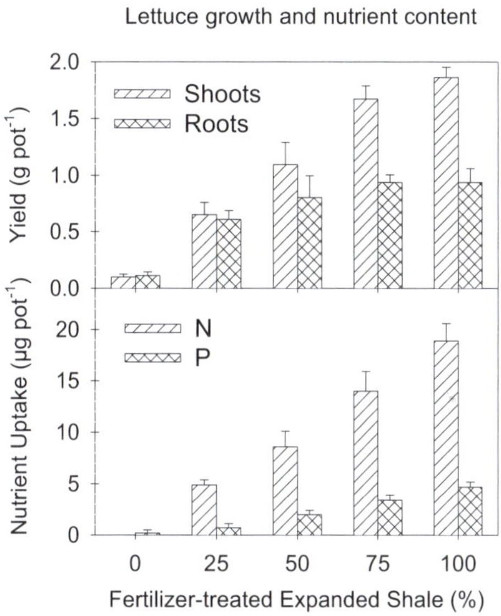

Fig. 5. Lettuce yields and total nutrient uptake for lettuce grown in fertilizer-treated expanded shale

Results for the shoot and root biomass yield as well as total nitrogen and phosphorus content of the aboveground tissue are shown in Fig. 5. The quantity of aboveground lettuce shoots increased linearly as the percentage of fertilizer-treated EXSH in the growing medium was increased (Fig. 5). Root mass also increased with increasing proportion of fertilizer- treated EXSH, but reached a maximum at the 75% fraction. Total uptake of nitrogen and phosphorus also increased linearly with increasing proportion of fertilizer-treated EXSH (Fig. 5). The yield and nutrient uptake data demonstrate that nutrients in the fertilizer-treated EXSH medium were taken up and assimilated by plants. The level of fertility was sufficient to support lettuce growth for approximately 45 days with no additional fertilization. Following harvest of the lettuce plants, additional lettuce seeds were planted into these same EXSH growing media treatments, but no fertilizer was added. This second planting of lettuce failed to thrive in the same way as the first planting (data not shown) suggesting that the ability of EXSH to provide a reservoir of abundant plant nutrients is limited without additional fertilizer applications. Other studies (Nash et al., 1990; Smalley et al., 1993; Sloan et al., 2002) have also reported good plant growth in growing media containing EXSH as a major ingredient. Together these observations suggest that expanded shale can retain and release sufficient amounts of nutrients for healthy plant growth either alone or in a mixture, but additional fertilization will occasionally be needed depending on the intensity of plant growth and the frequency of leaching events. EXSH is therefore suitable as an amendment for such growing media as soil, potting mixes, and green roof substrates.

4. Conclusions

Expanded Shale (EXSH) used in this study had small amounts of Ca, Mg, K, and Na oxides. This was confirmed by measured pH values of >8 in both water and 0.01 mol L^{-1} $CaCl_2$. However, a low CCE of 0.2-0.3% suggests that EXSH is unlikely to significantly affect pH when added to a more buffered materials such as soil, sphagnum peat moss or compost. The concentrations of water soluble Na, K, Mg and Ca were small and consistent with the measured EC of 1.6 dS m^{-1}. Acid extraction of these elements from EXSH yielded higher amounts with Mg and Ca concentrations increasing by 20-fold and 10-fold, respectively, suggesting that when the product is added to a growing medium, small quantities of plant-available Na, K, Mg and Ca can be provided initially and followed later by larger amounts of Mg and Ca in a slow release fashion. However, the concentration of these elements in EXSH is insufficient to affect plant nutrition positively or negatively. The release of NO_3-N and NH_4-N from nutrient-loaded EXSH was rapid and controlled by mainly by diffusion mechanisms, although cation exchange capacity was also responsible for retention of NH_4-N. The movement of PO_4-P on the other hand was influenced by both diffusion and the presence of calcium and magnesium oxides in the EXSH. The presence of these oxides enhanced the reversible processes of PO_4-P adsorption and precipitation in the EXSH medium. In practical terms, the ability of EXSH to reversibly adsorb fertilizer P suggests EXSH may be a suitable material for use as a slow-release phosphorus fertilizer. In this study, both the shoot and root mass of romaine lettuce increased with increasing proportions of fertilizer treated EXSH in the growing medium. This demonstrates that nutrients were released by the fertilizer treated EXSH and assimilated by the lettuce to support growth. The linear relationship between total nitrogen and phosphorus uptake and the proportion of fertilizer impregnated EXSH in the growing medium confirms the ability of EXSH to release nutrients for plant uptake.

5. References

Ampim, P.A.Y., J.J. Sloan, R.I. Cabrera, D.A. Harp, and F.H. Jaber. 2010. Green roof growing substrates: Types, ingredients, composition, and properties. J. Environ. Hort. 28:244-252.

AOAC International. 2005. Agricultural Liming Materials. Official Methods of Analysis of AOAC International, 17th Edition. OAC International, 481 North Frederick Ave, Suite 500, Gaithersburg, MD 20877-2417, United States of America.

ASTM D4972–01. 2007. Standard Test Method for pH of Soils. ASTM International, West Conshohocken, PA, 2003, DOI: 10.1520/D4972-01R07. www.astm.org.

Bunt, A.C. 1988. Media and mixes for container-grown plants. Unwin Hyman. London.

Chapman, H.D. 1965. Cation-Exchange Capacity. *In* C.A. Black et al. (eds.) Methods of Soil Analysis: Chemical and Microbiological Properties. Part 2. Agronomy No. 9. American Society of Agronomy. Madison, Wisconsin.

Dunnett, N. and N. Kingbury. 2008. Planting green roofs and living walls. Timber Press Inc. Portland, OR.

ESCSI (Expanded Shale, Clay and Slate Institute). 2002. Introducing the friendly material: Rotary kiln produced lightweight soil conditioner. Publication No. 8600.

Evanylo, Greg and Mike Goatley Jr. 2011. Chapter 9. Organic and Inorganic Soil Amendments. *In* M. Goatley Jr. and K. Hensler (eds.) Urban Nutrient Management Handbook. College of Agriculture and Life Sciences, Virginia Polytechnic Institute and State University. PUBLICATION 430-350. (Available online at: http://pubs.ext.vt.edu/430/430-350/430-350.html). Verified on 07-Sep-2011.

Ferguson, B.K. 2005. Porous pavements. CRC Press. Boca Raton.

Fischer, P. and E. Meinken. 1995. Expanded clay as a growing medium-Comparison of different products. Acta Hort. 401: 115-120.

Fischer, P. and F. Penningsfeld. 1979. Hydrokultur - Beurteilung verschiedener Blähtonherkünfte. Gb + Gw 79:106-108.

Forbes, M.G., K.R. Dickson, T.D. Golden, P. Hudak and R.D. Doyle. 2004. Dissolved phosphorus retention of light-weight expanded shale and masonry sand used in subsurface flow treatment wetlands. Environ. Sci. Technol. 38:892-898.

Forbes, M.G., K.L. Dickson, F. Saleh and W.T. Waller. 2005. Recovery and fractionation of phosphorus retained by lightweight expanded shale and masonry sand used as media in subsurface flow treatment wetlands. Environ. Sci. Technol. 39: 4621-4627.

Gajdos, R. 1997. Effects of two compost and seven commercial cultivation media on germination and yield. Compost Sci. and Util. 5:16-37.

Gibson, M.J. and J.G. Farmer. 1986. Multi-step sequential chemical extraction of heavy metals from urban soils. Environ. Pollut. 11:117-135.

Lemaire, F., A. Dartigues and L.-M. Rivière. 1985. Properties of substrate made with spent mushroom compost. Acta Hort. 172:13-29.

Mengel, K. and E.A. Kirkby. 1982. Principles of Plant Nutrition (3rd ed.). International Potash Institute. P.O. Box CH-3048, Worblaufen-Bern/ Switzerland.

Nash, M.A., T.P. Brubaker and B.W. Hipp. 1990. Expanded shale as a potting medium component for bedding plants. HortScience 25:1163.

Rund, R.C. 1984. Agricultural Liming Material. p.1. In S. Williams (ed.) Official Methods of Analysis of the Association of Official Analytical Chemists, 14th ed.

Sloan, J.J., R.I. Cabrera, P.A.Y. Ampim, S.A. George, and W.A. Mackay. 2010. Performance of ornamental plants in alternative organic growing media amended with increasing rates of expanded shale. HortTechnology 20:594-602.

Sloan, J.J., S.W. George, W.A. Mackay, P. Colbaugh and S. Feagley. 2002. The suitability of expanded shale as amendment for clay soils. HortTechnology 12:646-651.

Smalley, T.J., G.L. Wade, W.L. Corley and P.A. Thomas. 1993. Soil amendment increases growth of *Salvia splendens* and *Catharanthus roseus*. HortScience 28:540

Sposito, G. 1989. The Chemistry of Soils. Oxford University Press. New York, NY.

(TXI) Texas Industries Inc. 2009. Lightweight aggregate as a soil amendment. Retrieved 2 July 20009. http://www.txi.com/TXI-products/TXI-trugro.html.

Wright, R.D. 1986. Pour through nutrient extraction procedure. HortScience 21:227-229.

4

Poultry Litter Fertilization Impacts on Soil, Plant, and Water Characteristics in Loblolly Pine (*Pinus taeda* L.) Plantations and Silvopastures in the Mid-South USA

Michael A. Blazier[1], Hal O. Liechty[2], Lewis A. Gaston[1] and Keith Ellum[2]
[1]*Louisiana State University Agricultural Center,*
[2]*University of Arkansas Monticello*
USA

1. Introduction

Increasing global human populations and wealth have resulted in increased demands for animal protein and widespread use of confined animal feeding operations to meet added animal protein consumptive demands. Disposal of animal wastes from these operations can be ecologically and environmentally problematic (Kellogg et al., 2000; Roberts et al., 2004; Shober & Sims, 2003). Poultry production is an important source of this protein and is a major agricultural industry in the United States. The United States is the world's largest producer and second largest exporter of poultry meat (UDSA Economic Research Service, 2009). Four-fifths of the United States poultry industry is comprised of broiler meat production. Broiler meat production is largely concentrated in Southeastern states (Alabama, Arkansas, Florida, Georgia, Kentucky, Louisiana, Mississippi, North Carolina, Oklahoma, South Carolina, Tennessee, Texas and Virginia), with 82% of U.S. broiler production occurring in these states (National Agricultural Statistic Service, 2008).

Broiler production results in the generation of massive amounts of litter, a mixture of feces, feed, feathers and bedding materials such as straw, peanut or rice hulls, and wood shavings (Gupta et al., 1997; Weaver, 1998). The U.S.A. poultry industry produces more than 11 million Mg of litter per year (Cabrera & Sims, 2000). Broiler poultry litter contains several plant macro- and micronutrients (Table 1), which makes it desirable as an agricultural fertilizer (Sistani et al., 2008). Following removal from poultry production facilities, litter is commonly applied to nearby pastures, hay meadows, and agricultural crops such as corn and cotton to increase crop production and quality (Harmel et al., 2004; Sims & Wolf, 1994). Applications of poultry litter ranging from 4.5 to 11.2 Mg ha^{-1} yr^{-1} are common to supplement or replace inorganic annual fertilizer additions to pastures (Adams et al., 1994). Thus, poultry litter application is an efficient and potentially cost-effective method for improving forage production within the vicinity of production facilities, which helps to sustain non-poultry related agriculture economies in poultry producing regions. Substitution of broiler litter for inorganic fertilizers continues to increase in the southeastern U.S.A. as prices of inorganic fertilizers escalate (Funderberg, 2009).

Element	Nutrient Concentration	
	(g kg^{-1})	(mg kg^{-1})
C	280-320	
N	31-49	
P	4-13	
K	2-28	
Ca	2-28	
Mg	0.4-6	
Fe		1950-2395
Mn		277-424
Cu		263-332
Zn		252-404
B		45-55

Table 1. Ranges of reported nutrient concentrations in broiler poultry litter on an oven-dry basis. Adapted from Eichhorn, 2001; Ekinci et al., 2000; Kingery et al., 1994; Mitchell & Donald, 1995; Pote et al., 2003; Sauer et al., 2000; Sims, 1986; Williams et al., 1999

Poorly planned, excessive, or long-term applications of broiler litter to pastures and other agroecosystems can result in excessive nutrient losses, reductions in surface water quality, and potential risks to human health. Poultry litter is typically applied as a nitrogen fertilizer, but N availability from litter is relatively difficult to predict because only one-third of the N in litter is in exchangeable forms such as NH_4-N and NO_3-N. Two-thirds of N in litter is in organic form, which must be mineralized before it is plant available. Mineralization of N in litter varies from 40 to 90% with edaphic and environmental conditions, particularly conditions at the time of litter application (Mitchell & Donald, 1995). Gaseous losses of N from litter via volatilization can vary from 5 to 20% of total N, which reduces the amount of N available for plant use (Mitchell & Donald, 1995). While the amounts and forms of N in litter can vary considerably, those of other nutrients, particularly P, are relatively stable. As a result, if litter is applied at rates that supply sufficient N to meet crop demand soils can become saturated with P as well as K, Ca, Mg, Cd, Cu, Mn, and Zn (Edmeades, 2003; Kingery et al., 1994). Surface water runoff or soil water leaching associated with these nutrient-saturated soils can reduce water quality in watersheds (Friend et al., 2006; Gallimore et al., 1999; Gaston et al., 2003; Kellogg et al., 2000; Sauer et al., 1999; Sims & Wolf, 1994). Excess nutrients are transported to surface waters via runoff either in particulate forms or sorbed to soil particles which are suspended in surface runoff. Soluble P and C, NO_3-N, NH_4-N, and some organic N species have been demonstrated to be transported by runoff as a solution. NH_4-N and P are often sorbed to soil particles and conveyed by runoff through erosion, and organic C, P, and N have been shown to be moved by runoff in particulate form (Edwards & Daniel, 1992). Repeated applications of poultry litter can lead to accumulations of N and P in soil as well as elevated levels of one or both of these nutrients in surface runoff and subsurface water (McLeod and Hegg, 1984; Sharpley and Menzel, 1987; Kingery et al., 1994). The potential for P saturation and leaching may be particularly high for highly-fertilized and sandy soils (Breeuwsma and Silva, 1992; Nair and Graetz, 2004). Large or chronic accumulations of N and P can contribute to accelerated eutrophication of water bodies, impairing their use and potentially leading to fish mortality and growth of algae (Schindler, 1978; Lemunyon and Gilbert, 1993). Elevated concentrations of N and P in surface water and eutrophication of water bodies have been found in areas with high levels of confined poultry and other animal production (Daniel et al., 1998; Sharpley, 1999; Fisher et al., 2000). State and federal environmental protection agencies have responded to these environmental concerns by implementing regulations requiring poultry operations to develop nutrient management plans, which will frequently reduce the allowable amounts of litter that can be applied (Friend et al., 2006). It is estimated that 50% of the litter produced from areas with high concentrations of poultry production facilities cannot be applied to grasslands and croplands in these same areas due to environmental or economic constraints, which has led to surpluses of manure N and P production in some parts of the southeastern U.S. (Kellogg et al., 2000).

The environmental impacts of poultry litter fertilization could be reduced by applying surplus litter to terrestrial ecosystems other than pasture and cropland where nutrient levels in soils are low and which have a low risk of nutrient transport. Disposal of poultry litter to these ecosystems could increase the area of litter dispersal, reduce the spatial concentration of poultry litter application, and decrease risks to water quality in watersheds. Forests may be a viable alternative to pastures and croplands for broiler litter application. Similar to agroecoystems, forests are often limited by soil N and P supplies (Elser et al., 2007). Forests also have a high potential for nutrient uptake (O'Neill & Gordon, 1994) and have been successfully used to mitigate environmental impacts of municipal waste, municipal effluent, and mill waste disposal (Henry et al., 1993; Polglase et al., 1995; Falkiner & Polglase, 1997; Jackson et al., 2000). In addition, infiltration rates are much higher in forested landscapes than many agricultural landscapes, which could increase potential retention of nutrients and reduce losses through surface runoff in comparison to agricultural crops.

Loblolly pine (*Pinus taeda* L.) has been identified as a practical species to receive poultry litter application (Beem et al., 1998; Friend et al., 2006; Samuelson et al., 1999). Much of the poultry-producing regions of the southeastern U.S.A. are within the natural range of loblolly pine (*Pinus taeda* L.), so transportation of litter to land with loblolly pine is likely minimal (Friend et al., 2006). Loblolly pine is a prevalent and economically important species in the southeastern U.S.A.; it is used within the region to produce 18% of the world's supply of industrial timber (Allen et al., 2005; Prestemon and Abt, 2002). Loblolly pine growth is often limited by soil supplies of N and P (Binkley et al., 1999), and tree- and forest-level growth responses to N and P fertilization have been well demonstrated (Blazier et al. 2006; Colbert et al., 1990; Haywood et al., 1997; Murthy et al., 1997; Vose & Allen, 1988). Single applications of poultry litter, ranging between 2 to 23 Mg ha^{-1}, have been shown to increase loblolly pine growth rates (Dickens et al., 2004; Friend et al., 2006; Lynch and Tjaden, 2004; Roberts et al., 2006; Samuelson et al., 1999) and economic value (Dickens et al., 2004) of N- and P-deficient loblolly pine. Loblolly pine forests have a high capacity for fertilizer retention due to high plant biomass and soil organic matter. Will et al. (2006) reported that 90 to 100% of annually applied N and P was sequestered in aboveground biomass, soil organic matter, or the uppermost 10 cm of soil of a loblolly pine plantation. Due to this high fertilizer retention capacity, minimal offsite movement of nutrients associated with poultry litter application is expected. Furthermore, water runoff potential of forests is lower than that of grasslands and croplands due to their relatively higher infiltration (Zimmermann et al., 2006) and evapotranspiration rates (Farley et al., 2005). Friend et al. (2006) found that nutrients from an application of 4.6 Mg ha^{-1} of poultry litter (on a dry matter basis) was substantially contained within a loblolly pine forest and did not impair water quality.

The principal limitation to fertilizing loblolly pine with poultry litter is the relatively poor accessibility and maneuverability within dense plantations by ground-based fertilizer application equipment. Conventional manure-spreading equipment (manure trucks, tractor-drawn spreaders) cannot be driven through typical pine plantations due to close tree spacing, dense understory vegetation, and/or stumps high enough to cause equipment damage. These issues likely reduce the number of spreader contractors willing to operate within forests (Dickens et al., 2003).

Silvopasture is an alternative land management system that would allow the application of poultry litter to loblolly pine trees by largely circumventing maneuverability limitations of conventional loblolly pine plantations. Silvopasture management systems consist of forage grasses established and cultivated beneath trees in order to simultaneously produce timber and livestock (Clason & Robinson, 2000; Clason & Sharrow, 2000). Silvopasture regimes are currently the most popular form of agroforestry in the southeastern U.S. (Clason & Sharrow, 2000; Zinkhan & Mercer, 1996). Silvopastures are created by either planting trees in pastures (Robinson & Clason, 2002) or by establishing forage crops in forests (Clason & Robinson, 2000). Forage management in silvopastures is conducted similarly to conventional grasslands in the southeastern U.S.A.; herbicides and/or prescribed burning are used to reduce herbaceous and woody competition and fertilization is carried out to optimize forage yields. Due to land ownership and use patterns, there is high potential for conversion between agriculture and forestry in the southeastern U.S.A. (USDA SCS, 1989). Clason (1995) determined that loblolly pine was compatible with several forage crops in silvopasture systems. The relatively wide spacing of trees and forage understory in silvopastures make navigation of manure-spreading equipment possible (Figure 1).

Fig. 1. Applying broiler poultry litter to a silvopasture at the Louisiana State University Agricultural Center Hill Farm Research Station in northwest Louisiana, U.S.A. Picture by Terry Clason, USDA Natural Resource Conservation Service

A pine plantation in which straw is harvested is another management system where applications of poultry litter could increase commodity production. Pine straw mulch has emerged as a substantial commercial product for horticultural crops and landscaping in urban and suburban areas (Duryea and Edwards, 1989). Adding straw harvesting to conventional timber management regimes has been shown to markedly increase profits, with straw revenue potentially exceeding that of traditional forest products (Haywood et al., 1998; Lopez-Zamora et al. 2001; Roise et al,. 1991). These plantations are typically designed to allow access by conventional agronomic equipment to harvest the straw and are thus well suited for application of the poultry litter by small to mid-size manure or litter spreaders. Harvests of straw on large plantations are usually performed using a hay or pine straw rake, tractor, and mechanical baler (Mills and Robertson, 1991). Understory biomass is typically suppressed in straw harvesting management regimes to improve straw quality by eliminating woody and herbaceous debris (Mills and Robertson, 1991). Coarse and fine woody debris is also removed from the forest floor prior to baling to improve the economic value of baled pine straw (Minogue et al., 2007). This suppression of vegetation and woody debris removal between rows of trees fosters navigation of the plantations with tractor-drawn straw raking and baling equipment (Figure 2) as well as poultry litter application equipment.

Fig. 2. Mechanically baled straw in a 19-year-old loblolly pine plantation at the Louisiana State University Agricultural Center Calhoun Research Station in northeast Louisiana. Inset: Tractor-drawn straw baler used for mechanically baling straw in the plantation. Pictures by Keith Ellem, University of Arkansas Monticello

Poultry litter can be highly beneficial when applied to plantations in which straw is harvested because it can replenish nutrients lost in straw harvesting. The nutrient content in pine needles is substantial, and repetitive harvesting of pine straw removes significant amounts of nutrients from the soil. One metric ton of harvested straw contains approximately 21.3 kg nitrogen (N), 1.8 kg phosphorus (P), 4.5 kg potassium (K), 9.0 kg calcium (Ca), and 1.8 kg magnesium (Mg) (Pote & Daniel, 2008). Since fallen leaves are major sources of nutrient inputs to soils, repeated raking can reduce soil nutrient availability, particularly N, unless nutrients are replenished through management activities (Jorgenson & Wells, 1986; Lopez-Zamora et al., 2001). As such, periodic fertilization has been recommended to remedy nutrient removals that can occur with straw harvesting (Haywood et al., 1998; Lopez-Zamora et al., 2001).

Since poultry litter contains organic matter, it could potentially replenish some of the organic matter removed by pine straw raking. Fallen pine straw is a prominent source of organic matter in the soil organic horizon of pine forests, and it is the major reservoir of labile carbon used by soil microbes in the synthesis of new cells, a process that also mineralizes N (Pritchett & Fisher, 1987; Sanchez et al., 2006; Wagner & Wolf, 1999). Soil microbial biomass and activity are highly sensitive to changes in soil organic matter and are thus used as indicators of soil quality and sustainability (Fauci & Dick, 1994; Harris, 2003; Powlson & Brookes, 1987). Removal of the soil organic horizon decreased soil microbial biomass carbon (C_{mic}) due to reduced substrate availability in a study simulating organic matter removals associated with tree harvesting and site preparation in a boreal forest (Tan et al., 2005). Activities other than soil organic matter removal associated with straw harvesting may also impact soil biological properties. The suppression of understory vegetation prior to straw raking can reduce microbial biomass and activity because understory vegetation provides rhizodeposition important to soil microbes (Donegan et al., 2001; Gallardo and Schlesinger, 1994; Högberg et al., 2001). Inorganic fertilizers do not replenish organic matter essential as microbial substrates and may exacerbate soil microbial biomass and activity declines caused by organic matter removal (Blazier et al., 2005). In contrast, fertilization with poultry litter can increase soil microbial biomass and activity (Canali et al., 2004; Plaza et al., 2004). Soil organic matter also in part determines soil water availability and temperature (Attiwill & Adams, 1993), and poultry litter has been shown to increase soil water content and available water holding capacity and reduce soil temperature (Agbede et al., 2010; Warren & Fonteno, 1993).

Due to the potential influences of poultry litter on soil and tree nutrition, soil microbes, and tree growth, a series of experiments were conducted in the mid-South region of the U.S.A. This chapter will provide a review of the key results of these trials from 1996 through 2011. The focus of this chapter will be on the changes in soil nutrition, physical properties and microbes, tree nutrition and growth, and water nutrient contents in loblolly pine plantations and silvopastures in response to fertilization with conventional fertilizer and poultry litter.

2. Study descriptions

Results of five studies conducted in the mid-South U.S.A are described in this chapter. At least one treatment in each study received surface application of broiler litter as fertilizer, and loblolly pine was the tree component of each study. All studies occurred in the Western Gulf Coastal Region within areas identified by Friend et al. (2006) as having high occurrence of poultry production and southern pine forests. Two of the studies (SILVO, SWITCH) included poultry litter applied to silvopastures. The silvopasture in the SILVO study consisted as bahiagrass (Paspalum notatum Flüggé) established under thinned loblolly pine, and the silvopasture in the SWITCH study was comprised of switchgrass (Panicum virgatum L.) established under thinned loblolly pine. Two of the studies (AR-FORvsPAST, LA-FORvsPAST) included a comparison of broiler litter in loblolly pine and pastures. The STRAW study included poultry litter applied to a loblolly pine plantation in which straw was annually harvested.

The SILVO and SWITCH studies were conducted at the Louisiana State University Agricultural Center Hill Farm Research Station in Homer, Louisiana, U.S.A. The STRAW and LA-FORvsPAST trials were carried out at the Louisiana State University Agricultural Center Calhoun Research Station in Calhoun, Louisiana, U.S.A. The AR-FORvsPAST study was conducted at the University of Arkansas Southwest Research and Extension Center near Hope, Arkansas, U.S.A. Average annual precipitation of the region in which the studies were carried out is 120 cm, and average temperature is 18°C (Bailey, 1995). Primary study site characteristics are described in Table 2.

Study	Geographical Coordinates	Vegetation	Tree Age (years)	Tree Density (trees ha^{-1})	Soil Classification
SILVO	32°44′N, 93°03′W	loblolly pine-bahiagrass silvopasture	12	247	Loamy, siliceous, thermic Arenic Paleudults
SWITCH	32°44′N, 93°03′W	loblolly pine-switchgrass silvopasture	17	124	Loamy, siliceous, thermic Arenic Paleudults
STRAW	32°31′N, 92°21′W	loblolly pine plantation	10	618	Fine-loamy siliceous thermic Typic Fragiudults
LA-FORvsPAST	32°31′N, 92°21′W	loblolly pine, bermudagrass as vegetation type treatments	5	1586	Fine-loamy siliceous thermic Typic Fragiudults
AR-FORvsPAST	33°42′N, 93°32′W	loblolly pine, bahiagrass as vegetation type treatments	26	201	Fine-loamy, siliceous thermic Typic Fraigudults; Clayey, mixed, thermic Aquic Hapludults

Table 2. Location, vegetation, tree and density at study initiation, and soil characteristics for studies of poultry litter fertilization of loblolly pine in the mid-South U.S.A.

Treatments (Table 3) were replicated four times each in the STRAW, LA-FORvsPAST studies, three times in the SILVO and AR-FORvsPAST studies, and six times in the SWITCH study. Treatments were applied as a one-way treatment structure in the STRAW, LA-FORvsPAST, SWITCH, and SILVO studies. Treatments were applied as a split-plot treatment structure in the AR-FORvsPAST study, with vegetation type (pasture, forest) as a whole-plot treatment and fertilization as a sub-plot treatment. The experimental design was a randomized complete block design for all studies. In statistical analyses of all variables assessed in these treatments, differences among treatments were determined by analysis of variance at $\alpha = 0.05$; correlation among variables was assessed at $\alpha = 0.05$ as well.

Study	Treatment[1]	Treatment Description
SILVO	CONTROL	No treatment
	IF	Inorganic fertilizer mixture (diammonium phosphate, ammonium nitrate, muriate of potash to annually supply 114 kg N ha^{-1}, 39 kg P ha^{-1}, 20 kg K ha^{-1})
	PL5	Poultry litter applied at 5 Mg ha^{-1} that supplied N, P, K, Ca, Mg, Fe, Mn, Cu, Zn, B at 112, 36, 78, 106, 23, 9, 2, 1, 1.5, 0.2 kg ha^{-1}, respectively
	PL10	Poultry litter applied at 10 Mg ha^{-1} that supplied N, P, K, Ca, Mg, Fe, Mn, Cu, Zn, B at 224, 73, 157, 211, 45, 18, 3, 2, 3, 0.3 kg ha^{-1}, respectively
SWITCH	CONTROL	No treatment
	IF80	Ammonium nitrate applied that supplied 80 kg N ha^{-1}
	IF160	Ammonium nitrate applied that supplied 160 kg N ha^{-1}
	PL1.5	Poultry litter applied at 1.5 Mg ha^{-1} to supply N, P, K, Ca, Mg, Fe, Mn, Cu, Zn, and B at 80, 42, 90, 54, 15, 2, 1.5, 0.15, 1, and 0.1 kg ha^{-1}, respectively.
	PL3	Poultry litter applied at 3 Mg ha^{-1} to supply N, P, K, Ca, Mg, Fe, Mn, Cu, Zn, and B at 160, 84, 180, 108, 30, 4, 3, 0.3, 2, and 0.2 kg ha^{-1}, respectively.
STRAW	CONTROL	No treatment
	RAKE	Straw harvesting
	RAKE-IF	Straw harvesting, diammonium phosphate and urea inorganic fertilizers that supplied N and P at 193 and 102 kg ha^{-1}, respectively
	RAKE-PL	Straw harvesting, poultry litter applied at 8 Mg ha^{-1} that supplied N and P at 193 and 102 kg ha^{-1}, respectively. Other nutrients added by poultry litter not tested due to budget constraints.
LA-FORvsPAST	CONTROL	No treatment
	PL5	Poultry litter applied at 5 Mg ha^{-1} that supplied N, P, K, Ca, Mg, Fe, Mn, Cu, Zn, B at 112, 109, 92, 159, 34, 9, 2, 3, 2, 0.2 kg ha^{-1}, respectively
	PL10	Poultry litter applied at 10 Mg ha^{-1} that supplied N, P, K, Ca, Mg, Fe, Mn, Cu, Zn, B at 224, 218, 184, 318, 68, 18, 3, 6, 3, 0.3 kg ha^{-1}, respectively
	PL20	Poultry litter applied at 20 Mg ha^{-1} that supplied N, P, K, Ca, Mg, Fe, Mn, Cu, Zn, B at 448, 436, 368, 636, 136, 36, 6, 12, 6, 0.6 kg ha^{-1}, respectively
	FOREST	Loblolly pine plantation
	PASTURE	Bermudagrass pasture
AR-FORvsPAST	FOREST	Loblolly pine plantation
	PASTURE	Bahiagrass pasture
	CONTROL	No treatment
	PL9	Poultry litter at 9 Mg ha^{-1} that supplied N, P, K at 30, 14, 22 kg ha^{-1}. Other nutrients added by poultry litter not tested due to budget constraints.

Table 3. Treatments conducted in studies of poultry litter fertilization of loblolly pine conducted in the mid-South U.S.A. [1]Italicized treatments were applied as sub-plot treatments, underlined treatments were applied in all possible combinations to whole plots, all other treatments were applied as whole-plot treatments

Soil and/or plant responses to treatments were observed in the studies (Table 4). In the SILVO, SWITCH, and LA-FORvsPAST studies, grass clippings were randomly collected within quadrats either at the end of growing seasons (in the SWITCH study) or multiple times during the season and averaged (in the SILVO and LA-FORvsPAST studies) to determine forage yields. In the SILVO, STRAW, and LA-FORvsPAST studies loblolly pine basal area was measured by converting diameter at breast height measurements into basal area for all trees in measurement plots; basal area measures were summed for each plot to estimate stand level basal area. In the SILVO study soil was sampled by a tractor-mounted auger to the bottom of the B_t horizon and separated into A, E, and B_t horizons, which had average depths of 0.15, 0.48, and 0.59 m, respectively. In the SWITCH study, soil was sampled with punch augers to a 15-cm depth for labile C determination and 30 cm for nutrient analyses. Soil in the STRAW study was sampled with punch augers to a 15-cm depth. In the LA-FORvsPAST study, soil was sampled to a 15-cm depth with punch augers pre-treatment and sampled to 0-15, 15-30, 30-45, 60-80, and 80-100 cm depths post-treatment. Soil was sampled to a 15 cm depth in the AR-FORvsPAST study using punch augers. In the SILVO and STRAW studies, loblolly pine foliage was sampled from the upper third of crowns. Organic matter in soil samples was quantified by the Walkley-Black method (Walkley, 1947) in the SILVO, LA-FORvsPAST, and SWITCH studies and by the loss on ignition method (Ball, 1964) in the STRAW and AR-FORvsPAST studies. Soil pH was determined by pH meters in a 2:1 mixture of deionized water to soil in the SILVO and SWITCH studies. Phosphorus in the samples was extracted by Bray 2 P (Bray & Kurtz, 1945) in the SILVO and LA-FORvsPAST studies and by Mehlich 3 (Mehlich, 1984) in the AR-FORvsPAST and SWITCH studies. Nutrients other than P were extracted by ammonium acetate (K, Ca, Mg, Na) and DTPA (Cu, Fe, Mn, Zn) in the SILVO and LA-FORvsPAST studies (Gambrell, 1996; Helmke & Sparks, 1996). Mehlich 3 was used to extract K, Ca, Mg, S, Cu, and Zn in the SWITCH study (Mehlich, 1984). All nutrients from soil samples were quantified via ICP spectrometry (Jones & Case, 1990) in all studies in which soil was analyzed for nutrient concentration. Exchangeable N (NH_4-N, NO_3-N) was extracted by KCl extraction (Mulvaney, 1996) in the SILVO and AR-FORvsPAST studies and measured colorimetrically on Bran Luebbe (Bran-Luebbe, Inc, Delavan, WI) and Lachat autoanalyzers (Lachat Instruments, Loveland, CO, U.S.A.) in the SILVO and AR-FORvsPAST studies, respectively. In the STRAW and SWITCH studies soil labile C was measured by sequential fumigation incubation (Zou et al., 2005). In the STRAW and SWITCH studies microbial biomass C was measured by fumigation incubation (Jenkinson and Powlson, 1976*a,b*) and microbial activity was measured by an assay of dehydrogenase activity (Lenhard, 1956; Alef, 1995). In the STRAW study N mineralization and nitrification was measured using the buried bag method (Eno, 1960). In the AR-FORvsPAST study, potential N mineralization and nitrification (Hart et al., 1994) were assessed in samples aerobically incubated in the laboratory for 28 days; NH_4-N and NO_3-N used to determine mineralization and nitrification in this procedure were measured by the cadmium reduction method (Mulvaney, 1996) using a Lachat autoanalyzer. Total N in soil samples from the AR-FORvsPAST study was determined by dry combustion using an Elementar Vario MAX CN analyzer (Elementar Analysesysteme GmbH, Hanau, Germany).

All nutrients except N in foliage were analyzed by nitric acid digestion and ICP spectrometry; N in these samples was measured by Dumas combustion and thermal conductivity detection using a Leco N/protein analyzer (Leco Inc., St. Joseph, MI, U.S.A) (Helmke & Sparks, 1996; Tate, 1994; Zarcinas et al., 1987). In the SILVO study, N concentrations of bahiagrass samples

Study	Treatment or Measurement[1]	1996	1997	1998	1999	2000	2001	2002	2003	2004	2005	2006	2007	2008	2009	2010	2011
SILVO	Fertilization			■	■	■	■										
	Forage yield				■												
	Soil pH, OM, nutrients		■		■	■		■									
	Soil NH_4-N, NO_3-N		■					■									
	Pine foliage nutrients				■					■							
	Pine basal area		■	■	■	■		■									
SWITCH	Fertilization															■	
	Forage yield															■	
	Soil pH, OM, nutrients															■	■
	Soil LABC, CMIC, ACT															■	
STRAW	Straw raking					■	■	■	■	■	■	■	■	■	■		
	Fertilization								■	■	■	■	■	■	■		
	Soil STR, P, BD, WHC											■					
	Soil OM, LABC, NMIN, exchangeable N												■				
	Soil CMIC, ACT										■	■					
	Pine foliage nutrients												■				
	Pine basal area							■			■		■				
LA-FORvsPAST	Fertilization		■	■	■	■	■										
	Forage yield		■	■	■	■	■										
	Soil pH, OM, nutrients	■	■	■	■	■	■										
	Pine basal area	■	■	■	■	■	■	■									
	Runoff nutrients						■										
AR-FORvsPAST	Fertilization									■	■	■					
	Forage yield									■	■	■					
	Soil NH_4-N, NO_3-N									■	■	■	■				
	Soil potential NMIN										■	■					
	Soil OM, N, P									■	■	■	■				

Table 4. Timeline of treatments and measurements in studies of fertilization of loblolly pine with poultry litter in the mid-South U.S.A. Shaded cells designate years in which treatments or measurements occurred. [1]OM = organic matter, STR = strength, BD = bulk density, WHC = water holding capacity, LABC = labile C, NMIN = N mineralization, CMIC = microbial biomass C, ACT = microbial dehydrogenase activity

were ascertained by Kjeldahl method; other nutrients in the samples were determined by Dumas combustion and nitric acid digestion and ICP spectrometry (Helmke & Sparks, 1996; Horneck & Miller, 1998; Tate, 1994; Zarcinas et al., 1987).

Bulk density, porosity, soil moisture content, and air-filled porosity of soil samples collected in the STRAW study were analyzed using procedures of Blake & Hartge (1986) and Danielson & Sutherland (1986). Available water holding capacity was determined in the STRAW study using soil moisture retention curves (Brye, 2003; Gee et al., 1992). Soil strength in the STRAW study to 15- and 30-cm depths was measured with a Scout SCT compaction meter (Spectrum Technologies, Inc., Plainfield, IL, USA) (Bradford, 1986).

Soil water was collected to a 30-cm depth using tension lysimeters in the AR-FORvsPAST study; NO_3-N and PO_4-P in the water samples was analyzed by ion chromatography and NH_4-N was measured with a Lachat autoanalyzer. Water samples were also digested using a Kjeldahl digestion procedure and analyzed for total Kjeldahl nitrogen (TKN) and total Kjeldahl phosphorus (TKP) using the Lachat spectrophotometer. In the LA-FORvsPAST study, water was collected from runoff troughs after every rain event. Water samples were analyzed for total P by acid persulfate digestion and ICP spectrometry and for dissolved P by ICP spectrometry (Clesceri et al., 1998; Pote & Daniel, 2000).

3. Plant biomass and nutrition

Increases in forage yields were observed in response to poultry litter in the SILVO, LA-FORvsPAST, AR-FORvsPAST, and SWITCH studies. In the SILVO study poultry litter increased bahiagrass yields, but the magnitude of response was rate-dependent. The PL10 treatment had greater bahiagrass yields than all other treatments, and the PL5 and IF treatments had greater bahiagrass yields than the CONTROL treatment (Evans, 2000). The PL10 treatment also led to greater P, Zn, and Cu concentrations in bahiagrass relative to the CONTROL and IF treatments (Evans, 2000). These results indicated that poultry litter increased yields and nutritional quality of bahiagrass. Gaston et al. (2003) similarly found in the LA-FORvsPAST study that bermudagrass yields increased with increasing litter application rate. In the AR-FORvsPAST study, bahiagrass yields of the PL9 treatment were ~1.5 times greater than those of the CONTROL treatment in the first two years of fertilization. Switchgrass yield response to poultry litter in the SWITCH study was not rate-dependent as in the SILVO and LA-FORvsPAST studies, because both application rates led to comparable yields.

Loblolly pine growth was also improved by poultry litter in the SILVO and AR-FORvsPAST studies. In the SILVO study, tree- and stand-level basal area growth was increased by poultry litter at the 10 Mg ha^{-1} rate (Blazier et al., 2008a). As with forage yields, litter application rate affected the level of growth response. After four annual litter applications, the 10PL treatment had greater annual basal area growth per tree than that of all other treatments, and the 5PL treatment had greater annual basal area growth than the CONTROL treatment. Stand-level basal area growth of the 10PL treatment was greater than that of the CONTROL and IF treatments. All fertilizer treatments led to greater foliage N concentrations than the CONTROL treatment, and both poultry litter treatments had greater foliage P concentrations than the CONTROL treatment. These results, which are consistent with other studies (Dickens et al., 2004; Friend et al., 2006; Roberts et al., 2006), show that loblolly pine growth can be increased with poultry litter amendments. The levels of growth responses were somewhat surprising, because all foliage nutrient concentrations were above critical levels (Allen, 1987; Blazier et al., 2008a; Jokela, 2004). Due to the relatively low density of trees in silvopastures, trees may have more readily responded to fertilization by virtue of larger crown mass (which provides a larger nutrient sink per tree) and less competition for applied nutrients compared to that in typical pine plantations. The similarities in growth responses and N and P application rates of the 5LIT and INO treatments suggests that although the 5LIT treatment supplied more K and a wider array of nutrients than the INO treatment, N and P were likely the primary limiting nutrients in the stand (Blazier et al., 2008*a*). In the AR-FORvsPAST study, annual loblolly pine basal area growth in response to the PL9 treatment was 10.9% greater than that of the CONTROL treatment.

In the STRAW and LA-FORvsPAST studies, no significant loblolly pine growth responses to treatments were observed (Gaston et al., 2003). Before the studies were established, the land was intensively managed for forage production. As such, the decades of fertilization application at these locations had resulted in high nutrient availability. Foliage P and S concentrations were increased by the RAKE-PL treatment relative to the other treatments in the STRAW study, but these increases appeared to have been luxury consumption since these nutrient increases were not accompanied by increased loblolly pine growth.

4. Soil physical properties and organic matter

In the STRAW study, all treatments that included straw harvesting induced evidence of soil compaction by significantly increasing bulk densities (Table 5) to levels 0.6 to 3.3% greater than the 1.75 g cm^{-3} bulk density defined as a growth-limiting threshold for forests grown on loamy soils (Daddow and Warrington, 1983), whereas soil in the CONTROL treatment remained below this threshold. These bulk density increases were also associated with significant declines in porosity in all treatments that included straw harvesting (Table 5). These findings suggest that annual straw harvesting had potential to reduce tree growth through reduced rooting volume and aeration. Nevertheless, no decreases in loblolly pine growth were observed in response to raking, as described above. It is likely that equipment traffic and increased exposure of mineral soil to rainfall associated with straw harvesting led to these increases in bulk density. Similarities in bulk density and porosity among the RAKE treatment and treatments that included raking and fertilization suggest that the additional trafficking from fertilization equipment each season did not appreciably compact the soil and that straw harvesting was the predominant cause of soil compaction (Blazier et al, 2008*b*).

Treatment	Bulk Density (g cm^{-3})	Porosity (g kg^{-1})	Air-filled Porosity	Moisture (g kg^{-1})	Soil Strength (MPa)	Organic Matter (g kg^{-1})	Available Water Holding Capacity (g kg^{-1})
CONTROL	1.67 b	369 a	99 a	270 a	1.25 b	27.8 a	427 a
RAKE	1.81 a	318 b	51 b	268 ab	2.31 a	25.8 ab	367 b
RAKE-IF	1.76 a	334 b	86 a	248 b	2.45 a	19.0 b	353 b
RAKE-PL	1.78 a	329 b	48 b	281 a	0.99 b	25.8 ab	384 ab

Table 5. Soil physical properties and organic matter in response to pine straw harvesting and fertilization with inorganic fertilizer and poultry litter in a loblolly pine plantation in north central Louisiana, U.S.A. Means within columns followed by different letters differ at $P < 0.05$. Adapted from Blazier et al. (2008b)

The RAKE-PL treatment appeared to have ameliorated some of the soil physical impacts of the raking since soil strength, organic matter, and moisture in the RAKE-PL treatment were similar to that in the CONTROL treatment (Table 5). However, the RAKE-PL treatment was characterized by lower air-filled porosity than the CONTROL treatment (Table 5), so there may have been a compaction potential associated with the application of the poultry litter (Tekeste et al., 2007). Poultry litter did not alter soil physical properties in a manner similar to inorganic fertilizers. The RAKE-PL treatment was characterized by soil moisture content, strength, organic matter concentrations, and available water holding capacity similar to the CONTROL treatment (Table 5). The RAKE-PL treatment may have replenished some organic matter lost through straw harvesting and accelerated decomposition associated with increased nutrient levels, because broiler poultry litter typically consists of 44% organic matter (Adeli et al., 2006; Dick et al., 1998). These results suggest that use of poultry litter as a fertilizer source in an annual straw harvest regime was superior to inorganic fertilizers in sustaining soil physical quality.

In contrast with the RAKE-PL treatment, the RAKE and RAKE-IF treatments had detrimental effects on some soil physical properties. The RAKE and RAKE-IF treatments both had soil strengths 46% greater than the CONTROL treatment. Soil strengths of the RAKE and RAKE-IF treatments also exceeded the 2 MPa soil strength threshold defined as highly compacted because of demonstrated root growth restrictions (Taylor & Gardner, 1963; Tiarks & Haywood, 1996). Available water holding capacity was also reduced by the RAKE and RAKE-IF treatments relative to the CONTROL treatment. These findings suggest that the RAKE and RAKE-IF treatments made soil less amenable for root growth in the uppermost 5 cm of soil, which is the predominant zone in which tree roots, particularly fine roots, grow (Gilman, 1987). Relative to the CONTROL treatment, only the RAKE-IF treatment had greater soil strength, reduced moisture content, and reduced soil organic matter concentrations (Table 5). Repeated fertilization with inorganic nitrogen has been shown to reduce soil organic matter concentrations by increasing decomposition rates (Khan et al., 2007). Increased soil strength in response to the RAKE-IF treatment may have been due to the reductions in soil organic matter concentrations caused by this treatment. Soil strength tends to increase with decreasing soil organic matter concentrations because soil organic matter serves as an organic aggregate binding and bonding material (Munkholm et al., 2002). The relatively lower moisture content and available water holding capacity of the RAKE-IF treatment is consistent with its lower soil organic matter content because organic matter fosters soil moisture retention (Plaza et al., 2004; Powers et al., 2005).

There were no differences in soil organic matter among treatments in the SILVO, SWITCH, LA-FORvsPAST, and AR-FORvsPAST studies (Blazier et al., 2008a; Liechty et al., 2009), in which litter was not removed. As such, increases in forage and/or tree yields from fertilization in these studies were not associated with concomitant increases in soil organic matter. In the SILVO and SWITCH studies, the lack of declines in organic matter in response to inorganic fertilizer application as seen in the STRAW study was likely due to the straw raking done in tandem with fertilization in the STRAW study. As organic matter supply was drastically reduced by annual straw harvesting, stimulating decomposition with inorganic fertilizer led to significant declines in soil organic matter. Additionally, the increases in forage understory biomass of the SILVO and SWITCH studies may have been less prone to lead to increases in organic matter, as evidenced in the LA-FORvsPAST study. In that study no differences in organic matter were found among treatments in the pasture despite the increases in bermudagrass yields described above, whereas organic matter in the loblolly pine plantation differed among treatments as PL20 > PL10, PL5 > CONTROL.

5. Soil labile C, microbial biomass C, and microbial activity

Annual application of inorganic fertilizer had a profound effect on microbial biomass and activity in the STRAW study (Table 6). Microbial biomass C of the RAKE-IF treatment was lower than that of the CONTROL and RAKE treatments, and dehydrogenase activity of the RAKE-IF treatment was lower than all other treatments. The reductions in microbial biomass C and activity were apparently not a result of lower substrate supply, because labile C was similar among treatments (Table 6). Consequently, the higher potential turnover rate of the RAKE-IF treatment relative to all others is likely a result of reduced microbial biomass and activity rather than relatively high recalcitrance of organic matter.

The reductions in microbial biomass C and activity in the RAKE-IF treatment were likely associated with the lower pH of this treatment relative to all others. It has been well-demonstrated that intensive fertilization with inorganic N reduces soil pH and that declining pH is associated with reductions in soil microbial biomass and activity (Anderson and Domsch, 1993; Baath et al., 1995; Blazier et al., 2005). These results thus showed that microbial biomass and activity were reduced by declines in pH from inorganic fertilizer, whereas annual raking and fertilization with poultry litter had no such effects. In contrast to inorganic fertilizer, poultry litter tends to increase soil pH because litter contains calcium carbonate originating from poultry rations (Hue, 1992; Kingery et al., 1993). Although litter did not significantly increase pH in the STRAW study, litter sustained pH at levels comparable to the CONTROL treatment, which fostered microbial biomass C and activity levels comparable to the CONTROL treatment as well.

As in the STRAW study, inorganic fertilizer led to declines in microbial biomass C relative to the CONTROL treatment (Table 6) in the SWITCH study. Fertilization has been shown to reduce soil microbial biomass C in forest soils (Rifai et al., 2010; Wallenstein et al., 2006). Rifai et al. (2010) identified several possible mechanisms for soil microbial biomass declines in response to fertilization, including (1) pH reduction caused by nitrate leaching induced by application of high rates of NH_4NO_3, and (2) inhibition of organic compound decomposition from excess N that reduces organic matter available to soil microbes. In the SWITCH study there were no declines in pH among treatments consistent with declines in soil microbial biomass C, although pH of the inorganic fertilizer treatments were lower than those of the poultry litter treatments. Dehydrogenase activity decreased as fertilizer application rates increased for both fertilizer types. Since N was the sole nutrient added by inorganic fertilizer treatments in this study, these dehydrogenase activity trends suggest that excess N perturbed microbial decomposition of organic matter in this loblolly pine and switchgrass system. However, potential C turnover rate was shorter for the lower rate of inorganic fertilizer (IF80) relative to the lower rate of poultry litter (PL1.5) despite the equivalent N rate of the two treatments. Since labile C supply, microbial biomass C, and dehydrogenase activity were similar for the IF80 and PL1.5 treatments, the reason for the higher potential C turnover rate of the PL1.5 treatment was unclear and merited further study.

	Treatment				
STRAW	**CONTROL**	**RAKE**	**RAKE-IF**	**RAKE-PL**	
Labile C (mg kg^{-1})	475.1 a	522.3 a	457.0 a	582.5 a	
Potential C turnover rate (days)	46.0 b	53.2 b	92.9 a	62.8 ab	
Microbial biomass C (mg kg^{-1})	169.2 a	157.2 a	75.3 b	143.5 ab	
Dehydrogenase activity (μg g^{-1})	50.6 a	71.0 a	25.8 b	44.5 a	
pH	4.9 a	4.9 a	4.3 b	5.1 a	
	CONTROL	**IF80**	**IF160**	**PL1.5**	**PL3**
SWITCH					
Labile C (mg kg^{-1})	835.6 a	585.2 a	718.1 a	878.7 a	836.7 a
Potential C turnover rate (days)	29.7 ab	24.7 b	30.4 ab	43.8 a	37.3 ab
Microbial biomass C (mg kg^{-1})	410.9 a	341.6 b	348.4 b	320.4 ab	377.4 ab
Dehydrogenase activity (μg g^{-1})	11.0 ab	24.2 a	9.9 b	24.3 a	5.9 b
pH	5.5 bc	5.4 c	5.4 c	5.6 ab	5.7 a

Table 6. Soil labile C, microbial, and pH responses to fertilization in an annually raked loblolly pine plantation (STRAW) and a loblolly pine and switchgrass silvopasture (SILVO) in the mid-South U.S.A. For each study, means within columns followed by different letters differ at $P < 0.05$. Adapted in part from Blazier et al. (2008b)

6. Soil nutrients

6.1 Nitrogen

In all studies in which exchangeable soil N was measured, NO_3-N amounts or proportions of in soil increased in response to poultry litter application (Table 7). In the AR-FORvsPAST study, NO_3-N significantly increased in the loblolly pine plantation and in pasture relative to the CONTROL treatment following two years of poultry litter application. The proportion of NO_3-N to total exchangeable N was also greater in response to poultry litter than without litter application (Liechty et al., 2009). There was no difference in NO_3-N concentrations among treatments in the SILVO study, but as in the AR-FORvsPAST study the ratio of NO_3-N to total exchangeable N increased in response to poultry litter additions. This increase in the proportion of NO_3-N in the SILVO study occurred in response to both rates of broiler litter tested; no such increase was observed in response to the inorganic fertilizer mixture (Blazier et al., 2008a). Results similar to the SILVO study were also found in the STRAW study; NO_3-N increased in response to the treatment regime that included poultry litter, whereas no such increase was observed in response to non-fertilized treatments and the treatment regime that included a mixture of inorganic fertilizers (Liechty et al., 2009).

Increases in soil NO_3-N in response to poultry litter were attributable to greater nitrification rates (Table 7). Soil in plots treated with broiler litter had greater N mineralization rates in the AR-FORvsPAST study, and a greater proportion of mineralized N was nitrified. There was also a significant positive correlation between NO_3-N in soil and nitrification rates (Liechty et al., 2009). Similar results were observed in the STRAW study, in which both rates of poultry litter had greater N mineralization and nitrification than CONTROL and IF treatments (Blazier et al, 2008b). The greater nitrification and NO_3-N of poultry litter treatments relative to CONTROL treatments in both studies was likely predominately a function of the addition of N to soil by litter. Relatively high NO_3-N in soil after fertilization is in part indicative of low plant sequestration of applied N (Adeli et al., 2006), so consecutive applications of litter at the rates in these studies likely exceeded loblolly pine, bermudagrass, and bahiagrass N demand. The higher nitrification rates seen in response to poultry litter in these studies relative to inorganic fertilizer, even when both fertilizer sources were applied to provide the same N rates, was likely due to the differences in the effects of the fertilizer sources on soil pH. In the SILVO and STRAW studies, soil pH declined in response to inorganic fertilization applications relative to all other treatments (Tables 6 and 7). Likewise, soil pH of the poultry litter treatments in the SWITCH study was greater in response to broiler litter than to CONTROL and inorganic fertilizer treatments (data not shown). Ellum (2010) found in the STRAW study that nitrification was significantly and positively correlated with pH. Nitrification rates have been shown to decline with decreasing pH due to reductions in populations and activity of nitrifying bacteria (Aune & Lal, 1997).

	Treatment			
SILVO – 5 years post-treatment	**CONTROL**	**IF**	**PL5**	**PL10**
NO_3-N (mg kg^{-1})	7.2 a	0.1 a	4.8 a	16.2 a
Total exchangeable N (mg kg^{-1})	34.8 a	18.6 a	29.6 a	71.0 a
% NO_3-N	31.0 b	11.1 c	51.7 a	55.6 a
pH	4.9 a	4.5 a	5.0 a	5.0 a
STRAW – 5 years post-treatment	**CONTROL**	**RAKE**	**RAKE-IF**	**RAKE-PL**
NO_3-N (mg kg^{-1})	0.6 c	0.8 c	1.4 b	10.0 a
Total exchangeable N (mg kg^{-1})	6.5 b	5.1 c	6.3 b	14.4 a
% NO_3-N	15.9 c	17.8 c	25.2 b	65.6 a
N mineralization (mg kg^{-1})	23.6 b	18.5 b	13.7 b	51.2 a
N nitrification (mg kg^{-1})	23.2 b	17.4 b	17.3 b	48.2 a
% N nitrified	98.3 a	94.1 a	126.3 a	94.1 a
AR-FORvsPAST – 2 years post-treatment	**PASTURE-CONTROL**	**PASTURE-PL9**	**FOREST-CONTROL**	**FOREST-PL9**
NO_3-N (mg kg^{-1})	2.1 b	4.1 a	0.1 b	15.3 a
Total exchangeable N (mg kg^{-1})	8.9 b	11.0 b	6.2 b	24.4 a
% NO_3-N	21.4 bc	37.6 ab	1.1 c	57.1 a
N mineralization (mg kg^{-1})	14.3 b	23.6 a	7.3 c	14.3 b
N nitrification (mg kg^{-1})	15.6 b	26.3 a	3.4 d	13.3 c
% N nitrified	109.0 b	111.4 a	46.5 d	93.3 c

Table 7. Soil exchangeable N, mineralization, nitrification, and pH in response to fertilization in loblolly pine plantations, silvopasture, and bahiagrass pasture in a series of trials conducted in the mid-South U.S.A. Means within rows followed by different letters differ at $P < 0.05$. Adapted in part from Liechty et al. (2009)

Although differences in soil NO_3-N and nitrification between loblolly pine and bahiagrass pasture in the AR-FORvsPAST study in part reflected the differences in pH and C:N ratios of the soils in these two land uses (Richardson 2006), they also reflected the differences in uptake and use of available N forms by the loblolly pine and pastures. Although N mineralization and nitrification was greater in pasture when fertilized with poultry litter, the increase in NO_3-N remaining in soil per unit increase in potential net nitrification was greater in loblolly pine plantation than in pasture by the second application of poultry litter (Liechty et al., 2009). Conifer tree roots have been shown to preferentially absorb NH_4-N rather than NO_3-N (Kronzucker et al., 1997), whereas NO_3-N is preferentially taken up by forage (Blevins and Barker, 2007). Thus, loblolly pine plantation had a greater propensity to retain a higher proportion of NO_3-N than pasture. Given this tendency of loblolly pine to retain proportionally greater NO_3-N, it is likely that less poultry litter should be applied to such plantations than to pastures to minimize NO_3-N pollution in surface and subsurface water.

Annual raking in the STRAW study reduced total exchangeable N, and fertilization, regardless of source, replaced at least a portion of the lost N and increased total exchangeable N (Table 7). Interestingly, although both fertilizers increased exchangeable N, poultry litter increased exchangeable N to a greater extent than the inorganic fertilizer, although both fertilizers were applied at the same N rate (Ellum, 2010). The higher exchangeable N concentrations in the RAKE-PL reflected the increases in NO_3-N levels in the RAKE-PL treatment. The NO_3-N concentrations were nearly 7 and 17 times greater in this treatment than those in the RAKE-IF and CONTROL treatments, respectively. This result provides evidence of the propensity of loblolly pine plantations to accumulate NO_3-N in response to annual applications of broiler litter, even when exchangeable N is reduced by annual straw raking. To safeguard against such NO_3-N accumulation, it is likely necessary to fertilize a raked loblolly pine plantation with broiler litter less frequently and at lower rates than in the STRAW study.

6.2 Phosphorus

Soil test P accumulation, determined as the annual difference in soil test P concentrations from pre-treatment concentrations, increased in the uppermost soil horizon in all studies in which soil test P was measured (Table 8). In the SILVO study, both litter treatments had significantly greater soil test P accumulation in the uppermost soil horizon than the CONTROL and IF treatments. After the first application, soil test P accumulation was similarly increased by both litter rates. After four annual applications, the PL10 treatment had greater soil test P accumulation than all other treatments. The IF treatment did not result in a significant accumulation of soil test P at any point in the study (Liechty et al, 2009). Soil test P accumulation also increased in the SWITCH study in response to a single application of litter at both rates. In the LA-FORvsPAST and AR-FORvsPAST studies, soil test P accumulation increased in response to broiler litter in loblolly pine plantation and in pasture (Liechty et al., 2009). Increases in soil test P in surface soil in response to litter application have been similarly found in agricultural (Mitchell & Tu, 2006; Sharpley et al., 1993; Sistani et al., 2004) and forest (Friend et al., 2006) soils. In addition to these increases in upper soil horizons, soil test P accumulation was increased to the B_t horizon (an average depth of 0.59 m) by the 10PL treatment after four applications in the SILVO study (Blazier et al., 2008*a*). Additional evidence of increasing soil test P in lower soil profile was found in the LA-FORvsPAST study, in which soil test P concentrations of the 20PL treatment exceeded that of all others in the 30 to 45 cm depth in loblolly pine and bermudagrass soil in the seventh year of the study (data not shown). These increases in soil test P in surface and subsurface soil in response to annual litter applications suggest that vegetation P demands and soil P sorption capacity were exceeded at all sites irrespective of vegetation type and stand conditions.

Study	Treatment	Depth (cm)	Year after treatment 1	2	3	4	5	6	7
SILVO	CONTROL	0–15[a]	13.0 *b*	------	16.9 *b*	10.5 *c*	------	------	------
	IF		19.0 *b*	------	22.1 *b*	24.1 *c*	------	------	------
	PL5		36.7 *a*	------	68.5 *a*	87.2 *b*	------	------	------
	PL10		48.8 *a*	------	84.2 *a*	146.5 *a*	------	------	------
	CONTROL	15-48[b]	3.8 *a*	------	-0.4 *a*	-5.2 *b*	------	------	------
	IF		0.9 *a*	------	-2.0 *a*	-5.8 *b*	------	------	------
	PL5		3.3 *a*	------	1.1 *a*	0.9 *b*	------	------	------
	PL10		2.8 *a*	------	4.4 *a*	80.2 *a*	------	------	------
	CONTROL	48-59[c]	2.2 *a*	------	-1.4 *a*	-4.0 *b*	------	------	------
	IF		-0.4 *a*	------	-4.5 *a*	-6.6 *b*	------	------	------
	PL5		-0.5 *a*	------	-3.5 *a*	-5.7 *b*	------	------	------
	PL10		-1.8 *a*	------	-3.5 *a*	44.6 *a*	------	------	------
SWITCH	CONTROL	0-15	0.2 *c*	------	------	------	------	------	------
	IF80		0.1 *b*	------	------	------	------	------	------
	IF160		0.1 *b*	------	------	------	------	------	------
	PL90		0.5 *a*	------	------	------	------	------	------
	PL180		0.5 *a*	------	------	------	------	------	
LA-FORvsPAST[d]	CONTROL	0-15	-19.9 *b*	-8.67 *a*	-24.1 *b*	-11.9 *c*	------	-11.0 *c*	-19.6 *c*
	PL5		3.2 *b*	9.3 *a*	-14.2 *b*	26.4 *bc*	------	162.5 *bc*	82.9 *bc*
	PL10		32.2 *ab*	29.7 *a*	65.3 *a*	103.2 *b*	------	328.2 *b*	210.0 *b*
	PL20		71.2 *a*	43.5 *a*	80.8 *a*	243.4 *a*	------	760.0 *a*	447.6 *a*
AR-FORvsPAST[d]	CONTROL	0-15	------	10.1 *b*	------	------	------	------	------
	PL9		------	47.2 *a*	------	------	------	------	------

Table 8. Soil test P accumulation (mg kg^{-1}) in response to fertilization with poultry litter and inorganic fertilizer in the mid-South U.S.A. For each study site and soil depth, means within columns followed by different letters differ at $P < 0.05$. [a]Average depth of soil samples subdivided into the A horizon, [b]average depth of soil samples subdivided into the E horizon, [c]average depth of soil samples subdivided into the B_t horizon, [d]soil test P accumulation reported for study is an average of loblolly pine plantation and pasture soils because analyses did not reveal a treatment x land use type interaction. Adapted in part from Blazier et al. (2008*a*) and Liechty et al. (2009)

Land use type and rate affected soil test P trends in response to broiler litter in the LA-FORvsPAST study. Soil test P accumulation in the loblolly pine plantation averaged over all treatments exceeded that of the pasture for six years of the study (Figure 3). Initial soil test P concentrations of the pasture were 1.5 times greater than that of the loblolly pine plantation (data not shown), but in the first three years of treatment soil test P accumulation of the pasture was negative whereas soil test P accumulation of the loblolly pine plantation ranged from 51 to 76 mg kg^{-1} $year^{-1}$ over the same period. Until the final fertilization, soil test P increased more markedly in the loblolly pine plantation than in the pasture. These differences in soil test P accumulation trends between land use types may have been indicative of lower P demand by loblolly pine than bermudagrass, which led to a greater P accumulation in the soils of the loblolly pine plantations than pastures. Litter application rate also influenced soil P accumulation in both land use types in the LA-FORvsPAST study

(Table 8). Annual applications of litter at 5 Mg ha^{-1} did not significantly increase soil test P relative to the CONTROL treatment during the study. Soil test P accumulation was greater in response to the 20 Mg ha^{-1} litter application rate relative to the CONTROL and PL5 treatments throughout the study and greater relative to the 10 Mg ha^{-1} rate by the fourth annual fertilization.

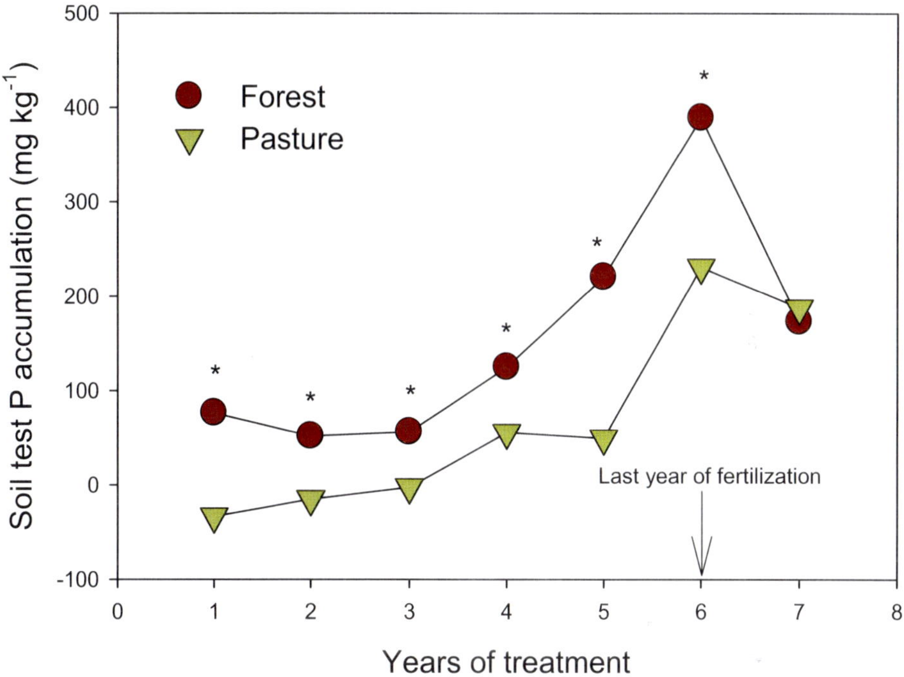

Fig. 3. Soil test P accumulation (0 to 15 cm) as affected by annual fertilization with poultry litter in a loblolly pine plantation and a bermudagrass pasture in the mid-South U.S.A. Asterisks denote years in which soil test P accumulation differed among land use types at $P < 0.05$

6.3 Other nutrients

Soil K concentrations were increased by broiler litter in the SILVO study (Table 9). A single application of the 10PL treatment increased K concentrations in the A horizon, and subsequent applications led to increases in K concentrations in the E horizon. Increases in K concentrations in lower soil depths have also been observed in response to annual litter fertilization of pastures and agricultural crops on sandy soils (Kingery et al., 1994; Mitchell & Tu, 2006). A similar increase in soil K concentrations in the uppermost 15 cm of soil in response to a single application of broiler litter was found in the SWITCH study (data not shown). In that study soil K increased more in response to the PL3 treatment than all others, and K concentrations of all other fertilizer treatments exceeded that of the CONTROL treatment. Results of both studies indicate that poultry litter can lead to increases in soil K concentrations in these silvopastures, even after a single application.

Although soil K concentrations increased in both the SILVO and SWITCH studies, the amount of poultry litter required to increase the concentrations differed between the two types of silvopastures. An application of only 1.5 Mg ha^{-1} of litter was needed to increase K concentrations in the loblolly pine-switchgrass silvopasture while in the loblolly pine-bahiagrass silvopasture K concentrations were observed only after two annual applications of 10 Mg ha^{-1} of poultry litter. Since the soil type was identical for these two studies, these results suggest that loblolly pine-bahiagrass silvopasture had a greater K demand than the loblolly pine-switchgrass silvopasture. The higher demand of the loblolly pine and bahiagrass pasture was likely due in part to loblolly pine density that was nearly double that in the loblolly pine and switchgrass silvopasture. The switchgrass also likely had a lower K demand than bahaiagrass , because switchgrass is characterized by relatively low nutrient demand despite its relatively high biomass growth potential (Tilman et al., 2006). Nevertheless, annual broiler application at 10 Mg ha^{-1} apparently exceeded vegetation K demand and sorption capacity of the A horizon in the loblolly pine and bahiagrass silvopasture as indicated by increased in K concentrations in the E horizon after four annual applications.

			Date			
Nutrient	**Horizon**	**Treatment**	**1997**	**1998**	**2001**	**2002**
K	A	CONTROL	42.1 a	30.1 b	23.3 a	30.7 a
		IF	33.9 a	42.9 b	23.7 a	26.0 a
		PL5	33.2 a	44.3 b	31.4 a	30.1 a
		PL10	39.8 a	62.6 a	34.9 a	36.1 a
	E	CONTROL	22.8 a	22.2 a	21.3 c	34.1 b
		IF	22.7 a	27.2 a	28.8 bc	37.8 b
		PL5	21.7 a	34.2 a	39.7 ab	43.8 b
		PL10	30.8 a	36.8 a	51.5 a	60.0 a
Mg	A	CONTROL	30.5 b	33.8 b	107.7 bc	96.2 c
		IF	32.0 a	26.9 bc	103.7 c	89.3 c
		PL5	34.8 a	38.4 ab	113.5 ab	100.9 b
		PL10	35.2 a	44.4 a	120.6 a	114.8 a
	E	CONTROL	26.6 a	25.8 b	107.8 b	100.4 b
		IF	29.8 a	25.8 b	112.4 b	104.4 b
		PL5	34.8 a	28.2 ab	126.6 ab	114.3 ab
		PL10	57.6 a	34.9 a	145.1 a	137.6 a
Ca	A	CONTROL	184.6 a	194.6 a	134.4 c	70.2 c
		IF	177.2 a	157.2 a	89.6 c	20.8 c
		PL5	171.2 a	196.0 a	162.9 b	95.5 b
		PL10	186.0 a	226.0 a	229.9 a	174..3 a

Table 9. Effects of annually fertilizing a loblolly pine and bahiagrass silvopasture with poultry litter and inorganic fertilizer on soil K and Mg in the A and E soil horizons and on Ca in the A horizon. For each nutrient and horizon, means within a column followed by a different letter differ at $P < 0.05$. Adapted from Blazier et al. (2008a)

As with K, soil Mg concentrations were increased in the A and E horizons by repeated applications of litter in the SILVO study (Table 9; Blazier et al., 2008a). After two applications soil Mg in the A and E horizons was increased by the 10 Mg ha^{-1} rate relative to the CONTROL and IF treatments, and after four applications the 5 Mg ha^{-1} rate led to greater soil K concentrations in the A horizon than in the CONTROL and IF treatments. However, the 5 Mg ha^{-1} did not increase soil K concentrations in the E horizon and did not increase soil K concentrations to levels in the A horizon comparable to that of the 10 Mg ha^{-1} rate after the fourth applications. By the fourth application, soil Ca concentrations in the A horizon were also increased by the poultry litter treatments, with that of the PL10 treatment exceeding all other treatments and that of the PL5 treatment greater than the CONTROL and IF treatments.

7. Water nutrients

Poultry litter applications led to increases in NO_3-N in soil water in the AR-FORvsPAST study. Total N concentrations in soil water were greater for pastures than the loblolly pine plantation and greater for the PL9 treatment than the CONTROL treatment; differences in NO_3-N accounted for the majority of the total N differences between land use types and treatments. In both pasture and loblolly pine plantation, NO_3-N concentrations increased in response to poultry litter application (Figure 4). Soil water NO_3-N concentrations were

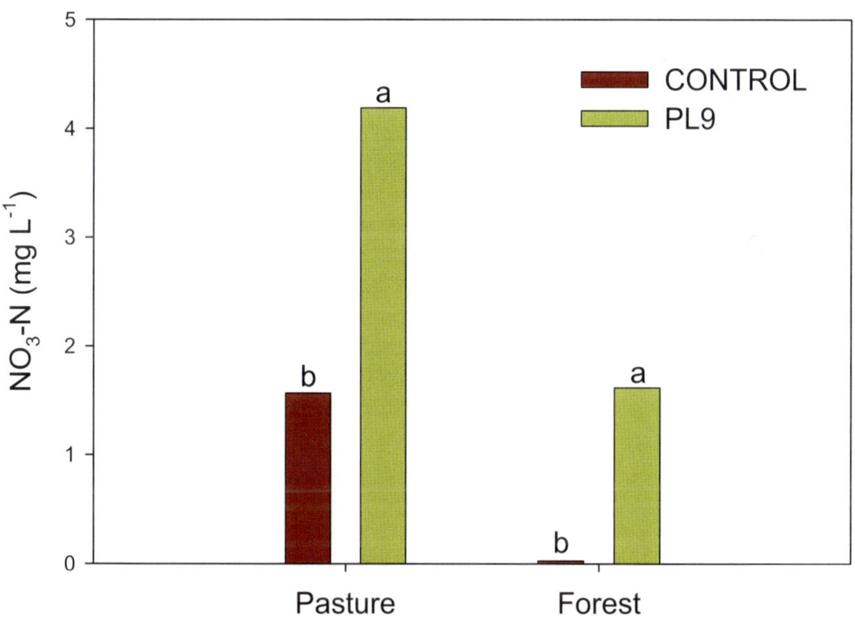

Fig. 4. Mean bi-weekly soil water (30 cm) NO_3-N concentrations in bermudagrass pasture and loblolly pine plantation treated with poultry litter. For each land use type, means headed by different letters differ at $P < 0.05$. Adapted from Liechty et al. (2009)

significantly positively correlated with potential nitrification rates. Although bi-weekly NO_3-N concentrations in soil water never exceeded the 10 mg L^{-1} drinking water standard of the U.S. Environmental Protection Agency in the loblolly pine plantation, this standard was exceeded in two or more sampling periods in pasture plots fertilized with poultry litter. Soil water N increased 51% more in pastures than in loblolly pine plantation, which suggests the potential for N pollution of water is greater for pastures fertilized with poultry litter than for loblolly pine plantations fertilized with poultry litter. However, because forest soils have an apparently greater propensity than pastures to retain proportionally greater NO_3-N in soil (described above), with long-term litter applications N losses in soil water from forests could be greater than in pastures (Liechty et al., 2009).

Repeated fertilization with poultry litter led to increases in total and dissolved P concentrations in runoff in pasture and loblolly pine plantation in the LA-FORvsPAST study. Total and dissolved P concentrations increased with increasing litter application rate in both land use types, although the P concentrations increased more markedly to 10 and 20 Mg ha^{-1} rates in pasture than in loblolly pine plantation. Total and dissolved P concentrations in runoff were positively correlated with Bray P concentrations in soil. These results indicate potential for losses of P in runoff in response to litter application in pasture and loblolly pine plantation, with modest evidence that P loss potential in loblolly pine plantation was lower. In the AR-FORvsPAST study, there were no significant differences in total P concentrations in soil water among treatments and land use types.

8. Conclusions

Poultry litter was a beneficial fertilizer for loblolly pine plantations and silvopastures in this series of studies. Unlike with inorganic fertilizer, soil pH did not decrease with poultry litter application, which sustained microbial biomass and activity at levels comparable to non-fertilized soil. Poultry litter application to soils that had annual pine straw harvesting maintained soil strength, organic matter, and soil moisture similar to those without straw harvesting, whereas applying inorganic fertilizer to soils with straw harvesting negatively impacted these soil attributes. Loblolly pine trees in plantations and silvopastures, as well as the grasses in silvopastures, responded to poultry litter fertilization with increased growth and nutrient concentrations. These increases in plant growth and nutrition provided some buffering against increasing soil nutrient concentrations when these plantations and silvopastures were annually fertilized with poultry litter. Nevertheless, poultry litter was more prone to lead to accumulation of NO_3-N and P in soil than inorganic fertilizer. Loblolly pine plantations were also more prone to increases in soil NO_3-N and P than pastures. Accumulations in soil NO_3-N and P were also associated with increased NO_3-N and P concentrations in soil water and runoff, respectively. As such, poultry litter fertilization of these loblolly pine plantations and silvopastures had the potential to contaminate soil water with N and P. Any poultry litter fertilization regimes for loblolly pine plantations and silvopastures must account for the greater tendencies of N and P accumulation in soil and water of these ecosytems; lower rates and/or frequencies than those used in these trials will likely be necessary for ecologically sustainable fertilization with poultry litter.

9. References

Adams, P.L., Daniel, T.C., Edwards, D.R., Nichols, D.J., Pote, D.H. & Scott, H.D. (1994). Poultry litter and manure contributions to nitrate leaching through the vadose zone. *Soil Science Society of America Journal,* Vol. 58 (No. 4): 1206-1211.

Adeli, A., Rowe, D.E. & Read, J.J. (2006). Effects of soil type on bermudagrass response to broiler litter application. *Agronomy Journal,* Vol. 98 (No. 1): 148-155.

Agbede, T.M., Oladitan, T.O., Alagha, S.A., Ojomo, A.O. & Ale, M.O. (2010). Comparative evaluation of poultry manure and NPK fertilizer on soil physical and chemical properties, leaf nutrient concentrations, growth and yield of yam (*Dioscorea rotundata* Poir) in southwestern Nigeria. *World Journal of Agricultural Sciences,* Vol. 6 (No. 5): 540-546.

Alef, K. (1995). Dehydrogenase activity, In: *Methods in applied soil microbiology and biochemistry,* Alef, K. & Nannipieri, P. (eds.). pp. 228-231. Academic Press, San Diego, CA.

Allen, H.L. (1987). Forest fertilizers: nutrient amendment, stand productivity, and environmental impact. *Journal of Forestry,* Vol. 85 (No. 2): 37-46.

Allen, H.L., Fox, T.R. & Campbell, R.G. (2005). What is ahead for intensive pine plantation silviculture in the South? *Southern Journal of Applied Forestry,* Vol. 29 (No. 2): 62-69.

Alikhani, H.A., Saleh-Rastin, N. & Antoun, H. (2006). Phosphate solubilization activity of rhizobia native to Iranian soils. *Plant and Soil,* Vol. 287 (No. 1-2): 35-41.

Anderson, T.H. & Domsch, K.H. (1993). The metabolic quotient for CO, (qCO_2) as a specific activity parameter to assess the effects of environmental conditions, such as pH, on the microbial biomass of forest soils. *Soil Biology and Biochemistry,* Vol. 25 (No. 3): 393-395.

Attiwill, P.M. & Adams, M.A. (1993). Tansley review no. 50: Nutrient cycling in forests. *New Phytologist,* Vol. 124: 561-582.

Aune, J.B. & Lal, R. (1997). Agricultural productivity in the tropics and critical limits of properties of oxisols, ultisols, alfisols. *Tropical Agriculture,* Vol.74 (No. 2): 96-103.

Baath, E., Frostegard, A., Pennanen, T., & Fritze, H. (1995). Microbial community structure and pH response in relation to soil organic matter quality in wood-ash fertilized, clear-cut or burned coniferous forest soils. *Soil Biology and Biochemistry,* Vol. 27 (No. 2): 229-240.

Bailey, R.G. (1995). *Description of the ecoregions of the United States* (2nd edition), USDA Forest Service Miscellaneous Publication No. 1391, USDA Forest Service, Washington, D.C.

Ball, D.F. (1964). Loss on ignition as an estimate of organic matter and organic carbon in non-calcareous soils. *Journal of Soil Science,* Vol. 15 (No. 1): 84.92.

Beem, M., Turton, D.J., Barden, C.J. & Anderson, S. (1998). *Application of poultry litter to pine forests,* OSU Extension Factsheet F-5037, Oklahoma Cooperative Extension Service, Oklahoma State University, Stillwater, OK.

Binkley, D., Burnham, H. & Allen, H.L. (1999). Water quality impacts of forest fertilization with nitrogen and phosphorus. *Forest Ecology and Management,* Vol. 121 (No. 3): 191-213.

Blake, G.R. & Hartge, K.H. (1986). Bulk density, In: *Methods of Soil Analysis, Part 1: Physical and Mineralogical Methods.* (2nd Edition), Klute, A. (ed.). pp. 363-375. Soil Science Society of America, Madison, WI.

Blazier, M.A., Hennessey, T.C. & Deng, S.P. (2005). Effects of fertilization and vegetation control on microbial biomass carbon and dehydrogenase activity in a juvenile loblolly pine plantation. *Forest Science*, Vol. 51 (No. 5): 449-459.

Blazier, M.A., Hennessey, T.C., Dougherty, P.M. & Campbell, R. (2006). Nitrogen accumulation and use by a young loblolly pine plantation in southeast Oklahoma: Effects of fertilizer formulation and date of application. *Southern Journal of Applied Forestry*, Vol. 30 (No. 2): 66-78.

Blazier, M.A., Gaston, L.A., Clason, T.R., Farrish, K.W., Oswald, B.P. & Evans, H.A. (2008*a*). Nutrient dynamics and tree growth of silvopastoral systems: impact of poultry litter. *Journal of Environmental Quality*, Vol. 37 (No. 4): 1546-1558.

Blazier, M.A., Hotard, S.L. & Patterson, W.B. (2008*b*). Straw harvesting, fertilization, and fertilizer type alter soil microbiological and physical properties in a loblolly pine plantation in the mid-South U.S.A. *Biology and Fertility of Soils*, Vol. 45 (No. 2): 145-153.

Blevins, D.G. & Barker, D.J. (2007). Nutrients and water in forage crops, In: *Forages: The Science of Grassland Agriculture. Volume II* (6th edition), Barnes, R.R., Nelson, C.J., Moore, K.J. & Collins, M. (Eds.). pp 67-80. Blackwell Publishing, Ames, IA.

Bradford, J.M. (1986). Penetrability, In: *Methods of Soil Analysis, Part 1: Physical and Mineralogical Methods*. (2nd edition), Klute A. (Ed.). pp. 463-478. Soil Science Society of America, Madison, WI.

Bray, R.H. & Kurtz, L.T. (1945). Determination of total, organic, and available forms of phosphorus in soils. *Soil Science*, Vol. 59 (No. 1): 39-45.

Breeuwsma, A. & Silva, S. (1992). *Phosphorus fertilization and environmental effects in the Netherlands and the Po region (Italy)*, Report 57, Winand Staring Centre for Integrated Land, Soil and Water Research, Wageningen, The Netherlands.

Brye, K.R. (2003). Long-term effects of cultivation on particle size and water-retention characteristics determined using wetting curves. *Soil Science*, Vol. 168 (No. 7): 459-468.

Cabrera, M.L. & Sims, J.T. (2000). Beneficial use of poultry by-products: challenges and opportunities, In: *Land application of agricultural, industrial, and municipal by-products* (1st edition), Power, J.F. & Dick, W.A. (Eds.). pp. 425-450. Soil Science Society of America, Madison, WI.

Canali, S., Tinchera, A., Intrigliolo, F., Pompili, L., Nisini, L., Mocali, S. & Torrisi, B. (2004). Effect of long term addition of composts and poultry manure on soil quality of citrus orchards in Southern Italy. *Biology & Fertility of Soils*, Vol. 40 (No. 3): 206-210.

Clason, T.R. (1995). Economic implications of silvipastures on southern pine plantations. *Agroforestry Systems*, Vol. 29: 227-238.

Clason, T.R. & Robinson, J.L. (2000). From a pine forest to a silvopasture system, Agroforestry Note 18, USDA Forest Service, USDA Natural Resource Conservation Service, Washington, D.C.

Clason, T.R. & Sharrow, S.H. (2000). Silvopastoral practices, In: *North American Agroforestry: An Integrated Science and Practice*, Garrett, H.E., Rietveld, W.J. & Fisher, R.F. (Eds.). pp. 119-147. Agronomy Society of America, Madison, WI.

Clesceri, L.S., Greenberg, A.E. & Eaton, A.D. (Eds.). (1998). *Standard methods for the examination of water and wastewater* (20th edition), American Public Health Association, Washington, DC.

Colbert, S.R., Jokela, E.J. & Neary, D.G. (1990). Effects of annual fertilization and sustained weed control on dry matter partitioning, leaf area, and growth efficiency of juvenile loblolly and slash pine. *Forest Science*, Vol. 36 (No. 4): 995-1014.

Daddow R.L. & Warrington, G.E. (1983). *Growth-limiting soil bulk densities as influenced by soil texture*, Watershed Systems Development Group Report WSDG-TN-00005, USDA Forest Service, Fort Collins, CO.

Daniel, T.C., Sharpley, A.N. & Lemunyon, J.L. (1998). Agricultural phosphorus and eutrophication: a symposium overview. *Journal of Environmental Quality*, Vol. 27 (No. 2): 251-257.

Danielson, R.E. & Sutherland, P.L. (1986). Porosity, In: *Methods of Soil Analysis, Part 1: Physical and Mineralogical Methods* (2nd edition). Klute, A. (Ed.). pp. 443-462. Soil Science Society of America, Madison, WI.

Dick, W.A., Eckert, D.J. & Johnson, J.W. (1998). *Land application of poultry litter*, Ohio State University Cooperative Extension Fact Sheet ANR-4-98, Ohio State University, Columbus, OH.

Dickens, E.D., Bush, P.B., & Morris, L.A. (2003). Poultry litter application recommendations in pine plantations. Warnell School of Forestry and Natural Resources, College of Agricultural and Environmental Sciences, University of Georgia, Athens, GA, Retrieved from http://www.bugwood.org/fertilization/PLARPP.html

Dickens, E.D., Richardson, B.W. & McElvany, B.C. (2004). Old-field thinned loblolly pine plantation fertilization with diammonium phosphate plus urea and poultry litter: 4 year growth and product class distribution results, In: *Proceedings of the 12th Biennial Southern Silvicultural Research Conference*, General Technical Report SRS-48, Outcalt, K. (Ed.), pp. 395-397. USDA Forest Service, Southern Research Station, Asheville, NC.

Donegan, K.K., Watrud, L.S., Seidler, R.J., Maggard, S.P., Shiroyama, T., Porteous, L.A. & DiGiovanni, G. (2001). Soil and litter organisms in Pacific Northwest forests under different management practices. *Applied Soil Ecology*, Vol. 18 (No. 2): 159-175.

Duryea, M.L. & Edwards, J.C. (1989). *Pine-straw management in Florida's forest*, Florida Cooperative Extension Service Institute of Food and Agricultural Science Circular 831, University of Florida, Gainsville, FL.

Edmeades, D.C. (2003). The long-term effects of manures and fertilizers on soil productivity and quality: a review. *Nutrient Cycling in Agroecosystems*, Vol. 66 (No. 2): 165-180.

Edwards, D.R. & Daniel, T.C. (1992). Environmental impacts of on-farm poultry waste disposal: a review. *Bioresource Technology*, Vol. 41 (No. 1): 9-33.

Eichhorn, M.M. (2001). *Impact of best management practices and organic wastes on water quality and crop production: Poultry litter application demonstration project*, Louisiana Department of Environmental Quality Projects CFMS514283 and CFMS554784 Final Report, Hill Farm Research Station, Louisiana Agricultural Experiment Station, Baton Rouge, LA.

Ekinci, K., Keener, H.M. & Elwell. D.L. (2000). Compositing short paper fiber with broiler litter and additives Part 1: Effects of initial pH and carbon/nitrogen ratio on ammonia emissions. *Compost Science & Utilization*, Vol. 8 (No. 2): 160-172.

Ellum, K. (2010). *Pine straw raking and fertilizer source impacts on nitrogen mineralization, pine needle gas exchange, and tree water stress in a loblolly pine plantation*, M.S. thesis, School of Forest Resources, University of Arkansas, Monticello, AR.

Elser, J. J., Bracken, M.E.S., Cleland, E.E., Gruner, D.S., Harpole, W.S., Hillebrand, H., Ngai, J.T., Seabloom, E.W., Shurin, J.B. & Smith, J.E. (2007). Global analysis of nitrogen and phosphorus limitation of primary producers in freshwater, marine, and terrestrial ecosystems. *Ecology Letters*, Vol. 10 (No. 12): 1135–1142.

Eno, C.F. (1960). Nitrate production in the field by incubating the soil in polyethylene bags. *Soil Science Society of America Proceedings*, Vol. 24: 277–279.

Evans, H.A. (2000). *Application of poultry litter and commercial fertilizer in a loblolly pine-bahiagrass silvopasture*, M.S. thesis, College of Forestry, Stephen F. Austin University, Nacogdoches, TX.

Falkiner, R.A. & Polglase, P.J. (1997). Transport of phosphorus through soil in an effluent-irrigated tree plantation. *Australian Journal of Soil Research*, Vol. 35: 385-398.

Farley, K.A., Jobbágy, E.G. & Jackson, R.B. (2005). Effects of afforestation on water yield: a global synthesis with implications for policy. *Global Change Biology*, Vol. 11 (No. 10): 1565-1576.

Fauci, F. & Dick, R.P. (1994). Microbial biomass as an indicator of soil quality: effects of long-term management and recent soil amendments, In: *Defining soil quality for a sustainable environment* (1st edition). Doran, J.W., Coleman, D.C., Bezdicek, D.F. & Stewart, B.A. (Eds). pp. 229-234. Soil Science Society of America, Madison, WI.

Fisher, D.S., Steiner, J.L., Endale, D.M., Stuedemann, J.A., Schomberg, H.H., Franzluebbers, A.J. & Wilkinson, S.R. (2000). The relationship of land use practices to surface water quality in the Upper Oconee Watershed in Georgia. *Forest Ecology and Management*, Vol. 128 (No. 1-2): 39-48.

Friend, A.L., Roberts, S.D., Schoenholtz, S.H., Mobley, J.A. & Gerard, P.D. (2006). Poultry litter application to loblolly pine forests: growth and nutrient containment. *Journal of Environmental Quality*, Vol. 35 (No. 3): 837-848.

Funderberg, E. (2009). Poultry litter as fertilizer. In: *Ag News and Views, January 2009, Soil & Crops*. The Samuel Roberts Noble Foundation. Available from: http://www.noble.org/ag/Soils/PoultryLitter/index.html.

Gallardo, A. & Schlesinger, W.H. (1994). Factors limiting microbial biomass in the mineral soil and forest floor of a warm-temperate forest. *Soil Biology & Biochemistry*. Vol. 26 (No 10): 1409-1415.

Gallimore, L.E., Basta, N.T., Storm, D.E., Payton, M.E., Huhnke, R.H. & Smolen, M.D. (1999). Water treatment residual to reduce nutrients in surface runoff from agricultural land. *Journal of Environmental Quality*, Vol. 28 (No. 5): 1474-1478.

Gambrell, R.P. (1996). Manganese, In: *Methods of soil analysis, part 3: Chemical methods* (3rd edition). Bartels, J.M. (Ed.) pp. 665-682. Soil Science Society of America, Madison, WI.

Gaston, L.A., Clason, T.R. & Cooper, D. (2003). Poultry litter fertilizer on pasture, silvopasture, and forest soils. *Louisiana Agriculture*, Vol. 46 (No. 3): 22-23.

Gaston, L.A., Drapcho, C.M., Tapadar, S. & Kovar, J.L. (2003). Phosphorus runoff relationships for Louisiana Coastal Plain soils amended with poultry litter. *Journal of Environmental Quality*, Vol. 32 (No. 4): 1422-1429.

Gee, G.W., Campbell, M.D., Campbell, G.S. & Campbell, J.H. (1992). Rapid measurement of low soil water potentials using a water activity meter. *Soil Science Society of America Journal*, Vol. 56 (No. 4): 1068-1070.

Gilman, E.F. (1987). *Where are tree roots?* Florida Cooperative Extension Service, Institute of Food and Agricultural Sciences, University of Florida, Extension Bulletin ENH137, Retrieved from http://edis.ifas.ufl.edu/pdffiles/WO/WO01700.pdf

Gupta, G., Borowiec, J. & Okoh, J. (1997). Toxicity identification of poultry litter aqueous leachate. *Poultry Science,* Vol. 76 (No. 10): 1364-1367.

Harmel, R.D., Torbet, H.A., Haggard, B.E., Haney, R. & Dozier, M. (2004). Water quality impacts of converting to a poultry litter fertilization strategy. *Journal of Environmental Quality,* Vol. 33 (No. 6): 2229-2242.

Harris, J.A. (2003). Measurements of the soil microbial community for estimating the success of restoration. *European Journal of Soil Science,* Vol. 54 (No. 4): 801-808.

Hart, S.C., Stark, J.M., Davidson, E.A. & Firestone, M.K. (1994). Nitrogen mineralization, immobilization, and nitrification, In: *Methods of Soil Analysis. Part 2. Microbiological and Biochemical Properties.* Weaver, R.W., Angele, S., Bottomly, P. (Eds.), pp. 985-1018. Soil Science Society of America. Madison, WI.

Haywood, J.D., Tiarks, A.E. & Sword, M.A. (1997). Fertilization, weed control, and pine litter influence loblolly pine stem productivity and root development. *New Forests,* Vol. 14 (No. 3): 233-249.

Haywood, J.D., Tiarks, A.E., Elliott-Smith, M.L. & Pearson, H.A. (1998). Response of direct seeded *Pinus palustris* and herbaceous vegetation to fertilization, burning, and pine straw harvesting. *Biomass Bioenergy,* Vol. 14 (No. 2): 157-167.

Helmke, P.A. & Sparks, D.L. (1996). Lithium, sodium, potassium, rubidium, and cesium, In: *Methods of soil analysis, part 3. Chemical methods,* Bartels, J.M. (Ed.), pp. 551-574. Soil Science Society of America, Madison, WI.

Henry, C.L., Cole, D.W., Hinckley, T.M. & Harrison, R.B. (1993). The use of municipal and pulp and paper sludges to increase production in forestry. *Journal of Sustainable Forestry,* Vol. 1 (No. 3): 41-55.

Högberg, P., Nordgren, A., Buchmann, N., Taylor, A.F.S., Ekblad, A., Högberg, M.N., Nyberg, G., Ottoson-Löfvenius, M. & Read, D.J. (2001). Large-scale forest girdling shows that current photosynthesis drives soil respiration. *Nature,* Vol. 411: 789-792.

Horneck, D.A. & Miller, R.O. (1998). Determination of total nitrogen in plant tissue, In: *Handbook of reference methods for plant analysis,* Yask, P. (Ed.), pp. 75-83. CRC Press, Boca Raton, FL.

Hue, N.V. (1992). Correcting soil acidity of a highly weathered ultisol with chicken manure and sewage sludge. *Communications in Soil Science and Plant Analysis,* Vol. 23 (No.3-4): 241-264.

Jackson, M.J., Line, M.A., Wilson, S. & Hetherington. S.J. (2000). Application of composted pulp and paper mill sludge to a young pine plantation. *Journal of Environmental Quality,* Vol. 29 (No. 2): 407-414.

Jenkinson, D.S. & Powlson, D.S. (1976*a*). The effects of biocidal treatments on metabolism in soil-I. Fumigation with chloroform. *Soil Biology & Biochemistry,* Vol. 8 (No. 3): 167-177.

Jenkinson, D.S. & Powlson, D.S. (1976*b*). The effects of biocidal treatments on metabolism in soil-V: A method for measuring soil biomass. *Soil Biology & Biochemistry,* Vol. 8 (No. 3): 209-213.

Jokela, E.J. (2004). Nutrient management for southern pines, In: *Slash pine: still growing and growing! Proceedings of the slash pine symposium,* General Technical Report SRS-76,

Dickens, E.D., Barnett, J.P., Hubbard, W.G. & Jokela, E.J. (Eds.), pp. 27-35. U.S. Department of Agriculture, Forest Service, Southern Research Station, Asheville, NC.

Jones, J.B. & Case, V.W. (1990). Sampling, handling, and analyzing plant tissue samples. In: *Soil testing and plant analysis* (3rd edition), Westerman, R.L. (Ed.), pp. 389-447. Soil Science Society of America, Madison, WI.

Jorgensen, J.R. & Wells, C.G. (1986). *Foresters' primer in nutrient cycling: a loblolly pine management guide,* General Technical Report SE-37, USDA Forest Service, Southeastern Forest Experiment Station, Asheville, NC.

Kellogg, R.L., Lander, C.H., Moffitt, D.C. & Gollehon, N. (Eds.). (2000). *Manure nutrients relative to the capacity of cropland and pastureland to assimilate nutrients: Spatial and temporal trends for the United States.* USDA Natural Resources Conservation Service Publication NPS 00-0579, GSA National Forms and Publication Center, Fort Worth, TX.

Khan, S.A., Mulvaney, R.L., Ellsworth, T.R. & Boast, C.W. (2007). The myth of nitrogen fertilization for soil carbon sequestration. *Journal of Environmental Quality,* Vol. 36 (No. 6): 1821-1832.

Kingery, W.L., Wood, C.W., Delaney, D.P., Williams, J.C. & Mullins, G.L. (1993). Implications of long-term application of poultry litter on tall fescue pastures. *Journal of Production Agriculture,* Vol. 6 (No. 3): 315-395.

Kingery, W.L., Wood, C.W., Delaney, D.P., Williams, J.C. & Mullins, G.L. (1994). Impact of long-term land application of broiler litter on environmentally related soil properties. *Journal of Environmental Quality,* Vol. 23 (No. 1): 139-147.

Kronzucker, H.J., Yaeesh Siddiqi, M. & Glass, A.D.M. (1997). Conifer root discrimination against soil nitrate and the ecology of forest succession. *Nature,* Vol. 385: 59-61.

Lemunyon, J. & Gilbert, R. (1993). The concept and need for a phosphorus assessment tool. *Journal of Production Agriculture,* Vol.6 (No. 4): 483-486.

Lenhard, G. (1956). The dehydrogenase activity in soil as a measure of the activity of soil microorganisms. *Z. Pflanzenernäh Düng Bodenkd,* Vol. 73:1-11.

Liechty, H.O., Blazier, M.A., Wight, J.P., Gaston, L.A., Richardson, J.D. & Ficklin, R.L. (2009). Assessment of repeated application of poultry litter on phosphorus and nitrogen dynamics in loblolly pine: implications for water quality. *Forest Ecology and Management,* Vol. 258 (No. 10): 2294-2303.

Lopez-Zamora, I., Duryea, M.L., McCormac, W.C., Comerford, N.B. & Neary, D.G. (2001). Effect of pine needle removal and fertilization on tree growth and soil P availability in a *Pinus elliottii* Engelm. var. *elliottti* stand. *Forest Ecology and Management,* Vol. 148 (No 1-3): 125-134.

Lynch, L. & Tjaden, R. (2004). Would forest landowners use poultry manure as fertilizer? *Journal of Forestry,* Vol. 102 (No. 5): 40-45.

McLeod, R.V. & Hegg, R.O. (1984). Pasture runoff quality from application of inorganic and organic nitrogen sources. *Journal of Environmental Quality,* Vol. 13 (No. 1): 122-126.

Mehlich, A. (1984). Mehlich 3 soil test extractant: A modification of the Mehlich 2 extractant. *Communications in Soil Science and Plant Analysis,* Vol. 15 (No. 12): 1409-1416.

Mills, R. & Robertson, D.R. (1991). *Production and marketing of Louisiana pine straw,* Louisiana Cooperative Extension Service Publication 2340, Louisiana State University Agricultural Center, Baton Rouge, LA.

Minogue, P.J., Ober, H.K. & Rosenthal, S. (2007). *Overview of Pine Straw Production in North Florida: Potential Revenues, Fertilization Practices, and Vegetation Management Recommendations,* School of Forest Resources and Conservation Department, Florida Cooperative Extension Service, Institute of Food and Agricultural Sciences Publication 125. University of Florida, Gainesville, FL.

Mitchell, C.C. & Donald, J.O. (1995). *The value and use of poultry manures as fertilizers,* Alabama Cooperative Extension System Circular ANR-224, Auburn University, Auburn, AL.

Mitchell, C.C. & Tu, S. (2006). Nutrient accumulation and movement from poultry litter. *Soil Science Society of America Journal,* Vol. 70 (No. 6): 2146-2153.

Mulvaney, R.L. (1996). Nitrogen-inorganic forms, In: *Methods of Soil Analysis. Part 3. Chemical methods,* Sparks, D.L. (Ed.), pp. 1123-1184. Soil Science Society of America, Madison, WI.

Munkholm, L.J., Schjønning, P., Debosz, K., Jensen, H.E. & Christensen, B.T. (2002). Aggregate strength and mechanical behavior of a sandy loam under long-term fertilization treatments. *European Journal of Soil Science,* Vol. 53 (No. 1): 129-137.

Murthy, R., Zarnoch, S.J. & Dougherty, P.M. (1997). Seasonal trends of light-saturated net photosynthesis and stomatal conductance of loblolly pine trees grown in contrasting environments of nutrition, water, and carbon dioxide. *Plant, Cell, and Environment,* Vol. 20 (No. 5): 558-568.

Nair, V.D. & Graetz, D.A. (2004). Agroforestry as an approach to minimizing nutrient loss from heavily fertilized soils: the Florida experience. *Agroforestry Systems,* Vol. 60: 269-279.

National Agricultural Statistics Service. (2008). *Poultry-Production and Value 2007 Summary,* United States Department of Agriculture National Agricultural Statistics Service, Washington, D.C.

O'Neill, G.J. & Gordon. A.M. (1994). The nitrogen filtering capacity of Carolina poplar in an artificial riparian zone. *Journal of Environmental Quality,* Vol. 23 (No. 6): 1218-1223.

Plaza, C., Hernández, D., García-Gil, J.C. & Polo, A. (2004). Microbial activity in pig slurry-amended soils under semiarid conditions. *Soil Biology & Biochemistry,* Vol. 36 (No. 10): 1577-1585

Polglase, P.J., Tompkins, D. & Falkiner, R.A. (1995). Mineralization and leaching of nitrogen in an effluent-irrigated pine plantation. *Journal of Environmental Quality,* Vol. 24 (No. 5): 911-922.

Pote, D.H. & Daniel, T.C. (2000). Analyzing for total phosphorus and total dissolved phosphorus in water samples. In: *Methods of Phosphorus Analysis for Soils, Sediments, Residuals, and Waters,* Southern Cooperative Series Bulletin No. 396., Pierzynski, G.M. (Ed.), pp. 91-93. Retrieved from: http://www.sera17.ext.vt.edu/Documents/Methods_of_P_Analysis_2000.pdf

Pote, D.H. & Daniel, T.C. (2008). Managing pine straw harvests to minimize soil and water losses. *Journal of Soil and Water Conservation,* Vol. 63 (No.1): 27-28.

Pote, D.H., Kingery, W.L., Aiken, G.E., Han, F.X., Moore Jr., P.A. & Buddington, K. (2003). Water-quality effects of incorporating poultry litter in perennial grassland soils. *Journal of Environmental Quality,* Vol. 32 (No. 6): 2392-2398.

Powers, R.F., Scott, D.A., Sanchez, F.G., Voldseth, R.A., Page-Dumroese, D., Elioff, J.D. & Stone, D.M. (2005). The North American long-term soil productivity experiment:

findings from the first decade of research. *Forest Ecology and Management,* Vol. 220 (No. 1-3): 31-50.

Powlson, D.S. & Brookes, P.C. (1987). Measurement of soil microbial biomass provides an early indication of changes in total soil organic matter due to straw incorporation. *Soil Biology & Biochemistry,* Vol. 19 (No. 2): 159-164.

Prestemon, J.P. & Abt, R.C. (2002). The southern timber market to 2040. *Journal of Forestry,* Vol. 100 (No. 7): 16-22.

Pritchett, W.L. & Fisher, R.F. (1987). *Properties and management of forest soils* (2nd edition), John Wiley & Sons, Inc., New York, NY.

Richardson, J. (2006). *Effects of poultry litter applied to pine plantations and pastures on water quality and soil nitrogen mineralization.* M.S. thesis, University of Arkansas at Monticello.

Rifai, S.W., Markewitz, D. & Borders, B. (2010). Twenty years of intensive fertilization and competing vegetation suppression in loblolly pine plantations: impacts on soil C, N, and microbial biomass. *Soil Biology & Biochemistry,* Vol. 42 (No. 5): 713-723.

Roberts, S.D., Friend, A.L. & Gerard, P.D. (2004). The effect of large applications of nutrients from organic waste on biomass allocation and allometric relations in loblolly pine, In: *Proceedings of the 12th Biennial Southern Silvicultural Research Conference,* General Technical Report SRS-48, Outcalt, K. (Ed.), pp. 398-402. USDA Forest Service, Southern Research Station, Asheville, NC.

Roberts, S.D., Friend, A.L. & Schoenholtz, S.H. (2006). Growth of precommercially thinned loblolly pine four years following application of poultry litter, In: *Proceedings of the 12th Biennial Southern Silvicultural Research Conference,* General Technical Report SRS-92, Outcalt, K. (Ed). pp. 139-142. USDA Forest Service, Southern Research Station, Asheville, NC.

Roise, J.P., Chung, J. & Lancia, R. (1991). Red-cockaded woodpecker habitat management and longleaf pine straw production: an economic analysis. *Southern Journal of Applied Forestry,* Vol. 15 (No. 2): 88-92.

Samuelson, L.J., Wilhoit, J. Stokes, T. & Johnson, J. (1999). Influence of poultry litter fertilization on 18-year-old loblolly pine stand. *Communications in Soil Science and Plant Analysis,* Vol. 30 (No. 3-4): 509-518.

Sanchez, F.G., Scott, D.A. & Ludovici, K.H. (2006). Negligible effects of severe organic matter removal and soil compaction on loblolly pine growth over 10 years. *Forest Ecology and Management,* Vol. 227 (No. 1): 145-154.

Sauer, T.J., Daniel, T.C., Nichols, D.J., West, C.P., Moore, P.A. & Wheeler, G.L. (2000). Runoff water quality from poultry litter-treated pasture and forest sites. *Journal of Environmental Quality,* Vol. 29 (No. 2): 515-521.

Schindler, D. (1978). Factors regulating phytoplankton production and standing crop in the word's freshwaters. *Limnology and Oceanography,* Vol. 23 (No. 3): 478-486.

Sharpley, A.N. (1999). Agricultural phosphorus, water quality, and poultry production: are they compatible. *Poultry Science,* Vol. 78 (No. 5): 660-673.

Sharpley, A.N. & Menzel, R.G. (1987). The impact of soil and fertilizer P on the environment. *Advances in Agronomy,* Vol. 41: 297-324.

Sharpley, A.N., Smith, S.J. & Bain, W.R. (1993). Nitrogen and phosphorus fate from long-term poultry litter applications to Oklahoma soils. *Soil Science Society of America Journal,* Vol. 57 (No. 4): 1131-1137.

Shober, A.L. & Sims, J.T. (2003). Phosphorus restrictions and land application of biosolids: current status and future trends. *Journal of Environmental Quality,* Vol. 32 (No. 6): 1955-1964.

Sims, J.T. (1986). Nitrogen transformations in a poultry manure amended soil: Temperature and moisture effects. *Journal of Environmental Quality,* Vol. 15 (No. 1): 59-63.

Sims, J.T. & Wolf, D.C. (1994). Poultry waste management: Agricultural and environmental issues. *Advances in Agronomy,* Vol. 52: 1-83.

Sistani, K.R., Adeli, A., McGowen, S.L., Tewolde, H. & Brink, G.E. (2008). Laboratory and field evaluation of broiler litter nitrogen mineralization. *Bioresource Technology,* Vol. 99 (No. 7): 2603-2611.

Sistani, K.R., Brink, G.E., Adeli, A., Tewolde, H. & Rowe, D.E. (2004). Year-round soil nutrient dynamics from broiler litter application to three bermudagrass cultivars. *Agronomy Journal,* Vol. 96 (No. 2): 525-530.

Tan, X., Chang, S.X. & Kabzems, R. (2005). Effects of soil compaction and forest floor removal on soil microbial properties and N transformations in a boreal forest long-term soil productivity study. *Forest Ecology and Management,* Vol. 217 (No. 2-3): 158-170.

Tate, D.F. (1994). Determination of nitrogen in fertilizer by combustion: Collaborative study. *Journal of AOAC International,* Vol. 77 (No. 4): 829-839.

Tekeste, M., Hatzhghi, D.H. & Stroonsnijder, L. (2007). Soil strength assessment using threshold probability approach on soils from three agro-ecological zones in Eritrea. *Biosystems Engineering,* Vol. 98 (No. 4): 470-478.

Tilman, D., Hill, J. & Lehman, C. (2006). Carbon-negative biofuels from low-input high-diversity grassland biomass. *Science,* Vol. 314 (No. 5805): 1598-1600.

United States Department of Agriculture (USDA) Economic Research Service. (2009). *Poultry and Eggs: Background.* Retrieved from: http://www.ers.usda.gov/Briefing/Poultry/Background.htm

United States Department of Agriculture Soil Conservation Service (USDA SCS). (1989). *Soil Survey of Claiborne Parish, Louisiana.* USDA SCS, Washington, D.C.

Vose, J.M. & Allen, H.L. (1988). Leaf area, stemwood growth, and nutrition relationships in loblolly pine. *Forest Science,* Vol. 34: 547-563.

Wagner, G.H. & Wolf, D.C. (1999). Carbon transformations and soil organic matter formation, In: *Principles and Applications of Soil Microbiology,* Sylvia, D.M., Fuhrmann, J.J., Hartel, P.G. & Zuberer, D.A. (Eds.). pp. 218-258. Prentice Hall Inc., Upper Saddle River, NJ.

Walkley, A. (1947). A critical examination of a rapid method for determining organic carbon in soils: Effect of variations in digestion conditions and of inorganic soil constituents. *Soil Science,* Vol. 63: 251-263.

Wallenstein, M.D., McNulty, S., Fernandez, I.J., Boggs, J. & Schlesinger, W.H. (2006). Nitrogen fertilization decreases forest soil fungal and bacterial biomass in three long-term experiments. *Forest Ecology and Management,* Vol. 222 (No. 1-3): 459-468.

Warren, S.L. & Fonteno, W.C. (1993). Changes in physical and chemical properties of a loamy sand soil when amended with poultry litter. *Journal of Environmental Horticulture,* Vol. 11 (No. 4): 186-190.

Weaver, T. (1998). Managing poultry manure nutrients. *Agricultural Research,* Vol. 46: 12-13.

Will, R.E., Markewitz, D., Hendrick, R.L., Meason, D.F., Crocker, T.R. & Borders, B.E. (2006). Nitrogen and phosphorus dynamics for 13-year-old loblolly pine stands receiving complete competition control and annual N fertilizer. *Forest Ecology and Management,* Vol. 227 (No. 1-2): 155-168.

Williams, C.M, Barker, J.C & Sims, J.T. (1999). Management and utilization of poultry wastes. *Reviews of Environmental Contamination & Toxicology,* Vol. 162: 105-157.

Zarcinas, B.A., Cartwright, B. & Spouncer, L.R. (1987). Nitric acid digestion and multi-nutrient analysis of plant material by inductively coupled plasma spectrometry. *Communications in Soil Science and Plant Analysis,* Vol. 18 (No. 1): 131-146.

Zimmermann, B., Elsenbeer, H. & DeMoraes, J.M. (2006). The influence of land-use changes on soil hydraulic properties: Implications for runoff generation. *Forest Ecology and Management,* Vol. 222 (No. 1-3): 29-38.

Zinkhan, F.C. & Mercer, D.E. (1996). An assessment of agroforestry systems in the southern U.S.A. *Agroforestry Systems,* Vol. 35: 303-321.

Zou, X.M., Ruan, H.H., Fu, Y., Yang, X.D. & Sha, L.Q. (2005). Estimating soil labile organic carbon and potential turnover rates using a sequential fumigation-incubation procedure. *Soil Biology & Biochemistry,* Vol. 37 (No. 10): 1923-1928.

5

Long-Term Effects of Residue Management on Soil Fertility in Mediterranean Olive Grove: Simulating Carbon Sequestration with RothC Model

O.M. Nieto[1,2], J. Castro[1] and E. Fernández[2]
[1]*IFAPA Centro Camino de Purchil, Junta de Andalucía, Granada,*
[2]*Dpto. Edafología y Química Agrícola, Facultad de Ciencias, Universidad de Granada, Granada,*
Spain

1. Introduction

Olive orchards are widely cultivated throughout the semiarid Mediterranean region. During olive growth a large quantity of vegetable residues are produced, either from the biannual pruning or from the olive-fruit cleaning in the oil mill, where the olive fruit is separated from the leaves, twigs and soil. These residues are generally discarded and the pruning debris is usually burned *in situ* or used for energy. Such practices not only release a large quantity of CO_2 into the atmosphere, but also fail to return to the soil the elements taken up by the tree. The use of crop residues is being widely debated today because of its impact on the soil degradation (Lal, 2008).

In Andalusian olive orchards, conventional agricultural practices such as tillage or non-tillage with bare soil also reduces the incorporation of plant remains into the soil, thus changing the quantity of soil organic carbon (SOC) in a variable way and accelerating erosive processes (Pastor, 2004). Nevertheless in recent years, the technique of shredding pruning debris and spreading the material over the orchard is becoming generalized as an alternative to burning. The residues left from fruit cleaning in the oil mill prior to extracting the oil, composed of leaves, green twigs, and superficial soil, can also be spread on the soil surface, returning to the soil the elements previously taken up by the tree. These new soil-management systems are an alternative for improving the soil quality and fertility in sustainable agricultural system (Ordóñez et al., 2001; Rodríguez-Lizana et al., 2008; Nieto et al., 2010).

Many studies on agricultural ecosystems, as reviewed by Jarecki & Lal (2003), have documented the changes in soil properties when the soil management shifts from tillage to cover crop, mainly the increase of SOC and nitrogen (N). The management of crop residues is an important aspect of conservation systems (Six et al., 1999; Paustian, 2000; Lal, 2008), since proper distribution on the ground surface reduces water losses and thus discourages soil erosion (Schomberg et al., 1999). Water and erosion constitute especially serious issues in zones that have a Mediterranean climate and can be only partially solved by recycling the crop debris (Rodríguez-Lizana et al., 2008).

When the soil includes great quantities of fresh plant material, it is necessary to separate the soil organic matter in order to quantify the SOC which is truly fixed. Some authors recommend methods to separate the water-floatable organic matter (FOM) from soil fractions according to size, using physical procedures such as ultrasound, together with a mixture of physical and chemical methods (Buyanovsky et al., 1994; Hevia et al., 2003). These latter procedures of fractionation appear to be more adequate when the residues added provide certain quantities of soil together with the plant debris.

According to Ingram & Fernandes (2001), the factors determining the current level of carbon in agricultural soils are the losses of soil and clay by erosion, the decline in plant debris, and the elimination of this. Franzluebbers (2002) proposed that soil quality is correlated with the stratification of the SOC. High SOC levels on the soil surface mitigate the direct impact of raindrops, protecting against sealing and the disruption of the soil structure (Hernanz et al., 2002).

Cultivation practices that improve soil quality and fertility, such as the use of crop residues, progressively change the physical and chemical properties of the soil (Rhoton et al., 1993; Ordóñez et al., 2001; Hernández et al., 2005). In addition to SOC and N, other nutrients have been made evaluated in this sense such as K^+ (Thomas et al., 2007), as has soil properties such as cation-exchange capacity (Oorts et al., 2003), or soil-water content (Rawls et al., 2003; Bescansa et al., 2006). The impact of different soil-management systems on soil properties have been studied for olive orchards (e.g. Hernández et al., 2005; Soria et al., 2005; Castro et al., 2008; Gómez et al., 2009) but only a few works have evaluated the effect of shredded olive-pruning debris (Ordóñez et al., 2001, Sofo et al., 2005; Rodríguez-Lizana et al., 2008).

Recently, agricultural soils have been identified as the major carbon pool in the context of its global cycle. Some authors (Jarecki & Lal, 2003; Hernández et al. 2005; Smith et al., 2008) have reported that strategies based upon changes in soil management in agricultural soils are potentially important in increasing carbon sequestration by the soil and in reducing the atmospheric CO_2 concentration. The main processes responsible for lowering current carbon levels in agricultural soils include erosion, tillage, and low inputs of agricultural residues (Lal, 2008; Álvaro-Fuentes et al., 2009). Some authors have emphasized that intensive tilling accelerates the decomposition of organic matter as result of the break-up of soil aggregates (Balesdent et al., 2000; Paustian et al., 2000) and contributes considerably to soil loss through erosion (Rodríguez-Lizana et al., 2008). Soil-management techniques that combine a restriction on tillage and the addition of organic residues are considered to be one potential way for improving soil properties and diminishing atmospheric CO_2 concentrations by storing carbon in the form of organic matter (IPCC, 2000; Jarecki & Lal, 2003).

The present work describes the effect on the soil after the spreading of olive-pruning debris together with the residues of the olive-fruit cleaning in two predominant soils in Andalusian olive orchards. In specific, study was made of the content, distribution, and stabilization of SOC (floatable and non-floatable in water) and N, the soil potential for carbon sequestration, as well as the effect of SOC in the K^+ content, bulk density (ρb), pH, cation-exchange capacity (CEC), and soil-water content (SWC) at -33 and -1500 kPa.

2. Material and methods

2.1 Field description

The study plot was located in the Cortijo El Empalme (Villacarrillo, Jaén), south-eastern Spain (38.175°N, 3.15°W) and is 812 m a.s.l. The climatic characteristics of the area are given in Table 1 (MAPA, 1989). The average annual rainfall was 550 mm, and average annual maximum and minimum temperatures of 37.0°C and 2.8°C, respectively. The natural vegetation is a perennial, sclerophyllous Holm oak (*Quercus ilex* L.) forest typical of the Mediterranean basin.

The orchard was comprised of adult olives (cv. picual) with 2-3 trunks and planting density of 82 trees ha^{-1}. The average slope is 3%. The orchard has underground drip irrigation and no fertilizers are applied to the soil. Following the WRBSR (FAO, 2006), the soils studied are classified as Chromic Calcisols (CLcr) and Calcic Vertisols (VRcc). The parent material is limestone in the CLcr and marls in the VRcc. The colour of the dry bare soils was dull brown (7.5 YR 5/4) in the CLcr and light grey (2.5 Y 7/1) in the VRcc.

Month	Temperature (°C)			Rainfall	ETo
	Maximum	Minimum	Mean	(mm/month)	(mm/month)
January	10.9	3.7	7.3	69	33
February	13.3	4.5	8.9	74	46
March	17.9	5.6	11.8	68	87
April	20.3	8.6	14.5	59	110
May	25.8	13.4	19.6	52	154
June	30.1	18.6	24.4	25	170
July	35.5	20.6	28.1	6	211
August	34.1	20.3	27.2	8	182
September	28.6	17.0	22.8	24	122
October	22.2	11.6	16.9	53	81
November	14.8	6.4	10.6	47	43
December	11.3	4.0	7.7	67	30

Table 1. Maximum, minimum and average monthly air temperature, monthly rainfall and potential evapotranspiration (ETo) for the study area (MAPA, 1989)

Before the experiment, the soil-management system of the orchard was conventional tillage (T), consisting of two or three passes (0.20 m deep) with a disc harrow and cultivator, twice a year to control weeds. This tillage was applied only to the open gaps between the trees (50% of the total area of the grove). Under the tree canopy (UC), the soil was completely cleared every year using pre- and post-emergence herbicides. Dead leaves, dried fruit and twigs were removed by manual blowers and a mechanical sweeper without breaking the surface crust. The trees were pruned every 2 yr and the debris was burned.

The soil-management system was changed in 1996 on the CLcr and in 2000 on the VRcc to cover crop, whereby shredded olive-pruning and the residues from the olive-fruit cleaning (PD+CR) were spread between the trees. The ground was not tilled and all these residues remained on the surface. The biomass input was quantified by the use of a 30 x 30 cm metal frame tossed at random 40 times between trees. The mean annual input was 23.9 ± 14.3 Mg

C ha^{-1} yr^{-1}. This area was studied by Soria (2002) under traditional tillage in 1997, just before to the experiment was started. The clay types are given in Table 2.

Soil type	Illite	Montmorillonite	Kaolinite
Chromic Calcisols	77	13	10
Calcic Vertisols	45	36	19

Table 2. Clay type (%) for 0-30 cm depth, in each soil type according to Soria (2002)

2.2 Sampling and analytical methods

Random soil samples were taken in two different areas: (i) between trees in PD+CR soil after removing the superficial plant-residue layer and (ii) UC area where the soil was completely bare. In addition, to establish the time-zero conditions for the experiment, two neighbouring tilled olive groves were sampled. A trench of 50 x 100 x 50 cm was opened and the samples were taken at depth intervals of 0-2, 2-5, 5-10, 10-15, and 15-30 cm. Three replicate plots per type of soil and area were sampled. Soil samples were also taken from the pits to determine ρb, following the method of Blake & Hartge (1986), using a set of cylinders of 2, 3, and 5 cm high specifically manufactured for this purpose.

The soil samples were dried and sieved (2-mm grid size). In the fine-earth fraction, the following analyses were performed: the textural analysis was made by the pipette method of Robinson (Soil Conservation Service, 1972); the SWC at field capacity was extracted in a pressure plate at -33 kPa, and the moisture at the wilting point was measured at -1500kPa (Cassel & Nielsen, 1986); the assimilable K^+ was extracted with NH_4OAc 1M; and the CEC was determined by saturation in sodium and, prior to washing with alcohol, extraction by sodium adsorbed with NH_4OAc 1M (Soil Conservation Service, 1972); the pH was measured in a soil suspension in distilled water (1:2.5).

For the determination of the SOC, N, and $CaCO_3$ equivalent, the sample was ground again (0.125 mm). The content of the SOC was determined using the method of Tyurin (1951); water-floatable organic matter (FOM) and non-water-floatable organic matter (NFOM) was separated following the method described by Hevia et al. (2003); for the total nitrogen, the Kjeldahl method was used (Bremner, 1965); and the $CaCO_3$ equivalent was determined by a manometric method (Williams, 1948). For plant remains, carbon and N were determined by the same methods. The SOC, N, and clay contents per hectare were computed by multiplying the soil mass (i.e. bulk-density) by the depth and the SOC, N, and clay concentrations, respectively. The CO_2 emissions from burning residues were determined from the values of carbon concentration in pruning debris using a molecular-weight ratio (1.00 g C = 3.67 g CO_2) (IPCC, 2000).

2.3 RothC model

A detailed description of the model is given in Coleman & Jenkinson (1996). In brief, the RothC model separates the SOC into four active compartments and a small amount of inert organic matter (IOM). Plant residues reintroduced to the soil (Figure 1) are divided into decomposable plant materials (DPM) and resistant plant materials (RPM), both undergoing decomposition to produce microbial biomass (BIO), humified organic matter (HUM) and CO_2 (lost from the system). The clay content of the soil determines the proportions that go to

CO_2 or to BIO + HUM. Each compartment, except for IOM, undergoes decomposition by first-order kinetics at its own characteristic rate, which is determined by using modifiers for soil moisture, temperature and plant cover.

The climatic input parameters include monthly average air temperature, monthly precipitation and monthly open-pan evaporation. Other input parameters are soil clay content, monthly carbon input from plant residues or farmyard manure and monthly information on soil cover, whether the soil is bare or covered by plants. As no data for open-pan evaporation were available, the average values for monthly potential evaporation (converted to open-pan evaporation) were used (Coleman & Jenkinson, 1996). The IOM was calculated using the equation proposed by Falloon et al. (1998). The turnover time was calculated as the total organic carbon content except IOM divided by the annual input of carbon into the soil (Jenkinson & Rayner, 1977).

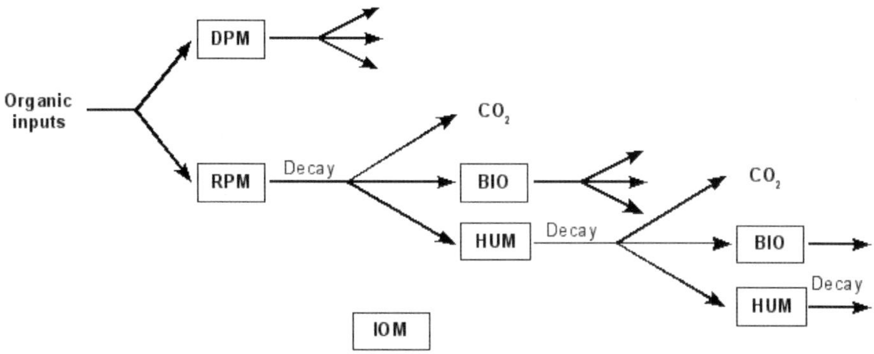

Fig. 1. Structure of the Rothamsted carbon model (from Coleman & Jenkinson, 1996)

RothC was designed to run in two modes: 'forward', in which known inputs are used to calculate changes in soil organic matter; and 'inverse', when inputs are calculated from known changes in soil organic matter. This model performed well for changing the carbon inputs to fit measured SOC values in other studies (Falloon & Smith, 2002). In our work, we used this model to assess the changes in the soil carbon content when soil management was changed from tillage to cover crop with pruning debris and the residues from the olive-fruit cleaning.

To run RothC at equilibrium (Coleman & Jenkinson, 1996), we needed to assume that the soils were in equilibrium (more than 30 yr with the same management). To determine the SOC value for the soil equilibrium under tillage, we sampled two neighbouring areas (one for each soil type) that ploughed the soils over the last 30 years. For the tilled olive grove, the soil was assumed to be in equilibrium. In this case, the RothC model was run iteratively in reverse to calculate how much organic carbon needs to enter a soil annually to give the measured amount of SOC. This value for the annual input of organic carbon was chosen to optimise the fitting between the modelled and measured data.

2.4 Statistical analysis

Data were analysed using SPSS v.10.0. Effects of the location and soil type for each variable were determined by one and two-way analysis of variance (ANOVA) at a confidence level of 95%. Tukey tests was performed for *post-hoc* comparisons between levels within each factor considered. Bartlett and Shapiro–Wilk tests were applied to check homoscedasticity and normality, respectively, to ensure that assumptions of the model were met. Spearman correlation was used to determine the degree of dependence between the SOC and other variables.

3. Results

3.1 Textural analysis, ρd, $CaCO_3$ and pH

Table 3 presents the results from the textural analysis, the percentage of gravel in the soils studied, ρb, $CaCO_3$ content, and pH. The gravel content increased in depth, with values greater in the CLcr. Despite differences in clay and sand percentages, both soils presented a loamy-clayey texture. The ρb diminished significantly in the uppermost 5 cm of the PD+CR with respect to the UC soils, with values equal to or lower than 1; beyond this depth, the values rose and tended to be equal. The $CaCO_3$ content was greater than 200 g kg^{-1} in both soils, with higher values in VRcc. However, the concentration of this element increased in depth until reaching a maximum of 706 g kg^{-1} in the last depth under the tree canopy in the CLcr. Both soils studied had basic pH values, which were significantly lower at the first two depths of the cover crop.

3.2 SOC, N and C:N

The spreading of the pruning debris and the cleaning residues significantly increased the SOC and N content in both soils, with maximum values in the uppermost 10 cm (Table 4). The differences were significant for depth, location and their interaction. No differences were registered in the SOC and N for each soil type. Under the canopy, both soils presented similar SOC and N contents, with values slightly higher in the upper 5-10 cm.

The C:N ratio reached maximum values in the uppermost 5 cm of PD+CR, being higher than 20 (Table 4). The differences were significant also for the location and type of soil, with higher values in CLcr than in VRcc. Under the canopy and in the last cm of the cover crop, all the C:N values were close to or lower than 10. In all cases, this value diminished in depth. The percentages of FOM and NFOM are represented in Figure 2. The presence of fresh plant residues increased the FOM in the uppermost cm of the soil; under the canopy and in depth, the FOM decreased.

The addition of debris in PD+CR increased the SOC and N content in the bulk soil, from 26.4 ± 1.2 and 27.1 ± 1.0 Mg C ha^{-1} to 158.0 ± 11.6 and 113.6 ± 17.8 Mg C ha^{-1} in CLcr and VRcc, respectively (Table 5). Under the tree, these values were intermediate for both soils. The stratification was similar in both soil types, higher than 10 for the cover crop and minimum (~1) in conventional tilled soil.

3.3 K^+, CEC and SWC

The CEC and K^+ contents were high for the PD+CR cover of both soils, diminishing over the profile (Table 6). The strongest differences were presented in the first layers of VRcc, with 30 $cmol_c$ kg^{-1} for CEC and K^+ values greater than 2 $cmol_c$ kg^{-1}. The presence of the plant residues on the soil surface also changed the SWC, with values significantly higher in the uppermost 2 or 5 cm of PD+CR and at all depths for the VRcc.

Depth (cm)	Gravel ($g\ kg^{-1}$)	Texture: Sand ($g\ kg^{-1}$)	Texture: Silt ($g\ kg^{-1}$)	Texture: Clay ($g\ kg^{-1}$)	ρb ($Mg\ m^{-3}$)	$CaCO_3$ ($g\ kg^{-1}$)	pH (1:2.5)
			Chromic Calcisols – Pruning debris + cleaning-residues cover				
0-2	188 (35)	405 (31)	256 (3)	339 (34)	0.9 (0.1)	234 (36)	7.8 (0.1)
2-5	184 (24)	384 (34)	276 (26)	340 (37)	1.0 (0.1)	228 (21)	7.8 (0.1)
5-10	317 (176)	464 (29)	235 (30)	301 (18)	1.7 (0.1)	340 (112)	8.3 (0.2)
10-15	238 (155)	496 (35)	215 (51)	288 (19)	1.8 (0.1)	344 (136)	8.3 (0.2)
15-30	173 (102)	433 (72)	246 (31)	322 (47)	1.7 (0.1)	399 (117)	8.4 (0.2)
			Chromic Calcisols – Under canopy				
0-2	147 (94)	458 (59)	203 (52)	339 (30)	1.6 (0.2)	353 (73)	8.3 (0.1)
2-5	167 (97)	480 (41)	181 (28)	339 (18)	1.6 (0.3)	374 (76)	8.3 (0.1)
5-10	160 (49)	483 (25)	184 (23)	333 (14)	1.6 (0.1)	422 (3)	8.3 (0.1)
10-15	163 (124)	500 (9)	177 (84)	322 (28)	1.6 (0.1)	524 (57)	8.3 (0.1)
15-30	276 (104)	521 (40)	201 (28)	278 (16)	1.6 (0.1)	706 (118)	8.4 (0.1)
			Calcic Vertisols – Pruning debris + cleaning-residues cover				
0-2	81 (36)	345 (58)	330 (13)	325 (45)	0.9 (0.2)	325 (40)	7.7 (0.1)
2-5	70 (41)	304 (88)	365 (24)	331 (81)	0.9 (0.1)	413 (91)	7.8 (0.1)
5-10	138 (47)	243 (18)	395 (51)	362 (21)	1.3 (0.1)	523 (49)	8.1 (0.1)
10-15	112 (37)	251 (21)	385 (17)	365 (16)	1.4 (0.1)	597 (16)	8.3 (0.1)
15-30	120 (35)	249 (19)	393 (20)	358 (11)	1.4 (0.1)	585 (17)	8.3 (0.1)
			Calcic Vertisols – Under canopy				
0-2	87 (17)	225 (13)	407 (15)	368 (3)	1.4 (0.1)	588 (6)	8.3 (0.1)
2-5	56 (11)	226 (7)	401 (15)	372 (17)	1.4 (0.1)	586 (11)	8.4 (0.1)
5-10	64 (39)	231 (25)	400 (14)	369 (14)	1.3 (0.1)	576 (26)	8.4 (0.1)
10-15	85 (18)	250 (59)	407 (28)	344 (34)	1.4 (0.1)	616 (74)	8.4 (0.1)
15-30	115 (105)	304 (53)	393 (45)	302 (14)	1.3 (0.1)	645 (54)	8.4 (0.1)
FACTOR				ANOVA *p*-value			
Depth	0.160	0.349	0.961	0.522	**<0.001**	**<0.001**	**<0.001**
Location	**<0.001**	**<0.001**	**<0.001**	0.822	**<0.001**	**<0.001**	**<0.001**
Soil type	**0.002**	**<0.001**	**<0.001**	**<0.001**	**<0.001**	**<0.001**	**0.009**
L x D	0.224	0.115	0.327	0.150	**<0.001**	0.058	**<0.001**
L x S	**<0.001**	**<0.001**	**<0.001**	0.985	**<0.001**	0.166	**0.001**

Table 3. Values for contents in gravel (>2 mm), sand (2-0.05 mm), silt (0.05-0.002 mm) and clay (<0.002 mm); bulk density (ρb); $CaCO_3$, and pH, in each soil type and location at different depths. For each value, the standard error is shown in parenthesis. Factors L: location, D: depth, S: soil type

Depth (cm)	SOC (g kg^{-1})	N (g kg^{-1})	C:N
	Chromic Calcisols – Pruning debris + cleaning-residues cover		
0-2	124.7 (9.0)	5.8 (1.0)	21.8 (3.1)
2-5	122.3 (15.5)	5.6 (1.0)	21.9 (1.3)
5-10	60.0 (13.9)	3.9 (0.5)	15.5 (3.1)
10-15	13.3 (2.1)	1.2 (0.3)	11.7 (3.9)
15-30	9.7 (1.5)	1.2 (0.3)	8.1 (0.9)
	Chromic Calcisols – Under canopy		
0-2	18.3 (10.1)	1.9 (1.0)	9.4 (1.9)
2-5	14.3 (4.9)	1.4 (0.5)	10.0 (0.1)
5-10	14.3 (4.0)	1.5 (0.5)	10.0 (1.8)
10-15	12.0 (1.0)	1.5 (0.5)	8.2 (1.8)
15-30	8.7 (1.2)	1.0 (0.2)	8.7 (0.3)
	Calcic Vertisols – Pruning debris + cleaning-.residues cover		
0-2	118.7 (1.5)	6.2 (0.7)	19.4 (1.9)
2-5	90.7 (0.6)	4.2 (0.9)	22.3 (5.1)
5-10	33.7 (7.6)	2.7 (0.3)	12.4 (1.3)
10-15	14.7 (1.2)	1.9 (0.4)	8.0 (1.2)
15-30	11.7 (2.5)	1.6 (0.2)	7.3 (2.0)
	Calcic Vertisols – Under canopy		
0-2	15.0 (1.0)	2.4 (0.2)	6.2 (0.7)
2-5	15.0 (3.0)	1.9 (0.3)	7.9 (1.9)
5-10	12.7 (3.8)	1.7 (0.6)	7.7 (2.6)
10-15	8.7 (1.5)	1.5 (0.6)	7.1 (4.7)
15-30	6.7 (1.5)	1.1 (0.3)	6.4 (0.9)
FACTOR		ANOVA *p*-value	
Depth	**<0.001**	**<0.001**	**<0.001**
Location	**<0.001**	**<0.001**	**<0.001**
Soil type	0.226	0.101	**<0.001**
L x D	**<0.001**	**<0.001**	**<0.001**
L x S	0.366	0.086	**<0.001**

Table 4. Values of soil organic carbon (SOC), nitrogen (N) concentration, and C:N ratio, in each soil type and location at the different depths. For each value, the standard error is shown in parenthesis. Factors L: location, D: depth, S: soil type

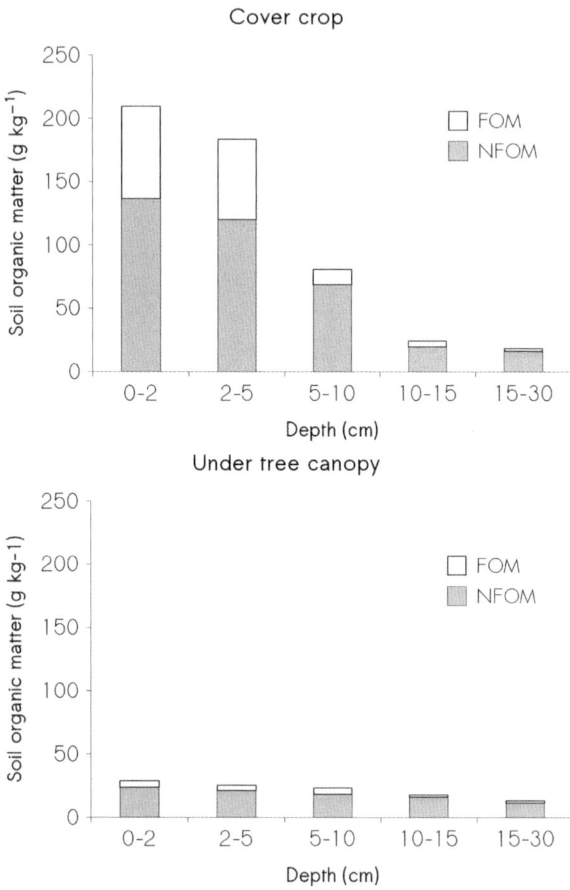

Fig. 2. Contents in floatable (FOM) and non-floatable (NFOM) organic matter in water for each location at the different depths

Soil type		SOC (Mg ha^{-1})	N (Mg ha^{-1})	Stratification ratio of SOC
Chromic Calcisols	Cover crop	158.0 (11.6)a	11.2 (1.1)a	13.1 (1.6)a
	Under canopy	55.4 (11.8)b	6.2 (1.1)b	2.0 (0.9)b
	Tillage	26.4 (1.2)c	3.4 (0.6)c	0.9 (0.1)b
	ANOVA (*p*-value)	**<0.001**	**<0.001**	**<0.001**
Calcic Vertisols	Cover crop	113.6 (17.8)a	10.1 (0.6)a	10.5 (2.2)a
	Under canopy	37.9 (4.5)b	5.4 (2.1)b	2.3 (0.6)b
	Tillage	27.1 (1.0)b	3.8 (0.3)b	1.2 (0.1)b
	ANOVA (*p*-value)	**0.001**	**0.006**	**<0.001**

Table 5. Soil organic carbon (SOC), nitrogen (N), and stratification ratio of SOC for 0-30 cm depth, in each soil type and location. For each value, the standard error is shown in parenthesis

3.4 Correlations with SOC

The correlations between SOC and other soil parameters are summarized in Table 7. The SOC was correlated negatively with the silt content in all cases, and positively with the clay content only in soils without carbon addition (UC). As expected, the SOC correlated negatively with ρb, $CaCO_3$ and pH, and positively with N. The K^+ and CEC correlated

Depth (cm)	K^+ ($cmol_c\ kg^{-1}$)	CEC ($cmol_c\ kg^{-1}$)	Soil-water content ($m^3\ m^{-3}$) -33 kPa	-1500 kPa
		Chromic Calcisols – Pruning debris + cleaning-residues cover		
0-2	1.5 (0.8)	22.5 (8.6)	0.34 (0.02)	0.22 (0.01)
2-5	1.5 (0.7)	25.0 (7.0)	0.35 (0.02)	0.25 (0.02)
5-10	1.5 (0.4)	19.6 (3.5)	0.24 (0.03)	0.15 (0.03)
10-15	0.9 (0.4)	17.2 (3.3)	0.17 (0.01)	0.09 (0.01)
15-30	0.6 (0.2)	15.9 (2.2)	0.19 (0.02)	0.11 (0.02)
		Chromic Calcisols – Under canopy		
0-2	1.4 (0.5)	15.1 (0.6)	0.19 (0.04)	0.10 (0.01)
2-5	1.1 (0.5)	17.5 (6.3)	0.17 (0.03)	0.09 (0.01)
5-10	0.8 (0.2)	13.0 (0.6)	0.17 (0.01)	0.10 (0.01)
10-15	0.4 (0.1)	12.0 (1.4)	0.17 (0.01)	0.10 (0.01)
15-30	0.4 (0.1)	10.5 (2.1)	0.17 (0.02)	0.10 (0.01)
		Calcic Vertisols – Pruning debris + cleaning-residues cover		
0-2	2.6 (0.8)	30.0 (7.2)	0.35 (0.03)	0.30 (0.06)
2-5	1.9 (0.5)	25.1 (8.7)	0.31 (0.03)	0.23 (0.05)
5-10	1.3 (0.3)	19.6 (4.8)	0.30 (0.01)	0.18 (0.02)
10-15	0.5 (0.3)	15.9 (2.9)	0.29 (0.02)	0.17 (0.02)
15-30	0.4 (0.1)	15.4 (2.6)	0.29 (0.02)	0.17 (0.02)
		Calcic Vertisols – Under canopy		
0-2	0.8 (0.3)	16.7 (2.9)	0.30 (0.01)	0.18 (0.02)
2-5	0.6 (0.2)	18.0 (1.5)	0.29 (0.01)	0.19 (0.03)
5-10	0.3 (0.1)	16.4 (0.7)	0.29 (0.01)	0.17 (0.02)
10-15	0.2 (0.2)	16.7 (4.8)	0.29 (0.01)	0.17 (0.03)
15-30	0.1 (0.1)	12.5 (2.6)	0.28 (0.02)	0.16 (0.03)
FACTOR		ANOVA *p*-value		
Depth	**<0.001**	**<0.001**	**<0.001**	**<0.001**
Location	**0.008**	**<0.001**	**<0.001**	**<0.001**
Soil type	0.470	**<0.001**	**<0.001**	**<0.001**
L x D	**0.013**	**0.003**	**<0.001**	**<0.001**
L x S	0.053	**<0.001**	**<0.001**	**<0.001**

Table 6. Potassium (K^+), cation-exchange capacity (CEC) and soil-water content at -33 and -1500 kPa in each soil type and location at the different depths. For each value, the standard error is shown in parenthesis. Factors L: location, D: depth, S: soil type

positively with the organic fractions (SOC, NFOM, and FOM), especially in VRcc. Under canopy, the correlation coefficients lowered the significance value, and were not significant for FOM. The correlations between the SWC and the organic fractions were positive and significant in the cover crop of the CLcr; the degree of significance diminishing in the VRcc. No firm correlations were found for the soil fine fraction and SWC.

	All data	CLcr		VRcc	
	(n=60)	PD+CR	UC	PD+CR	UC
			SOC (Mg ha^{-1})		
Silt	-0.94 **	-0.77 **	-0.64 *	-0.68 **	-0.69 **
Clay	0.20	-0.04	0.94 **	0.19	0.89 **
			SOC (%)		
ρb	-0.43 **	-0.82 **	0.07	-0.78 **	0.27
$CaCO_3$	-0.50 **	-0.63 *	-0.36	-0.84 **	-0.44
pH	-0.64 **	-0.80 **	-0.13	-0.90 **	-0.22
N	0.86 **	0.92 **	0.87 **	0.97 **	0.63
K^+	0.80 **	0.58 *	0.65 **	0.94 **	0.77 **
CEC	0.63 **	0.61 *	0.49	0.83 **	0.64 *
SWC -33 kPa	0.53 **	0.84 **	0.61 *	0.64 **	0.04
SWC -1500 kPa	0.56 **	0.86 **	0.51	0.81 **	0.23
			NFOM		
K^+	0.79 **	0.54 *	0.74 **	0.90 **	0.72 **
CEC	0.63 **	0.62 *	0.45	0.80 **	0.70 **
SWC -33 kPa	0.50 **	0.87 **	0.53 *	0.68 **	0.07
SWC -1500 kPa	0.53 **	0.87 **	0.30	0.74 **	0.28
			FOM		
K^+	0.68 **	0.67 **	0.42	0.81 **	0.46
CEC	0.52 **	0.56 *	0.31	0.65 **	0.04
SWC -33 kPa	0.52 **	0.81 **	0.57 *	0.52 *	0.09
SWC -1500 kPa	0.55 **	0.81 **	0.51	0.71 **	-0.17

Table 7. Correlation coefficients of soil organic fractions with the properties studied, for the entire dataset (all data) and for each soil type and location. CLcr: Chromic Calcisols; VRcc: Calcic Vertisols; PD+CR: cover crop with pruning debris and the fruit-cleaning residues; UC: under the tree canopy. * Significant at $P < 0.05$; ** Significant at $P < 0.01$ according to Spearman's test

3.5 RothC

The results for changing soil management from conventional tillage to cover crop are summarized in Table 8. At first, we calculated the annual carbon input for the tillage of olive trees from the SOC concentration measured in neighbouring areas, assuming a steady state. For both soils, the annual input modelled was 1.0 Mg C ha^{-1} yr^{-1}. During the experiment, we measured a carbon input at the soil-surface in PD+CR cover of 23.9 ± 14.3 Mg C ha^{-1} yr^{-1}. After 10 and 6 yr of change in the soil-management system, the annual carbon inputs needed to reach the SOC values between trees as estimated by the model were very similar to those measured in both soils. The turnover time decreased from 26 yr in T soils to 6 and 5 yr for CLcr and VRcc, respectively.

Scenario	Soil type	SOC measured (Mg C ha^{-1})	IOM (Mg C ha^{-1})	Input C modelled (Mg C ha^{-1} yr^{-1})	Input C measured (Mg C ha^{-1} yr^{-1})	Turnover time (yr)
T	CLcr (equ)	26.4	2.0	1.0	-	26
	VRcc (equ)	27.1	2.1	1.0	-	26
PD+CR	CLcr (10 yr)	158.0	2.0	25.3	23.9	6
	VRcc (6 yr)	113.6	2.1	23.6	23.9	5

Table 8. Measured and modelled data for the turnover of organic carbon in olive-grove soils under conventional tillage (T) and mulched with residues from pruning debris and olive-fruit cleaning (PD+CR) (from Nieto et al., 2010)

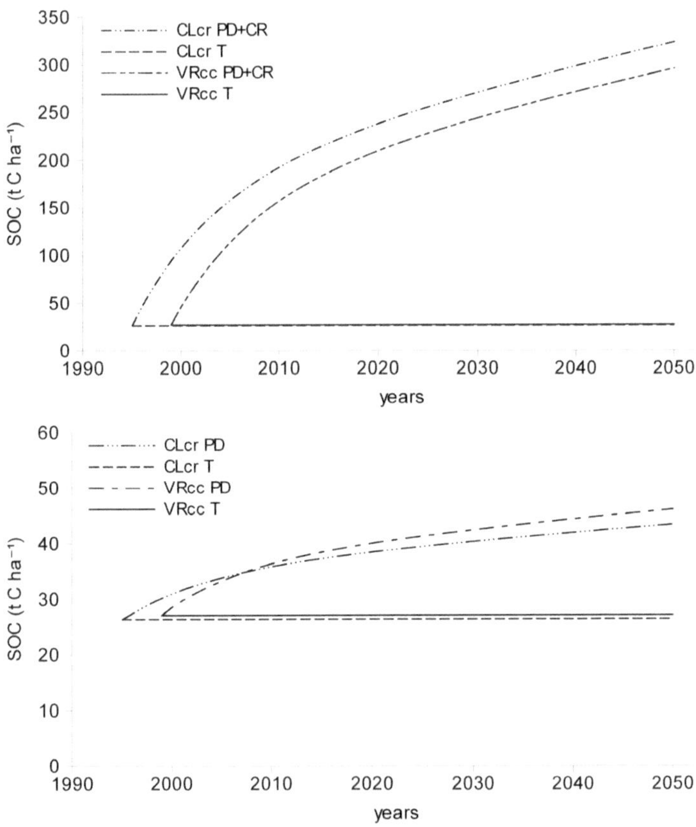

Fig. 3. SOC modelled for CLcr and VRcc under conventional tillage (T), shredded olive-pruning debris cover (PD) and mulching with residues from olive-fruit cleaning and pruning debris (PD+CR) (from Nieto et al., 2010)

The C sequestration was calculated as the difference between CO_2 emissions in each type of management. Under tillage, the cleaning residues were stored in areas close to the oil mill, so that we could not include them in the CO_2 emission balance. Although the pruning debris had a different use in the two cases, under tillage they were burnt and in the cover crop they were spread on the soil.

Figure 3 shows SOC modelled by RothC for 50 yr in both soils under three management systems: cover with PD+CR, only pruning debris (PD), and tillage (Nieto et al., 2010). When mulched with PD+CR the model predicted a continuous increase in SOC, but without reaching a state of equilibrium. No differences could be discerned between either type of soil, due mainly to the large quantities of organic carbon added. When the input was that of PD alone, the model gave a lower SOC content than in the previous case.

CO_2 emissions are given in Table 9 when the PD were burnt in tillage, and when shredded and spread on the ground. As a result of the addition of PD to the soil, the CO_2 that was previously released into the atmosphere during the years of tilled management was reduced more than 55% for both soils. For the decrease in CO_2 emission, the RothC model estimated a potential carbon sequestration of 0.5 and 0.6 Mg C ha^{-1} yr^{-1} for CLcr and VRcc (Table 10). We did not model carbon sequestration after mulching with CRs because this waste matter is normally discarded. Nevertheless, the total carbon content in the soil registered an increase during the experiment of 13.2 and 14.4 Mg C ha^{-1} yr^{-1} for each soil type.

Scenario		CO_2 released to the atmosphere (Mg CO_2 ha^{-1})	
		CLcr (10 yr)	VRcc (6 yr)
Tillage	Burn pruning debris	24.0	14.4
	CO_2 lost from soil	12.0	7.2
	Total	36.0	21.6
Pruning debris	Burn pruning debris	-	-
	CO_2 lost from soil	15.9	9.2
	Total	15.9	9.2

Table 9. CO_2 released into the atmosphere after 10 and 6 years of conventional tillage and mulching with shredded pruning debris, for both types of soil (from Nieto et al., 2010)

Soil type	SOC increase (Mg C ha^{-1} yr^{-1})	CO_2 reduction (Mg CO_2 ha^{-1} yr^{-1})	C sequestration (Mg C ha^{-1} yr^{-1})
CLcr	13.2	2.0	0.5
VRcc	14.4	2.1	0.6

Table 10. SOC increase as result of changes in soil management from conventional tillage to mulching with residues from pruning debris and olive-fruit cleaning (PD+CR). CO_2 reduction and carbon sequestration after the addition of only pruning debris (from Nieto et al., 2010)

4. Discussion

4.1 Soil organic carbon and related soil properties

The two types of soils studied were close together and are widely represented in the areas dedicated to olive cultivation in the Mediterranean region. The parent material, limestone in the CLcr and marls in the VRcc, conditioned the development of different profiles with respect to such characteristics as colour, and content, as well as type of clay, sand, and silt.

Large quantities of soil from nearby zones were applied with the plant residues from olive-fruit cleaning. For this reason, such properties as texture and $CaCO_3$ content presented similar values for the first cm of PD+CR in both soils types. The pH values changed with the SOC content (Thomas et al., 2007).

As opposed to the findings of Fontaine et al. (2004), the addition of fresh organic matter did not cause a negative balance in the SOC, since we began with degraded soils of very low SOC contents. Many authors have also found a rise in SOC values in agricultural soils within at least the uppermost 10 cm in depth, after using conservation practices (Angers et al., 1997; Hernanz et al., 2002; Jarecki & Lal, 2003). In olive orchards, some authors as Hernández et al. (2005), Sofo et al. (2005), Castro et al. (2008) and Gómez et al. (2009), have reported increases in the SOC and N content after applying plant residues. Our values were higher than those reported by these authors because, together with the shredded pruning debris, a major quantity of residues composed of soil and plant debris were brought from the olive-cleaning processes, reaching a biomass accumulation greater than that indicated by these authors. In addition, the soil from the cleaning process originated from the soil under the tree canopy, with small aggregates, fine material and a high SOC content.

Under the tree canopy, despite that the soil was maintained free of weeds with herbicides and free of plant debris by sweeping, SOC values reached 64.7 and 43.3 Mg C ha^{-1}, higher than tilled soils between rows (Table 5). This was due to the greater presence of roots in this zone and the continual dropping of olive leaves, which fell under the canopy, where they remained until the annual cleanup before harvest (Ordóñez et al., 2001; Soria et al., 2005).

The high values of the SOC found in PD+CR indicate the effectiveness of the treatment in terms of storage, reaching values of up to 158.0 Mg ha^{-1} after 10 years of management. These values, together with those of N were far higher than reported by other researchers in agricultural areas (Hernanz et al., 2002; Hernández et al., 2005) but similar to those found by Jarecki & Lal (2005) for forest soils in Ohio.

Texture plays a major role in SOC accumulation in the soil. In many works, the losses of SOC from agricultural soils were lower in clayey soils, since these tended to accumulate this fraction more rapidly and retain it longer (Percival et al., 2000; Arrouays et al., 2006). In our work, a correlation between the SOC and the clay was found only in under the canopy (Table 6), since the input of plant debris changed this relationship. Similar results have been reported by Castro et al. (2008).

According to Hevia et al. (2003) the NFOM presented high correlations with the clay content only in the soils UC (Figure 4). This showed that the addition of plant debris and other residues with high carbon content alter the relation between SOC and the fine-size particle.

The C:N relationship indicates the rate of the mineralization of the soil organic matter. According to Giménez and Bratos (1985), C:N values higher than 15 indicate a very low N release. Our results show high values at the uppermost soil levels of PD+CR, and thus humification processes predominated. At greater depths and in UC, mineralization processes predominating (values lower than 10). It bears mentioning that a high C:N relationship does not necessarily signify N deficiencies in soil, as pointed out by Rhoton et al. (1993), as the progressive increase in organic matter of the soil augments the availability of many nutrients, including N. In this case, the high superficial values cause SOC to act as a protector and store of N, releasing this nutrient little by little into the soil.

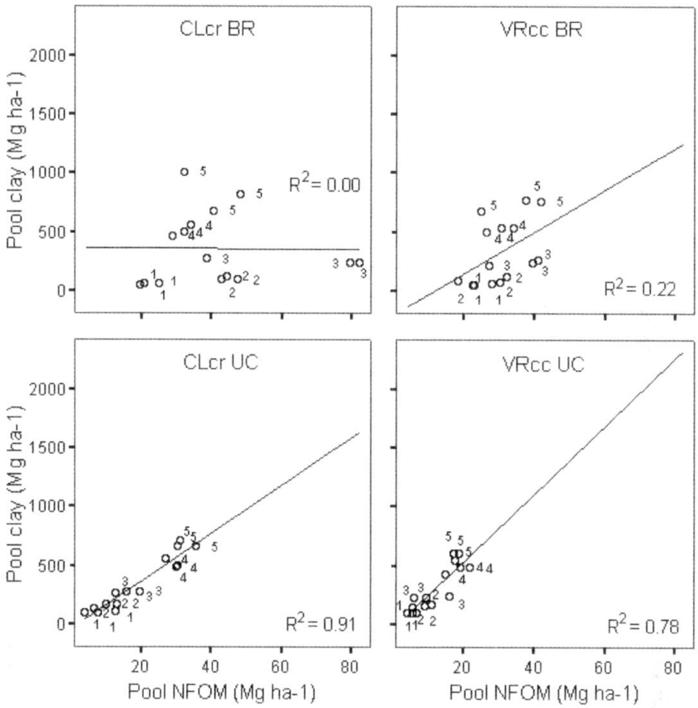

Fig. 4. Linear regression between clay and NFOM (non-floatable organic matter) for each soil type and location (1: 0-2 cm, 2: 2-5 cm, 3: 5-10 cm, 4: 10-15 cm, 5: 15-30 cm)

One of the most important characteristics of this management is the continued non-tillage, which affects the distribution of certain properties such as SOC throughout the soil profile. Franzluebbers (2002) showed that SOC stratification ratios higher than 2 are infrequent under degradation conditions. The data compiled in our study, lower than 2 in conventional tillage and higher in PD+CR cover, confirm the increase in soil quality as well as a control of erosion and degradation processes.

The high values of ρb UC indicated a compaction process related to cultivation (Soria et al., 2005). Between rows, the spreading of olive-pruning debris and the maintenance of the material on the surface (non-tillage) generated a mulching effect (Bescansa, 2006; Lal, 2008) that protected the soil effectively with respect to nutrient loss and sediment production (Rodríguez-Lizana et al., 2008). Thus, ρb diminished in the uppermost 2 cm of PD+CR by

44% in CLcr and by 31% in VRcc, with respect to UC. This difference between the two soil types is related to the SOC content, determined by the application time of the plant residues. Similar results have been reported by Ordóñez et al. (2001) after 6 years of applying olive-pruning debris.

The K^+ content raised with the SOC (Rhoton et al., 1993; Thomas et al., 2007) by 3.4-fold in the VRcc and by 1.5-fold in the CLcr over the entire depth studied with respect to UC. This higher increase observed in the VRcc was linked to the greater content in smectite clays (Bhonsle et al., 1992; Ghosh & Singh, 2001).

According to Oorts et al. (2003) organic matter can be responsible for as much as 85% of the CEC of the soil. The correlation between CEC and SOC found in the present work (Table 6) was similar to those documented by Caravaca et al. (1999) for calcareous soils. The higher values of CEC were detected in VRcc, with lower duration of the experiment and greater fine-fraction content than CLcr (Table 3). These results coincide with those of Leinweber et al. (1993), who indicated that the order of factors that affect the CEC is: particle size of the fraction, management, and duration of the experiment.

The water-storage capacity of the soil changed too with the organic matter content since the pore size and distribution changes (Rawls et al., 2003; Bescansa et al., 2006). However, the effect of SOC on water retention was high in sandy soils and marginal in fine-textured soils (Bauer & Black, 1981). Our results match those of Rawls et al. (2003) in the regression tree for the SWC at -33 and -1500 kPa and in the variation of the moisture with the change in organic-carbon content, which, for high SOC contents, increased for all the textures but in a higher proportion with greater sand contents, as occurred in our case in the CLcr.

4.2 Soil carbon sequestration

The effectiveness of changing the management from traditional tillage to cover crop between trees is manifest in this study in the resulting high SOC and N values. The SOC rose from 34.0 and 46.2 Mg ha^{-1} to 158.0 and 113.6 Mg ha^{-1} for CLcr and VRcc, respectively, indicating greater in soil fertility.

The RothC model estimated the carbon input into the soil of the tilled olive grove in equilibrium as being 1.0 Mg C ha^{-1} yr^{-1}, which falls within the range of 1-2 Mg C ha^{-1} yr^{-1} estimated by Jenkinson & Rayner (1977). Romanyà et al. (2000) registered similar results for a vineyard in the Mediterranean area, with an annual carbon input of 1.4 Mg C ha^{-1} yr^{-1}. The only estimates of carbon input for olive groves were reported by Sofo et al. (2005), who registered an annual input as senescent leaves of 0.4 Mg C ha^{-1} yr^{-1} but did not account for other inputs such as root turnover and rhizodeposition.

In our work, the SOC content modelled for the VRcc soil (Figure 3) was higher than that predicted for the CLcr after eight years of mulching. This might be explained by the higher percentage of clay in the VRcc soil, which is masked by the large quantities of organic matter added in the form of PD+CR. The relationships between fine soil fractions and organic-carbon sequestration have been addressed by other authors such as Paustian et al. (2000). It has been noted that carbon is physically protected against biodegradation when it is contained in clay- or silt-sized micro-aggregates (Balesdent et al., 2000).

The RothC model fitted carbon turnover satisfactorily for the change from tillage to PD+CR cover due to the absence of erosion (Gottschalk et al., 2010). During the 10 and 6 years that the experiment lasted, we registered a much higher annual input of residues into the soil between trees than that reported by other authors, due to the fact that our management system included the addition of cleaning residues as well as pruning debris. For this scenario, RothC predicts an annual input of 25.3 and 23.6 Mg C ha^{-1} yr^{-1} for CLcr and VRcc soils, respectively (Table 8), thus confirming that the model closely fits this kind of management.

Turnover time is defined as the migration of organic carbon through a given volume of soil (Jenkinson & Rayner, 1977). The high values found in tillage (Table 8) indicate carbon stabilization in the soil, signifying that carbon migrates slowly from one pool to another. Our results are higher than those modelled for Kenyan savanna (~16 yr) or dry forest in Zambia (~8 yr) by Jenkinson et al. (1999). For PD+CR the turnover times were lower than those found by these latter authors. This showed rapid migration of carbon and thus a soil that is not close to equilibrium.

During the first years of olive-tree establishment, CO_2 is distributed preferentially in the permanent structures and the root system, but in mature olives trees, fixed CO_2 is located to a greater extent in the leaves and fruit, and consequently also in the pruning debris (Sofo et al., 2005). Thus pruning debris is an important carbon reservoir that can be returned to the soil rather than the atmosphere in the form of CO_2 (Figure 3). When the pruning debris is shredded and spread on the ground, the RothC model predicts a decrease in CO_2 emission of ~2 Mg CO_2 ha^{-1} yr^{-1} for each soil (Table 10). This value is within the range of the data collected by Smith et al. (2008) for warm-dry zones, who registered an emission reduction of 3.45 Mg CO_2 ha^{-1} yr^{-1} when degraded lands are restored and 0.33 Mg CO_2 ha^{-1} yr^{-1} for croplands under tillage and residue management.

Abandoning tillage in favour of using organic waste to cover the ground is considered to be an efficient way of increasing carbon sequestration in agricultural soils (cf. Smith et al., 2008; Lal, 2008). In our experiment, we measured a carbon sequestration of 0.5 and 0.6 Mg C ha^{-1} yr^{-1} with pruning debris cover in CLcr and VRcc soils, respectively (Table 10). Fairly wide ranges have been estimated for carbon sequestration in agricultural soils. Álvaro-Fuentes et al. (2009) measured a carbon sequestration of 0.46 and 0.15 Mg C ha^{-1} yr^{-1} for continuous barley and barley-fallow rotation in Mediterranean area, respectively. For the European area, Smith et al. (2000) estimated a carbon sequestration of 0.7 Mg C ha^{-1} yr^{-1} with crop residues and 0.4 Mg C ha^{-1} yr^{-1} with no tillage. Using organic residues Hutchinson et al. (2007) registered average rates of potential carbon gain from 0.1 to 0.5 Mg C ha^{-1} yr^{-1}. After changes in cropland use, and introduction of the best series of management techniques for every land use and climate zone, IPCC (2000) suggested a carbon sequestration potential of 0.3 Mg C ha^{-1} yr^{-1}. Our results coincide with these values, supporting the idea that the recycling of pruning debris in olive groves is an effective way of storing carbon in the soil. As with SOC, carbon sequestration was greater in VRcc soil, with its higher clay content.

Apart from the carbon sequestration from the recycling of pruning debris, the addition of the fruit-cleaning residues and the absence of tilling resulted in an increase in SOC (Figure 3). Although it may be difficult to introduce the reuse of the large quantities of CR + PD described in this work as a commonplace practice in olive-grove management, merely recycling the pruning, which is produced in the grove itself, together with a policy of zero tillage, will increase the SOC content considerably compared to T soils.

5. Conclusions

This work shows the improvement in soil quality and fertility after the soil management in olive orchards was changed from conventional tillage to non-tillage with plant residues cover. The application of shredded olive-pruning debris and the plant residues and soil from olive-fruit cleaning increased the organic fraction in both soils, CLcr and VRcc, with respect the tillage soils. The changes affected the uppermost 10 cm of the soils although the SOC content was greater in the CLcr, where the management spanned a longer time period.

With the change in SOC, some soil properties were affected. The spreading of plant residues lowered the pH and the ρb in the uppermost soil depths between trees, where there was an effect of mulching with respect to under canopy soils, and the N content increased. The slow release of this N ensured plant nutrition and the soil fertility despite the high values of the C:N relationship.

The non-tillage generated a highly fertile surface layer in PD+CR, which also protected the soil physically, as demonstrated by the high stratification observed, similar to that described in forest systems. In addition, the NFOM percentage, although higher under the tree canopy, was high in both types of soils, indicating the stability of the material applied. The addition of plant debris also increased the K^+ content, CEC and water-storage capacity, improving the soil quality.

When the soil management was changed from conventional tillage to cover crop with PD+CR, carbon storage in the soil improved considerably together with its general quality. Carbon turnover in Mediterranean olive-grove with PD+CR cover was quite accurately predicted by the RothC model. Over the long term, the carbon sequestration was higher in soils with greater quantities of clay. A soil-management system that abandons tillage in favour of reusing both pruning debris and the residue from cleaning the olives to cover the ground, constitutes the most effective way of increasing soil quality and diminishing CO_2 emissions in one of the most extensive agricultural enterprises in the entire Mediterranean area.

6. Acknowledgments

We would like to thank F. J. Garcia de Zuñiga, the farm owner, for his collaboration. The work has been financed with investigation projects INIA RTA2007-010-C3-01.

7. References

Angers, D.A., Bolinder, M.A., Carter, M.R., Gregorich, E.G., Drury, C.F., Liang, B.C., Voroney, R.P., Simard, R.R., Donald, R.G., Beyaert, R.P. & Martel, J. (1997). Impact of tillage practices on organic carbon and nitrogen storage in cool, humid soils of eastern Canada. *Soil and Tillage Research,* Vol. 41, pp. 191-201.

Álvaro-Fuentes, J., López, M.V., Arrúe, J.L., Moret, D. & Paustian, K. (2009). Tillage and cropping effects on soil organic carbon in Mediterranean semiarid agroecosystems: Testing the Century model. *Agriculture, Ecosystems and Environment,* Vol. 134, pp. 211-217.

Arrouays, D., Saby, N., Walter, C., Lemercier, B. & Schvartz, C. (2006). Relationships between particle-size distribution and organic carbon in French arable topsoils. *Soil Use and Management,* Vol. 22, pp. 48-51.

Balesdent, J., Chenu, C. & Balabane, M. (2000). Relationship of soil organic matter dynamics to physical protection and tillage. *Soil and Tillage Research,* Vol. 53, pp. 215-230.

Bauer, A. & Black, A.L. 1981. Soil carbon, nitrogen, and bulk density comparisons in two cropland tillage systems after 25 years and in virgin grassland. *Soil Science Society of America Journal,* Vol. 45, pp. 1166-1170.

Bescansa, P., Imaz, M.J., Virto, I., Enrique, A. & Hoogmoed, W.B. (2006). Soil water retention as affected by tillage and residue management in semiarid Spain. *Soil and Tillage Research,* Vol. 87, pp. 19-27.

Bhonsle, N.S., Pal, S.K. & Sekhon, G.S. (1992). Relationship of K forms and release characteristics with clay mineralogy. *Geoderma,* 54, pp. 285-293.

Blake, G.R. & Hartge K.H. (1986). Bulk density. In: Klute, A. (Ed.), *Methods of Soil Analysis.* Part 1. Physical and Mineralogical Methods. 2nd ed. ASA, SSSA Monograph No. 9, Madison, pp. 363-375.

Bremner, J.M. (1965). Nitrogen availability indexes. In: Black, C.A., Evans, D.D., Esminger, T.E., Clark, F.E. (Eds.), *Methods of Soil Analysis.* Part. 2. Chemical and Microbiological Properties. American Society Agronomy, Madison, pp. 1324-1345.

Buyanovsky, G.A., Aslam, M. & Wagner, G.H. (1994). Carbon turnover in soil physical fractions. *Soil Science Society of America Journal,* Vol. 58, pp. 1167-1174.

Caravaca, F., Lax, A. & Albaladejo, J. (1999). Organic matter, nutrient contents and cation exchange capacity in fine fractions from semiarid calcareous soils. *Geoderma,* 93, pp. 161-176.

Cassel, D.K. & Nielsen, D.R. (1986). Fields capacity and available water capacity. In: Klute, A. (Ed.), *Methods of Soil Analysis.* Part 1. Physical and Mineralogical Methods. 2nd ed. ASA, SSSA Monograph No. 9, Madison, pp. 901-926.

Castro, J., Fernández-Ondoño, E., Rodríguez, C., Lallena, A.M., Sierra, M. & Aguilar, J. (2008). Effects of different olive-grove management systems on the organic carbon and nitrogen content of the soil in Jaén (Spain). *Soil and Tillage Research,* Vol. 98, pp. 56-67.

Coleman, K. & Jenkinson, D.S. (1996). RothC-26.3 - a model for the turnover of carbon in soil. En D.S. Powlson, P. Smith, J.U. Smith, eds. Evaluation of soil organic matter models using existing, long-term datasets. Springer-Verlag, Berlin, pp. 237-246.

Falloon, P., Smith, P., Coleman, K. & Marshall, S. (1998). Estimating the size of inert organic matter pool from total soil organic carbon content for use the Rothamsted Carbon Model. *Soil Biology and Biochemistry,* Vol. 30, pp. 1207-1211.

Falloon, P. & Smith, P. (2002). Simulating SOC changes in long-term experiments with RothC and CENTURY: model evaluation for a regional scale application. *Soil Use and Management,* Vol. 18, pp. 101-111.

FAO (2006). *World Reference Base for Soil Resources.* IUSS - ISRIC - FAO, Roma, 117 p.

Fontaine, S., Bardoux, G., Abbadie, L. & Mariotti, A. (2004). Carbon input to soil may decrease soil carbon content. *Ecology Letters,* Vol. 7, pp. 314-320.

Franzluebbers, A.J. (2002). Water infiltration and soil structure related to organic matter and its stratification with depth. *Soil and Tillage Research,* Vol. 66, pp. 197-205.

Ghosh B.N. & Singh, R.D. (2001). Potassium release characteristics of some soils of Uttar Pradesh hills varying in altitude and their relationship with forms of soil K and clay mineralogy. *Geoderma,* Vol. 104, pp. 135-144.

Giménez, M. & Bratos, J. (1985). *Análisis de suelos.* Ed. ERT. Spain.

Gómez, J.A, Sobrinho, T.A., Giráldez, J.V. & Fereres, E. (2009) Soil-management effects on runoff, erosion and soil properties in an olive grove of Southern Spain. *Soil and Tillage Research,* Vol. 102, pp. 5-13.

Gottschalk, P., Bellarby, J., Chenu, C., Foereid, B., Smith, P., Wattenbach, M., Zingore, S. & Smith, J. (2010). Simulation of soil organic carbon response at forest cultivation sequences using ^{13}C measurements. *Organic Geochemistry,* Vol. 41, pp. 41-54.

Hernández, A.J., Lacasta, C. & Pastor, J. (2005). Effects of different management practices on soil conservation and soil water in a rainfed olive orchard. *Agricultural Water Management*, Vol. 77, pp. 232-248.

Hernanz, J.L., López, R., Navarrete, L. & Sánchez-Girón, V. (2002). Long-term effects of tillage systems and rotations on soil structural stability and organic carbon stratification in semiarid central Spain. *Soil and Tillage Research*, Vol. 66, pp. 129-141.

Hevia, G.G., Buschiazzo, D.E., Hepper, E.N., Urioste, A.M. & Antón, W.L. (2003). Organic matter in size fractions of soils of the semiarid Argentina. Effects of climate soil texture and management. *Geoderma*, Vol. 116, pp. 265-277.

Hutchinson, J.J., Campbell, C.A. & Desjardins, R.L. (2007). Some perspectives on carbon sequestration in agriculture. *Agricultural and Forest Meteorololgy,* Vol. 142, pp. 288-302.

Ingram, J.S. & Fernandes, E.C.M. (2001). Managing carbon sequestration in soils: concepts and terminology. *Agriculture, Ecosystems and Environment*, Vol. 87, pp. 111–117.

IPCC (2000). Special report on land use, land-use change and forestry. Cambridge University Press, Cambridge.

Jarecki, M.K. & Lal, R. (2003). Crop management for soil carbon sequestration. *Critical Reviews in Plant Sciences,* Vol. 22, No. 5, pp. 471-502.

Jarecki, M.K. & Lal, R. (2005). Soil organic carbon sequestration rates in two long-term no-till experiments in Ohio. *Soil Science*, Vol. 170 No. 4, pp. 280-291.

Jenkinson, D.S. & Rayner, H. (1977). The turnover of soil organic matter in some of the Rothamsted classical experiments. *Soil Science,* Vol. 123, No. 5, pp. 298-305.

Jenkinson, D.S., Meredith, J., Kinyamario, J.I., Warren, G.P., Wong, M.T.F., Harkness, D.D., Bol, R. & Coleman, K. (1999). Estimating net primary production from measurements made on soil organic matter. *Ecology,* Vol. 80 No. 8, pp. 2762-2773.

Lal, R. (2008). Soils and sustainable agriculture. A review. *Agronomy for Sustainable Development,* Vol. 28, pp. 57-64.

Leinweber, P., Reuter, G. & Brozio, K. (1993). Cation exchange capacities of organo-mineral particle-size fractions in soils from long-term experiments. *Journal of Soil Science,* Vol. 44, pp. 111-119.

Ministerio de Agricultura, Pesca y Alimentación - MAPA (1989). Caracterización agroclimática de la provincia de Jaén. Dirección General de Producción Agraria, Madrid.

Nieto, O.M., Castro, J., Fernández, E. & Smith, P. (2010). Simulation of soil organic carbon stocks in a Mediterranean olive grove under different soil-management system using the RothC model. *Soil Use and Management*, Vol. 26, pp. 118-125.

Oorts, K., Vanlauwe, B. & Merckx, R. (2003). Cation exchange capacities of soil organic matter fractions in a Ferric Lixisol with different organic matter inputs. *Agriculture, Ecosystems and Environment*, Vol. 100, pp. 161-171.

Ordóñez, R., Ramos, F.J., González, P., Pastor, M. & Giráldez, J.V. (2001). Influencia de la aplicación continuada de restos de poda de olivo sobre las propiedades físico-químicas de un suelo de olivar. In: López, J.J., Quemada, M. (Eds.), *Temas de Investigación en la Zona no Saturada*, Vol. 5. Univ. Pública Navarra, Spain.

Pastor, M. (2004). Sistemas de manejo del suelo. In: Barranco, D., Fernández-Escobar, R., Rallo, T. (Eds.), *El cultivo del olivo*. Mundi-Prensa, Spain, pp. 231-285.

Paustian, K., Six, J., Elliott, E.T. & Hunt, H.W. (2000). Management options for reducing CO_2 emissions from agricultural soils. *Biogeochemistry,* Vol. 48, pp. 147-163.

Percival, H.J., Parfitt, R.T. & Scott, A.N. (2000). Factors controlling soil carbon levels in New Zealand grasslands: is clay content important? *Soil Science Society of America Journal,* Vol. 64, pp. 1623-1630.

Rawls, W.J., Pachepsky, Y.A., Ritchie, J.C., Sobecki, T.M. & Bloodworth, H. (2003). Effect of soil organic carbon on soil water retention. *Geoderma,* Vol. 116, pp. 61-76.

Rhoton, F.E., Bruce, R.R., Buehring, N.W., Elkins, G.B., Langdale, C.W. & Tyler, D.D. (1993). Chemical and physical characteristics of four soil types under conventional and no-tillage systems. *Soil and Tillage Research,* Vol. 28, pp. 51-61.

Rodríguez-Lizana, A., Espejo-Pérez, A.J., González-Fernández, P. & Ordóñez- Fernández, R. (2008). Pruning residues as an alternative to traditional tillage to reduce erosion and pollutant dispersion in olive groves. *Water, Air and Soil Pollution,* Vol. 193, pp. 165-173.

Romanyà, J., Cortina, J., Falloon, P., Coleman, K. & Smith, P. (2000). Modelling changes in soil organic matter after planting fast-growing Pinus radiata on Mediterranean agricultural soils. *European Journal of Soil Science,* Vol. 51, pp. 627-641.

Schomberg, H.H. & Steiner, J.L. (1999). Nutrient dynamics of crop residues decomposing on a fallow no-till soil surface. *Soil Science Society of American Journal,* Vol. 63, pp. 607-613.

Six, J., Elliott, E.T. & Paustian, K. (1999). Aggregate and SOM dynamics under conventional and no-tillage system. *Soil Science Society of American Journal,* Vol. 63, pp. 1350-1358.

Smith, P., Nabuurs, G.J., Janssens, I.A., Reis, S., Marland, G., Soussana, J.F., Christensen, T.R., Heath, L., Apps, M., Alexeyev, V., Fang, J., Gattuso, J.P., Guerschman, J.P., Huang, Y., Jobbagy, E., Murdiyarso, D., Ni, J., Nobre, A., Peng, C., Walcroft, A., Wang, S.Q., Pan, Y. & Zhou, G.S. (2008). Sectoral approaches to improve regional carbon budgets. *Climatic Change,* Vol. 88, pp. 209-249.

Sofo, A., Nuzzo, V., Palese, A. M., Xiloyannis, C., Celano, G., Zukowskyj, P. & Dichio, B. (2005). Net CO_2 storage in Mediterranean olive and peach orchards. *Scientia Horticulturae,* Vol. 107, pp. 17-24.

Soil Conservation Service (SCS). 1972. *Soil Survey Laboratory Methods and Procedures for Collecting Soil Samples, Soil Survey Report 1.* US Department of Agriculture: Washington, DC.

Soria, L. (2002). *Fertilización y riego en el olivar de la provincia de Jaén: comarcas de La Loma y Sierra Morena.* Ph.D. thesis, University of Jaén, Spain.

Soria, L., Fernández, E., Pastor, M., Aguilar, J. & Muñoz, J.A. (2005). Impact of olive-orchard cropping systems on some soil physical and chemicals properties in southern Spain. In: Faz, A., Ortiz, R, Mermut, A.R. (Eds.), *Advances in GeoEcology,* Catena Verlag, Vol. 36, pp. 428-435.

Thomas, G.A., Dalal, R.C. & Standley, J. (2007). No-till effects on organic matter, pH, cation exchange capacity and nutrient distribution in a Luvisol in the semi-arid subtropics. *Soil and Tillage Research,* Vol. 94, pp. 295-304.

Tyurin, I.V. (1951). *Analytical procedure for a comparative study of soil humus.* Trudy. Pochr. Inst. Dokuchaeva 38.

Williams, D.E. (1948). A rapid manometric method for the determination of carbonate in soils. Soil Science *Society* American *Proceeding,* Vol. 13, pp. 27-129.

Comparison of the Effects of Saline and Alkaline Stress on Growth, Photosynthesis and Water-Soluble Carbohydrate of Oat Seedling (*Avena sativa L*)

Rui Guo[1], Ji Zhou[1], WeiPing Hao[1], DaoZhi Gong[1], SongTao Yang[2], XiuLi Zhong[1] and FengXue Gu[1]
[1]*Institute of Environment and Sustainable Development in Agriculture(IEDA), Chinese Academy of Agricultural Sciences (CAAS)/Key Laboratory of Dry land Agriculture, MOA, Beijing*
[2]*The College of Urban and Environment Sciences, Northeast Normal University, ChangChun*
China

1. Introduction

High-quality arable land is necessary for human socio-economic development, but up to 60% (0.9 x 10^9 ha) of the world's arable land is considered too saline and/or too alkaline for agricultural use, it severely affect agricultural productivity (Richards 1990; Allakhverdiev et al., 2000; Läuchli and Lüttge 2002). The main ions in salt soil are Na^+, K^+, Cl^- and NO_3^- in the salt soil, the saline stress usually had been defined two distinct kinds of stresses: saline stress and alkaline stress (Shi and Yin 1993). Saline stress generally involves osmotic stress and ion injury, but alkaline stress not only included osmotic stress and ion injury, but also has high pH effect (Yang et al., 2009a). The high pH environment surrounding the roots can directly cause mineral elements availability significantly decreased and Ca^{2+}, Mg^{2+} and H PO_3^- to precipitate, which possible inhibit ion uptake and disrupt the ion homeostasis of plant cells (Shi and Wang, 2005; Xue and Liu, 2008).

Oat is one of the major economic and ecological food crops in the world, and it is a kind of low sugar, high protein, fat and energy foods, it's exposed to multiple environment stresses, including drought, salinity etc (Shtangeeva and Ayrault, 2007). But salinity and alkalinity are significant constraints on the productivity of oat, a major staple cereal crop with global socioeconomic importance (Gale, 2005). The tolerance of oat has been widely studied; its physiological characteristics enable it to tolerate saline stress, however there were only a few researches on alkaline stress. In this study, the aim was to analyze and compare the growth, photosynthesis and water soluble carbohydrate in oat seedlings under saline and alkaline stresses, to determine and elucidate how high pH levels cause damage to oat seedlings tissues and how oat adapt to overcome alkaline stress. The goal can be achieved either by improving the soil or planting crops that are tolerant to both saline and alkaline stress.

Abbreviations

FW - fresh weight: *DW* - dry weight; RGR – relative growth rate; *WC* - water content; *P*n - net photosynthetic rate; *g*s -stomatal conductance; *E* -transpiration rate; *WUE* -water use efficiency.

2. Materials and methods

2.1 Design of simulated saline and alkaline conditions

Saline conditions were simulated by mixing the neutral salts NaCl and Na_2SO_4 in a 9:1 molar ratio and applying them to germinating wheat plants at five concentrations: 20, 40, 60, 80 and 100 mM. Similarly, alkaline stress conditions were simulated by mixing $NaHCO_3$ and Na_2CO_3 in a 9:1 molar ratio and applying them to germinating wheat plants at the same five concentrations. The electrical conductivity (EC), pH and osmotic potential of the stress treatment solutions were measured using a conductivity meter (DDG-2080-S, Anhui, China), PSH-3C and a water potential meter (Psypro Wescor Corporation, US), respectively.

2.2 Plant materials and growth

Seedling growth was tested by sowing 20 seeds in 17-cm diameter plastic pots containing 2.5 kg of washed sand. The seedlings were watered daily with 0.5 time Hoagland nutrient solution. All pots were placed outdoors and sheltered from rain, with a day/night temperature range of 21.0–25.5°C/18.5–21.0°C. After 4 weeks, 60 pots containing uniformly growing seedlings were selected and divided randomly into 12 sets, each comprising five pots. One set was used as the untreated control (watered with Hoagland nutrient solution), a second set was used to determine growth parameters at the beginning of treatment (see below), and the 10 remaining sets were used for the stress treatments, which were applied daily for 16 days.

2.3 Measurement of physiological indices

Relative growth rate (RGR) was determined by RGR = [(ln Dry Weight (DW) at the end of stress treatment) - (ln DW at the start of stress treatment)] / total treatment duration. The water content (%WC) was calculated using the formula (FW-DW)/DW. The electrolyte leakage rate (%ELR) was determined by measuring the electrical conductivity of 1 g fresh leaf material (after washing three times with deionized water to remove surface electrolytes) which was placed in a closed cuvette containing 20 ml deionized water and incubated at 25°C on a rotary shaker for 4 h, and comparing it to the electrical conductivity of the same sample after autoclaving at 120°C for 20 min. The activity of the root system was determined by incubating fresh root tissue for 60 min at 37°C in 0.04% triphenyl tetrazolium chloride (TTC) in phosphate buffer (pH 7.0), extracting the red product in ethyl acetate and determining is absorbance at 485 nm by spectrophotometry. The leaves were measured by portable photosynthesis system (LI - 6400, USA) in their locations to determined net photosynthetic rate (P_n), stomatal conductance - (*g*s) and transpiration rate (*E*) of leaves, the water use efficiency (*WUE*) has been determined by $WUE = P_n / E$.

2.4 Measurement of carbohydrates

The shoots and roots of oat seedlings were oven-dried at 80°C separately and then ground into small pieces. The fructan, sucrose, glucose and fructose content were determined using a high performance liquid chromatography diode (HPLC, Shimadu Class - VP, Japan), using columns (Sugar - PAK™ I) produced by the Waters company, and the concentration of the mobile phase was 0.1 mmol·L^{-1} $EDTANa_2Ca$, with the flow velocity controlled at 0.3

mL·min^{-1} at 70°C. The retention times of fructan, sucrose, glucose and fructose were 9.058, 13.545, 16.740 and 20.349 min, respectively. According to the chromatographic peak area, the carbohydrate content was determined (Rovia et al., 2008). The total carbohydrate content was determined according to the chromatographic peak area (Corradini et al., 2004; Rovio et al., 2008) using the phenol–sulfuric acid method (Somani et al., 1987).

2.5 Statistical analyses

Statistical analysis included one-way analysis of variance (ANOVA) in SPSS (Version 13.0, SPSS, Chicago, IL, USA) and Duncan's method to detect differences in physiological parameters in plants under saline and alkaline stress ($P < 0.01$).

3. Results

3.1 Electrical conductivity (EC) and osmotic potential of stress treatment solutions

The results given in Table 1 show that, with increased salinity, the EC increased (Saline stress: 3.67 to 11.23 dS/m; Alkaline stress: 3.01 to 9.98 dS/m), while the osmotic potential of the stress treatment solution decreased (Saline stress: -0.18 to -0.59 MPa; Alkaline stress: -0.13 to -0.52 MPa). The pH value decreased slightly with increasing saline concentration (Saline stress: 6.51 to 6.40), but it increased in alkaline stress solution (Alkaline stress: 8.30 to 9.56).

Salinity (mM)		EC (dS/m)	pH	OP (MPa)
Control		1.60	6.70	-0.05
Salt-stress	20	3.67	6.51	-0.18
	40	6.10	6.55	-0.25
	60	8.30	6.46	-0.38
	80	10.68	6.41	-0.51
	100	11.23	6.40	-0.59
Alkali-stress	20	3.01	8.30	-0.13
	40	4.97	8.66	-0.21
	60	6.48	8.69	-0.30
	80	8.21	8.84	-0.47
	100	9.98	9.56	-0.52

Table 1. The electrical conductivity (EC), pH and osmotic potential (OP) of stress treatment solutions

3.2 Growth

The patterns of RGR of shoot and root decreased with increasing salinity concentrations, but the decreased trend of alkaline stress was significantly than saline stress (Fig.1 A and B). The pattern of *WC* was similar as the pattern of RGR, the extent of reduction under alkaline stress was greater than those under saline stress (Fig. 1 C and D). Saline stress caused only minor changes in ELR and RST, but alkaline stress led to a significantly increase in ELR and RST declined dramatically (Fig. 1 E and F).

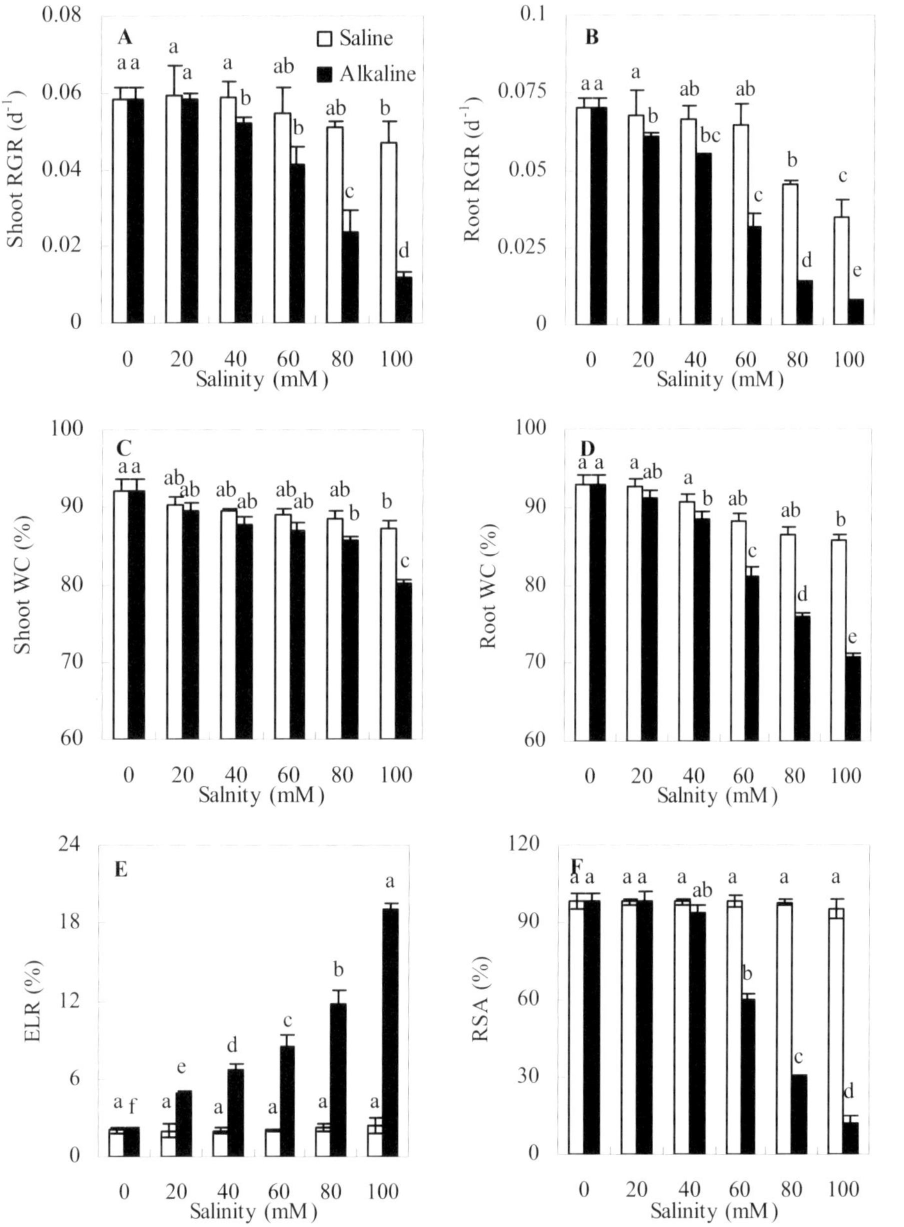

Fig. 1. Effects of salt and alkali stresses on under and up ground relative growth rate (RGR), water content (WC), Electrolyte leakage rate (ELR) and Root system activity (RSA). Salt-stress: NaCl:Na_2SO_4=9:1; Alkaline-stress: $NaHCO_3$:Na_2CO_3=9:1, the values are the means of five replicates. Means followed by different letters in the same stress type are significantly different at P <0.01 according to Duncan's methods

3.3 Photosynthesis

Under saline stress, P_n decreased slightly with increasing saline concentrations, however, under alkaline stress, the P_n value decreased dramatically with alkaline concentration increased (Fig. 2 A). The change of *gs* had a little increased at 40mM saline stress and then beginning to decreased, however, gs decreased sharply under alkaline stress (Fig. 2 B). There was no any change of *E* under saline stress, but under alkaline stress *E* decresed significantly with increasing stress, especially at 60mM (Fig. 2 C). No significant changes in *WUE* of oat seedlings was treated with saline stress, compare with saline stress, *WUE* had a decreased trend under alkaline stress (Fig. 2 D).

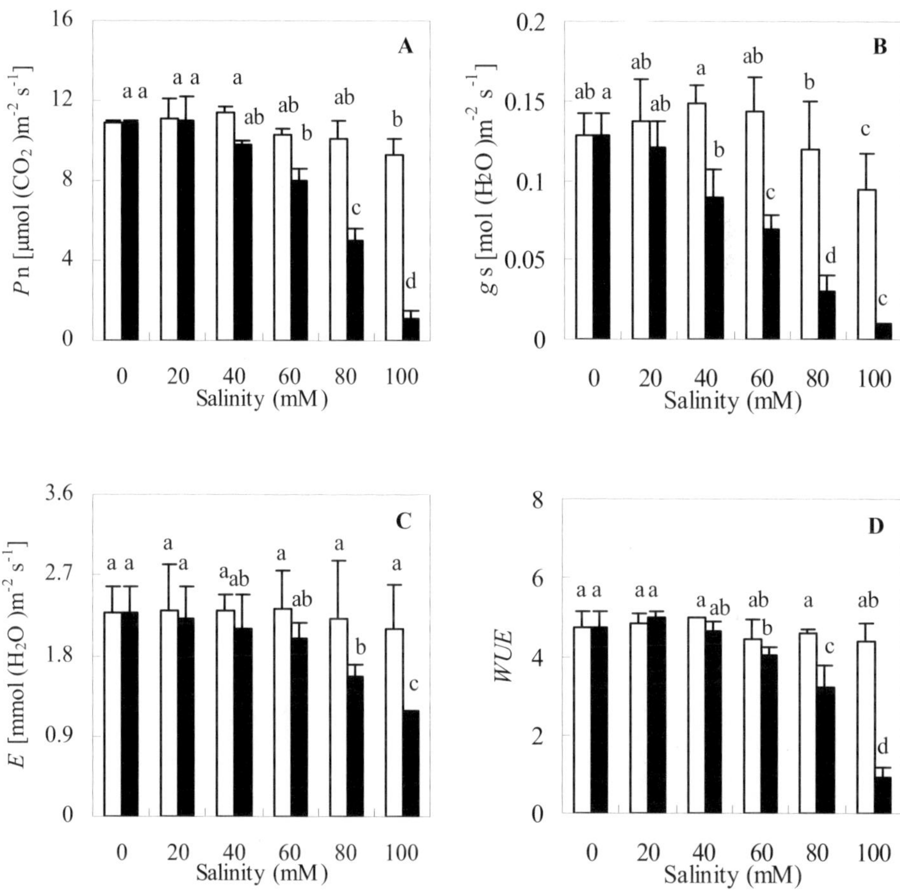

Fig. 2. Effects of salt and alkali stresses on net photosynthesis - *Pn* (A), stomatal conductance - *gs* (B), transpiration rate - *E* (C) Water Use Efficiency – *WUE* (D). Salt-stress: $NaCl:Na_2SO_4=9:1$; Alkaline-stress: $NaHCO_3:Na_2CO_3=9:1$, the values are the means of five replicates. Means followed by different letters in the same stress type are significantly different at $P<0.01$ according to Duncan's methods

3.4 Carbohydrates

There was little change in TCC in the shoot of oat seedlings with two stresses, it increased slightly with higher salinity and alkalinity (Fig 3 A_1). The TCC of root of the control was relatively high under saline stress and it increased with increasing saline stress after 40mM; but the change in TCC declined significant reduction under alkaline stress (Fig 3 A_2). The fructan content of shoot increased under saline stress with increasing salinity, but under alkaline stress it was higher than in the control and appeared to show decline trend with salinity increasing (Fig 3 B_1). The fructan contents in root of oat seedling under saline stress showed increased change trend with increasing salinity concentration; but the fructan contents decreased with alkaline stress increased (Fig. 3 B_2). Under two stresses sucrose content of shoot and root were no significant changed compared with the control (Fig. 3 C_1 and C_2), and the trends of change in glucose were same as the patterns of the fructan (Fig. 3 D_1 and D_2), however it nearly no accumulated in root under two stresses (Fig. 3 D). Under 40mM saline stress, saline stress promoted fructose synthesis and then decreased with increasing salinity; which was higher under alkaline stress than in the control and appeared to have no significant dependence on salinity (Fig. 3 E_1); however, it nearly no accumulated in root under two stresses (Fig. 3 E_2).

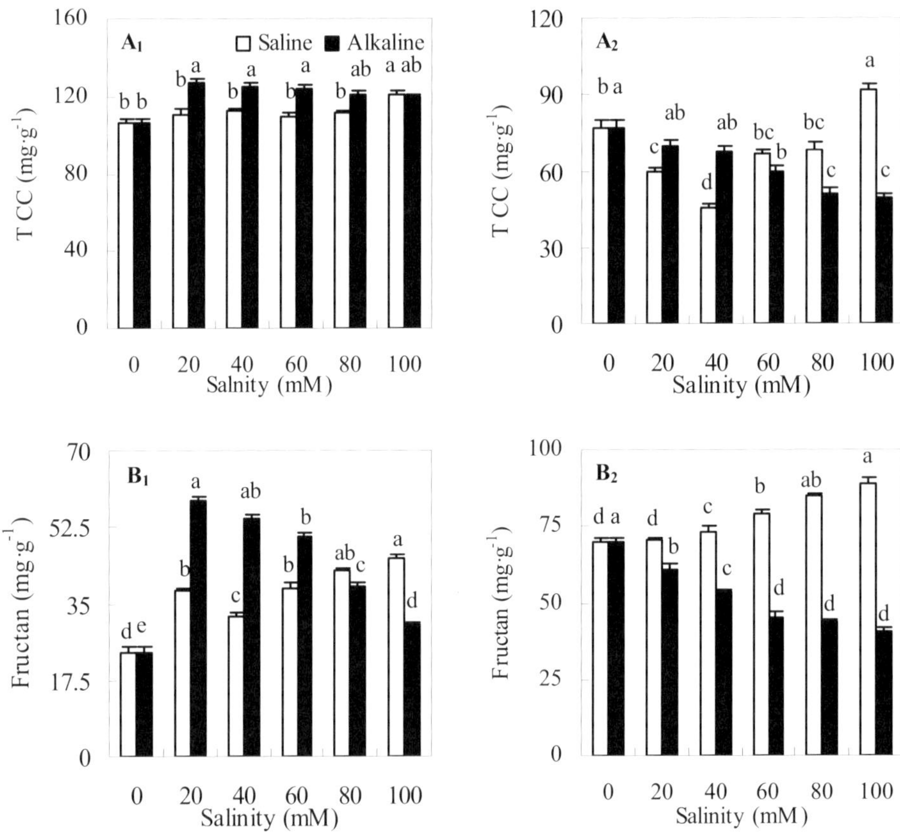

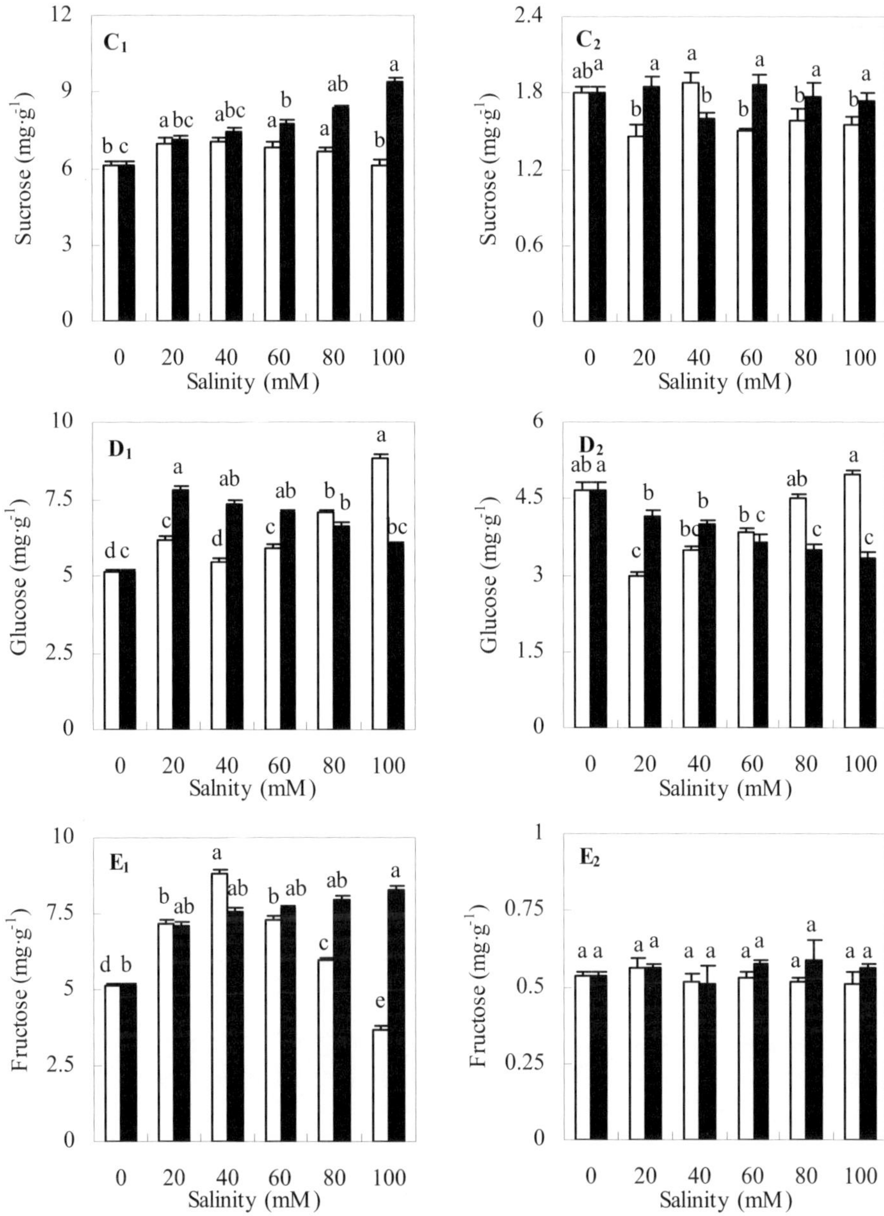

Fig. 3. Effects of salt and alkali stress on the levels of total carbohydrate content (TCC) (A1), fructan (B_1), sucrose (C_1), glucose (D_1) and fructose (E_1) in the shoots of oat seedlings, and of total carbohydrate content (TCC) (A_2), fructan (B_2), sucrose (C_2), glucose (D_2) and fructose (E_2) in the roots of oat seedlings. Salt-stress: $NaCl:Na_2SO_4=9:1$; Alkaline-stress: $NaHCO_3:Na_2CO_3=9:1$, the values are the means of five replicates. Means followed by different letters in the same stress type are significantly different at $P<0.01$ according to Duncan's methods

4. Discussion

4.1 Growth

The RGR of plant is considered to be an important index for the degree of stress, our results show that the RGR decreased in oat seedlings with increasing stress intensity for both stresses; however, the decrease for alkaline stress was greater than saline stress (Lissner et al., 1999). The different injuriousness of two stresses may result from their different mechanisms of action; the injurious effects of salinity are commonly thought to be from low water potentials and ion toxicities (Munns, 2002). Figure 1 F indicated that saline stress did not affect oat seedling root system activity, whereas alkaline stress led to a sharp decrease and even to the death of root cells. We suspect that there are the direct effect (root damage) and indirect effect (nutritional disorder) in alkaline condition; it's maybe the reasons why alkaline stress is more harmful to oat than saline stress. WC as approach to osmotic adjustment reflects osmotic stress, in this experiment compare with saline stress, alkaline stress cause WC severe reduced. Moreover, there nearly no effects of saline stress on the ELR, however, it sharply increased under alkaline stress. The results revealed that alkaline stress might damage root functions, including, the absorption of water and ions. These may be the main reasons why the oat RGR was lower for alkaline than saline stress (Yang et al., 2008).

4.2 Photosynthesis

A regression analysis between *Pn* and saline concentration is shown in Table 2, the *Pn* decreased by 3.525 for each 10 mM increase in saline stress and by 19.552 for each 10 mM increase in alkaline stress (Table 2). The photosynthesis of oat seedling were significantly lower under alkaline stress than that under saline stress, suggesting that alkaline stress might not only damage root functions, but also it further affects contents of photosynthetic area, photosynthetic pigments and the membrane system. Under both stresses, the environment water potential and WC of oat seedlings (Fig. 1 C and D) decreased with increasing stress intensity, the g_s was closely correlated with the change of environment water potential. The g_s of oat seedlings were lower than that of control, especially under alkaline stress. The *WUE* fell sharply under alkaline stress (Table 2), it is likely that the reduction in intracellular CO_2 partial pressure caused by stomatal closure helped to reduce net photosynthetic activity, and that the high pH in the leaf had a direct effect on the photosynthetic apparatus and on water potential, resulting in the differing results for water use efficiency (Bethke and Drew, 1992; Yang et al., 2009a,b). Our results showed clearly that saline stress had only a marginal effect on photochemical processes as previously reported (Larcher et al., 1990; Everard et al., 1994; Lu et al., 2002). In contrast, alkaline stress had a significant impact on chlorophyll a fluorescence, probably because the massive Na^+ influx and the high pH combine to encourage photoinhibition, and thus reduce the activity of *PSII*.

4.3 Carbohydrates

The accumulation and secretion of carbohydrates are a physiological response of plants to various stresses (Rosa-Ibarra and Maiti, 1995; Pan 2001). There is a different in the accumulation of carbohydrates in shoot and root respond to saline and alkaline stress, the results indicated that carbohydrates were accumulated in shoots under both saline and alkaline stresses; however, the extent of accumulation under alkaline stress was much higher than under saline stress. In the process of adapting to two stresses in root, oat seedlings with different adaptive pathways will have different carbohydrates metabolic regulating. The research indicated that the different part of origination of oat seedlings might have different physiological pathways under saline and alkaline stress, which should be further investigated. The results of the component of carbohydrates showed that fructan was the dominant component in oat seedlings. Fructan is considered to play a key role in stress-induced metabolic processes, which appears to be advantageous for plants under low temperature, drought and anoxia (Galiba et al., 1997). Our data implied that under saline stress, oat seedling accumulated fructan to resist saline stress, however, the concentrations of fructan was decreasing with salinity concentration increase under alkaline stress, suggesting that alkaline stress was more harmful for oat seedling than saline stress. Oat seedling might enhance fructan synthesis to remedy the osmotic stress and to maintain a stable ion balance to avoid ion injury; but the high pH of alkaline stress might inhibit the fructan synthesis. The material and energy demands of fructan synthesis are much greater. Sucrose is the main carbohydrates transport form and fructose is one of monosaccharides and had a significantly high concentration in drought sensitive in plant and it participate the fructan metabolism (Martin et al., 1993, kerepesi et al., 1998). In this research, there are no significant change and differences in sucrose and fructose content under two stresses in oat seedlings root, it reveal saline and alkaline no affect they synthesis. The change trend of glucose was followed the fructan, it indicated that under saline stress glucose content increased with salinity increased, which could synthesis more glycosides, oligosaccharides and polysaccharides to increase osmotic potential in intracellular, and enhance the salt tolerance of cells (Ghasempour et al., 1998). But the alkaline stress inhibited or destroyed glucose synthesis.

	Regression equation	R^2	Decrease in *Pn* and *WUE* for 10 mM increment in salinity
Pn	Y_{Salt}= -0.3525x + 11.745	0.699	3.525
	Y_{Alkali}= -1.9552x + 14.477	0.883	19.552
WUE	Y_{Salt}= -0.0904x + 4.984	0.5445	0.904
	Y_{Alkali}= -0.7142x + 6.2631	0.7702	7.142

Table 2. Regression equations between *Pn* and *WUE*

5. Conclusion

In summary, the germination of oat was inhibited by both saline and alkaline stress. The inhibit impact of alkaline stress was significantly greater than those of saline stress, reflecting the specific detrimental effects of a high pH environment. Alkaline stress might caused roots' physiological structure been destroyed, reducing root system activity, and finally inhibit oat seedlings growth. The photosynthesis of oat was not significantly lower than that of control until the intensity of saline stress reached 80 mM, but decreased sharply under alkaline stress with increasing salinity. This implied that saline stress and alkaline stress were actually not only two distinct stresses, but also that the resistance of oat to saline stress was stronger than that to alkaline stress, indicating intracellular photosynthesis were interfered with by alkaline stress (high pH). The content of carbohydrate increased to a greater extent in response to alkaline stress than saline stress. These data suggest that plants may initially sense high pH environments as an extreme form of salinity stress, such that the response is initially similar to the response to saline stress but more pronounced. However, because these tolerance mechanisms are not sufficient to counteract the specific toxic effects of high-pH environments, additional and more specific mechanisms such as reduced fructan synthesis. These results provide useful data that will facilitate the development of strategies for the creation of engineered oat varieties that are more tolerant towards alkaline stress.

6. Acknowledgements

This work supported by grants from the Project of the National Natural Science Foundation of China (No. 30870238, 30871447, 50709040, 31070398). The basic research special fund operations (No. BSRF 200901, BSRF200803), the international scientific and technological cooperation projects (No. 2010DFB30550), the youth foundation of northeast normal university of China (20090501).

7. References

Allakhverdiev, S.I., Sakamoto A., Nishiyama, Y., Inaba ,M., Murata, N., 2000. Ionic and osmotic effects of NaCl-induced inactivation of photosystems I and II in *Synechococcus* sp. Plant Physiol. 123, 1047-1056.

Bethke, P.C., Drew, M.C., 1992. Stomatal and non-stomatal components to inhibition of photosynthesis in leaves of Capsicum annuum during progressive exposure to NaCl salinity. Plant Physiol. 99, 219-226.

Corradini, C.F., Bianchi, D., Matteuzzi, A., Amoretti, M.R., Zanoni, S., 2004. High-performance anion-exchange chromatography coupled with pulsed amperometric detection and capillary zone electrophoresis with indirect ultra violet detection as powerful tools to evaluate prebiotic properties of *fructooligosaccharides* and *inulin*. Journal of Chromatography. 1054, 165-173.

Everard, R., Gucci, S.C., Kann, J.A., Loescher, W.H., 1994. Gas exchange and carbon partitioning in the leaves of celery (Apium graveolens L.) at various levels of root zone salinity. Plant Physiol. 106, 281–292.

Gale, K.R., 2005. Diagnostic DNA markers for quality traits in oat. Journal of Cereal Science. 41, 181–192.

Galiba, G., Kerepesi, I., Snape, J.W., Sutka, J., 1997. Location of a Gene Regulating Cold-Induced Carbohydrate Production on Chromosomes 5A of Oat. Theor Appl Genet. 95, 265-270.

Ghasempour, H.R., Gaff, D.F., Williams, R.P.W., Gianello, R.D., 1998. Contents of Sugars in Leaves of Drying Desiccation Tolerant Flowering Plants, Particularly Grasses, Plant Growth Regul. 24, 185–191.

Kerepesi, I., Galiba, G., and Ba´nyai, E., 1998. Osmotic and Salt Stresses Induced Differential Alteration in Water-Soluble Carbohydrate Content in Oat Seedlings. J. Agric. Food Chem. 46, 5347-5354.

Läuchli, A., Lüttge, U., 2002. Salinity: Environment - Plants - Molecules. Springer. 552.

Larcher, W., Wagner, J., Thammathaworn, A., 1990. Effects of superimposed temperature stress on in vivo chlorophyll fluorescence of *Vigna unguiculata* under saline stress. Plant Physiol. 136, 92–102.

Lissner, J., Schierup, H.H., Comın, F.A., Astorga, V., 1999. Effect of Climate on the Salt Tolerance of Two *Phragmites Australis* Populations, I. Growth, Inorganic Solutes, Nitrogen Relations and *Osmoregulation*. Aquat Bot. 64, 317–333.

Lu, C.M., Qiu, N.W., Lu, Q.T., Wang, B.S., Kuang, T.Y., 2002. Does saline stress lead to increased susceptibility of photosystem II to photoinhibition and changes in photosynthetic pigment composition in halophyte *Suaeda salsa* grown outdoors. Plant Science. 163, 1063-1068.

Martin, M., Miceli, F., Morgan, J.A., Scalet, M., Zerbi, G., 1993. Synthesis of Osmotically Active Substrates in Winter Oat Leaves as Related to Drought Resistance of Different Genotypes. J. Agric. Crop Sci. 171, 176-184.

Munns, R., 2002. Comparative physiology of saline and water stress. Plant Cell Environ. 25, 239–250.

Pan, R.C., 2001. Higher Education press. Plant Physiology. 4nd edn. (In Chinese).

Richards, J.F., 1990. Land transformation. *The Earth as transformed by human action*. (ed. by B.L. Turner, W.C., Clark, R.W., Kates, J.F., Richards, J.T., Mathews., W.B.Meyer) Cambridge University Press, Cambridge.163-178.

Rosa-Ibarra, M., Maiti, R.K., 1995. Biochemical Mechanism in glossy sorgum lines for resistance to salinity stress. J. Plant Physiol. 146, 515-519.

Rovio, S., Simolin, H., Koljonen, K., Sir´en, H., 2008. Determination of Monosaccharide Composition in Plant Fiber Materials by Capillary Zone Electrophoresis, Journal of Chromatography A. 1185, 139–144.

Shi, D.C., Yin, L.J., 1993. Difference between salt (NaCl) and alkaline (Na_2CO_3) stresses on *Puccinellia tenuiflora* (Griseb.) Scribn et Merr. plants. Acta Bo. 3, 144–149.

Shi, D.C., Wang, D., 2005. Effects of Various Salt–Alkaline Mixed Stresses on *Aneurolepidium Chinense* (Trin.) Kitag. Plant Soil. 271, 15–26.

Shtangeeva., I., Ayrault, S., 2007. Effects of Eu and Ca onYield and Mineral Nutrition of Oat (Triticum aestivum) Seedlings, Environmental and Experimental Botany. 49-58.

Somani, B. L., J. Khanade, Sinha, R., 1987. A modified anthrone-sulfuric acid method for the determination of fructose in the presence of certain proteins. Analytical Biochemistry. 167, 327-330.

Xue, Y.F., Liu, Z.P., 2008. Effects of NaCl and Na_2CO_3 stresses on photosynthesis and parameters of chlorophyll fluorescence in *Helianthus Tuberosus* seedlings, Journal of Plant Ecology. 32, 161-167.

Yang, C., Wang, P., Li, C., Shi, D.C., Wang, D., 2008. Comparison of effects of salt and alkali stresses on the growth and photosynthesis of oat. Photosynthetica. 46, 107–114.

Yang, C., Xu, H.H., Wang, L., Liu, J., Shi, D.C., Wang, D., 2009a. Comparative effects of salt-stress and alkali-stress on the growth, photosynthesis, solute accumulation, and ion balance of barley plants. Photosynthetica. 47, 79–86.

Yang, C., Zhang, M., Liu, J., Shi, D.C., Wang, D., 2009b. Effects of buffer capacity on growth, photosynthesis, and solute accumulation of a glycophyte (oat) and a halophyte (*Chloris virgata*), Photosynthetica. 47, 55-60.

7

Physiological and Biochemical Mechanisms of Plant Adaptation to Low-Fertility Acid Soils of the Tropics: The Case of Brachiariagrasses

T. Watanabe[1], M. S. H. Khan[2], I. M. Rao[3], J. Wasaki[4], T. Shinano[5], M. Ishitani[3], H. Koyama[6], S. Ishikawa[7], K. Tawaraya[8], M. Nanamori [1], N. Ueki [8] and T. Wagatsuma[8]

1. Introduction

Brachiaria species are the most widely planted tropical forage grasses in the world (Miles et al., 2004). For example, in Brazil alone, about 80 million hectares are planted to *Brachiaria* pastures (Macedo, 2005). They increase animal productivity by 5 to 10 times with respect to native savanna vegetation in the tropical areas of Latin America, thus representing a significant contribution to farmer's income (Rao et al., 1993). Although their origin is from the tropical areas of Africa, they are also used for livestock production in South-East Asia and Australia. Among them, *Brachiaria decumbens* cv. Basilisk, *Brachiaria brizantha* cv. Marandu, and *Brachiaria ruziziensis* cv. Kennedy have been more commonly utilized for livestock production in the tropics (Miles et al., 2004). Among the three grasses, *B. decumbens* is highly adapted to infertile acid soils, i. e., high level of tolerance to high aluminium (Al) saturation, low phosphorus (P) and low calcium (Ca) supply in soil (Louw-Gaume et al., 2010 a,b; Rao et al., 1995, 1996; Wenzl et al., 2001, 2003), but also highly sensitive to a major insect, spittlebugs (Miles et al., 2006) and produces mycotoxin after infection with *Pithomyces chartarum* (Andrade et al., 1978). *B. brizantha* cv. Marandu is highly resistant to spittlebugs, adapted to seasonal drought stress, highly responsive to fertilizer application but is not well adapted to low fertility acid soils (Miles et al., 2004, 2006). *B. ruziziensis* cv. Kennedy is sensitive to spittlebugs, performs better in well-drained fertile soils, has high forage quality but poorly adapted to low fertility acid soils (Ishigaki, 2010; Miles et al., 2004). *B. decumbens* and *B. brizantha* are generally tetraploid, apomicts while *B. ruziziensis* is diploid, sexual (Miles et al., 2004, 2006).

---

[1]*Graduate School of Agriculture, Hokkaido University, Kita-ku, Sapporo, Japan*
[2]*Department of Soil Science, HMD Science and Technology University, Dinajipur, Bangladesh*
[3]*Centro Internacional de Agricultura Tropical (CIAT), A.A.6713, Cali, Colombia*
[4]*Graduate School of Biosphere Science, Hiroshima University, Higashi-Hiroshima, Japan*
[5]*National Agricultural Research Center for Hokkaido Region, Sapporo, Japan*
[6]*Faculty of Applied Biological Sciences, Gifu University, Gifu, Japan*
[7]*National Institute for Agro-Environmental Science, Tsukuba, Japan*
[8]*Faculty of Agriculture, Yamagata University, Tsuruoka, Japan*

CIAT and its collaborators have an on-going breeding program to combine the desirable attributes from the three grasses (Miles et al, 2004, 2006). *B.* hybrid cv. Mulato is the product of three generations of crosses between *B. ruziziensis*, *B. decumbens* and *B. brizantha*. This grow well in low P, low fertility acid soils in both wet and dry seasons (Rao et al., 1998), and produces a large numbers of panicles with well synchronized flowering and good caryopsis formation, which leads to good-quality seed. Defining the specific physiological and biochemical mechanisms that are associated with greater adaptation to low fertility acid soils will contribute to developing rapid and reliable methods to select the phenotypes and to develop molecular markers for marker assisted breeding of brachiariagrasses. Developing superior *Brachiaria* hybrids from the on-going breeding programs that combine the desirable attributes including adaptation to major biotic and abiotic constraints, forage quality, and seed production will facilitate sustainable intensification of crop-livestock systems in the tropics (Miles et al., 2004, 2006; Rao, 2001 a,b). This chapter reviews the progress made in defining the physiological and biochemical mechanisms of adaptation of brachiariagrasses to low fertility acid soils.

There is limited knowledge on the comparative differences in Al resistance among *B. decumbens, B. brizantha, B. ruziziensis* and *B.* hybrids (Mulato and Mulato II) grown in hydroponic system. Identification of plant attributes that contribute to greater ability to acquire nutrients under low pH, low P and high Al conditions is critical to develop brachiariagrases that are productive and persistant under infertile acid soil conditions.

There have been some discussions on the validity of short-term screening technique that uses simple solution of Al and Ca to test the effect of Al on relative root elongation of young seedlings: whether the results obtained by this short-term screening technique can apply to the behaviour of older plants is under discussion (Ryan et al., 2011). However, significant positive correlations were observed on Al resistance of 15 cultivars of sorghum (*Sorghum bicolor* Moench) and 10 cultivars of maize (*Zea mays* L.) with data obtained using short-term (1 day) screening and long-term screening technique using hydroponic system (Akhter et al., 2009). Similar results were also observed with 8 rice cultivars (unpublished data). In this chapter, we consider short-term vs long-term responses of several plant species including brachiariagrasses that differed widely in their level of Al resistance.

B. decumbens is known for very high level of Al resistance, however, the mechanisms responsible for this high level of Al resistance was not associated with exudation capacity of organic acid anions from root tips (Wenzl et al., 2001). It is important to define the specific mechanism(s) contributing to the high level of Al resistance in *B. decumbens*.

Higher level of Al exclusion was found in *B. decumbens*, however, the specific mechanisms related to Al exclusion are not known. Several mechanisms other than organic acid anion exudation were found and the related Al resistance genes have been reported (Huang et al., 2009; Yamaji et al., 2009; J.L. Yang et al., 2008; Z-B Yang et al., 2010). We found a new mechanism for higher level of Al tolerance of an Al-resistant rice cultivar based on higher abundance of sterols in plasma membrane (PM) lipids of root-tip cells (Khan et al., 2009). Rice is also known for its greater level of Al resistance than other cereal crops and its level of resistance was also not related to organic acid anion exudation (Ishikawa et al., 2000; Ma et al., 2002). Therefore, it is crucial to test the role of sterols in PM lipids of root-tip cells of brachiariagrasses.

High concentrations of phenolic compounds have been reported as one of the promising mechanisms to explain higher level of Al resistance of a forage legume, *Lotus pedunculatus* Cav. (Stoutjesdijk et al., 2001) and several common woody plants (Ofei-Manu et al., 2001).

This can be explained by higher complexing abilities of phenolic compounds with Al ions (Cornard & Merlin, 2002; Yoneda & Nakatsubo, 1998). Phenolic compounds have also been reported to be solubilized into lipid layer (Boija et al., 2006, 2007). Existence of phenolic compounds in lipid layer was found to make PM less fluid (Arora et al., 2000). Lipid layer with less fluidity will make PM less permeable even in the presence of Al ions (Khan et al., 2009). The contribution of phenolic compounds in root-tip portion to high level of Al resistance in *B. decumbens* is not known.

A major constraint to agriculture on tropical and subtropical soils is P deficiency (Fairhust et al., 1999). Applying large amounts of fertilizer to correct P deficiency is not feasible for most resource-poor farmers in developing countries. Thus the agricultural productivity becomes limited in the near future. Moreover, P fertilizer is receiving more attention as a nonrenewable resourse (Cordell et al., 2009; Steen, 1998). For sustainable P management in agriculture on tropical and subtropical soils, it is essential to define the mechanisms involved in making plants more efficient in P acquisition and use (Lynch, 2011; Ramaekers et al., 2010; Rao et al., 1999).

P is a constituent of phospholipids (PL), nucleic acids, nucleosides, coenzymes, and phosphate esters in plants. P helps regulate plant metabolisms by controlling enzymatic activity through phosphorylation and/or dephosphorylation. To overcome P deficiency, plants develop several strategies, including the well-known one of secreting acid phosphatase (APase), ribonuclease (RNase), and organic acids into the rhizosphere to improve P availability in the soil (Duff et al., 1994; Green, 1994; Jones, 1998; Rao et al., 1999; Tadano et al., 1993; Wasaki et al., 2003a). Besides from acquiring P from the outside of the plant, APase is also recognized for its role in efficient utilization of absorbed P for metabolism (Duff et al., 1989, 1991). One of the mechanisms to increase P recycling ability is dependent on the activity of APase (Duff et al., 1991, 1994) and RNase (Howard et al., 1998) in the cell. The increase of both enzymes is reported and these enzymes are considered to utilize those P compounds that are stored in vacuole. The other mechanism is bypassing several metabolic pathways to reduce the usage of P molecule. Theodorou and Plaxton (1993) showed that P deficiency induces some glycolytic enzymes, such as phosphoenolpyruvate carboxylase (PEPC) and phosphoenolpyruvate phosphatase (PEPP). These catalyze the bypass reaction of pyruvate kinase (PK), which is responsible for regulating carbon flow from glycolysis to the TCA cycle. PEPC replenishes intermediates of the TCA cycle, and may help regulate both carbon and nitrogen metabolism, and P recycling under P deficiency. Kondracka and Rychter (1997) observed that, in P-deficient bean leaves, the rate of malate synthesis increases, and the accumulation of aspartate and alanine (products of PEP metabolism) is also enhanced. In early stages of P deficiency, the increased activity of PEPC and use of PEP in amino acid synthesis are probably the most important reactions for P recycling in bean leaves during photosynthesis. Thus, PEP metabolism by PEPC and PEPP, or PEP transport via PPT (which transports PEP from the cytosol into chloroplasts in leaves) may affect carbon distribution within the plant under P deficiency.

Under P-deficient conditions, the brachariagrasses improve their P acquisition by enhancing root growth, uptake efficiency, and ability to use poorly available plant P (Louw-Gaume et al., 2010b; Rao et al., 1999, 2001a). Although they have much lower internal requirements for P than do other grasses, they also show interspecific differences (Rao et al., 1996).

Recent advances in post-genomic studies have indicated that transcriptomic analysis is a useful tool for understanding gene expression networks. Some transcriptomic studies of low-P adaptation strategies have also been carried out using cDNA arrays (Hammond et al., 2003; Misson et al., 2005; Ramaekers et al., 2010; Uhde-Stone et al., 2003; Wang et al., 2002; Wasaki et al., 2003b, 2006; Wu et al., 2003). Although some aspects of plant strategies for coping with P-deficient conditions are understood, the majority are not; a broad view of gene expression is necessary to fully elucidate all of the mechanisms involved (Ramaekers et al., 2010). In this chapter, we review the progress in understanding the transcriptomic changes by P deficiency in rice plants, which is a model of Gramineae plants and also relatively tolerant of low P and low pH conditions. We use qRT-PCR for quantification of the effects of P deficiency on phosphohydrolases and carbon metabolism in rice leaves. We also review the progress in defining the physiological and biochemical bases of improved P use efficiency in *B.* hybrid (cv. Mulato).

2. Materials and methods

We randomly selected seven plant species, i.e., rice (*Oryza sativa* L. cv. Sasanishiki), maize (*Zea mays* L. cv. Pioneer 3352), pea (*Pisum sativum* L. cv. Kinusaya), barley (*Hordeum vulgare* L. cv. Manriki), tea (*Camellia sinensis* L. cv. Yabukita), siclepod (*Cassia tora* L.) and *B. brizantha*. Seeds of rice, maize, pea were soaked in tap water under aeration for 3 to 24 h depending on plant species. Seeds of siclepod were notched by a razor to facilitate germination and soaked in tap water for 12 h. Seeds of barley and *B. brizantha* were not soaked. These seeds were germinated on a nylon screen that was put on a polypropylene container filled with tap water under aeration at 27°C in a growth room. Seeds of tea were germinated in quarts sand wetted with deionized water, and then seedlings with roots approximately 1 cm long were transferred to the container described above. All the seedlings with roots approximately 5 cm long were used in the following experiments. In the short-term experiment, ten seedlings of each plant species were pretreated in 600 mL volume of a solution with (Al treatment) or without (control) 50 μ M $AlCl_3$ containing 0.2 mM $CaCl_2$ at pH 4.7 for 1 h. After measurement of the root length with a ruler, all the seedlings were transferred into a Al-free 0.2 mM $CaCl_2$ solution at pH 4.7. Root length was measured again after 24 h. For the tea plant, the root length was measured after 3 d because of the slower root elongation than in the others. Relative growth (%) in the short-term treatment with Al was calculated as the ratio of net root re-elongation of the primary root in the Al treatment to that in the control. In the long-term experiment, twenty seedlings of each plant species were precultured in a 54-L volume of nutrient solution at pH 5.2. The nutrient solution was composed of 1.43 mM NH_4NO_3, 0.7 mM $NaNO_3$, 0.13 mM NaH_2PO_4, 0.78 mM K_2SO_4, 1 mM $CaCl_2$, 0.6 mM $MgSO_4$, 36 μM $FeSO_4$, 9 μ M $MnSO_4$, 0.08 μ M $CuSO_4$, 0.03 μ M $(NH_4)_6Mo_7O_{24}$, 18.5 μ M H_3BO_3, 1.5 μ M $ZnCl_2$. After 1 week, the seedlings were transferred into the nutrient solution with same composition as above (control) or the nutrient solution containing 100 μ M Al and 10 μ M P in soluble form (Al treatment) at pH 5.2 or 4.5, respectively. The solutions were renewed weekly. The concentrations of Al and P, and pH of the solution were monitored every day and adjusted if required. Two months after the Al treatment, the plants were harvested and dried in a draft oven (60°C). Relative growth (%) in the long-term treatment with Al was calculated as the ratio of the dry weight of whole plant in the Al treatment to that in the control. The concentrations of K, P, Mg, and Ca were determined by ICP-AES (inductively coupled plasma-atomic emission spectrometry) after digestion of the plant samples using an acid mixture (HNO_3: $HClO_4$ = 5: 3, v/v). Extraction and determination of the phenolic compounds were carried out as described by Ofei-Manu et al. (2001). Seedlings of *Brachiaria* hybrid (*B.* ruziziensis Ger. & Ev. clone 44-06 × *B.* brizantha (A. Rich.) Stapf CIAT 36061, also known as cv. Mulato), *Andropogon gayanus* Kunth (CIAT 621), barley (cv. Ryofu) were transferred to 36-L containers containing nutreint solution with or without 0.37 mM Al (as $Al_2[SO_4]_3$). At the end of 10 days of treatment, the roots of seedlings from each treatment were sampled, dried in a forced-air oven at 80°C for 72 h, and then weighed. The dried samples were ground and digested with H_2SO_4-H_2O_2 for Al analysis by ICP-AES. *Brachiaria* seedlings were prepared as described above, and transferred to 36-L containers carrying the standard nutrient solution, but with 2.8 mM Al at pH 3.7 added, and left to grow for 1 month. The much higher Al concentration was used to ensure clear peaks in the ^{27}Al NMR spectrum. Even so, the *Brachiaria* seedlings grew well (data not shown). After treatment, roots were removed from the seedlings and washed, first with tap water, then with deionized water. The roots were grouped into three: Fraction (a), roots given the water washings only, and used to determine total amounts of Al and organic acid anions; and Fraction (b), roots were also washed with 0.1 M HCl for 5 min to remove apoplastic, soluble or loosely bound, components. Each fraction of *Brachiaria* roots was placed in a 10-mm-diameter NMR tube. $AlCl_3$ (0.1 M) solution was used as an external reference to calibrate the chemical shift (0 ppm). ^{27}Al NMR spectra were recorded, using a Bruker MSL400 spectrometer at 104.262 MHz. The spectra were obtained by using a frequency range of 62.5 kHz, a pulse width of 12 μs, a delay time of 0.16 ms, a cycle time of 0.5 s, and 4000 scans. For estimation of low P tolerance, we have selected *Brachiaria* hybrid (cv. Mulato) and rice (*Oryza sativa* L. cv. Nipponbare). For the seedling growth nutrient solution was prepared with 2.12 mM N (NH_4NO_3), 0.77 mM K (K_2SO_4:KCl = 1:1), 1.25 mM Ca ($CaCl_2 \cdot 2H_2O$), 0.82 mM Mg ($MgSO_4 \cdot 7H_2O$), 35.8 μM Fe ($FeSO_4 \cdot 7H_2O$), 9.1 μM Mn ($MnSO_4 \cdot 4H_2O$), 46.3 μM B (H_3BO_3), 3.1 μM Zn ($ZnSO_4 \cdot 7H_2O$), 0.16 μM Cu ($CuSO_4 \cdot 5H_2O$), 0.05 μM Mo $(NH_4)_6Mo_7O_{24} \cdot 4H_2O$), with 6 μM P ($NaH_2PO_4 \cdot 2H_2O$) and pH was maintained at 5.2 for 1 week preculture. Then phosphorus level was changed as 0, 6 and 32 μM for 2 weeks. After sampling, samples were freeze-dried, then P fractionation was carried out based on Schmidt-Thannhauer-Schneider method. Samples were also used for measurements of APase and RNase activities. $^{14}CO_2$ was generated by adding 30% PCA into $NaH^{14}CO_2$ (18.5 kBq), then applied to plants for 5 min in a vinyl package under natural light condition. After sampling, samples were fractionized by using column method to obtain organic acids, amino acids, and sugars fraction, then ^{14}C content in each sample was determined by using scintillation counter. In transcriptomic analyzes and subsequent molecular analysis, we have selected rice (*Oryza sativa* L. cv. Michikogane) and the growth condition was same as shown before except for the P treatment was done using 0 or 32 μM levels. Total RNA was extracted from frozen samples using a sodium dodecyl sulfate (SDS)-phenol method. The real-time PCR was performed by the LightCycler™ system (Roche) with the LightCycler DNA Master SYBR Green I kit for PCR (Roche), and the TaqStart™ antibody (Clontech) was used for the repression of unspecific amplification.

2.1 Mechanisms of Al resistance in brachiariagrasses

2.1.1 Differences in Al resistance and nutrient acquisition among crops and brachiariagrasses

Differences in Al resistance among crops and *B. brizantha* are shown in Fig.1 (the left panel). Al resisance was ranked as follows: *B. brizantha* > rice > tea > maize > pea, siclepod > barley. The order of Al resistance in the short-term experiment was well correlated with that of the long-term experiment in spite of marked difference in the treatment conditions, i.e., duration of Al treatment, Al concentration, composition of co-existing nutrients, pH, etc. ($R^2 = 0.785$ [$p< 0.01$], right side figure of Fig.1) (Ishikawa et al., 2000). Although *B. brizantha* is relatively less adapted than *B.decumbens* to infertile acid soils, *B. brizantha* was found to be superior in its level of Al resistance to the other crops tested.

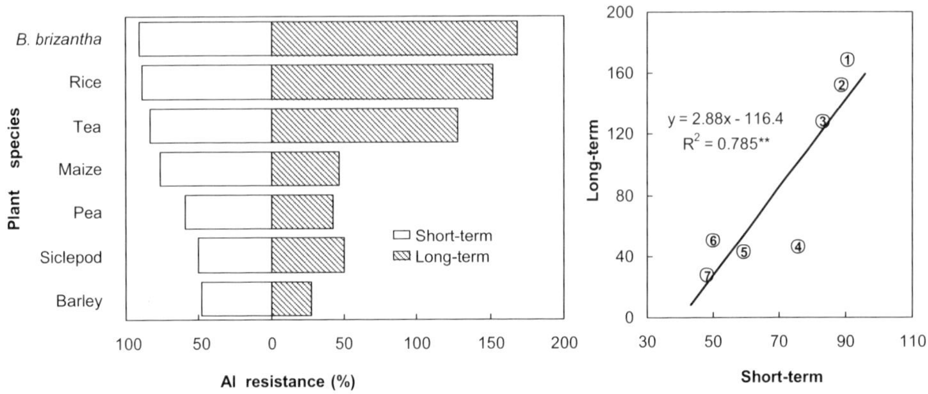

Fig. 1. Differences in Al resistance among crops and *Brachiaria brizantha*. The left panel shows Al resistance values in long-term experiment and short-term experiment. The right figure shows the relationship between Al resistance in long-term experiment and that in short-term experiment. ① *B. brizantha*, ② rice, ③ tea, ④ maize, ⑤ pea, ⑥ siclepod, ⑦ barley

We also compared nutrient acquisition ability of *B. brizantha* with that of crops. Nutrient acquisition ability was compared by quantifying relative nutrient status in shoots in Al treatment to that in low pH • low P treatment. Among the 5 species compared, *B. brizantha* was found to have superior nutrient acquisition abilities for K,P and Mg (Fig. 2). However, *B. brizantha* showed the lowest acquisition ability for Ca. *B. brizantha* can acquire highest amounts of K, P and Mg even under high Al, low P and low pH conditions. *B. brizantha* may have the lowest requirement of Ca for normal growth even in acid soil conditions. This greater ability of *B. brizantha* to acquire nutrients under simulated acid soil conditions could

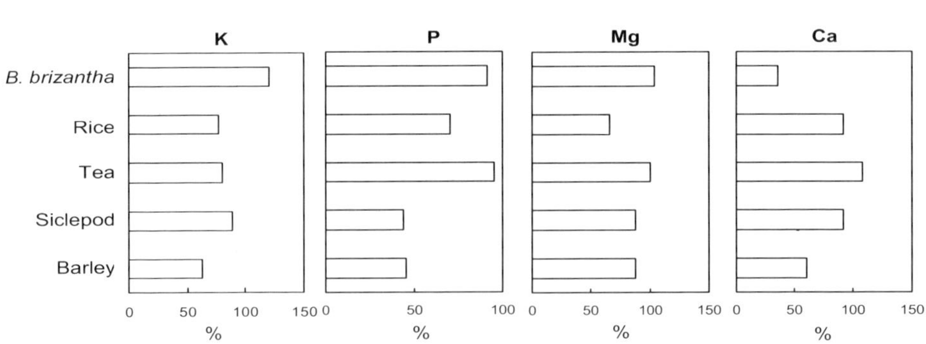

Fig. 2. Relative nutrient status of each nutrient concentration in Al treatment to that in control treatment

be responsible for its greater vigour in acid soil conditions during the pasture establishment phase (Rao et al., 1996).

2.1.2 Differences in Al resistance, Al accumulation and plasma membrane permeability among brachiariagrasses

Adaptive responses of several brachiariagrasses to infertile acid soils have been identified and described by previous research (Louw-Gaume et al., 2010 a,b; Rao et al., 1995, 1996; Wenzl et al., 2001, 2003). We compared Al resistance, Al accumulation (hematoxylin staining method [Wagatsuma et al., 1995] and PM permeability (FDA-PI fluorescence staining method [Ishikawa et al., 2001]) among 4 brachiariagrasses in short-term experimental conditions that were described in the former section, together with most Al-resistant rice cultivar Rikuu-132 (Khan et al., 2009) as a reference plant species. Al resistance was ranked as follows: *B. decumbens, B.* hybrid, *B. brizantha* > *B. ruziziensis* > Rikuu-132 (Fig. 3). Al resistance of *B.* hybrid (cv. Mulato) and *B. brizantha* was found to be comparable to that of *B.decumbens* which has been ranked as the most Al-resistant brachiariagrass (Wenzl et al., 2001). Although Al resistance of *B. ruziziensis* was found to be markedly lower than *B. decumbens* (Wenzl et al., 2001, 2003), its resistance level was higher than that of the most Al-resistant rice cultivar. It was suggested that the highest Al resistant phenotype of *B.* hybrid may be ascribed to the Al resistance genes from *B. decumbens* or *B. brizantha* and not from *B. ruziziensis*. Al accumulation was localized mainly within 1- mm root-tip portion and its concentration corresponds reversely to Al resistance order: the least Al accumulation was recognized for the most Al-resistant *B. decumbens*. PM lipid layer was less permeable to Al in brachiariagrasses than in the most Al-resistant rice cultivar and its less permeable PM characteristic was localized mainly within 1- mm root-tip portion (Fig.4).

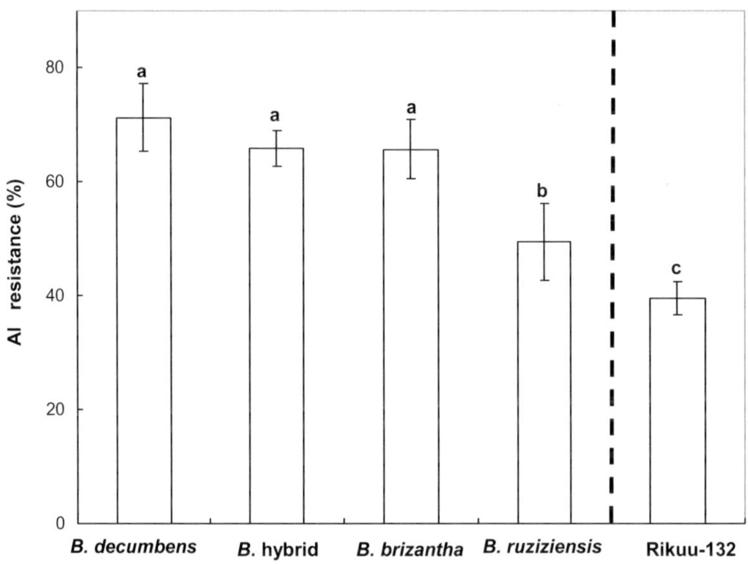

Fig. 3. Differences in Al resistance among brachiariagrasses and Al-resistant rice cultivar Rikuu-132

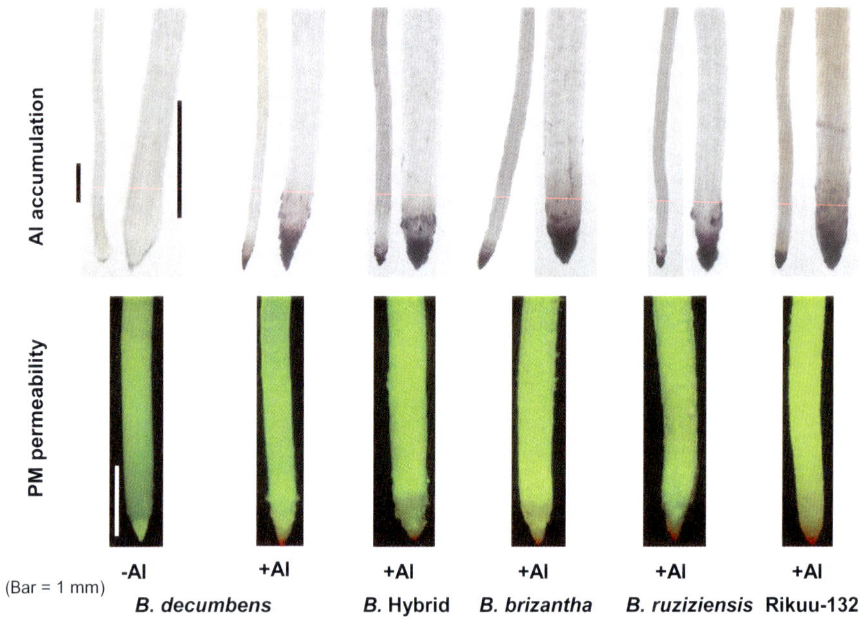

Fig. 4. Differences in Al accumulation and PM permeability among brachiariagrasses and Al-resistant rice cultivar (Rikuu-132)

Permeability of PM was negatively associated with Al resistance: the least PM permeabilization was observed with the most Al-resistant *B. decumbens*. The less permeability of PM and the lower Al accumulation in root-tip portion in Al-resistant brachiariagrass agree well with the former results which have been recognized in Al-resistant plant species, cultivars, or lines (Ishikawa et al., 2001; Ishikawa & Wagatsuma, 1998; Wagatsuma et al., 2005).

2.1.3 Lipid composition and phenolics concentration in root-tip portion of brachiariagrasses in relation to Al resistance

The lower ratio of PL to sterols (S) (PL was measured by molybdenum blue spectrophotometric method after extraction with isopropanol-chloroform-H_2O [2:2:1] ; S was measured by ortho-phthalaldehyde colorimetric method after extraction with dichloromethane-methanol [2:1] [Khan et al., 2009]) in root-tip portion was found to be beneficial for the less permeability of PM in the presence of Al, which agrees with the results of rice cultivars (Khan et al., 2009). In the more proximal root region (0-10 mm from root apex), the ratio of PL to S for *B. decumbens* was higher than that of *B. ruziziensis*, but on the contrary, it was lower in root-tip portion (0-2 mm from root apex) under Al treatment conditions (Fig.5).

The lower negativity of PM surface that was associated with the lower ratio of PL to S in root-tip portion could contribute to lower permeability of PM to Al. This is highly consistent with Gouy-Chapman-Stern model of Al rhyzotoxicity (Kinraide, 1999). In case of rice cultivars (Khan et al., 2009), wheat lines, triticale lines, maize cultivars (unpublished data), lipid compositional difference in connection with Al resistance were recognized in root-tip portion of 0-10 from root apex. However, in brachiariagrasses, lipid compositional difference was related with Al resistance only in root-tip portion of 0-2 mm from root apex. We suggest that the high level of Al resistance in brachiariagrass is extremely localized at the root tip.

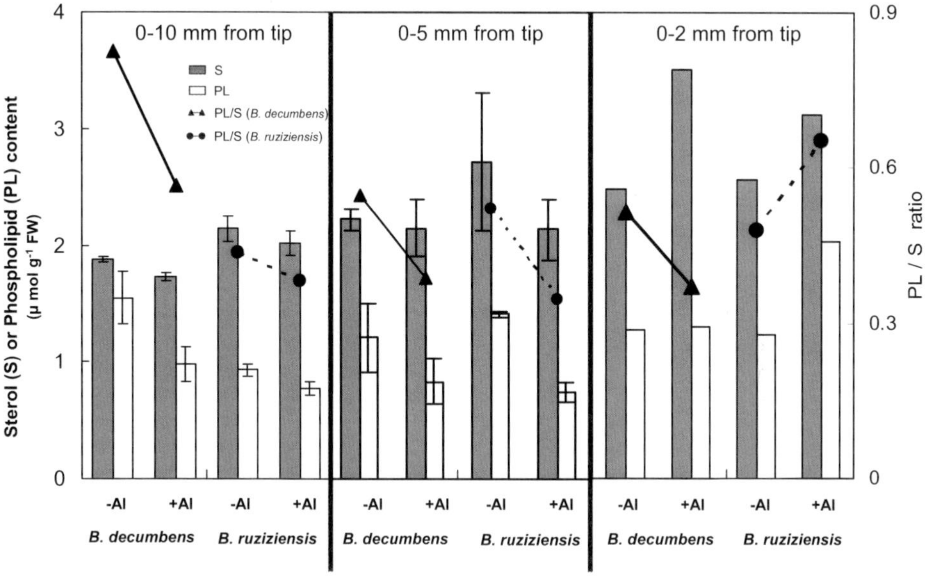

Fig. 5. Sterol or phospholipid content in the different segment of root of brachiariagrasses treated with or without Al in solution

It is known that phenolic compounds can be solubilized into lipid layer, and the lipid layer solubilized with phenolic compounds is transformed into the less fluid layer (Arora et al., 2000; Boija & Johansson, 2006). Higher concentration of phenolic compounds was detected in root-tip portion (0-5 mm and 0-2 mm from root apex) of *B. decumbens* than in *B. ruziziensis* (Fig.6).

The concentration of phenolic compounds was lower in the portion of 0-10 mm from root apex than that of the shorter part from root apex (data not shown). Phenolic compounds have been detected basically in the cell wall, vacuole, and to a small extent in the cytoplasm and nucleus (Hutzler et al., 1998). At around neutral pH of cytosol, the binding affinity to Al ions was significantly higher for phenolic compounds than for organic acids (Ofei-Manu et al., 2001). Higher concentration of phenolic compounds is considered to be more effective for greater detoxification of Al ions in cytosol of *B. decumbens*. Additionally, higher inclusion of phenolic compounds into PM lipid layer may be more favourable for making the PM less permeable in the presence of Al ions, although there are no reports on the inclusion of phenolic compounds in plant lipid layer. Several quantitative and qualitative changes in PM may contribute to superior level of Al resistance in *B. decumbens*. These include: higher proportion of S relative to PL, higher concentration of phenolic compounds in cytosol, and higher inclusion of phenolic compounds in PM lipid layer in root-tip portion. These changes may contribute to an extremely strong PM lipid layer which plays a key role in exclusion of Al and high level of Al resistance in *B. decumbens*. Direct demonstration of the existence of phenolic compounds in PM lipid layer will be an important task for the future research.

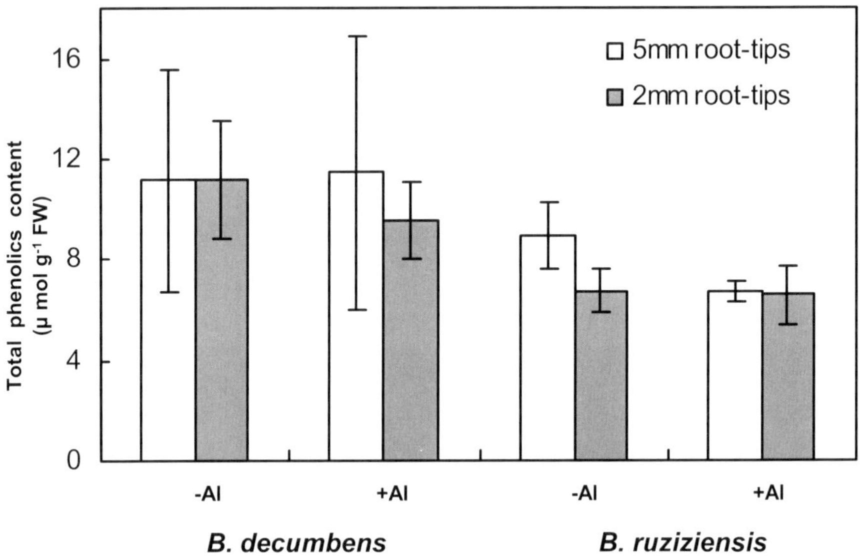

Fig. 6. Total phenolic compounds in root-tips of two brachiariagrasses treated with or without Al in solution

2.1.4 Mechanisms of high level of Al resistance in *B.* hybrid (cv. Mulato)

B. hybrid showed higher resistance to Al similar to *B. decumbens*. When *B.* hybrid seedlings were grown with an extremely high concentration of Al (0.37 mM) for 10 days, no growth inhibition was observed (Fig. 7). Moreover, Al application did not inhibit the uptake of nitrogen (N), P and K in *B.* hybrid. *Andropogon gayanus*, a poaceous pasture grass, is also very resistant to Al and Al application significantly increased Al concentration in both leaf and root of this species (Fig. 8). In *B.* hybrid, by contrast, significant increase in Al accumulation was also observed in root but not in leaf. This indicates that some mechanisms restricting Al translocation from roots to shoots should exist in *B.* hybrid. The ^{27}Al NMR spectrum obtained from intact roots showed several peaks downfield at 10-20 ppm (Fig.9a), suggesting that most of the soluble Al in roots makes complexes presumably with organic acid anions (Fatemi et al., 1992; Kerven et al., 1995). Since the ^{27}Al NMR spectrum did not change after removing soluble and/or loosely bound apoplastic Al, these Al complexes in roots were likely to be localized in the symplast of cells. In many Al-accumulator species, leaves and roots with high concentration of Al are detoxified by organic ligands, such as Al-oxalate in *Melastoma malabathricum* (Watanabe et al., 1998, 2005). The same mechanisms are considered possible in roots of *B.* hybrid. It has been reported that Cd translocation from roots to shoots is restricted by Cd isolation in root vacuoles (Miyadate et al., 2010). Al in the *B.* hybrid may also compartmentalize in root vacuoles and, thus, may not be translocated to shoots.

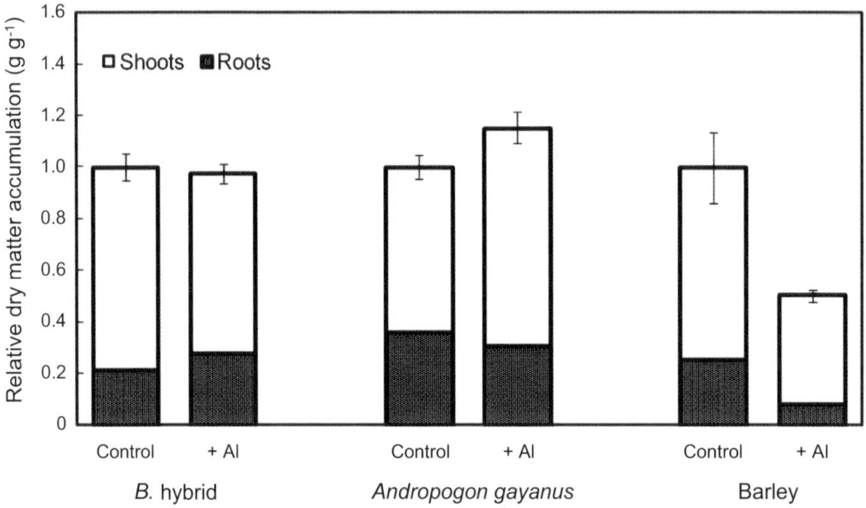

Fig. 7. Effects of Al toxicity on growth of *Brachiaria* hybrid, *Andropogon gayanus*, and barley. Growth was expressed as the relative dry matter accumulation (i.e. [dry weight after treatment - initial dry weight in each treatment]/[dry weight after treatment-initial dry weight in control treatment])

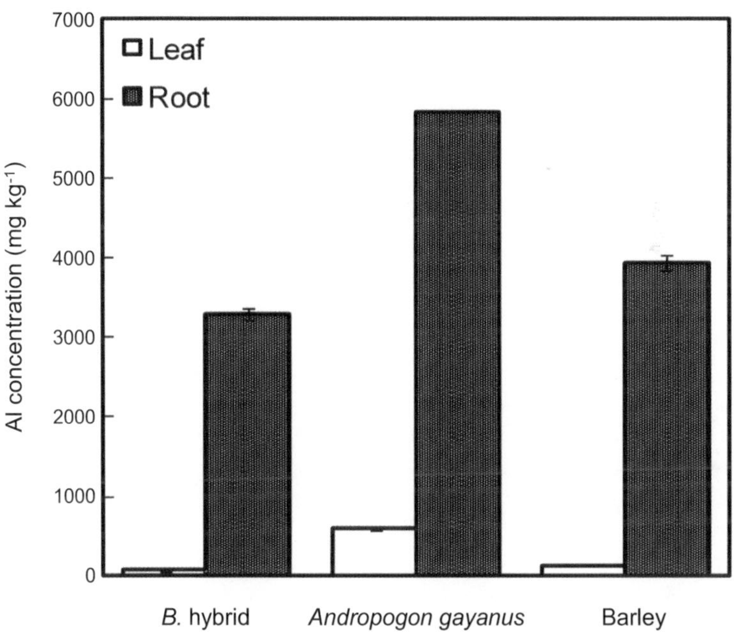

Fig. 8. Concetration of Al in *Brachiaria* hybrid, *Andropogon gayanus*, and barley after the Al treatment

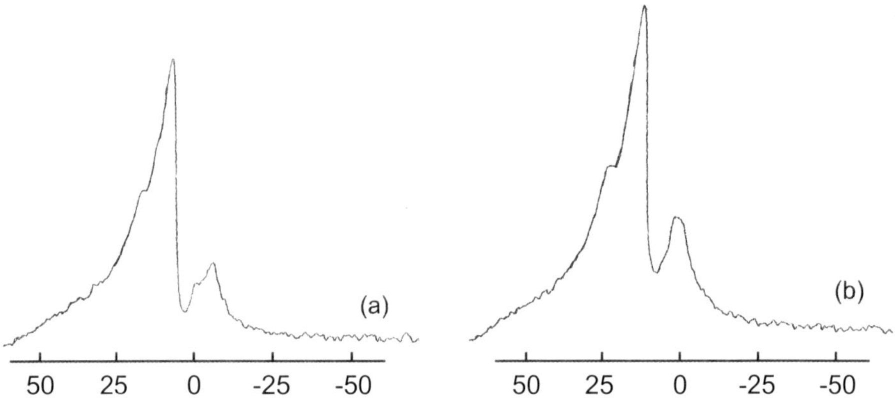

Fig. 9. ^{27}Al NMR spectra in intact roots (a) and roots after removing soluble and/or loosely bound apoplastic Al by 0.1 M HCl (b). $AlCl_3$ (0.1 mM) was used as an external reference to calibrate the chemical shift (0 ppm)

2.2 Mechanisms of low P tolerance in brachiariagrass comparing with those in rice

2.2.1 Low P tolerance of *B.* hybrid

As indicated above, *Brachiaria* species are well adapted to the low-fertility acid soils of the tropics because they are highly tolerant of high Al and low supplies of P and Ca (Louw-Gaume et al., 2010b; Rao et al., 1995, 1996, 2001b; Wenzl et al., 2001). They have lower internal requirements for P than other grasses because they are able not only to acquire P with their extensive root systems but also to use the acquired P more efficiently for growth and metabolism (Rao et al., 1996, 1999). However, mechanisms of P-use efficiency are relatively less known in plants, including *B.* hybrid. Because carbon metabolism is well known to be affected by the P status in plant tissue (Rao, 1996), we studied low-P-tolerance mechanisms, in terms of P recycling and carbon metabolism, in the *B.* hybrid comparing them with those of rice (Nanamori et al., 2004).

B. hybrid and rice plants were cultivated in nutrient solutions with or without 32 μM P. The data obtained on growth parameters and nutrient status are shown in Fig. 10. When P supply in the nutrient solution was low, root:shoot ratio increased, especially in the *B.* hybrid. We found that, for the *B.* hybrid, vigorous root growth is a mechanism for acquiring larger amounts of P from low P conditions. This finding was supported by the high levels of N concentration found in *B.* hybrid roots, while P concentration in *B.* hybrid leaves was significantly lower than that of rice leaves. Lower P concentration in *B.* hybrid leaves may indicate that the *B.* hybrid uses P more efficiently to sustain active metabolism for dry matter production. The P concentration of *B.* hybrid was quite low (0.44 and 0.56 mg-P/gDW in roots and leaves, respectively) and less than rice, which is also known as a low P tolerant plant. Results on the fractionation of P compounds indicated that acid-soluble Pi accounted for about half of the total P in the *B.* hybrid (Fig. 11). Results on the Pi:total P ratio in *B.* hybrid leaves under P deficiency indicate that the *B.* hybrid can survive with extreme low intracellular Pi concentration. This may be due to rapid turnover of other organic P pools under P-deficient conditions.

Chapin and Bieleski (1982) studied the impact of mild P stress on P fractions in relation to plant growth in barley and low-P-adapted barleygrass. They found that barleygrass had a higher proportion of Pi at each level of P supply. They explained this as a consequence of slower growth in barleygrass and higher P status rather than any inherent difference in mechanism. However, in our study, the higher Pi proportion in the *B.* hybrid, compared with that of rice, coincided with lower P concentrations, as explained above. We, therefore, speculate that recycling of internal organic P compounds could be an important mechanism of P-use efficiency in the *B.* hybrid.

Bosse and Köck (1998) have shown activities of APase and RNase were induced during P deficiency, and that this induction is associated with P turnover in plants. In our study, APase and RNase activities were both strongly induced in both rice and *B.* hybrid by P deficiency (Fig. 12). Induction of APase activity was markedly higher in roots under P-deficient conditions. Duff et al. (1994) reported the existence of extracellular APase in roots, where it is localized mainly in apical meristems and outer and surface cells. It is involved in hydrolyzing and mobilizing Pi from organic phosphates in the soil for plant nutrition. The induction of APase in roots may also be associated with excretion. Bosse and Köck (1998) suggested that the increase in activity of phosphohydrolases was a specific response to the decline of cellular available Pi in Pi-starved tomato seedlings. Although Pi in roots was lower than in shoots of both test crops, it was impossible to account for the difference of APase induction between roots and shoots only by the difference in intracellular Pi concentration. Thus, we suggest that some other signal transduction pathway must be operating between roots and shoots against P starvation in the cell. APase activity in shoots was greater in the *B.* hybrid than in rice, suggesting the possibility of rapid P turnover in the *B.* hybrid. This may enable the *B.* hybrid to survive under low P conditions. APase may not be a major mechanism for scavenging or acquiring P because differences in APase induction could not sufficiently account for the diverse growth response of genotypes of both common

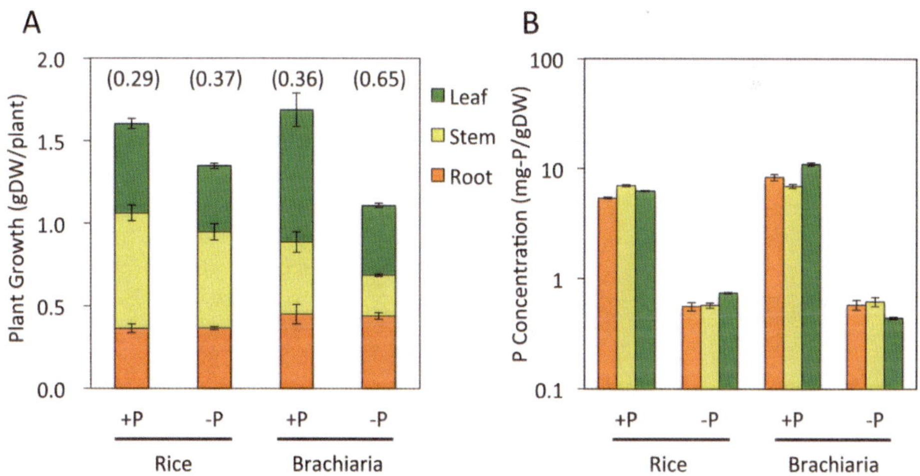

Fig. 10. Growth (A) and P concentration (B) of rice and *Brachiaria* hybrid plants grown under P sufficient and deficient conditions. Error bars indicate S.E. (n = 3)

bean and maize plants under P deficiency (Yan et al., 2001; Yun & Kaeppler, 2001). However, we observed in our study that APase activity was induced by P deficiency and the activity seems to be correlated well with P-use efficiency, as indicated by the lower value of total P concentration, so that the function of APase in adaptation to low-P conditions should not be underestimated. RNase activity was also high in roots under P-deficient conditions (Fig. 12). Nürnberger et al. (1990) and Löffler et al. (1992) showed that both extracellular and intracellular RNase activities were induced in tomato-cell culture under P deficiency. Extracellular RNase could help degrade the RNA from senescing cells that have been either damaged or lysed, and also help degrade any RNA that might be present in the rhizosphere. Thus, the high RNase activity in roots may be associated with secretion similar to APase. RNase activity in shoots was also greater in the *B.* hybrid than in rice, indicating that RNase also contributes to rapid P turnover. Glund et al. (1990) showed that, in the relationship between Pi concentration and RNase activity, induction of RNase under P starvation occurs when the intracellular content of P is very high.

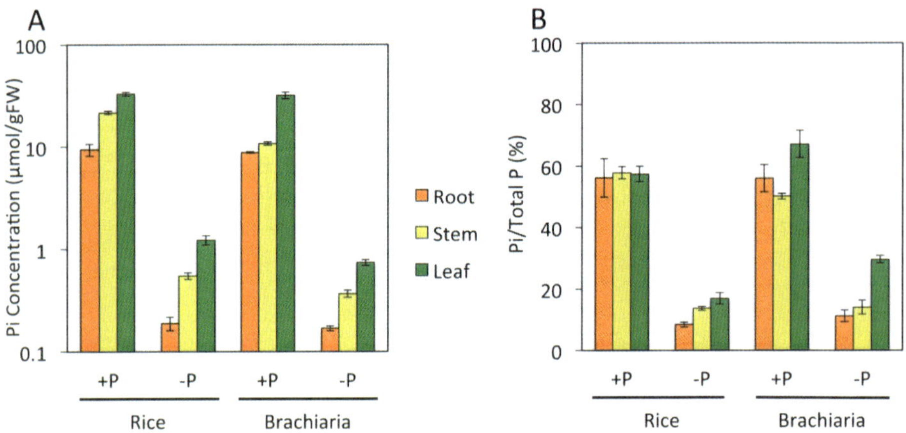

Fig. 11. Pi concentration (A) and ratio of Pi to total P (B) of rice and *Brachiaria* hybrid plants grown under P sufficient and deficient conditions. Error bars indicate S.E. (n= 3)

The above studies indicate that phosphohydrolases, such as APase and RNase, were induced by P deficiency as a P-recycling system. Coinciding with such a mechanism, it is possible that carbon metabolism could also be altered under P deficiency. We therefore studied photosynthate partitioning under P deficiency, tracing photosynthetically fixed ^{14}C in leaves. In rice, photosynthates mainly distributed to sugars, which consist of sucrose, indicating that rice enhanced the sucrose synthesis pathway (Fig. 13). The mRNA accumulation of sucrose phosphate synthase (SPS) also increased as mentioned previously. Hence, sucrose concentration in rice leaves was remarkably high (Fig. 13). The ^{14}C distribution proportion to sugars increased with P deficiency. Enhanced sucrose synthesis in rice leaves through P deficiency may contribute to P recycling because P is liberated during sucrose synthesis (Rao, 1996). However, sucrose catabolism was restricted because the ^{14}C distribution ratio to amino acids and organic acids decreased with P deficiency and with

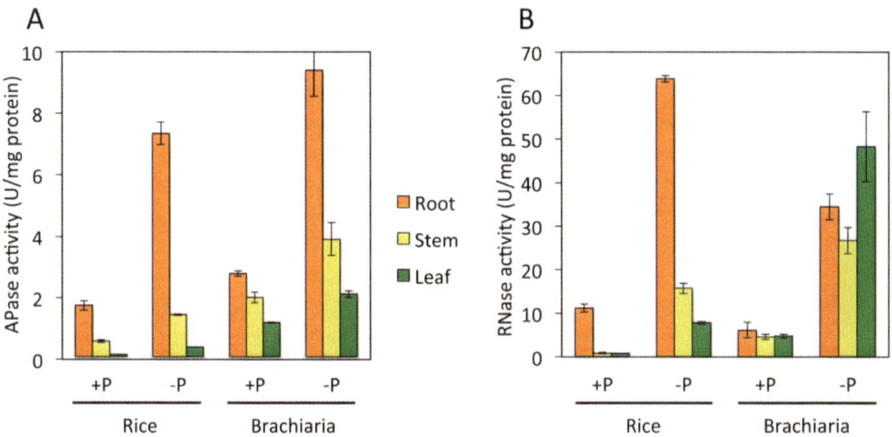

Fig. 12. Acid phosphatase (A) and ribonuclease (B) activities in leaves of rice and *Brachiaria* hybrid grown under P sufficient and deficient conditions. Error bars indicate S.E. (n = 3)

carbohydrate accumulation (Fig. 13). Sucrose synthesis may, therefore, not contribute efficiently to P recycling. However, the ^{14}C distribution proportion to sugars in the *B.* hybrid was not as marked as in rice (Fig. 13), and the effect of P deficiency was smaller. The ^{14}C distribution ratio to amino acids and organic acids in the *B.* hybrid was greater than in rice, and slightly affected by P deficiency. The decrease of total organic acids and carbohydrates in *B.* hybrid leaves under P deficiency suggests that the *B.* hybrid can sustain active amino acid and organic acid pathways with enhanced sugar catabolism, using P efficiently under P deficiency. PK and its bypassing enzymes catalyze the PEP-consuming reaction in leaves, with PEPP activity increasing by a factor of 5.6 to 6.0 with P deficiency. This induction of PEPP is likely to be associated with P recycling, as Duff et al. (1989) suggested. PK was also induced by P deficiency, but not significantly in the *B.* hybrid. PEPC activity was slightly induced by P deficiency in rice but not in the *B.* hybrid. The decrease of PEPC activity in *B.* hybrid leaves would result from reduced net photosynthesis under P deficiency. Kondracka and Rychter (1997) suggest that facilitating the PEP metabolism may be important in view of the P recycle. PEPC and PEPP are considered to function in P recycling as PK-bypass pathways. If these enzyme activities are induced in P recycling, then the carbon flow to the TCA cycle is expected to increase. The ^{14}C distribution ratio to amino acids and organic acids increased slightly in the *B.* hybrid with P deficiency (Fig. 13), indicating that these bypassing enzymes may function to facilitate carbon flow to the TCA cycle. However, in rice, the ^{14}C distribution ratio to amino acids and organic acids decreased with P deficiency. Therefore, the PK bypassing mechanism under P deficiency may not contribute to facilitating the carbon flow to the TCA cycle in rice. In addition to the PK-bypassing mechanism, carbon export from chloroplast to cytosol via the triose-phosphate translocator (TPT) may be a process that significantly affects carbon partitioning under P deficiency (Rao, 1996). When plants are starved for P, triose-P exports from chloroplast to cytosol via TPT, and subsequent sucrose synthesis in the cytosol is likely to be restricted (Rao, 1996). The ^{14}C distribution ratio to sugars and to residue, which mainly consists of sucrose and starch respectively, increased with P deficiency in both rice and *B.* hybrid (Fig. 13), indicating that restriction of triose-P exports from chloroplast to cytosol via TPT may not occur.

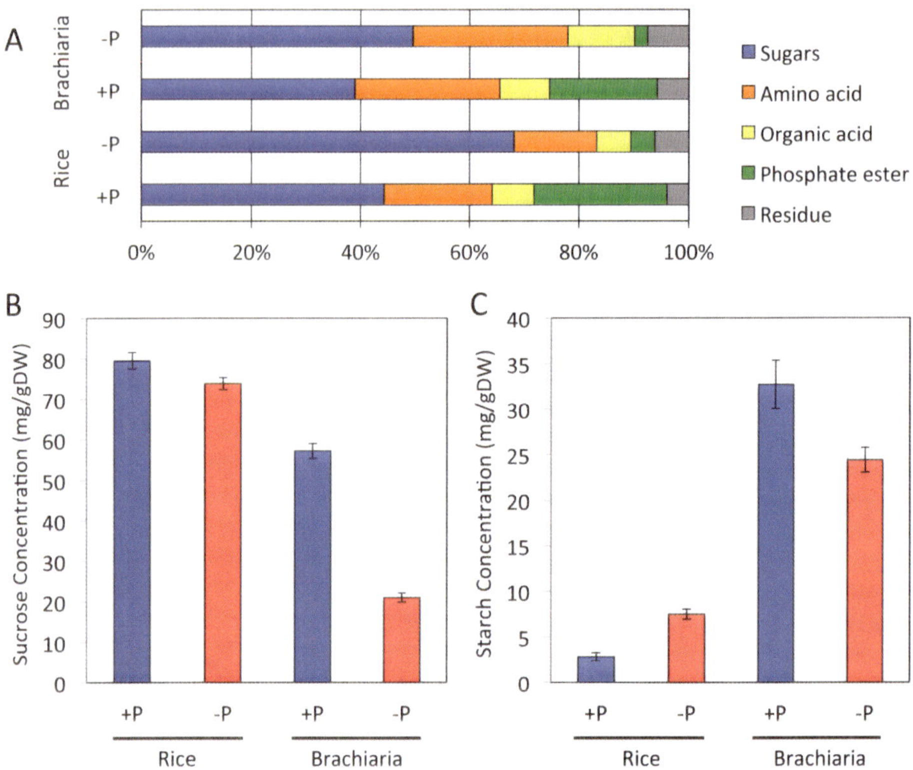

Fig. 13. Photosynthetically assimilated ^{14}C distribution (A), sucrose and starch concentration(B and C, respectively) in leaves of rice and *Brachiaria* hybrid. Error bars indicate S.E. (n = 3)

2.2.2 Transcriptomic analysis of P deficient rice plants

Rice (*Oryza sativa* L. ssp. japonica) plants were germinated and cultured in nutrient solutions containing 0 and 32 μM NaH_2PO_4 for -P and +P treatments, respectively. The seedlings were cultivated for 9 days after transplanting. Total RNA of leaves and roots was used for transcriptomic analyzes by using cDNA arrays (Wasaki et al., 2003b, 2006). As the response of rice roots, there were 15 up-regulated genes in the short-term (24 h) and 86 in the long-term (9 d) treatment with -P, whereas there were 23 and 97 down-regulated genes in the two treatments, respectively. The number of genes regulated (especially down-regulated genes) by the P deficiency was lower in leaves than in roots. There was one up-regulated gene in the short-term (24 h) and 48 genes in the long-term (9 days) -P treatments, whereas there were eight and four down-regulated genes in these two treatments, respectively. None of the genes were regulated in a similar manner between the short and long-term -P leaves. This result suggests that the responses in P-deficient rice leaves are different between short- and long-term treatments, whereas those of roots are relatively similar. *OsPI1* (*Oryza sativa* phosphate-limitation inducible gene 1; Wasaki et al. 2003c), showed the most significant increase in its transcription in the long-term -P treatment, in both the roots and leaves. This gene was classified as a member of *TPSI1/Mt4* family, which is the P-deficient responsive non-coding RNA. The *SqdX*-like gene, a homolog of sulfoquinovosyl diacylglycerol (SQDG) synthesis related genes, was up-regulated significantly in the -P roots. P deficiency enhances dynamic lipid reconstruction and causes SQDG or galactolipids accumulation and expression of a related gene in leaves (Essigmann et al., 1998; Nakamura et al., 2009). Because SQDG has the ability to substitute for PL, it was suggested that the increase of SQDG synthesis is available for the efficient use of P in the membrane (Essigmann et al., 1998). Four genes related to P metabolism were induced in leaves by long-term -P treatment. Inorganic pyrophosphatase and a phosphatase probably contributed to the maintenance of Pi concentration in the tissue by the direct production of Pi from organic phosphate compounds. It was concluded that the function of inorganic pyrophosphatase was common in both roots and leaves, because expression was induced in both organs by long-term -P treatment. Both bi-functional nuclease and S-like RNase expression levels were increased by the -P conditions; their contribution is to produce monomeric nucleotides as substrates for phosphatases (Duff et al., 1994; Green, 1994; Palma et al., 2000).

Many genes involved in polysaccharide metabolisms were up- and down-regulated in leaves by long-term -P and P re-supply treatments, respectively. It is probable that the up-regulation of ADP-glucose pyrophosphorylase, which is a key enzyme of starch synthesis, and starch synthetic enzymes such as starch branching enzyme and starch synthases, induces the accumulation of starch in leaves under -P conditions. In fact, there are many reports of the accumulation of starch in the chloroplasts of P-deficient rice and other plants (Ciereszko & Barbachowska, 2000; Fredeen et al., 1989; Nomura et al., 1995; Qui and Israel, 1992; Rao et al., 1993; Usuda & Shimogawara, 1991). We concluded that the starch accumulation in leaves grown under P-deficient conditions was caused by the disruption of the export of triose phosphate from the stroma by the Pi translocator (Nátr, 1992). Nátr (1992) also noticed the liberation of Pi by the enhancement of starch synthesis. Because starch synthesis and the induction of Pi utilizing enzymes are synchronized, it is a reasonable speculation that the starch accumulation in the P-deficient leaves is a result of the maintenance of the internal Pi concentration.

Fig. 14 shows a summary of metabolic changes based on the regulation of gene expression in the leaves and roots of rice exposed to -P stress. Some important metabolic changes in roots by -P are suggested, namely: (1) acceleration of carbon supply for organic acid synthesis through glycolysis; (2) alteration of lipid metabolism; (3) rearrangement of compounds for cell wall; and (4) changes of gene expression related to the response for metallic elements such as Al, Fe and Zn. The major responses in leaves were involved in internal P utilization. The response in leaves seems to be less dramatic than that in roots; however, it is probable that an important function is regulated in shoots, such as the regulation of the novel *TPSI1/Mt4* gene family (Burleigh & Harrison, 1999), which contains rice *OsPI1* (Wasaki et al., 2003c).

2.2.3 Bypass pathways in rice for P use efficiency in plant

From our previous study using microarray on P deficient rice, we found that several genes relating C and P metabolism in chloroplast changed their expression level. One of them is phosphoenolpyruvate/phosphate translocator (PPT), and it showed enhancement under phosphorus deficient condition. PPT transports PEP into the chloroplast and antiports Pi to cytosol (Hausler et al., 2000), the role of PPT under P deficient condition is considered to supply substrate for the shikimate pathway. There exit another phosphate transporter; triose phosphate translocater (TPT) on chloroplast membrane which loads triose phosphate into cytoplasm and antiports phosphate into chloroplast. These two phosphate translocators are considered to regulate the phosphate level in the chloroplast.

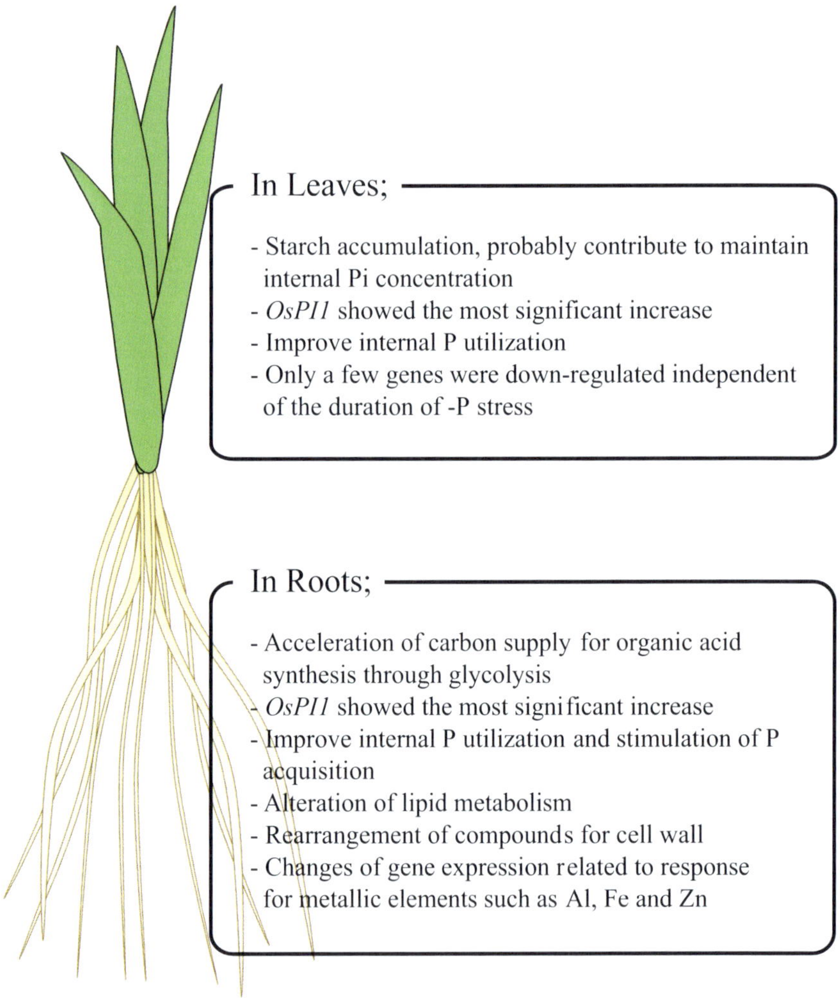

Fig. 14. Summary of plant responses to phosphorus deficiency in shoot and root tissue

As comprehensive analysis of each pathway using intact plant has not reported, we evaluated it by using quantitative real time PCR (qRT-PCR) to determine the expression level of each gene under P deficient condition (Shinano et al., 2005). While the expression level of mRNA is not simply representing the activity of those enzymes corresponding to the gene, obtained information will be very useful to consider plant response to P deficiency.

2.2.4 Gene coding key enzyme of sucrose synthesis

The synthesis of sucrose will liberate phosphate from intermediate compounds, thus it is expected that the level of mRNA for SPS increased with -P treatment. SPS exist in cytosol of mesophyll cell and the combined reaction of SPS and sucrose phosphate phosphatase is main route for sucrose synthesis. That is, during sucrose synthesis, one molecule of Pi is liberated from sucrose phosphate. The –P treated plants first uses Pi stored in vacuole but after they used up all Pi in vacuole, cytosolic Pi content became lower. Then the plants with lower Pi concentration may facilitate sucrose synthesis and excrete Pi from sugar phosphate to keep up the Pi concentration in cytosol of mesophyll cell. Sucrose content in phosphate starved plant varies with species. In common bean and sugar beet, leaf sucrose content increased by P deficiency (Ciereszko & Barbachowska, 2000; Rao et al., 1990), although in leaves of Arabidopsis it decreased. Our results indicate that rice increases sucrose synthesis with P deficiency.

2.2.5 Genes coding candidates for glycolytic bypass enzymes

NADP dependent glyceraldehyde 3-phosphate dehydrogenase (NADP-G3PDH) instead of NAD dependent G3PDH (NAD-G3PDH), and PEPC instead of PK are expected to play alternative pathways to regulate carbon flow under P deficient condition.

In rice leaves under P deprivation, we did not see any increase in relative expression of NADP-G3PDH, which is known as P starvation inducible bypassing enzyme for NAD-G3PDH in *Brassica nigra* (Duff et al., 1989). On the contrary, NAD-G3PDH relative expression was significantly high in –P plants at 21 days. In the level of gene expression argument, this result may suggest NADP-G3PDH is not working as glycolytic bypass in rice plant. The lack of induction of nonphosphorylating pathway was also seen in other plant species, such as *S. minutum* (Theodorou et al., 1991) and *A. brevipes* (Guerrini et al., 2000). Also PEPC and PK have the relationships of glycolytic bypass induced under P deficiency (Li & Ashiharam, 1990). Even though PEPC was thought to be catalyzed with the alternative pathway of PK under P deficiency (Li & Ashiharam, 1990), relative expression of both genes was increased by –P treatment. Increase of PEPC expression by P deficiency is also known in lupin.

2.2.6 Genes encoding chloroplast membrane transporters

Precise value of Pi concentration in cytosol and chloroplast is not known while it is suggested that the value is between 10 to 15 mM in cytosol (Mimura, 1999) and 20 to 35 mM in chloroplast (Diez & Heber, 1984). This indicates that higher requirement for maintaining Pi level in chloroplast rather than in cytosol, and low P condition is expected to increase the level of TPT and *in versa* in PPT. While the expression level of TPT was not changed by P deficient condition, the expression level of PPT increased dramatically. When one molecule of P is transported into chloroplast as PEP while exporting one Pi, the incorporated PEP is decomposed in the chloroplast thereby having no net change in the P level of the chloroplast. We assumed that the role of PPT is increasing the PEP metabolism and makes a cycle from primary photosynthate synthesized in chloroplast and metabolized in cytosol with glycolysis then re-enter chloroplast with PEP then decomposed to release Pi in the chloroplast. From the analysis of rice microarray, PKp (plastid type PK) and shikimate kinase expression were enhanced under P deficiency. These results indicate physiological adaptation to incorporate PEP into chloroplast to support photosynthetic carbon flow and synthesis of secondary metabolic compounds. Recently, another type of phosphate transporter (PHT2; 1, which has high homology with Na^+/Pi symporter of fungi) was reported (Versaw & Harrison, 2002). There is need to evaluate how these transporters are operated to regulate phosphate flux within these subcellular organs.

3. Conclusions

Brachiariagrasses are highly adapted to infertile acid soils, however, the physiological and biochemical mechanisms responsible for their superior adaptation have not yet been fully defined. This chapter summarizes the recent progress towards this objective. Comparative differences in Al resistance among 4 brachiariagrasses and 6 reference plant species were analyzed, and the following order of Al resistance was observed: *B. decumbens*, *B.* hybrid, *B. brizantha* > *B. ruziziensis* > rice (the most Al-resistant cultivar Rikuu-132) > tea (cv. Yabukita) > maize (cv. Pioneer 3352) > pea (cv. Kinusaya) > siclepod > barley (cv. Manriki). The order of Al tolerance in the short-term experiment with exposure to Al (1-h of 50 μ M $AlCl_3$ in 0.2 mM $CaCl_2$ followed by 24-h of Al-free 0.2 mM $CaCl_2$) was well correlated with that in the long-term exposure experiment (2 months of Al treatment with full nutrients) in spite of the differences in the treatment conditions, i.e., duration of Al treatment, Al concentration, composition of co-existing nutrients, and pH. Short-term Al resistance screening technique is accepted to be useful for the evaluation of Al resistance in spite of the simple composition of the treatment solution, considering the positive correlation data obtained formerly among 15 cultivars of sorghum, 10 cultivars of maize, and 8 cultivars of rice. Brachiariagrass showed greater abilities to acquire K, P and Mg, and to tolerate to lower concentration of Ca in shoots in the presence of high concentration of Al in the growing medium including low P at low pH conditions. The level of Al resistance of *B.* hybrid was ranked to be high as comparable to the most Al-resistant *B. decumbens*. It was suggested that the highest Al resistance phenotype of *B.* hybrid may be ascribed to the Al resistance genes from *B. decumbens* or *B. brizantha* but not from *B. ruziziensis*. Extremely high level of Al resistance found in *B. decumbens* was attributed to localized tip portion of less than 2 mm from root apex due to low amount of Al accumulation, low permeability of PM to Al, lower ratio of PL to S, and higher concentration of phenolic compounds in the tip portion of root as compared with other brachiariagrasses. Thus *B. decumbens* is considered to possess multiple physiological and biochemical mechanisms to resist high level of Al in soil solution, and its strategy may be extremely localized in the tip portion of the root apex. *B.* hybrid also exhibited good level of Al resistance. When an extremely high concentration of Al (0.37 mM) was included into the culture solution, significant increase in Al accumulation was observed only in root part. ^{27}Al NMR analysis suggested that the most part of Al in roots was likely to be localized in the cytosol of cells in organically complexed forms and this complexation may inhibit greater upward translocation of Al to shoots, which is beneficial to reduce Al toxicity in shoots.

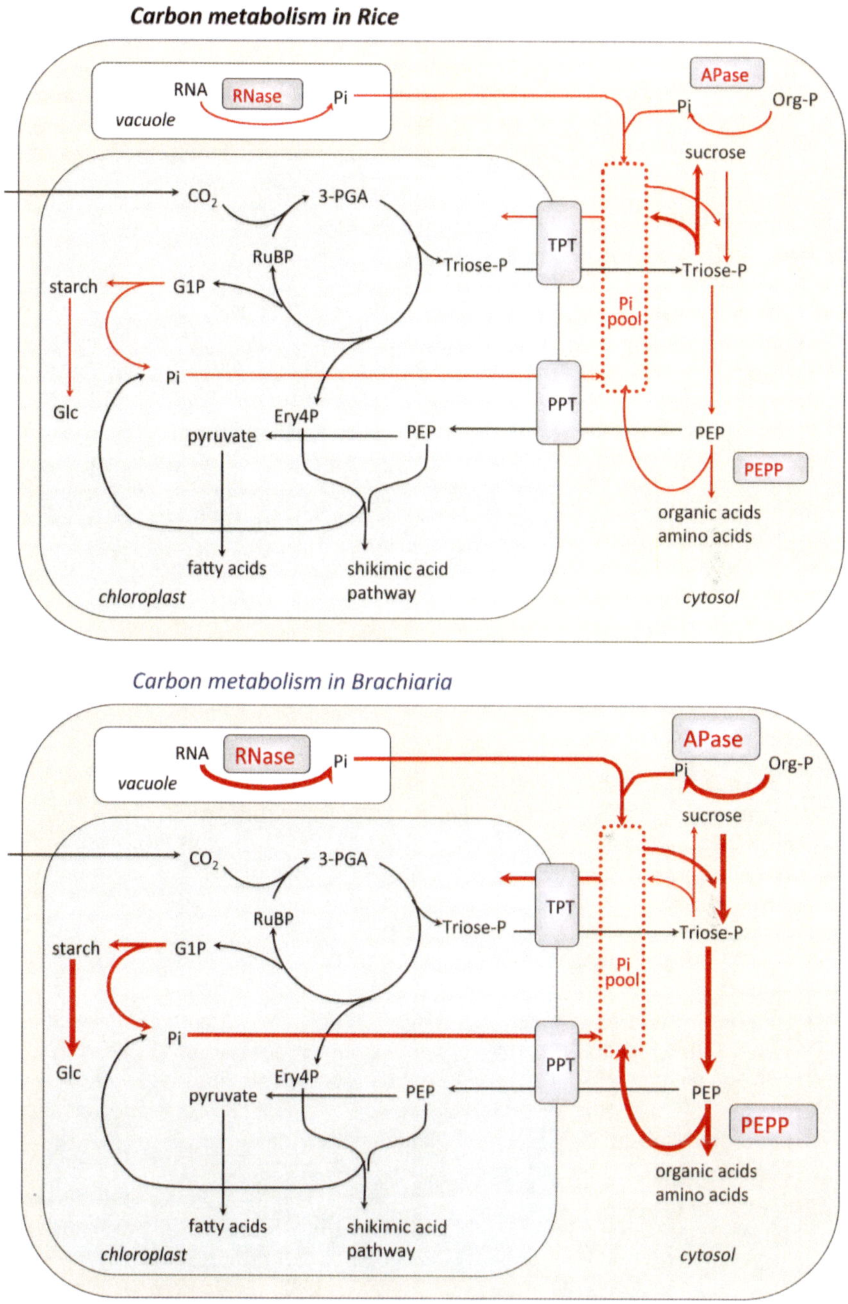

Fig. 15. Summary of metabolic changes in leaves and roots with P deficiency: A. Rice; B *Brachiaria* hybrid. Red arrows indicate P deficiency inducible pathways.

Our study shows that tolerance of low P in both rice and *B.* hybrid involved marked differences in P recycling and carbon metabolism. We summarized the proposed P recycling mechanisms involved in carbon metabolism of rice and *B.* hybrid in Fig. 15. For rice, strategies for low-P tolerance include (1) decreased carbon flow to amino acids and organic acids, and decreased N concentration; and (2) improved partitioning of photosynthates to sucrose, combined with restricted sugar catabolism. For the *B.* hybrid, low-P tolerance involved two major strategies under P deficiency: (1) increasing the ability to use P efficiently by inducing APase and RNase in shoots; and (2) enhancing sugar catabolism and subsequent synthesis of amino acids and organic acids in leaves. Brachiariagrasses also showed greater abilities to acquire K, P and Mg, and to tolerate low concentration of Ca in shoots in the presence of high Al concentration in the growing medium.

In summary, studies on physiological and biochemical mechanism of adaptation of brachiariagrasses grown under simulated conditions of low fertility acid soils indicated their higher level of resistance to Al and tolerance to low supply of P and Ca. This was mainly attributed to their greater ability of Al complexation and Al localization in roots, less upward translocation of Al to shoot tissue, improved P utilization efficiency due to high PPT, and greater acquisition efficiency of K, P and Mg. Identifying the genes responsible for these superior traits of brachiariagrasses is a major objective for future research.

4. Acknowledgment

This work was supported by a Grant-in-Aid for Scientific Research to Wagatsuma T (18208008 and 23380041) from the Japan Society for the Promotion of Science.

5. References

Akhter, A., Khan, M. S. H., Egashira, H., Tawaraya, K., Rao, I. M., Wenzl, P., Ishikawa, S & Wagatsuma, T. (2009). Greater Contribution of Low-nutrient Tolerance to Sorghum and Maize Growth under Combined Stress Conditions with High Aluminum and Low Nutrients in Solution Culture Simulating the Nutrient Status of Tropical Acid Soils. *Soil Science and Plant Nutrition*, Vol. 55, No. 3, (May 2009), pp. 394-406, ISSN 1747-0765

Andrade, S. O., da Sila Lopes, H. O., de Almeida Barros, M., Leite, G. G., Dias, S. M., Saueressig, M., Nobre, D & Temperini, J. A. (1978). Photosensitization in Cattle Grazing on Pasture of *Brachiaria decumbens* Stapf Infested with *Pithomyces chartarum* (Berk. & Curt.) M. B. Ellis. *Arquivos do Instituto Biologico*, Vol.45, No. 2, (April-June 1978), pp. 117-136, ISSN 0020-3653 (In Portuguese with English abstract)

Arora, A., Byrem, T. M., Nair, M. G & Strasburg, G. M. (2000). Modulation of Liposomal Membrane Fluidity by Flavonoids and Isoflavonoids. *Archives of Biochemistry and Biophysics*, Vol.373, No. 1, (Januray 2000), pp. 102-109, ISSN 0003-9861

Boija, E & Johansson, G. (2006). Interactions Between Model Membranes and Lignin-related Compounds Studied by Immobilized Liposome Chromatography. *Biochimica et Biophysica Acta*, Vol. 1758, No. 5, (May 2006), pp. 620-626, ISSN 0005-2736

Boija, E., Lundquist, A., Edwards, K. & Johansson, G. (2007). Evaluation of Bilayer Disks as Plant Cell Membrane Models in Partition Studies. *Analytical Biochemistry*, Vol. 364, No. 2 (May 2007), pp. 145-152, ISSN 0003-2697

Bosse, D. & Köck, M. (1998). Influence of Phosphate Starvation on Phosphohydrolases during Development of Tomato Seedlings. *Plant, Cell and Environment*, Vol.21, No.3, (March 1998), pp. 325-332, ISSN 0140-7791

Burleigh, S. H. & Harrison, M. J. (1999). The Down-regulation of *Mt4*-like Genes by Phosphate Fertilization Occurs Systemically and Involves Phosphate Translocation to the Shoots. *Plant Physiology,* Vol.119, No.1 (January 1999), pp. 241-248, ISSN 0032-0889

Chapin, F. S. III & Bieleski, R. L. (1982). Mild Phosphorus Stress in Barley and Related Low-phosphorus-adapted Barleygrass: Phosphorus Fractions and Phosphate Absorption in Relation to Growth. *Physiologia Plantarum,* Vol.54, No.3, (March 1982), pp. 309-317, ISSN 0031-9317

Ciereszko, I. & Barbachowska, A. (2000). Sucrose Metabolism in Leaves and Roots of Bean (*Phaseolus vulgaris* L.) during Phosphate Deficiency. *Journal of Plant Physiology,* Vol.156, No. 5/6, pp. 640-644, ISSN 0176-1617

Cordell, D., Drangert, J. O. & White, S. (2009). The Story of Phosphorus: Global Food Security and Food for Thought. *Global Environmental Change*, Vol.19, No.2, (May 2009), pp. 292-305, ISSN 0959-3780

Cornard, J. P. & Merlin, J. C. (2002). Specroscopic and Structural Study of Complexes of Quercetin with Al(Ⅲ). *Journal of Inorganic Biochemistry*, Vol. 92, No. 1, (September 2002), pp. 19-27, ISSN 0162-0134

Dietz, K. & Heber, U. (1984). Rate-limiting Factors in Leaf Photosynthesis. I. Carbon Fluxes in the Calvin Cycle. Biochimica et Biophysica Acta, Vol.767, No. 3, (December 1984), pp. 432-443, ISSN 0005-2728

Duff, S. M., Lefebvre, D. D. & Plaxton, W. C. (1989). Purification and Characterization of a Phospho*enol*pyruvate Phosphatase from *Brassica nigra* Suspension Cells. *Plant Physiology,* Vol.90, No. 2, (June 1989), pp. 734-741, ISSN 0032-0889

Duff, S. M. G., Plaxton, W. C. & Lefebvre, D. D. (1991). Phosphate-starvation Response in Plant Cells: *De novo* Synthesis and Degradation of Acid Phosphatases. *Proceedings of the National Academy of Science of the United States of America,* Vol.88, No.21, (November 1991), pp. 9538-9542, ISSN 0027-8424

Duff, S. M. G., Sarath, G. & Plaxton, W. C. (1994). The Role of Acid Phosphatase in Plant Phosphorus Metabolism. *Physiologia Plantarum,* Vol.90, No.4, (April 1994) pp. 791-800, ISSN 0031-9317

Essigmann, B., Güler, S., Narang, R. A., Linke D. & Benning C. (1998). Phosphate Availability Affects the Thylakoid Lipid Composition and the Expression of *SQD1*, a Gene Required for Sulfolipid Biosynthesis in *Arabidopsis thaliana. Proceedings of the National Academy of Sciences of United States of America,* Vol.95, No.5, (March 1998), pp. 1950-1955, ISSN 0027-8424

Fairhust, T., Lefrovy, R., Mutert, E. & Batjes, N. (1999). The Importance, Distribution and Causes of Phosphorus Deficiency as a Constraint to Crop Production in the Tropics. *Agroforestry Forum*, Vol.9, No.1, pp. 2-8, ISSN 0966-8616

Fatemi, S. J. A., Williamson, D. J. & Moore, G., R. (1992). A ^{27}Al NMR Investigation of Al^{3+} Binding to Small Carboxylic Acids and the Proteins Albumin and Transferrin. *Journal of Inorganic Biochemistry*, Vol. 46, No. 1, (April 1992) pp. 35-40, ISSN 0162-0134

Fredeen, A. L., Rao, I. M. & Terry, N. (1989). Influence of Phosphorus Nutrition on Growth and Carbon Partitioning in *Glycine max*. *Plant Physiology*, Vol. 89, No. 1, (January 1989), pp. 225-230, ISSN 0032-0889

Glund, K., Nürnberger, T., Abel, S., Jost, W., Preisser, J. & Komor, E. (1990). Intracellular Pi-compartmentation during Phosphate Starvation-triggered Induction of an Extracellular Ribonuclease in Tomato Cell Culture. In: *Progress in Plant Cellular and Molecular Biology*. H.J.J. Nijkamp, L.W.H. Van Der Plas, & J. Van Aartrijk, (Eds.), 338–342, Kluwer Academic Publishers, 978-0-7923-0873-7, Boston

Green, P. J. (1994). The Ribonuclease of Higher Plants. *Annual Review of Plant Physiology*, Vol.45, (Jun 1994), pp. 421-425, ISSN 0066-4294

Guerrini, F., Cangini, M. & Boni, L. (2000). Metabolic Responces of the Diatom *Achanthes brevipes* (Bacillariophyceae) to Nutrient Limitation. *Journal of Phycology*, Vol.36, Nol. 5, (October 2000), pp. 882-890, ISSN 0022-3646

Hammond, J. P., Bennett, M.J., Bowen, H.C., Broadley, M.R., Eastwood, D.C., May, S.T., Rahn, C., Swarup, R., Woolaway, K. E. & White , P. J. (2003). Changes in Gene Expression in *Arabidopsis* Shoots during Phosphate Starvation and the Potential for Developing Smart Plants. *Plant Physiology*, Vol. 132, No. 2, (June 2003), pp. 578-596, ISSN 0032-0889

Hausler, R. E., Baur, B., Scharte, J., Teichmann, T., Eicks, M., Fischer, K. L., Flugge, U. I., Schubert, S., Weber, A. & Fischer, K. (2000). Plastidic Metabooolite Transporters and Their Physiological Functios in the Inducible Crassulacean Acid Metabolism Plant *Messembryanthemum crystallinum*. *The Plant Journal*, Vol.24, No.3, (November 2000), pp. 285-296, ISSN 0960-7412.

Howard, C. J., LeBrasseur, N. D., Bariola, P. A. & Pamela, J. G. (1998). Control of Ribonuclease in Response to Phosphate Limitation: Induction of RNS1 in Arabidopsis. In: *Phosphorus in Plant Biology: Regulatory Roles in Molecular, Cellular, Organic, and Ecosystem Processes*, J.P. Lynch & J. Deikman (Eds.), American Society of Plant Physiology, ISBN 094-3088-38-0, Rockville, MD, U.S.A.

Huang, C. F., Yamaji, N., Mitani, N., Yano, M., Nagamura, Y. & Ma, J. F. (2009). A Bacterial-type ABC Transporter is Involved in Aluminum Tolerance in Rice. *The Plant Cell*, Vol.21, No. 2, (Feburuary 2009), pp. 655-667, ISSN 1040-4651

Hutzler, P., Fischbach, R., Heller, W., Jungblut, T. P., Reuber, S., Schmitz, R., Veit, M., Weissenbock, G & Schnitzler, J-P. (1998). Tissue Localization of Phenolic Compounds in Plants by Confocal Laser Scanning Microscopy. *Journal of Experimental Botany*, Vol. 49, No. 323, pp. 953-965, ISSN 0022-0957

Ishigaki, G. (2010). Studies on Establishment of Tissue Culture System and its Utilization for Breeding in Ruzigrass (*Brachiaria ruziziensis*). (in Japanese with English Tables, Figures, and Summary). Dissertation thesis, University of Miyazaki, Miyazaki, Japan, Retrieved from http://hdl.handle.net/10458/2755

Ishikawa, S. & Wagatsuma, T. (1998). Plasma Membrane Permeability of Root-tip Cells Following Temporary Exposure to Al Ions is a Rapid Measure of Al Tolerance among Plant Species. *Plant and Cell Physiology*, Vol. 39, No. 5, (May 1998), pp. 516-525, ISSN 0032-0781

Ishikawa, S., Wagatsuma, T., Sasaki, R. & Ofei-Manu, P. (2000). Comparison of the Amount of Citric and Malic Acids in Al Media of Seven Plant Species and Two Cultivars

Each in Five Plant Species. *Soil Science and Plant Nutrition*, Vol. 46, No. 3, (September 2000), pp. 751-758, ISSN 1747-0765

Ishikawa, S., Wagatsuma, T., Takano, T., Tawaraya, K. & Oomata, K. (2001). The Plasma Membrane Intactness of Root-tip Cells is a Primary Factor for Al Tolerance in Cultivars of Five Plant Species. *Soil Science and Plant Nutrition*, Vol. 47, No. 3, (September 2001), pp. 489-501, ISSN 1747-0765

Jones, D. L. (1998). Organic Acids in the Rhizosphere – A Critical Review. *Plant and Soil*, Vol. 205, No. 1, (August 1998), pp. 25-44, ISSN 0032-079X

Kerven, G. L., Larsen, P. L., Bell, L. C. & Edwards, D. G. (1995). Quantitative ^{27}Al NMR Spectroscopic Studies of Al(III) Complexes with Organic Acid Ligand and Their Comparison with GEOCHEM Predicted Values. *Plant and Soil*, Vol. 171, No. 1, pp. 35-39, ISSN 0032-079X

Khan, M. S. H., Tawaraya, K., Sekimoto, H., Koyama, H., Kobayashi, Y., Murayama, T., Chuba, M., Kambayashi, M., Shiono, Y., Uemura, M., Ishikawa, S. & Wagatsuma, T. (2009). Relative Abundance of Δ^5-sterols in Plasma Membrane Lipids of Root-tip Cells Correlates with Aluminum Tolerance of Rice. *Physiologia Plantarum*, Vol. 135, No. 1, (January 2009), pp. 73-83, ISSN 0031-9317

Kinraide, T. (1999). Interactions Among Ca^{2+}, Na^{+} and K^{+}in Salinity Toxicity : Quantitative Resolution of Multiple Toxic and Ameliorative Effects. *Journal of Experimental Botany*, Vol. 50, No. 338, (September 1999), pp. 1495-1505, ISSN 0022-0957

Kondracka, A. & Rychter, A. M. (1997). The Role of Pi Recycling Processes during Photosynthesis in Phosphate-deficient Bean Plants. *Journal of Experimental Botany*, Vol. 48, No. 7, (July 1997), pp. 1461-1468, ISSN 0022-0957

Li, X. N. & Ashiharam H. (1990). Effects of Inorganic Phosphate on Sugar Catabolism by Suspension-cultured *Catharanthus roseus*. *Phytochemistry*, Vol. 29, No. 2, (February 1990), pp. 497-500, ISSN 0031-9422

Löffler, A., Abel, S., Jost, W., Beintema, JJ. & Glund, K. (1992). Phosphate-regulated Induction of Intracellular Ribonulease in Cultured Tomato (*Lycopersicon esculentum*) Cells. *Plant Physiology*, Vol.98, No.4, (April 1992), pp. 1472-1478, ISSN 0032-0889

Louw-Gaume, A. E., Rao, I. M., Gaume, A. J. & Frossard, E. (2010a). A Comparative Study on Plant Growth and Root Plasticity Responses of Two *Brachiaria* Forage Grasses Grown in Nutrient Solution at Low and High Phosphorus Supply. *Plant and Soil*, Vol. 328, No. 1-2, (March 2010), pp.155-164, ISSN 0032-079X

Louw-Gaume, A., Rao, I. M., Frossard, E. & Gaume, A. (2010b). Adaptive Strategies of Tropical Forage Grasses to Low Phosphorus Stress: The Case of Brachiariagrasses. In: *Handbook of Plant and Crop Stress. Third Edition*, M. Pessarakli, Ed., Taylor & Francis Group, pp. 1111-1144, ISBN 1439813965, USA, Boca Raton

Lynch, J.P. (2011). Root Phenes for Enhanced Soil Exploration and Phosphorus Acquisition: Tools for Future Crops. *Plant Physiology*, Vol. 156, No. 3, (July 2011), pp. 1041-1049, ISSN 0032-0889

Ma, J. F., Shen, R., Zhao, Z., Mathias, W., Takeuchi, Y., Ebitani, T & Yano, M. (2002). Response of Rice to Al Stress and Identification of Quantitative Trait Loci for Al Tolerance. *Plant and Cell Physiology*, Vol. 43, No. 6, (Jun 2002), pp. 652-659, ISSN 0032-0781

Macedo, M. A. M. (2005). Pastagens no Ecosistema Cerrados: Evalucao das Pesquisas Para o Desenvolvimento Sustentavel. *Reuniao Anual da Siciedade Brasileira de Zootecnia,* Vol. 41, pp. 56-84. UFGO, SBZ, Goiania.

Miles, J.W.; do Valle, C.B.; Rao, I.M. & Euclides, V.P.B. (2004). Brachiariagrasses. In: *Warm-season (C4) grasses* (L. Moser, B. Burson & L.E. Sollenberger, Eds.). ASA-CSSA-SSSA, Madison, WI, USA, pp. 745-783, ISSN 0011-183X

Miles, J. W., Cardona, C. & Sotelo, G. (2006). Recurrent Selection in a Synthetic Brachiariagrass Population Improves Resistance to Three Spittlebug Species. *Crop Science,* Vol. 46, No. 3, (June 2006), pp. 1088-1093, ISSN 0011-183X

Mimura, T. (1999). Regulation of Phosphate Transport and Homeostasis in Plant Cells. In: *International Review of Cytology,* K.W. Jeon (Ed.), Vol. 191, pp. 149-200. ISBN 978-0123-645-95-1

Misson, J., Raghothama, K. G., Jain, A., Jouhet, J., Block, M. A., Bligny, R., Ortet, P., Creff, A., Somerville, S., Rolland, N., Doumas, P., Nacry, P., Herrerra-Estrella, L., Nussaume, L. & Thibaud, M. C. (2005). *Arabidopsis thaliana* Affymetrix Gene Chips Determined Plant Responses to Phosphate Deprivation. *Proceedings of the National Academy of Sciences of USA,* Vol.102, No.33, (August 2005), pp. 11934-11939, ISSN 0027-8424

Miyadate, H., Adachi, S., Hiraizumi, A., Tezuka, K., Nakazawa, N., Kawamoto, T., Katou, K., Kodama, I., Sakurai, K., Takahashi, H., Satoh-Nagasawa, N., Watanabe, A.,Fujimura, T. & Akagi, H. (2010). OsHMA3, P1B-type of ATPase Affects Root-to-shoot Cadmium Translocation in Rice by Mediating Efflux into Vacuoles. *New Phytologist,* Vol. 189, No. 1, (January 2010), pp. 190-199, ISSN 0028-646X

Nakamura, Y., Koizumi, R., Shui, G., Shimojima, M., Wenk, M.R., Ito, T. & Ohta, H. (2009). Arabidopsis Lipins Mediate Eukaryotic Pathway of Lipid Metabolism and Cope Critically with Phosphate Starvation. *Proceedings of the National Academy of Sciences of USA,* Vol. 106, No. 49, (December 2009), pp. 20978-20983, ISSN 0027-8424

Nanamori, M., Shinano, T., Wasaki, J., Yamamura, T., Rao, I.M. & Osaki, M. (2004). Low Phosphorus Tolerance Mechanisms: Phosphorus Recycling and Photosynthate Partitioning in the Tropical Forage Grass, *Brachiaria* Hybrid cultivar Mulato Compared with Rice. *Plant and Cell Physiology,* Vol. 45, No. 4, (April 2004), pp. 460-469, ISSN 0032-0781

Nátr, L. (1992). Mineral Nutrients - A Ubiquitous Stress Factor for Photosynthesis. *Photosynthetica,* Vol.27, No. 3, (September 1992), pp. 271-294, ISSN 0300-3604

Nomura, M., Imai, K. & Matsuda, T. (1995). Effects of Atmospheric Partial Pressure of Carbon Dioxide and Phosphorus Nutrition on the Ultrastructure of Rice (*Oryza sativa* L.) Chloroplasts. *Japanese Journal of Crop Science,* Vol.64, No. 4, pp. 784-793, ISSN 0011-1848

Nürnberger, T., Abel, S., Jost, W. & Glund, K. (1990). Induction of An Extracellular Ribonulease in Cultured Tomato Cells upon Phosphate Starvation. *Plant Physiology,* Vol. 92, No. 4, (April 1990), pp. 970-976, , ISSN 0032-0889

Ofei-Manu, P., Wagatsuma, T., Ishikawa, S. & Tawaraya, K. (2001). The Plasma Membrane Strength of the Root-tip Cells and Root Phenolic Compounds are Correlated with Al Tolerance in Several Common Woody Plants. *Soil Science and Plant Nutrition,* Vol. 47, No. 2, (June 2001), pp. 359-375, ISSN 1747-0765

Palma, D. A., Blumwald, E. & Plaxton, W. C. (2000). Upregulation of Vacuolar H^+-translocating Pyrophosphatase by Phosphate Starvation of *Brassica napus*

(Rapeseed) Suspension Cell Cultures. *FEBS Letter*, Vol. 486, No. 2, (December 2000), pp. 155-158, ISSN 0014-5793

Qui, J. & Israel, D. W. (1992). Diurnal Starch Accumulation and Utilization in Phosphorus-deficient Soybean Plants. *Plant Physiology*, Vol.98, No.1, (January 1992), pp. 316-323., ISSN 0032-0889

Ramaekers, L., Remans, R., Rao, I. M., Blair, M. W. & Vanderleyden, J. (2010). Strategies for Improving Phosphorus Acquisition Efficiency of Crop Plants. *Field Crops Research*, Vol. 117, No. 2-3, (June 2010), pp. 169-175.

Rao, I. M., Fredeen, A. L. & Terry, N. (1990). Leaf Phosphate Status, Photosynthesis and Carbon Partitioning in Sugar Beet. III. Diurnal Changes in Carbon Partitioning and Export. *Plant Physiology*, Vol. 92, No.1, (January 1990), pp. 29-36 , ISSN 0032-0889

Rao, I. M. (1996). The Role of Phosphorus in Photosynthesis. In : *Handbook of Photosynthesis*, M. Pessarakli, (Ed.), pp. 173-194, Marcel Dekker, New York

Rao, I. M. (2001a). Adapting Tropical Forages to Low-fertility Soils. In: *Proceedings of the XIX International Grassland Congress*. (J.A. Gomide.; W.R.S. Mattos. & S.C. da Silva. Eds.), Brazilian Society of Animal Husbandry, Piracicaba, Brazil, pp. 247-254, ISBN 857-1330-1-7

Rao, I. M. (2001b). Role of Physiology in Improving Crop Adaptation to Abiotic Stresses in the Tropics: The Case of Common Bean and Tropical Forages. In: *Handbook of Plant and Crop Physiology*, M. Pessarakli Ed., pp. 583-613, Marcel Dekker, ISBN 0824705467, Inc., New York, USA

Rao, I. M., Ayarza, M. A. & Garcia, R. (1995). Adaptive Attributes of Tropical Forage Species to Acid Soils. I . Differences in Plant Growth, Nutrient Acquisition and Nutrient Utilization among C_4 Grasses and C_3 Legumes. *Journal of Plant Nutrition*, Vol. 18, No. 10, (October 1995), pp. 2135-2155, ISSN 0176-1617

Rao, I. M., Friesen, D. K. & Osaki, M. (1999). Plant Adaptation to Phosphorus-limited Tropical Soils. In: *Handbook of Plant and Crop Stress*. (M. Pessarakli , Ed.), pp. 61-96, Marcel Dekker, Inc., ISBN 0824719484, New York, USA.

Rao, I. M.; Kerridge, P. C. & Macedo, M. C. M. (1996). Nutritional Requirements of *Brachiaria* and Adaptation to Acid Soils. In: *Brachiaria*: Biology, Agronomy, and Improvement (J.W. Miles., B. L. Maass. & C. B. Valle, Eds.), Centro Internacional de Agricultura Tropical, Cali, Colombia, pp. 53-71, ISBN 958-9439-57-8

Rao, I. M., Miles, J. W & Granobles, J. C. (1998). Differences in Tolerance to Infertile Acid Soil Stress Among Germplasm Accessions and Genetic Recombinants of the Tropical Forage Grass Genus, *Brachiaria*. *Field Crops Reseach*, Vol. 59, No. 1, (October 1998), pp. 43-52, ISSN 0378-4290

Rao, I. M., Zeigler, R. S., Vera, R. & Sarkarung, S. (1993). Selection and Breeding for Acid-soil Tolerance in Crops: Upland Rice and Tropical Forages as Case Studies. BioScience, Vol. 43, No. 7, (July 1993), pp. 454-465, ISSN 0006-3658

Ryan, P. R., Tyerman, S. D., Sasaki, T., Furuichi, T., Yamamoto, Y., Zhang, W. H. & Delhaize, E. (2011). The Identification of Aluminium-resistance Genes Provides Oppotunities for Enhancing Crop Production on Acid Soils. *Journal of Experimental Botany*, Vol. 62, No. 1, (January 2011), pp. 9-20, ISSN 0022-0957

Shinano, T., Nanamori, M., Dohi, M., Wasaki, J. & Osaki, M. (2005). Evaluation of Phosphorus Starvation Inducible Genes Relating to Efficient Phosphorus

Utilization in Rice. *Plant and Soil*, Vol. 269, No. 1-2, (February 2005), pp. 81-87, ISSN 1573-5036

Steen, I. (1998). Phosphorus Availability in the 21st Centrury. *Phosphorus &Potassium*, Vol. 217, (September-October 1998), ISSN 0031-8426

Stoutjesdijk, P. A., Sale, P. W. & Larkin, P. J. (2001). Possible Involvement of Condensed Tannins in Aluminium tolerance of *Lotus pedunculatus*. *Australian Journal of Plant Physiology*, Vol. 28, No. 11, (November 2001), pp. 1063-1074, ISSN 0310-7841

Tadano, T., Ozawa, K., Sakai, H., Osaki, M. & Matsui, H. (1993). Secretion of Acid Phosphatase by the Roots of Crop Plants under Phosphorus-deficient Conditions and Some Properties of the Enzyme Secreted by Lupin Roots. *Plant and Soil*, Vol. 155/156, No. 1, (October 1993), pp. 95-98, ISSN 1573-5036

Theodrou, M.E., Elrifi, I.R., Turpin, D.H. & Plaxton, W.C. (1991). Effect of Phosphorus Limitation on Respiratory Metabolism in the Green Alga *Selenastrum minutum*. *Plant Physiology*, Vol. 95, No. 4, (April 1991), pp. 1089-1095. , ISSN 0032-0889

Theodorou, M. E. & Plaxton, W. C. (1993). Metabolic Adaptations of Plant Respiration to Nutritional Phosphate Deprivation. *Plant Physiology*, Vol. 101, No.2, (February 1993), pp. 339-344, ISSN 0032-0889

Uhde-Stone, C., Zinn, K. E., Ramirez-Yáñez, M., Li, A., Vance, C. P. & Allan, D. L. (2003b). Nylon Filter Arrays Reveal Differential Gene Expression in Proteoid Roots of White Lupin in Response to Phosphorus Deficiency. *Plant Physiology*, Vol. 131, No. 3, (March 2003), pp. 1064-1079, ISSN 0032-0889

Usuda, H. & Shimogawara, K. (1991). Phosphate Deficiency in Maize. II. Enzyme Activities. *Plant and Cell Physiology*, Vol. 32, No. 8, (December 1991), pp.1313-1317, ISSN 0032-0781

Versaw, W. K. & Harrison, M. I. (2002) A Chloroplast Phosphate Transporter, PHT2; 1, Influences Allocation of Phosphate within the Plant and Phosphate-starvation Responses. *The Plant Cell*, Vol. 14, No. 8, (August 2002), pp. 1751-1766, ISSN 1040-4651

Wagatsuma, T., Ishikawa, S., Obata, H., Tawaraya, K. & Katohda, S. (1995). Plasma Membrane of Younger and Outer Cells is the Primary Specific Site for Aluminum Toxicity in Roots. *Plant and Soil*, Vol. 171, No. 1, (April 1995), pp. 105-112, ISSN 0032-079X

Wagatsuma, T., Ishikawa, S., Uemura, M., Mitsuhashi, W., Kawamura, T., Khan, M. S. H. & Tawaraya, K. (2005). Plasma Membrane Lipids are the Powerful Components for Early Stage Aluminum Tolerance in Triticale. *Soil Science and Plant Nutrition*, Vol. 51, No. 5, pp. 701-704, (September 2005), ISSN 1747-0765

Wang, Y. -H., Gravin, D. F. & Kochian, L. V. (2002). Rapid Induction of Regulatory and Transporter Genes in Response to Phosphorus, Potassium, and Iron Deficiencies in Tomato Roots. Evidence for Cross Talk and Root/Rhizosphere-mediated Signals. *Plant Physiology*, Vol. 130, No. 3, (November 2002), pp. 1361-1370, ISSN 0032-0889

Wasaki, J., Shinano, T., Onishi, K., Yonetani, R., Yazaki, J., Fujii, F., Shimbo, K., Ishikawa, M., Shimatani, Z., Nagata, Y., Hashimoto, A., Ohta, T., Sato, Y., Miyamoto, C., Honda, S., Kojima, K., Sasaki, T., Kishimoto, N., Kikuchi, S. & Osaki, M. (2006). Transcriptomic Analysis Indicates Putative Metabolic Changes Caused by Manipulation of Phosphorus Availability in Rice Leaves. *Journal of Experimental Botany*, Vol. 57, No. 9, (September 2006), pp. 2049-2059, ISSN 0022-0957

Wasaki, J., Yamamura, T., Shinano, T. & Osaki, M. (2003a). Secreted Acid Phosphatase is Expressed in Cluster Roots of Lupin in Response to Phosphorus Deficiency. *Plant and Soil*, Vol. 248, No. 1-2, (January 2003), pp. 129-136, ISSN 1573-5036

Wasaki, J., Yonetani, R., Kuroda, S., Shinano, T., Yazaki, j., Fujii, F., Shimbo, K., Yamamoto, K., Sakata, K., Sasaki, T., Kishimoto, N., Kikuchi, S., Yamagishi, M. & Osaki, M. (2003b). Transcriptomic Analysis of Metabolic Changes by Phosphorus Stress in Rice Plant Roots. *Plant, Cell and Environment*, Vol. 26, No. 9, (September 2003), pp. 1515-1523, ISSN 0140-7791

Wsaki, J., Yonetani, R., Shinano, T., Kai, M. & Osaki, M. (2003c). Expression of the *OSPI1* Gene, Cloned from Rice Roots Using cDNA Microarray, Rapidly Responds to Phosphorus Status. *New Phytologist*, Vol. 158, No. 2, (May 2003), pp. 239-248, ISSN 0028-646x

Watanabe, T., Misawa, S. & Osaki, M. (2005). Aluminum Accumulation in the Roots of *Melastoma malabathricum* L., an Aluminum-accumulating Plant. *Canadian Journal of Botany*, Vol. 83, No. 11, (Nobember 2005), pp. 1518-1522, ISSN 0008-4026

Watanabe, T., Osaki, M., Yoshihara, T. & Tadano, T. (1998). Distribution and Chemical Speciation of Aluminum in the Al Accumulator Plant, *Melastoma marabathricum* L. *Plant and Soil*, Vol. 201, No. 2, pp. 165-173, ISSN 0032-079X

Wenzl, P., Mancilla, L. I., Mayer, J. E., Albert, R. & Rao, I. M. (2003). Simulating Infertile Acid Soils with Nutrient Solutions : The Effects on *Brachiaria* species. *Soil Science Society of America Journal*, Vol. 67, No. 5, (September-October 2003), pp. 1457-1469, ISSN 1747-0765

Wenzl, P., Patino, G. M., Chaves, A. L., Mayer, J. E. & Rao, I. M. (2001). The High Level of Aluminum Resistance in Signalgrass is not Associated with Known Mechanisms of External Detoxification in Root Apices. *Plant Physiology*, Vol. 125, No. 3, (March 2001), pp. 1473-1484, ISSN 0032-0889

Wu, P., Ma, L., Hou, X., Wang, M., Wu, Y., Liu, F. & Deng, X. W. (2003). Phosphate Starvation Triggers Distinct Alterations of Genome Expression in *Arabidopsis* Roots and Leaves. *Plant Physiology*, Vol. 132, No. 3, (July 2003), pp. 1260-1271, ISSN 0032-0889

Yamaji, N., Huang, C. F., Nagao, S., Yano, M., Sato, Y., Nagamura, y. & Ma, J. F. (2009). A Zinc Finger Transcription Factor ART 1 Regulates Multiple Genes Implicated in Aluminum Tolerance in Rice. The Plant Cell, Vol. 21, No. 10, (October 2009), pp. 3339-3349, ISSN 1040-4651

Yan, X., Liao, H., Trull, M. C., Beebe, S. E. & Lynch, J. P. (2001). Induction of a Major Leaf Acid Phosphatase Does not Confer Adaptation to Low Phosphorus Availability in Common Bean. *Plant Physiology*, Vol. 125, No. 4, (April 2001), pp. 1901-1911, ISSN 0032-0889

Yang, J. L., Li, Y. Y., Zhang, Y. J., Zhang, S. S., Wu, Y. R., Wu, P., Zheng, S. J. (2008). Cell Wall Polysaccharides Are Specifically Involved in the Exclusion of Aluminum from the Rice Root Apex. *Plant Physiology*, Vol. 146, No. 2, (February 2008), pp. 602-611, ISSN 0032-0889

Yang, Z-B., Eticha, D., Rao, I. M. & Horst, W. J. (2010). Alteration of Cell-wall Porosity is Involved in Osmotic Stress-induced Enhancement of Aluminium Resistance in Common Bean (*Phaseorus velgaris* L.). *Journal of Experimental Botany*, Vol. 61, (July 2010), pp. 3245-3258, ISSN 0022-0957

Yoneda, S. & Nakatsubo, F. (1998). Effect of the Hydroxylation of Patterns and Degrees of Polymerization of Condensed Tannins on Their Metal-chelating Capacity. *Journal of Wood Chemistry and Technology*, Vol. 92, No. 2, (April 2008), pp. 193-205, ISSN 0277-3813

Yun, S. J. & Kaeppler, S. M. (2001). Induction of Maize Acid Phosphatase Activities under Phosphorus Starvetion. *Plant and Soil*, Vol. 237, No. 1, (November 2001), pp. 109-115, ISSN 1573-5036

8

Modeling of the Interannual Variation in Ecosystem Respiration of a Semiarid Grassland

Tomoko Nakano[1] and Masato Shinoda[2]
[1]*Department of Economics, Chuo University, Higashi-nakano, Hachioji, Tokyo,*
[2]*Arid Land Research Center, Tottori University, Hamasaka, Tottori*
Japan

1. Introduction

The Mongolian steppe zone comprises a major part of the East Asian grasslands. The regional climate is typically continental and semiarid, with low precipitation and large diurnal and annual temperature fluctuations. Rainfall normally occurs during the warm season (i.e., summer), but large interannual variability in the amount and pattern of rain has been observed and the area has often suffered from droughts.

Respiratory CO_2 efflux from terrestrial ecosystems (ecosystem respiration, R_{eco}) is an important component of the global carbon cycle, and since R_{eco} includes soil respiration (both root respiration and heterotrophic microbial decomposition) as well as aboveground respiration, rates of R_{eco} depend on complex interactions between biotic and abiotic factors. In general, temperature has been recognized as the major regulator of R_{eco} (Fang & Moncrieff, 2001; Lloyd & Taylor, 1994). Although many temperature models have been proposed, the simplest is the so-called exponential Q_{10} relationship (van't Hoff model), whereby the parameter Q_{10} represents the factor by which the respiration rate increases with a 10°C temperature increase. Soil water is also an important variable controlling R_{eco}, especially in semiarid and arid regions (e.g., Davidson et al., 1998; Hunt et al., 2002; Wen et al., 2006; Xu & Qi, 2001). Studies have shown that both very low and very high water content reduces ecosystem and soil respiration via the direct inhibition of biological activity or the inhibition of oxygen diffusion, respectively (Davidson et al., 1998; Gaumont-Guay et al., 2006; Wen et al., 2006). An increasing number of studies have indicated that R_{eco} rates are positively correlated with features of plant productivity such as the photosynthetic rate, leaf area index, and dry weight of green biomass (e.g., Aires et al., 2008; Crain et al., 1999; Jassal et al., 2008; Sampson et al., 2007).

Grassland ecosystems in semiarid regions demonstrate a sensitive response to climate changes, especially to changes in precipitation (Miyazaki et al., 2004; Ni, 2003; Shinoda et al., 2010), suggesting that interannual climate variability can lead to major changes in the respiratory CO_2 release from the ecosystem. Studies have demonstrated that the ecosystem switches from a net carbon sink during non-drought years to a net carbon source in drought years (Aires et al., 2008; Meyers, 2001).

We conducted field measurements of CO_2 fluxes and environmental parameters in a semiarid grassland area of Mongolia since 2004 and have quantitatively documented the dependence of ecosystem respiration on environmental variables (Nakano & Shinoda, 2010; Nakano et al., 2008). Based on our findings, we constructed an empirical respiration model in which the rates of ecosystem respiration are calculated from the soil temperature, soil water content, and aboveground biomass (AGB) (Nakano & Shinoda, 2010). In the present study, interannual changes in ecosystem respiration during the growing seasons from 2007 to 2010 at the semiarid grassland site in Mongolia were estimated using this model. We assessed whether temperature, moisture, or plant biomass was the most important environmental parameter controlling interannual variability in ecosystem respiration.

2. Methods

2.1 Study site and field measurements

Our study sites were located in grassland areas in central Mongolia (Fig. 1), which contain typical steppe vegetation that is grazed by livestock. The plant community at the area consists of graminaceous perennial grasses (*Agropyron cristatum, Cleistogenes squarrosa, Stipa krylovii*), forbs (*Artemisia adamsii, Chenopodium* spp.), and small shrubs (*Caragana* spp.). Soils are classified as Kastanozems (IUSS Working Group WRB, 2006), which is widely distributed in the Mongolian steppes. The bulk density of the upper 5 cm of the soil profile is approximately 1.5 g cm^{-3}.

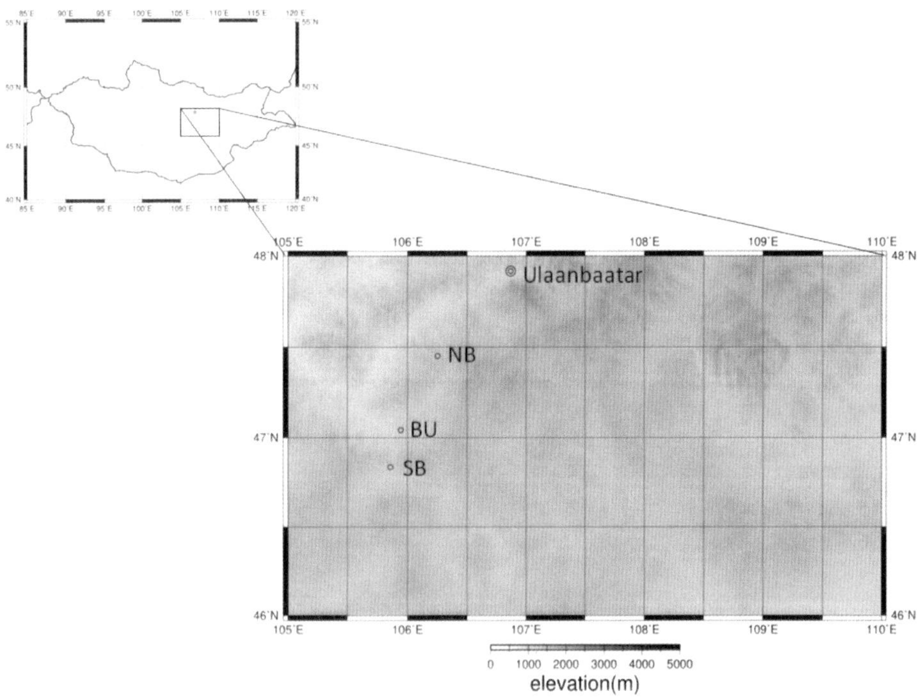

Fig. 1. Study site locations in semiarid grassland areas of central Mongolia

Site BU [47°02′37.2″N, 105°57′04.9″E, 1200 m above sea level (asl)] was established in 2004 in Bayan Unjuul county, which has an Institute of Meteorology and Hydrology of Mongolia (IMH) meteorological station. Data from the IMH station indicated that the mean annual air temperature and precipitation from 1995 to 2010 were 0.3°C and 155.8 mm, respectively. Most of the precipitation occurred between May and August. Two additional sites were established in 2009 and 2010: site NB, located approximately 60 km north of site BU (47°27′02.1″N, 106°15′09.2″E, 1507 m asl) and site SB, located approximately 30 km south of site BU (46°50′01.4″N, 105°51′31.2″E, 1296 m asl).

The rate of R_{eco} was measured using a closed-chamber technique during the growing seasons of 2004, 2005, 2006, 2009, and 2010 at site BU, in 2010 at site NB, and in 2009 and 2010 at site SB. We set up automated weather and ground observation systems at site BU in June 2004, and standard meteorological and soil parameters were continuously measured at 30-minute intervals. Meteorological and soil parameters were measured at sites NB and SB with a portable weather station and soil temperature and moisture sensors for each flux measurement period. Live AGB was measured by clipping green parts of the vegetation and oven-drying them at 80°C for 48 hours. The AGB was determined at each point where the closed-chamber measurements were made. The details and results of these measurements are described by Nakano et al. (2008) and Nakano & Shinoda (2010).

2.2 Respiration model

An empirical respiration model for which the rates of ecosystem respiration are calculated using the soil temperature, soil water content, and AGB, was constructed based on the field measurement results (Nakano & Shinoda, 2010). In the model, the rates of R_{eco} were calculated using the following equation (van't Hoff-type model):

$$R_{eco} = R_{20}Q_{10}^{(Ts-Tref)/10}, \tag{1}$$

where R_{20} is the standardized rate of respiration (μmol m^{-2} s^{-1}) at the reference temperature (20°C), Q_{10} is the factor by which the respiration rate increases for each 10°C increment increase in temperature (dimensionless), T_s is the soil temperature (°C) at 5 cm depth, and T_{ref} is the reference temperature of 20°C.

Nakano & Shinoda (2010) revealed that R_{20} values can be expressed as the linear model of AGB (Eq. 2), the residuals of which are calculated from the soil water content (Eq. 3),

$$R_{20} = a_1 AGB + a_2, \tag{2}$$

$$\text{Residual} + 2.0 = \exp(a_3W^2 + a_4W + a_5), \tag{3}$$

where a_1, a_2, a_3, a_4, and a_5 are parameters, AGB is the aboveground green biomass (g dry weight m^{-2}), and W is the soil volumetric water content at 3 cm depth (m^3 m^{-3}). Since the right side of Eq. 3 contained only positive values, the residuals were transformed by adding 2.0.

The temperature sensitivity of R_{eco} (Q_{10}) was expressed as a linear function of soil water content,

$$Q_{10} = a_6W + a_7. \tag{4}$$

Parameters for the model are listed in Table 1. The model was validated by comparing the observed and modeled values (r^2 = 0.754; Fig. 6 in Nakano & Shinoda, 2010). We adopted this model only for the growing season from May to September because ecosystem respiration during the winter and freeze–thaw cycles in the early spring and late fall might be controlled by different factors.

	Parameter	Value
Eq2	a_1	0.0289
	a_2	0.558
Eq3	a_3	-142
	a_4	31.9
	a_5	-0.403
Eq4	a_6	3.72
	a_7	1.51

Table 1. Parameters of the R_{eco} model. Values cited are from Nakano & Shinoda (2010)

In the present study, CO_2 effluxes during the growing season (May–September) from 2007 to 2010 at site BU were calculated from the observed soil temperature data at a depth of 5 cm and soil volumetric water content (VWC) at 5 cm depth, which were sampled every 30 minutes. We confirmed the absence of a significant difference between VWCs at 3 cm depth and those at 5 cm depth, and thus the values at 5 cm depth were used in the model.

Since AGB was not continuously measured at site BU, temporal changes in AGB were estimated from a remotely sensed vegetation index. The normalized difference vegetation index (NDVI) from the MODIS/Terra satellite was used in this study because NDVI is commonly applied to estimate the vegetation productivity of terrestrial ecosystems (Nemani et al., 2003; Running et al., 2004). We downloaded the 8-day Land Surface Reflectance (MOD09A1) data sets from the Land Processes Distributed Active Archive Center (LP DAAC), US Geological Survey (http://lpdaac.usgs.gov) and calculated NDVI values as follows:

$$NDVI = (\rho_{nir} - \rho_{red})/(\rho_{nir} + \rho_{red}), \quad (5)$$

where ρ_{nir} and ρ_{red} are the reflectances of near-infrared and red bands, respectively.

Green parts of the plants were clipped in four to ten 1 × 1-m quadrats at sites BU, NB, and SB in 2009 and 2010, and the AGB was determined by oven-drying and weighing the clippings. Average AGB values were calculated for each site and each year. Figure 2 shows the relationship between the average AGB and NDVI at the same times and locations as the field measurements. The AGB values showed a strong positive correlation with NDVI (r^2 = 0.849, $P < 0.001$). Using this relationship, temporal changes in AGB for every 8-day interval at site BU were estimated from the NDVI data, and daily values were interpolated from the 8-day values.

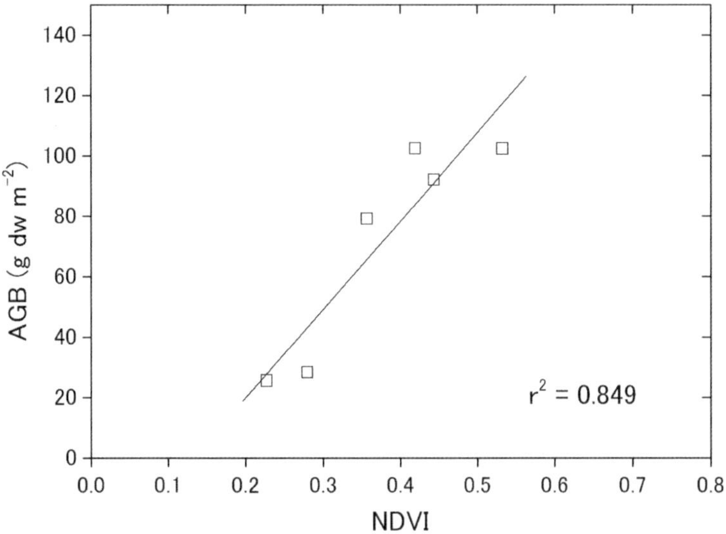

Fig. 2. Relationship between the aboveground biomass (AGB) measured at sites BU, NB, and SB and the normalized difference vegetation index (NDVI) obtained from satellite remote sensing data. The solid line is the linear regression line

3. Results and discussion

3.1 Meteorology and plant biomass

The monthly variations in air temperature, precipitation, and soil volumetric water content (VWC) values during 2007–2010 are shown in Fig. 3, and plant AGB derived from 8-day NDVI data is given in Fig. 4. Monthly mean air temperature was >0°C from April to October, reaching the maximum in July of every year. Average temperature during the growing season (May–September) was 17.8°C in 2007, 15.4°C in 2008, 15.5°C in 2009, and 16.4°C in 2010 (Table 2). The amount and timing of precipitation differed markedly among years; e.g., in 2007, the cumulative precipitation during the growing season was 79.3 mm, less than half of that in 2009 (178.1 mm) (Table 2). In June and July 2007, precipitation was considerably lower than in the other years, resulting in an extremely low AGB during the entire 2007 growth season (Fig. 4). A great deal of rain fell in the early 2008 growing season (June), with even more rain in the late 2009 growing season (August). This resulted in high VWC values for June 2008 (0.067 m^3 m^{-3}) and August 2009 (0.078 m^3 m^{-3}). Large amount of precipitation and high soil moisture in the early growing season of 2008 stimulated plant growth and increased AGB rapidly. As a consequence, the average AGB during the whole growing season was highest in 2008 (Table 2), although 2008 was not the wettest year. In 2010, the cumulative precipitation and average soil water content throughout the growing season were relatively low. However, a great deal of rain fell in July of that year, and subsequently, the amount of plant biomass rapidly increased in early July. Thus, the average AGB for 2010 was similar to that of 2009.

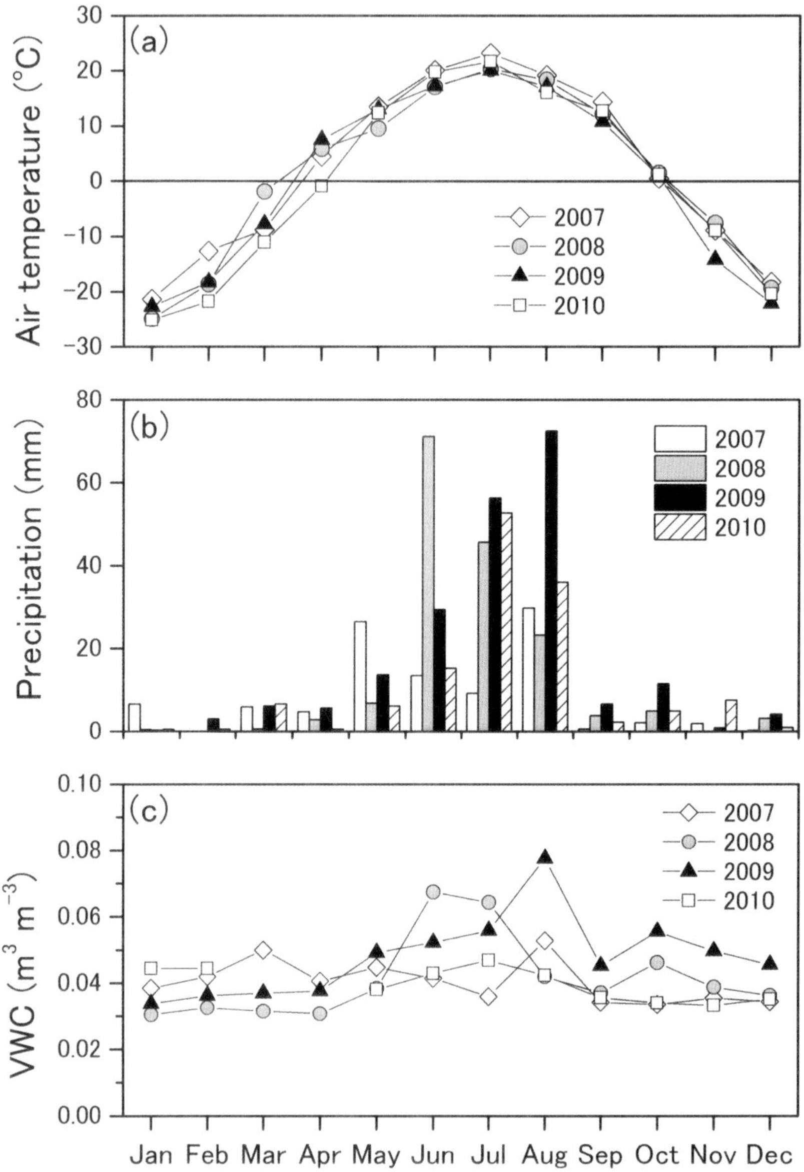

Fig. 3. Variation in (a) monthly mean air temperature, (b) monthly precipitation, and (c) monthly mean soil volumetric water content (VWC) from 2007 to 2010 at site BU

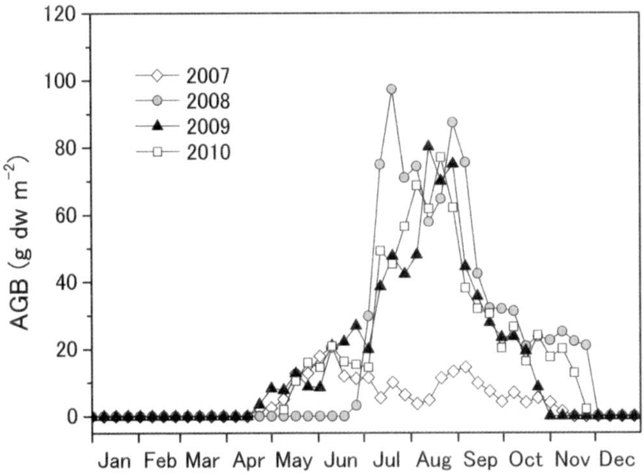

Fig. 4. Time course of live aboveground biomass (AGB) derived from the remotely sensed NDVI data

3.2 Estimation of ecosystem respiration

The half-hourly rate of R_{eco} was calculated using our empirical respiration model, and the daily flux was determined (Fig. 5). Variation in the daily rate of R_{eco} appeared to coincide with fluctuations in VWC caused by rain events. Yearly maxima of the daily R_{eco} were 2.7 g C m^{-2} d^{-1} in 2007, 4.9 g C m^{-2} d^{-1} in 2008, 4.7 g C m^{-2} d^{-1} in 2009, and 4.3 g C m^{-2} d^{-1} in 2010. Cumulative rates of ecosystem respiration during the growing season are listed in Table 2. The maximum and minimum amounts of ecosystem respiration within the 4 years were observed in 2009 and 2007, the wettest and driest years, respectively. Studies on ecosystem respiration in grassland ecosystems have indicated that respiratory efflux is generally larger in wet years than in dry years (Aires et al., 2008; Meyers, 2001).

	2007	2008	2009	2010
Air temperature (°C)	17.8	15.4	15.5	16.4
Precipitaion (mm)	79.3	150.5	178.1	112.1
AGB (g dw m^{-2})	14.0	39.9	37.7	37.3
VWC (m^3 m^{-3})	0.042	0.050	0.056	0.043
Soil temperature (°C)	21.2	17.9	18.4	19.1
R_{eco} (g C m^{-2})	155.4	286.6	295.0	222.8

Table 2. Average air temperature, cumulative precipitation, average aboveground green biomass (AGB), average soil volumetric water content (VWC), average soil temperature at 5 cm depth (T_s), and cumulative rate of ecosystem respiration (R_{eco}) during the growing season (i.e., May–September) from 2007 to 2010

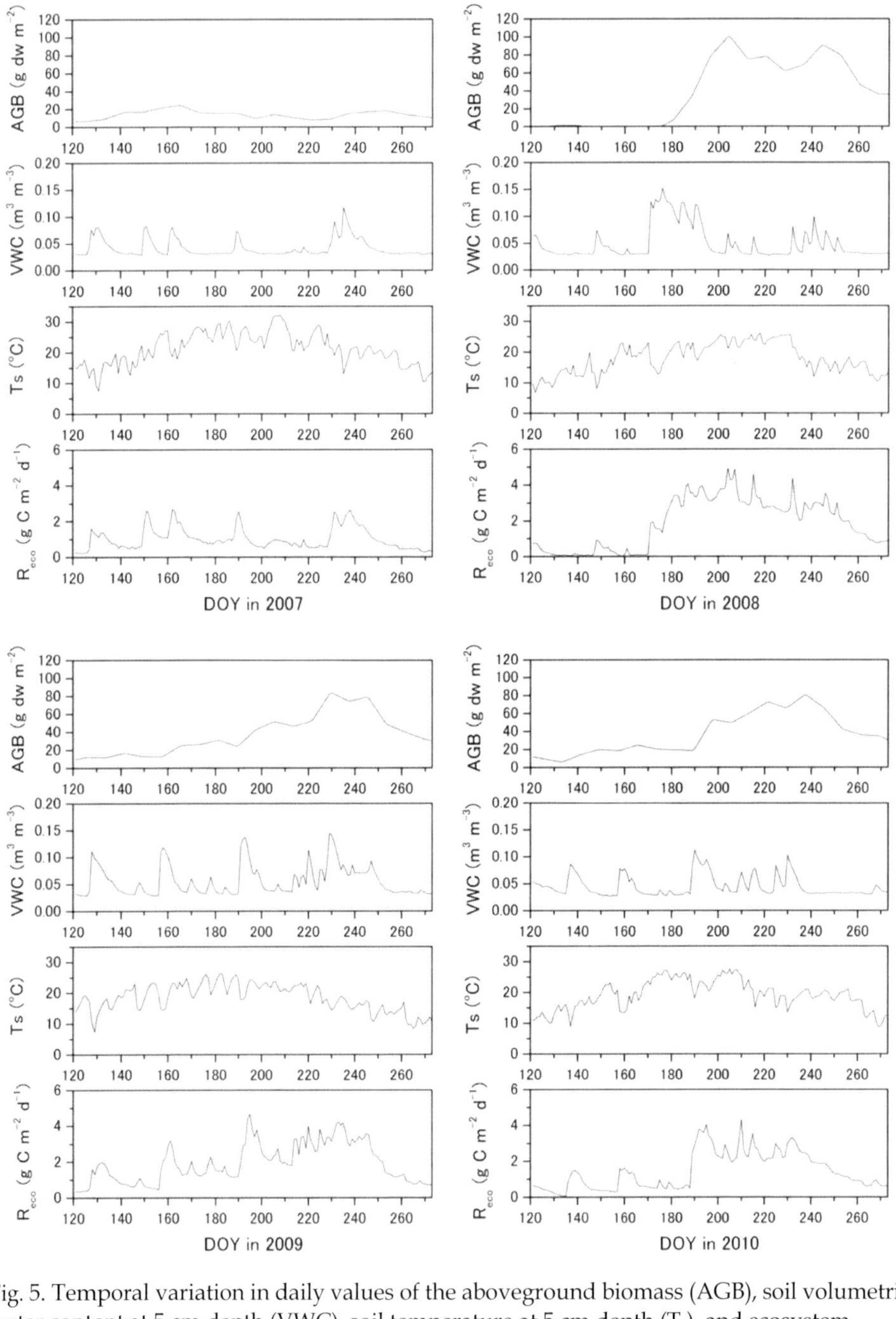

Fig. 5. Temporal variation in daily values of the aboveground biomass (AGB), soil volumetric water content at 5 cm depth (VWC), soil temperature at 5 cm depth (T_s), and ecosystem respiration (R_{eco}) during the growing season (i.e., May–September) from 2007 to 2010

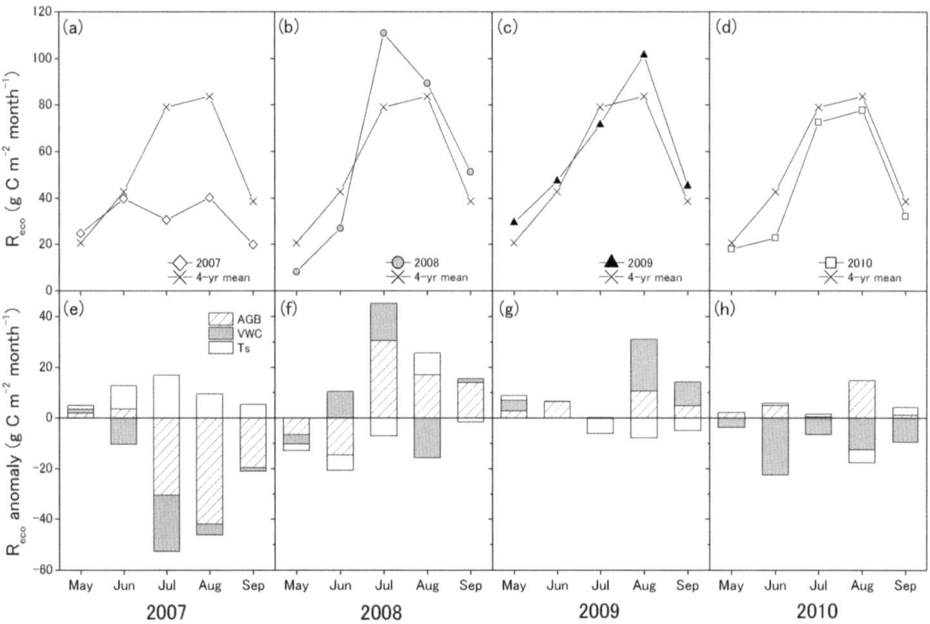

Fig. 6. (a-d) Monthly rate of ecosystem respiration (R_{eco}) and virtual R_{eco} calculated from the 4-year mean values of the aboveground biomass (AGB), soil volumetric water content at 5 cm depth (VWC), and soil temperature at 5 cm depth (T_s). (e-h) Components of the controlling factors of the anomalous R_{eco} values

To clarify the individual effects of AGB, VWC, and soil temperature (T_s) on the interannual variation in ecosystem respiration, we computed the average rate of R_{eco} using the 4-year mean values of AGB, VWC, and T_s measured every 30 minutes (Fig. 6a-d), and the significance of the anomalous values was assessed. To examine the effect of a single parameter, R_{eco} was calculated by setting the remaining parameters as equivalent to the mean values. This process allowed us to measure how ecosystem respiration responded to variations in AGB, VWC, and T_s. Fig. 6e-h shows the contribution of each parameter (AGB, VWC, and T_s) to the anomalous R_{eco} values. In 2007, the soil temperature was higher than in other years, and soil water content and plant biomass were lower. While the higher temperature increased the rate of ecosystem respiration, depression of R_{eco} by the drier soil and less plant biomass were much more significant. A decrease in AGB accounted for the greatest proportion of the R_{eco} depression, particularly in August and September. Consequently, the total amount of ecosystem respiration in 2007 was considerably lower than the average rate. In 2009, the high soil water content in August and September was correlated with the large amount of total ecosystem respiration during the growing season. Low soil water content decreased the R_{eco} in 2010, whereas plant biomass values were slightly higher and compensated for the depression in respiration influenced by the soil water content. In general, ecosystem respiration is considered to be a function of temperature. However, the results of this study suggest that variations in soil water content and plant biomass are more important determinants of the interannual variation in ecosystem respiration in semiarid grasslands.

Recent field-based studies have indicated the importance of plant photosynthetic productivity in the variability of carbon dynamics in grassland ecosystems. Aires et al. (2008) carried out the eddy-covariance measurements of CO_2 flux over a Mediterranean grassland in Portugal during dry and normal years and revealed that variations in ecosystem respiration were mainly controlled by canopy photosynthesis. Shimoda et al. (2009) suggested that large biomass contributes to the higher respired carbon loss over a temperate grassland in Japan. Craine et al. (1999) examined the role of a decreased carbohydrate source on soil respiration in an annually-burned grassland through manipulations of light intensity and the removal of aboveground biomass. They showed that shading and clipping grasses caused 40% and 19% reductions in soil CO_2 flux, respectively. A model study also demonstrated that a model using only climate drivers as predictors of R_{eco} failed to describe part of the temporal variability in the data and that dependency on grass primary production needed to be included as an additional driver of R_{eco} (Migliavacca et al., 2011). Several studies have conducted rainfall manipulation experiments to examine how the magnitude and timing of rainfall affect the ecosystem and soil respiration (e.g., Cable et al. 2008; Chou et al. 2008; Liu et al., 2002; Thomey et al., 2011). Chou et al. (2008) found that soil respiration was more sensitive to rainy season length than to the quantity of rainfall over a Californian annual grassland. Thomey et al. (2011) demonstrated that a single large rainfall led to a significant increase in soil respiration and the strength of these patterns differed between years in a desert grassland in New Mexico. In agreement with these studies, our results indicated that the effect of aboveground biomass and soil water content was larger than that of soil temperature on the interannual variation in ecosystem respiration and that contribution of plant AGB and soil water content to the respiration variability differed from year to year.

4. Conclusions

Interannual variations in ecosystem respiration over 4 years, including several hydrologically contrastive years, were computed by inputting observed meteorological parameters and remotely sensed plant biomass data into the empirical respiration model. The total amount of ecosystem respiration was highest in the wettest year (2009) and lowest in the driest year (2007). We examined the effects of plant AGB, soil water content, and soil temperature on the variation in ecosystem respiration. Ecosystem respiration is generally regarded as a function of temperature, whereas the results of this study suggest that the variations in soil water content and plant biomass are more important controls for interannual ecosystem respiration variation in semiarid grasslands.

5. Acknowledgment

We thank Ganbat Bavuudorj from the Institute of Meteorology and Hydrology of Mongolia for his help in field measurements. This research was funded by Chuo University (Grant for Special Research) and the Ministry of Education, Culture, Sports, Science and Technology, Japan (Grants-in-Aid nos. 20255001 and 22510017).

6. References

Aires, L.M.I., Pio, C.A., & Pereira, J.S. (2008). Carbon dioxide exchange above a Mediterranean C3/C4 grassland during two climatologically contrasting years. *Global Change Biol.*, Vol. 14, pp. 539–555.

Cable, J.M., Ogle, K., Williams, D.G., Weltzin, J.F., & Huxman, T.E. (2008). Soil texture drives responses of soil respiration to precipitation pulses in the Sonoran desert: implications for climate change. *Ecosystems*, Vol. 11, pp. 961–979.

Chou, W.W., Silver, W.L., Jackson, R.D., Thompson, A.W., & Allen-Diaz, B. (2008). The sensitivity of annual grassland carbon cycling to the quantity and timing of rainfall. *Global Change Biol.*, Vol. 14, pp. 1382–1394.

Craine, J.M., Wedin, D.A., & Chapin III, F.S. (1999). Predominance of ecophysiological controls on soil CO_2 flux in a Minnesota grassland. *Plant and Soil*, Vol. 207, pp. 77–86.

Davidson, E.A., Belk, E., & Boone, R.D. (1998). Soil water content and temperature as independent or confound factors controlling soil respiration in a temperate mixed hardwood forest. *Global Change Biol.*, Vol. 4, pp. 217–227.

Fang, C. & Moncrieff, J.B. (2001). The dependence of soil CO_2 efflux on temperature. *Soil Biol. Biochem.*, Vol. 33, pp. 78–90.

Gaumont-Guay, D., Black, T.A., Griffis, T.J., Barr, A.G., Jassal, R.S., & Nesic, Z. (2006). Interpreting the dependence of soil respiration on soil temperature and water content in a boreal aspen stand. *Agric. For. Meteorol.*, Vol. 140, pp. 220–235.

Hunt, J.E., Kelliher, F.M., McSeveny, T.M., & Byers, J.N. (2002). Evaporation and carbon dioxide exchange between the atmosphere and a tussock grassland during a summer drought. *Agric. For. Meteorol.*, Vol. 111, pp. 65–82.

IUSS Working Group WRB (2006). *World reference base for soil resources 2006*. 2nd edition. World Soil References Reports 103, FAO, Rome.

Jassal, R.S., Black, T.A., Novak, M.D., Gaumont-Guay, D., & Nesic, Z. (2008). Effect of soil water stress on soil respiration and its temperature sensitivity in an 18-year-old temperate Douglas-fir stand. *Global Change Biol.*, Vol. 14, pp. 1–14.

Liu, X., Wan, S., Su, B., Hui, D., & Luo, Y. (2002). Response of soil CO_2 efflux to water manipulation in a tallgrass prairie ecosystem. *Plant and Soil*, Vol. 240, pp. 213–223.

Lloyd, J. & Taylor, J.A. (1994). On the temperature dependence of soil respiration. *Funct. Ecol.*, Vol. 8, pp. 315–323.

Meyers, T.P. (2001). A comparison of summertime water and CO_2 fluxes over rangeland for well watered and drought conditions. *Agric. For. Meteorol*, Vol. 106, pp. 205-214.

Migliavacca, M., Reichstein, M., Richardson, A.D., Colombo, R., Sutton, M.A., Lasslop, G., Tomelleri, E., Wohlfahrt, G., Carvalhais, N., Cescatti, A., Mahecha, M.D., Montagnani, L., Papale, D., Zaehle, S., Arain, A., Arneth, A., Black, T.A., Carrara, A., Dore, S., Gianelle, D., Helfter, C., Hollinger, D., Kutsch, W.L., Lafleur, P.M., Nouvellon, Y., Rebmann, C., Rocha, H.R.Da., Rodeghiero, M., Roupsard, O., Sebasti, M.T., Seufert, G., Soussana, J.F., & Van Der Molen, M.K. (2011). Semiempirical modeling of abiotic and biotic factors controlling ecosystem respiration across eddy covariance sites. *Global Change Biol.*, Vol. 17, pp. 390-409.

Miyazaki, S., Yasunari, T., Miyamoto, T., Kaihotsu, I., Davaa, G., Oyunbaatar, D., Natsagdorj, L., & Oki, T. (2004). Agrometeorological conditions of grassland vegetation in central Mongolia and their impact for leaf area growth. *J. Geophys. Res.*, Vol. 109, D22106, doi:101029/2004JD005179.

Nakano, T., Nemoto, M., & Shinoda, M. (2008). Environmental controls on photosynthetic production and ecosystem respiration in semi-arid grasslands of Mongolia. *Agric. For. Meteorol*, Vol. 148, pp. 1456-1466.

Nakano, T. & Shinoda, M. (2010). Response of ecosystem respiration to soil water and plant biomass in a semiarid grassland. *Soil Sci. Plant Nutr.*, Vol. 56, pp. 773-781.

Nemani, R.R., Keeling, C.D., Hashimoto, H., Jolly, W.M., Piper, S.C., Tucker, C.J., Myneni, R.B., & Running S.W. (2003). Climate-driven increases in global terrestrial net primary production from 1982 to 1999. *Science*, Vol. 300, pp. 1560−1563.

Ni, J. (2003). Plant functional types and climate along a precipitation gradient in temperate grasslands, north-east China and south-east Mongolia. *J. Arid Environ.*, Vol. 53, pp. 501−516.

Running, S.W., Nemani, R.R., Heinsch, F.A., Zhao, M.S., Reeves, M., & Hashimoto, H. (2004). A continuous satellite-derived measure of global terrestrial primary production. *Bioscience*, Vol. 54, pp. 547−560.

Sampson, D.A., Janssens, I.A., Curiel Yuste, J., & Ceulemans, R. (2007). Basal rates of soil respiration are correlated with photosynthesis in a mixed temperate forest. *Global Change Biol.*, Vol. 13, pp. 2008−2017.

Shimoda, S., Lee, G., Yokoyama, T., Liu, J., Saito, M., & Oikawa, T. (2009). Response of ecosystem CO2 exchange to biomass productivity in a high yield grassland. *Environ. Exp. Bot.*, Vol. 65, pp. 425-431.

Shinoda, M., Nachinshonhor, G.U., & Nemoto, M. (2010). Impact of drought on vegetation dynamics of the Mongolian Steppe: A field experiment. *J. Arid Envioron.*, Vol. 74, pp. 63−69.

Thomey, M.L., Collins, S.L., Vargas, R., Johnson, J.E., Brown, R.F., Natvig, D.O., & Friggens, M.T. (2011). Effect of precipitation variability on net primary production and soil respiration in a Chihuahuan Desert grassland. *Global Change Biol.*, Vol. 17, pp. 1505−1515.

Wen, X.F., Yu, G.R., Sun, X.M., Li, Q.K., Liu, Y.F., Zhang, L.M., Ren, C.Y., Fu, Y.L., & Li, Z.Q. (2006). Soil moisture effect on the temperature dependence of ecosystem respiration in a subtropical Pinus plantation of southeastern China. *Agric. For. Meteorol.*, Vol. 137, pp. 166−175.

Xu, M. & Qi, Y. (2001). Soil-surface CO_2 efflux and its spatial and temporal variations in a young ponderosa pine plantation in northern California. *Global Change Biol.*, Vol. 7, pp. 667−677.

9

Soil Carbon Sequestration Under Bioenergy Crops in Poland

Magdalena Borzecka-Walker, Antoni Faber,
Katarzyna Mizak, Rafal Pudelko and Alina Syp
Institute of Soil Science and Plant Cultivation-State Research Institute
Poland

1. Introduction

Agriculture practices have an important role to play in mitigating climate change due to atmospheric enrichment of carbon dioxide, and other greenhouse gases (GHG). Land management can strongly influence soil carbon stocks and careful management can be used to sequestered soil carbon. It is important to propose contemporary management practises to farming, like the conversion from a tillage system to no-tillage, incorporation of cover crops and forages in the crop rotation, use of crop residues and biosolids e.g. mulch, implementation of biocrops, as well as integrated nutrient management which including compost/manures as well as the precision use of fertilizers and integrated pest management. Sustainable management in agriculture should reduce and avoid the introduction of carbon dioxide (CO_2) to the atmosphere, which is one of three most prevalent GHGs directly emitted by human activity. CO_2 is the most important anthropogenic GHG, and according to IPCC Fourth Assessment Report (2007), anthropogenic CO_2 emissions grew by about 80% between 1970 and 2004. Carbon sequestration is a process through which agricultural and forestry practices remove carbon dioxide (CO_2) from the atmosphere into a form that does not affect atmospheric chemistry (Lal, 2004a). A natural way to trap atmospheric CO_2 is by photosynthesis, where carbon dioxide is absorbed by plants and turned into carbon compounds, stored or fixed C as soil organic carbon (SOC). The SOC pool consist litter, humads and humus, which it is comprised of mixtures of plant and animal residues at various stages of decomposition along with microbial by-products (Lal, 2004a). Agriculture is responsible for 13.5% of global anthropogenic GHG emissions (IPCC, 2007), but if sustainable land management practices are implemented, agricultural soils could become a carbon sink (Dumanski et al., 1998). There are five principal global carbon pools. The oceanic pool (38 Gt) is the largest, followed by the geologic (5 Gt), pedologic (2.5 Gt), biotic (0.56 Gt), and the atmospheric pool (0.76 Gt). The soils beneath the oceans are the most important reservoir of carbon in the terrestrial biosphere and contain three times the amount as compared with those that are found in vegetation (Lal, 2004b; SEC, 2009). Soils contain more than twice the carbon that can be found in the atmosphere and the loss of carbon from soils can have a significant effect on atmospheric CO_2 concentrations, which can influence the climate (Smith, 2008). Many studies have examined the sequestration potential in agriculture and forestry in Europe (Smith et al., 1997; Smith et al., 2000; Vleeshouwers & Verhagen, 2002; Freibauer et al., 2004;

Smith, 2004;) and globally (Smith 2004; IPCC 2007; Lal, 2004a), as well as in other regions of the world such as North America (Dumanski et al., 1998; Franzluebbers & Follett, 2005) or Africa (Ringius, 2002). The potential for carbon sequestration in the European Union (EU) is approximately 90–120 Mt C/y, in the US cropland is 75–208 Mt C/y, in Canada is approximately 24 Mt C/y, to obtain this potential, optimal land management practices have to be implemented (Hutchinson et al., 2007). It is estimated that the global potential scale of carbon sequestration in soils used for agricultural purposes is around 0.3 t C/ha/y on arable lands, and around 0.5 - 0.7 t C/ha/y on grasslands (IPCC, 2000). The conducted researches indicate existence of a high potential for carbon sequestration in soils under agricultural crops. Depending on the used method for its evaluation and it range between from 0.15-0.22 t C/ha/y for willow (Bradley & King, 2004) up to 0.93 t C/ha/y for Miscanthus (Matthews & Grogan 2001). The net soil carbon sequestration simulated for biocrops in Poland was around 0.38 - 0.95 t C/ha/y Miscanthus crops and 0.22 - 0.39 t C/ha/y for willow coppice (Borzecka-Walker et al., 2008).

There are many policies, directives, standards, as well as norms in the EU designed to stimulate and support the reduction of GHG emission and to improve the carbon mitigation potential. The publication of a Green paper "Towards a European strategy for the security of energy supply" (2000) started a debate on energy security, which is considered a key element of politico-economic independence of the EU. It stressed the need to improve the organisation's strategic stocks of raw materials and coordinate its use. Additionally, the European Commission presented a White Paper that sets out the actions necessary to strengthen the Union's ability to adapt to a changing climate. To support the biofuels industry, the Energy Taxation Directive allows exemptions or reductions from energy taxation for biofuels (Directive, 2003/96/EC). The aims of the recently released European Parliament and the Council directive on the promotion of the use of energy from renewable sources amending and subsequently repealed Directives 2001/77/EC and 2003/30/EC (Directive 2009/28/EC); are to achieve by 2020 a 20% share of energy from renewable sources in the EU's final consumption of energy and a 10% share of energy from renewable sources in each member state's transport energy consumption. Moreover the GHG emission saving from the use of biofules and bioliguids shall be at least 35%, 50% in 2017 and 60% in 2018 yrs.

The aim of this review is an evaluation into the current knowledge of carbon sequestration and to present potential bioenergy crops for carbon sequestration in Poland.

2. The soil's organic matter balance

There are several methods and simulation models useful for defining the content of soil organic matter. The following work will present two methods which were applied for the Polish territory. It should be stressed that the results obtained by using both methods are comparable. Based on these results, it can be concluded that the coefficient method used in the assessment of carbon balance has a small error and gives equally reliable results as while using the soil profiles method.

2.1 The soil's organic matter balance based on the determination of soil profiles

The content of organic matter in soils of agricultural land is highly variable. The results of determinations carried out in Poland show that it varies in the arable layer within the limits of 0.5-10% with an average of 2.2%. According to the division used in Polish soil with low humus content (<1%) account for approximately 6% of agricultural land, the average (1.1-2.0%), approximately 50% of the area, but about the content > 2% of approximately 33%. The global balance of organic matter in Polish soils is negative in all regions (-0.06 to - 1.05 t C/ha/y), with the average for the country of -0.47 t C/ ha/y. This means that in large areas of Poland we note the CO_2 emissions from soil to the atmosphere (Terelak et al., 2001; Stuczynski et al., 2007). In the years 2000-2004, a preliminary analysis of soil humus content trends were carried out under repeated testing of standard profiles. Studies have shown a significant decline in humus, mainly in soils initially rich in organic matter. A decline in soil organic matter is associated with the change of soil water relations, i.e. more intensive use and drainage. In contrast, a large part of the light soils of the last 30 years recorded an increase of humus content associated with an increased level of fertilization and increase in quantity of crop residues (Stuczynski et al., 2007). Based on measurements taken in the years 1968-1983 and in 2003, the changes in soil organic matter and humus loss risk were able to be calculated. The results presented in figure 1 show both accumulation and humus loss in soil as well as soil organic matter balance. The highest losses of soil organic matter were calculated for the Kujawsko-pomorskie voivodeship, whilst the lowest was in the Malopolskie voivodeship. Voivodeships of the North Western part of the country have the lowest soil organic balance, and this indicates a greater share of soils with a higher risk of loss of function due to mineralisation of soil humus.

2.2 The soils organic matter balance based on coefficients

The amount of organic matter in soils is a key indicator for the quality, and is significant for their physicochemical properties such as sorption. Maintaining high humus content in soil is important because of its impact on soil carbon sequestration. Increasingly popular intensive use of soils, combined with a simplified crop rotation, increased predominance of cereal plants with reduced amounts of livestock, leads to a reduction in the amount of organic residues entering the soil, which in turn leads to reduced carbon sequestration in soil.

The basic principle of good farm management is to maintain a positive, or at least a sustainable balance of soil organic matter. This balance can be obtained by the selection of species of cultivation plants, their participation in the crop structure, and the quantity of manure and organic. The various species of crops leave different amounts of crop residue. The soil carbon in cropland can be increased by planting more forages, and increasing residue inputs from plants with high biomass potential. Approximately, it can be concluded that the weight of cereal crop residues is about 3-fold greater than the root, and legumes with grasses, by up to 6-fold. In addition, a different duration and degree of shading the soil surface and the number of tillage performed and care, which affects the mineralisation of humus.

The cultivated plants can be separated in three groups depending of the impact on the balance of humus in the soil.

The first group includes plants with a potential in enriching the soil with organic matter. Among them are primarily long-term forage legumes and their mixtures with grasses and grasses grown in the field as well as crops grown for energy sources like tall grasses, fast growing trees. In addition, legumes and intercrops ploughed as green fertilizers have little positive effect on the balance of humus. The reproduction rate of soil for this group of plants ranges from 0.21 to 2.10, depending on the type of soil they are grown on (Fotyma & Mercik, 1992).

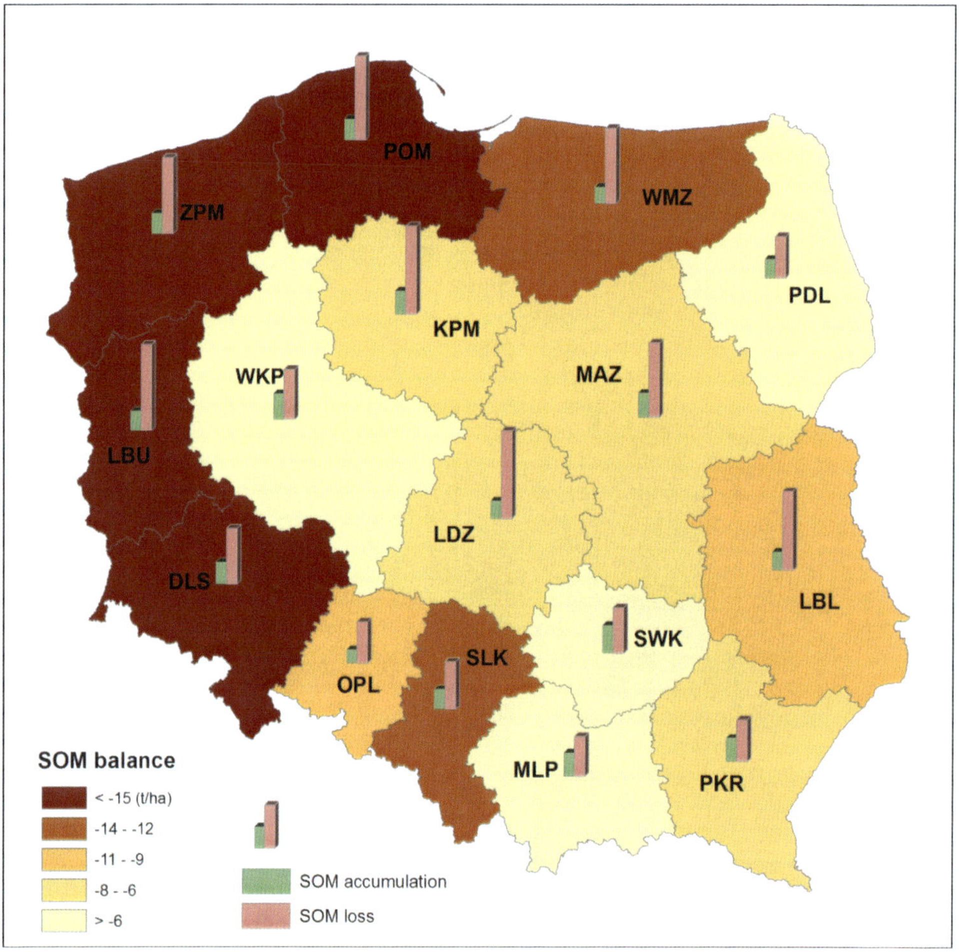

Fig. 1. Forecast loss of soil organic matter (SOM) from agricultural land[1].
Source: own work based on Stuczynski et al., (2007)

[1] Map code: Dolnośląskie (DLS), Kujawsko-pomorskie (KPM), Lubelskie (LBL), Lubuskie (LBU), Lodzkie (LDZ), Małopolskie (MLP), Mazowieckie (MAZ), Opolskie (OPL), Podkarpackie (PKR), Podlaskie (PDL), Pomorskie (POM), Slaskie (SLK), Swietokrzyskie (SWK), Warminsko-mazurskie (WMZ), Wielkopolskie (WKP), Zachodniopomorskie (ZPM)

The second group includes plants with a potential in degrading the soil organic matter. This group includes mainly root crops, root vegetables and corn. The characteristics for this group of crops is very little crop residue, seeded in wide rows, intercrops, heavy maintenance, and the short canopy (cover spacing) increases the distribution of humus resulting in increased erosion. The soil degradation rate for this group of plants ranges from -0.12 to -1.54, and is dependent on the type of soil they are grown on (Fotyma & Mercik, 1992). The mineralisation for these types of plants per year is about 1.0-1.5 t/ha of humus. To compensate for this loss about 15-16 t/ha manure should be used.

The third group include plants with a small negative or neutral impact on soil organic matter. This group of plants includes cereals and oilseeds. Cereals previously were treated as plants degrading the soil organic matter, but changes in agricultural techniques (density of straw, shortening of straw), and combine harvester collection, leaves a lot of crop residue that significantly reduces their negative impact on the balance of soil organic matter. It should be emphasized that the quality of cereal crop residues is worse due to the unfavourable ratio of carbon to nitrogen. The soil degradation rate for this group of plants ranging from -0.49 to -0.56, is dependent on the type of soil they are grown on (Fotyma & Mercik, 1992).

The coefficients values determine the amount of soil organic matter t/ha can enriched or depleted by following a one-year cultivation of the plants or through the application of 1t/ha dry matter of different natural and organic fertilizers. Using these coefficients can be simplified way to determine the soil organic matter balance for a farm, region or country. A positive result indicates a normal economy and organic matter, thus ensuring the long-term stabilisation of humus content at an optimum level. If the balance is negative then changes are necessary. This can be achieved by changing the crop structure (introduction of plants with positive coefficient), or increasing the dosage of organic fertilizers (ploughed straw) or intercrops cultivation for ploughing. Throughout the calculations the following formula was used (see Equation 1):

$$\text{Degradation coefficient} = \frac{\sum(\%\ \text{cereals area x} - 0.53) + (\%\ \text{root crops} \times -1.40) + (...)}{\text{sown area}\ (\%)} \tag{1}$$

The numerator is the sum of the ratio (the share of particular groups or species of plants in the crop structure multiplied by the coefficients for these species), while the denominator is the percentage of a sown area (where taking into account all the sown land as 100%). Based on those coefficients following the agricultural use of arable land, allows us to calculate the decreases in the soil organic matter amount by 0.39 to 0.66 t/ha for particular voivodeships and approximately 0.53 tonnes per one ha per year in Poland (Kus et al., 2006). Figure 2 presents carto-diagram which shows spatial differentiation, presented by standard deviation methods.

The best situation of soil organic matter was calculated for the voivodeships of Warminsko-mazurskie, Podlaskie followed by the Malopolskie, and their positive situation is associated with a high share of legumes or their mixtures with grasses. An adverse situation appears in the Dolnoslaskie and Opolskie voivodeships, where there is a large share of root crops and maize. To offset this loss, approximately six tons of manure should be applied on every hectare of arable land used. The calculations (Kus et al., 2006) show that the national average production of natural fertilizers (manure) was approximately 7.3 t/ha of sowing area (Fig. 3).

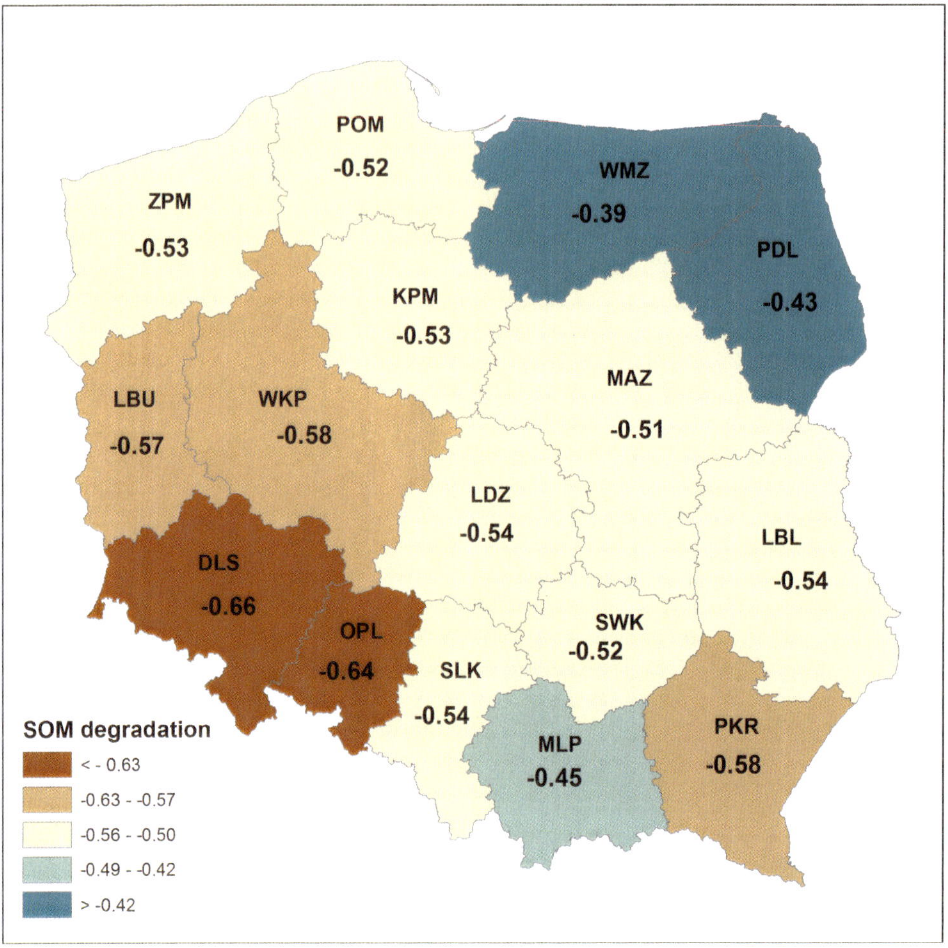

Fig. 2. The coefficient values of soil organic matter degradation for the individual regions calculated for the cropping system average for 2002-2005[1].
Source: own work based on Kus et al., (2006)

The highest amount of manure was produced in the Podlaskie voivodeship due to high livestock and low sowing area. This led to a significant surplus of soil organic matter in the Podlaskie voivodeship. The lowest production of manure was in the Dolnosląskie and Zachodniopomorskie voivodeships where the sowing area is high and livestock low.

In some voivodeships with a negative value of SOM, a balance can be best achieved through ploughing the straw. Particularly large quantities of straw can be ploughed in four voivodeships, (0.9 - 1.0 t/ha in Opolskie and Lubuskie, about 1.2 t/ha in Zachodniopomorskie and to 1.9 t/ha Dolnoslaskie (Fig. 4). However, in some voivodeships such as Lubelskie, only 0.2 tons of the straw per 1 ha of arable land can be ploughed. In total, across the country three million tonnes of straw, which is less than 12% of the collected straw, could be allocated for ploughing.

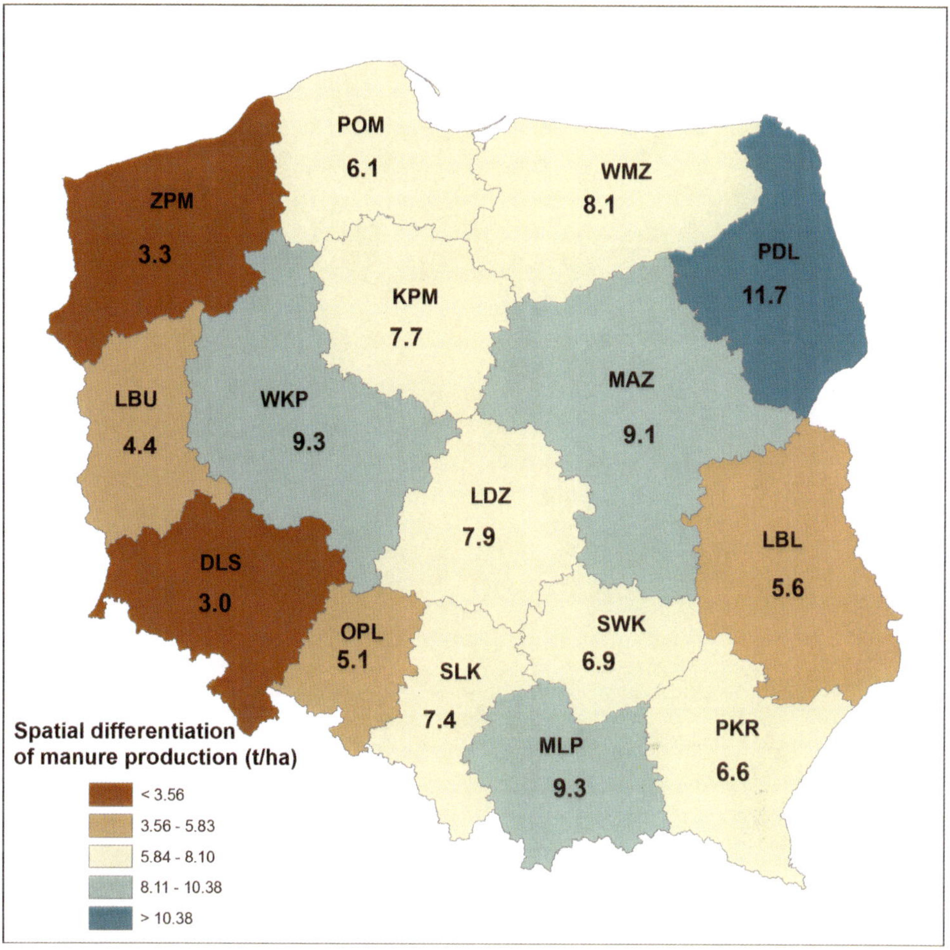

Fig. 3. The production of manure per 1 ha of sowing area (average 2002-2005)[1].
Source: own work based on Kus et al., (2006)

2.3 Assessment of C organic balances

As indicators for determining a positive, neutral or negative balance of carbon in the soil the scale developed by Korschens et al., (2004) can be used - presented in the table 1. The presented scale has a range between below -200 kg/ha/y and values above 300 kg/ha/y. Comparing the results with this scale allows judging the impact on soil functions and potential yield performance of plants. It is important to emphasize that the carbon balance of more than 300 increases high emissions of nitrogen. Therefore, soil carbon sequestration cannot be considered in isolation from the nitrogen emissions.

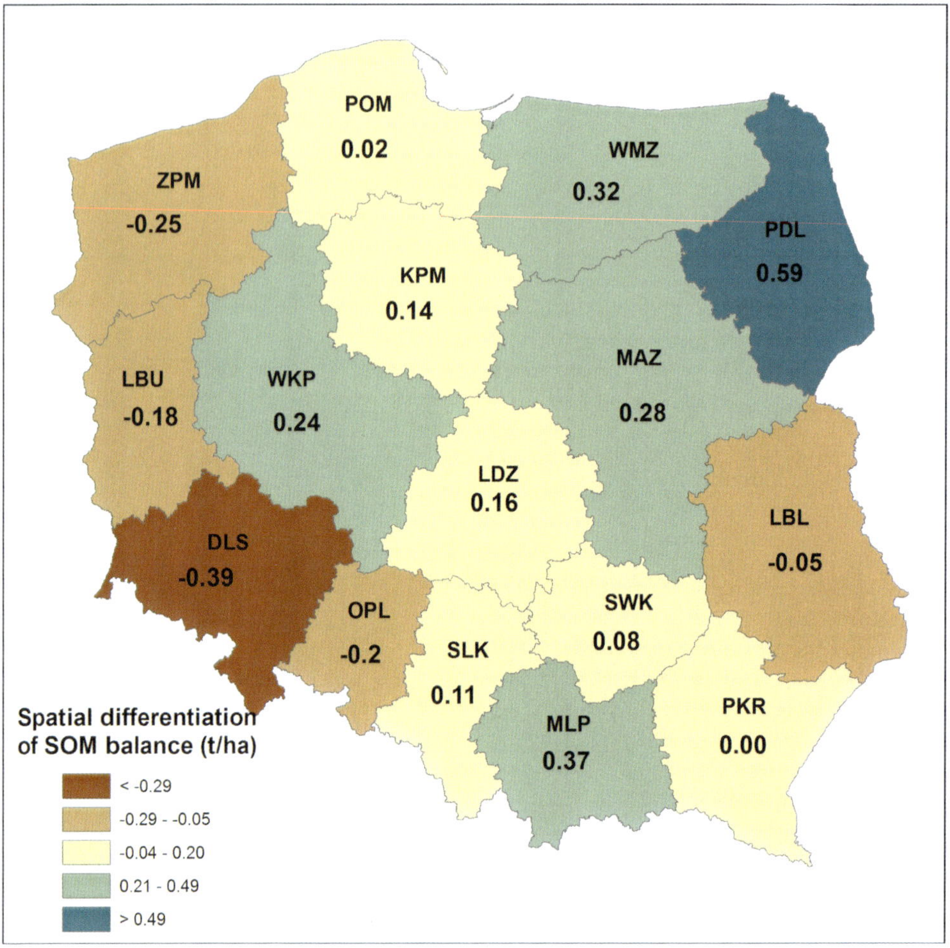

Fig. 4. Balance of soil organic matter (t/ha) in voivodeship[1].
Source: own work based on Kus et al., (2006)

Balance		Impact
kg C/ha/y	group	
<-200	very low	unfavourable influence of soil functions and yield performance
-200- -75	low	medium-term tolerable, especially on soils enriched with humus
-75- +100	optimal	optimal in terms of yield loss at low risk of long-term site-adapted setting humus content
100-+300	high	medium-term tolerable, especially on impoverished soils with humus
>300	Very high	increased risk of nitrogen loss, low efficiency

Table 1. Evaluation of the carbon organic balance (VDLUFA, 2004)

2.4 The potential land use change for sequestering carbon in soils

Land use change significantly affects soil carbon stock (Guo & Gifford, 2002). Most long-term research shows significant changes in SOC (Smith, 2008). Land use change can play a positive or negative role in mitigating global warming by sequestering carbon from the atmosphere into vegetation and soils. Many land use activities generate carbon sequestration and thus counteract the impact of emissions made in a different place. There are two components of estimated emissions from land use change: decomposition of vegetation and mineralisation/oxidation of humus or SOC (Lal, 2004b). The conversion of arable land to woodland may result in a substantial increase in soil carbon sequestration from 0.3 to 0.6 t C/ha/y. The conversion of arable land to grassland may result in a substantial increase in soil C sequestration from 1.2 to 1.7 t C/ha/y. The potential carbon sequestration rate in the conversion of woodland to arable land was -0.6 t C/ha/y, while the conversion of grassland to arable land was at a rate of between -1.0 and -1.7 t C/ha/y (Freibauer et al., 2004). In addition, Guo & Gifford (2002) in a long-term experiment had shown that a conversion of forestland or grassland to arable land caused a significant loss of SOC, whereas a conversion of forestry to grassland did not result in a loss for all cases. The largest potential decrease of SOC loss is in land use change on highly organic soil (Gronlund et al., 2008). Drainage and cultivation of peat soils stimulates soil organic matter (SOM) mineralisation, which substantially increases CO_2 emissions from soils. Because of this, the Directive 2009/28/EC prohibits the use of land with high carbon stock (i.e. wetlands, continuously forested areas, and peat land) for the production of biofuels.

2.5 Potential of management change for sequestering carbon in soils

Correct agricultural practices of the soil can have a significant influence for carbon sequestration. A change in conventional cultivation practises has an important role in improving the soil's structure. Implementing modern practises like reduced tillage, no tillage or conservation tillage can significantly improve the soil's organic matter. Conventional tillage is defined as the mechanical manipulation (ploughing, disking and harrowing) of the top soil that leaves no more than 15% of the ground cover with crop residues. Such tillaging tends to disrupt the soil structure, accelerating the decomposition of soil organic matter, and making the bared topsoil vulnerable to erosion by rain and wind (Hillel & Rosenzweig, 2009). The alternative for conventional tillage is conservation tillage. The European Conservation Agriculture Federation (ECAF) defines conservation tillage as soil management practices, which minimise the disruption of the soil's structure, composition and natural biodiversity, thereby also minimising erosion and degradation, and water contamination (ECAF, 2002). By avoiding deep ploughing, this can increase the sequestration rate by 1.4 to 4.1 t C/ha/y. There is a growing interest in the impact of conservation tillage practices on carbon sequestration in recent years. According to Holland (2004), agriculture can act both as a sink and a source of CO2 emission and the use of conservation practices by agriculture could decrease this emission. The coverage of the soil surface with straw and cover crops, increases biomass productivity and turns the soil into a tremendous carbon sink. Reducing the intensity of soil cultivation lowers energy consumption and the emission of carbon dioxide, while carbon sequestration is raised through the increase of soil organic matter (Holland, 2002). On the basis of long-term experiments, West & Post (2002) concluded that conversion of conventional tillage to no-till sequesters an average of 0.57±0.14 t C/ha/y. Long-term field experiments are the most reliable source information about GHG emissions from different agricultural systems. However, they are difficult to manage and limited by time and costs (Li et al., 2009). Reduced tillage, enhanced crop residue incorporation, and farmyard manure application each increased soil C-sequestration, increased N2O emissions, and had little effect on CH4 uptake. Over 20 years, increases in N2O emissions, which were converted into CO2-equivalent emissions with 100-year global warming potential multipliers, offset 75–310% of the carbon sequestered, depending on the scenario (Li et al., 2005). Simulation models provide an alternative method of assessment of agricultural practices effects (Farge et al., 2007). Many models have been developed to describe the responses of crop growth, soil water dynamics and soil biogeochemistry such as Roth C (Groenigen et al., 2010) for organic carbon turnover, CENTURY (Grant et al., 2004) or DNDC (Li et al., 1992; Giltrap et al., 2010) for carbon and nitrogen cycles. In order to calculate the management influence of carbon changes, we used 'Tool for Estimation of Changes in Soil Carbon Stocks associated with management Changes in Croplands and Grazing Lands based on IPCC Default Data'. At the moment Poland is assumed as having a cold temperate climate with both maritime and continental elements impact. The conversion from full tillage to reduced tillage under cold a temperate climate with maritime influence can cause the annual carbon stock to change by approximately 0.56 t C/ha, while the conversion from full tillage to no tillage 0.76 t C/ha depending on type of climate, soil, and the amount of fertiliser applied while under a cold temperate climate with a continental influence is 0.20 and 0.28 t C/ha respectively.

3. Potential of cultivation energy plants for sequestering carbon in soils

The soil carbon in croplands can be increased by planting more forages, and increasing residue inputs from plants with a high biomass potential. Hopes for increased soil carbon sequestration are associated with an increase in large-scale energetic crops cultivation. Energy crops are characterised by rapid growth and large amount of biomass produced and consequently a very large amount of crop residue left at the field. Annually about 30% of the senescent leaves and post harvest remnants are entering the soil (Matthews & Grogan 2001). One factor that is highly important for the amount of carbon sequestration in soil is that these perennial plants are grown for about 20 years on one field. The cultivation of energy crops is associated with greenhouse gas emissions (burning fuel, the production of fertilizers, crop protection). The assumption is that carbon sequestration of around 0.25 t C/ha/y resulting from energetic crops cultivation, allows biomass combustion to be neutral in terms of greenhouse gas emissions (Volk et al., 2004). Nevertheless, different authors found that the carbon sequestration rates for these cultivars are different. Bradley and King (2004), determined the carbon sequestration in forests and willows cultivations at 0.15-0.22 t C/ha/y, whereas in Miscanthus cultivation it was at 0.13-0.20 t C/ha/y. According to Matthews and Grogan (2001), carbon sequestration in the surface layer of the soil (0-23 cm) was at 0.31 for forests, and 0.41 for the cultivation of willow, whereas for Miscanthus it was measured at 0.93 t C/ha/y. Freibauer et al., (2004) and Smith (2004), determined the carbon sequestration in cultivations of energy crops at appropriately 0.60 and 0.62 t C/ha/y. Unfortunately for the carbon sequestration the liquidation of biocrops plantations causes large losses of accumulated carbon. This topic is not yet completely understood and requires further work from researchers.

3.1 Materials and methods

As mentioned before Poland has a temperate climate on the Western side of the Vistula River with larger influence of maritime climate than on the Eastern side of the river that has a larger influence of continental climate. For research on soil carbon sequestration dependant on the climate, we have divided Poland along the Vistula River into two regions of climatic influence. We have selected randomly ten experimental positions with available climate data from the MARS Database elaborated by the Joint Research Centre European Commission – 5 grids cells (50 km x 50 km) in each region.

For a simulation of total inflows of carbon into soil under willow and Miscanthus crops in a 19 year period, the DNDC model was used. The model was calibrated based on the data from experimental fields established in 2003 at two Experimental Stations of the Institute of Soil Science and Plant Cultivation, Puławy. The experimental fields are located in the Experimental Station Pulawy-Osiny on heavy black earth, and at the Experimental Station in Grabow on medium-heavy soil where five genotypes of Miscanthus and four clones of willow were planted.

3.2 Results and discussion

The results have shown that the significant difference in yield of Miscanthus grown under two different climate influences, as well as organic matter input, while soil organic matter did not denote a statistically significant difference. In comparing the case of willow, the yield was not significantly different, but the input of organic matter and soil organic matter was significantly different. This might be explained as the temperature had an impact in the case of Miscanthus but it did not influence willow plants. The opposite situation was found in the effect of precipitation. Borzęcka-Walker et al., (2008), did not find any significant differences between the yields of willow clones. The yield of willow grown in two different localisations, ranged from 11.1 - 13.7 t/ha/y. There were lower simulated yields for willow cultivations (13.8-18.1 t/ha/y) located on very good soils of Eastern Europe (Fischer et al., 2005). It can be assumed that the limited water influenced the experimental yield of willow. The dry matter yield of Miscanthus genotypes was significantly different within an average of 10.2 - 20.7 t/ha/y at both localisations. The yield for the first year of the experiment was low; this could be because it was the second year of cultivation when the plants are still not mature enough to obtain an economic yield (Clifton-Brown & Lewandowski 2000). In the second year of the experiment, the yield witnessed a high increase. In the third year of experiment this was characterised with very bad weather conditions, including a late spring ground-frost and long summer draught. The yields were approaching the presupposed simulated yields for Miscanthus cultivations (17.7-21.8 t/ha/y) located on very good soils of Eastern Europe (Fischer et al., 2005). It can be assumed with a high probability that the limited water in 2005 did not influence the experimental Miscanthus yield, but there was an influence from weather condition in 2006 (Borzecka-Walker et al., 2008).

	Miscanthus		Willow	
	maritime	continental	Maritime	continental
yield t C/ha/y	4.8	5.1	4.8	5.0
organic matter input t C/ha/y	5.4	6.1	2.9	3.5
soil organic carbon t C/ha/y	1.5	1.7	0.71	0.85

Table 2. Soil carbon (C) balance (t C ha y) under Miscanthus and Willow cultivation

The aboveground biomass has a high influence on the amount of carbon sequestration, which enters the soil usually in the form of senescent leaf mass and postharvest remnants. Kahle (2001) measured in Germany that about 3.0-7.5 t/ha/y aboveground biomass full to soil. Matthews and Grogan (2001) in Great Britain estimated the inflow of the organic matter at a level of 7.5 t/ha/y, while for Poland it was calculated at from 2.63 to 6.58 t/ha/y (Borzecka-Walker et al., 2008). The organic matter input table 2 shows a greater potential for carbon mitigation for Miscanthus than for willow. Despite the very different movement of C into the soil plantations of willow and Miscanthus, most of this element accumulates in the litter. This is labile to C fraction, which in total will be mineralised and released as CO_2 in a short time after the restoration of conventional land use. Moderately stable fractions of carbon in the form of living organisms (humads) is the second largest fraction of sequestrated carbon. In almost all cases it will be transformed into humus after a change of use in the plantation. The stable fraction C (humus) in the lifetime of the plantation rose to a negligible extent. So it can be concluded that an effective carbon sequestration expressed the sum of humus fractions and humads. For willow and Miscanthus (tab. 3), during the period of cultivation it was respectively at 0.20, 0.21 for the maritime climate and the 0.23 and 0.25 in the continental climate. The obtained values from the model are close to the obtained values by Bradley & King (2004), who have determined the carbon sequestration in forests and willows cultivations at 0.15-0.22 t C/ha/y, whereas in Miscanthus cultivations they were at 0.13-0.20 t C/ha/y. According to Matthews & Grogan (2001), carbon sequestration in the surface layer of the soil (0-23 cm) was at 0.31 for forests, and 0.41 for the cultivation of willow, whereas for Miscanthus it was measured at 0.93 t C/ha/y. The net soil carbon sequestration in Miscanthus crops was around 0.38-0.95 t C/ha/y and 0.22-0.39 t C/ha/y for coppice willow (Borzecka-Walker et al., 2008). Freibauer et al., (2004) and Smith (2004) have determined the carbon sequestration in cultivations of energy crops appropriately at 0.60 and 0.62 t C/ha/y. Much of the carbon mitigation potential associated with the use of SRC willow and Miscanthus as bioenergy crops arises from their indefinite capacities as 'carbon neutral' alternatives to fossil fuel combustion (Grogan & Matthews, 2001). The assumption is that carbon sequestration of around 0.25 t C/ha/y resulting from energy crops cultivation makes biomass combustion neutral in terms of greenhouse gas emissions (Volk et al., 2004). The new cultivations will result in changes in fossil-fuel use, agricultural inputs, and carbon emissions with fossil fuels and other inputs. Management practices that alter crop yields and land productivity can affect the amount of land use crop production with further significant implications for both emissions and sequestration potential (West & Marlnd, 2003; Schneider & Mccarl, 2003).

SOM	Litter	Humads	Humus
Maritime			
Willow	0.28	0.19	0.01
Miscanthus	0.86	0.20	0.01
Continental			
Willow	0.32	0.21	0.02
Miscanthus	0.91	0.22	0.03

Table 3. Soil organic matter (SOM) pools under Miscanthus and willow cultivation

The organic matter input (tab. 2) is basically consisting of litter and dead roots fractions. Miscanthus harvested in autumn delivers approximately 20% of leaf mass and some of underground biomass while 30 % of leaf mass and some underground biomass from spring harvest is entering the soil. In compare 100 % of willow leaf mass and some of underground biomass is entering the soil. Soil organic matter includes tree pools: very labile fraction of litter, labile humads, and passive humus (tab.3).

4. Conclusion

Agriculture practices have an important role to play in mitigating climate change due to atmospheric enrichment of CO_2 and other greenhouse gases. To improve the negative balance of soil carbon sequestration in Poland, corrective action should be taken. Land management can strongly influence soil carbon stocks and careful management can be used to increase soil carbon sequestration. It is important to propose contemporary management practises to farming like the conversion from tillage to no tillage systems, incorporation of cover crops, forages in crop rotation, as well as a liberal use of crop residues and biosolids like mulch. Special care should be taken of integrated nutrient management including compost/manures and precision use of fertilizers and integrated pest management. A very important role can be played by the implementation of biocrops which are characterised with very high potential of carbon sequestration and much lower GHG emission during the cultivation.

When considering carbon sequestration it should be mandatory to combine these analyses with nitrogen (N) as carbon and nitrogen move through terrestrial ecosystems coupled with biogeochemical cycles, and increasing C stocks in soils and vegetation which have an impact on the N cycle.

5. Acknowledgment

The work was done based on the results obtained within the research projects N N315 759240 and N N313 436839, funded by the Ministry of Science and Higher Education.

6. References

Borzecka-Walker, M.; Faber, A. & Borek, R. (2008). Evaluation of carbon sequestration in energetic crops (Miscanthus and coppice willow). *Int. Agrophysics*, No 22, 185-190.

Bradley, R.I. & King, J.A. (2004). A review of farm management techniques that have implications for carbon sequestration – validating an indicator. OECD Expert Meet. Farm Management Indicators and the Environment, 8-12 March, Palmerston North, Avaialble from http://webdomino1.oecd.org/comnet/agr/farmind.nsf/22afaebba539ba74c1256a3b004d5175/b3b8d25f219f4ae3c1256bd5004874f1/$FILE/Bradley1.pdf

Clifton-Brown, J.C.; & Lewandowski, I.; (2000). Overwintering problems of newly established Miscanthus plantations can be overcome by identifying genotypes with improved rhizome cold tolerance. *New Phytol.* No 148, 287-294.

COM, (2000) 769: Green Paper: Towards a European strategy for the security of energy supply

Directive, 2003/96/EC. (2003). The European Parliament and of The Council on restructuring the Community framework for the taxation of energy products and electricity.

Directive, 2009/28/EC. (2009). The European Parliament and of The Council on the promotion of the use of energy from renewable sources amending and subsequently repealing Directives 2001/77/EC and 2003/30/EC.

Dumanski, J.; Desjardins, R.L.; Tarnocal, C.; Monreal, EG.; Gregorich, E.G.; Kirkwood, V. &Campbell, C.A. (1998). Possibilities for future carbon sequestration in Canadian agriculture in relation to land use changes. *Climatic Change,* Vol. 40, 81-103.

ECAF (European Conservation Agriculture Federation). Observations on the Communication form the European Commission: Towards a Thematic Strategy for Soil Protection (COM (2002) 179 final) www.ecraf.org

Farge, P.K.; Ardő, J.; Olsson, L.; Rienzi, E.A. & Pretty, J.N. (2007). The potential for soil carbon sequestration in the three tropical dryland farming systems of Africa and Latin America: a modeling approach. *Soil Tillage Res,* Vol. 94, 457-472.

Fischer, G.; Prieler, S.; & van Velthuizen, H.; (2005). Biomass potentials of Miscanthus, willow and poplar: results and policy implications for Eastern Europe. Northern and Central Asia. *Biomass and Bioenergy.* No 28,119-132.

Fotyma, M. & Mercik, S. (1992). Współczynniki reprodukcji glebowej materii organicznej. In: *Chemiarolna,* PWN Warszawa.

Franzluebbers, A.J. & Follett, R.F. (2005). Greenhouse gas contributions and mitigation potential in agricultural regions of North America: Introduction. *Soil & Tillage Research,* Vol. 83, 1–8.

Freibauer, A.; Rounsevell, M.; Smith, P. & Verhagen A. (2004). Carbon sequestration in European agricultural soils. *Geoderma,* Vol. 122, 1–23.

Giltrap, D.L.; Li, Ch. & Saggar, S. (2010). DNDC: A process-based model of greenhouse gas fluxes from agricultural soils. *Agriculture Ecosystems and Environment,* Vol. 136, 292-300.

Grant, B.; Smith, W.N.; Desjardins, R.; Lemke, R. & Li, C. (2004). Estimated N_20 and CO_2 emissions as influenced by agricultural practices in Canada. *Climatic Change,* Vol. 65, 315-332.

Groenigen, K.J.; Hastings, A.; Forristal, D.; Roth, B.; Jones, M. & Smith, P. (2010). Soil C storage as affected by tillage and straw management: An assessment using field measurements and model predictions. *Agriculture Ecosystems and Environment,* Vol. 140, 218-225.

Grogan, P. & Matthews, R. (2001). Review of the potential for soil carbon sequestration under bioenergy crops in the U.K. Report for U.K. Dept. of Environment, Food, and Rural Affairs. Contract NF 0418.

Gronlund, A.; Hauge, A.; Hovde, A. & Rasse, D.P. (2008). Carbon loss estimates from cultivated peat soils in Norway: a comparison of three methods *Nutr Cycl Agroecosyst,* Vol. 81, 157–167.

Guo, L.B. & Gifford, R. M. (2002). Soil carbon stocks and land use change: a meta analysis. *Global Change Biology,* Vol. 8, 345-360.

Hillel, D. & Rosenzweig, S. (2009). Soil and carbon climate change. *CSA News,* Vol. 54, No 6, 4-11.

Holland, J.M. (2004). The environmental consequence of adopting conservation tillage in Europe: reviewing the evidence. *Agriculture Ecosystems and Environment,* Vol. 103, 1-25.

Hutchinson, J.J.; Campbell, C.A. & Desjardins, R.L. (2007). Some perspectives on carbon sequestration in agriculture. *Agricultural and Forest Meteorology,* Vol. 142, 288–302.

IPCC, (2000). IPCC special report on Impacts. Adaptation and Vulnerability. http://www.grida.no/climate /ipcc_t ar/wg2/index.htm

IPCC, (2007). Fourth Assessment Report. Synthesis Report. http://www.ipcc.ch/pdf/assessment-report/ar4/syr/ar4_syr.pdf

Kahle, P.; Beuch, S.; Boelcke, B.; Leinweber, P.; & Schulten, H.R.; 2001: Cropping of Miscanthus in Central Europe: biomass production and influence on nutrients and soil organic matter. *Eur. J. Agron.* No 15: 171-184.

Knowler, D. & Bradshaw, B. (2007). Farmers' adoption of conservation agriculture: A review and synthesis of recent research. *Food Policy,* Vol. 32, 25-48.

Kus, J.; Madej, A. & Kopinski, J. (2006). Bilans słomy w ujęciu regionalnym. *Studia i Raporty IUNG-PIB,* Wyd. IUNG, Pulawy, No 3, 211-225.

Lal, R. (2004a). Agricultural activities and the global carbon cycle. *Nutrient Cycling in Agroecosystems,* Vol. 70, 103–116.

Lal, R. (2004b). Soil carbon sequestration to mitigate climate change. *Geoderma,* Vol. 123, 1 - 22.

Li, C.; Frolking, S. & Butterbach-Bahl, K. (2005). Carbon sequestration in arable soils is likely to increase nitrous oxide emissions, offsetting reductions in climate radiative forcing. *Climatic Change,* Vol. 72, 321-338.

Li, C.; Frolking, S. & Frolking, T.A. (1992). A model of nitrous-oxide evolution from soil driven by rainfall events: model structure and sensitivity. *Journal of Geophysical Research–Atmospheres,* Vol. 97, 9759–9776.

Li, T.; Feng, Y. & Li, X. (2009). Predicting crop growth under different cropping and fertilizing management practices. *Agricultural and Forest Meteorology,* Vol. 149, 985-998.

Matthews, R.B. & Grogan, P. (2001). Potential C-sequestration rates under short-rotation coppiced willow and Miscanthus biomass crops: a modelling study. *Aspects Appl. Biol.,* Vol. 65, 303-312.

Ringius, L. (2002). Soil carbon sequestration and the CDM: opportunities and challenges for Africa. *Climatic Change,* Vol. 54, 471–495.

Schneider, U.A. & Mccarl, B.A. (2003). Economic potential of biomass based fuels for greenhouse gas emission mitigation. *Environmental and Resource Economics,* Vol. 24, 291–312.

SEC. Commission Staff Working Document. (2009). The role of European agriculture in climate change mitigation. Brussels, 23.7.2009, 1093 final.

Smith, P. (2004). Carbon sequestration in croplands: the potential in Europe and the global context. *Europ. J. Agronomy,* Vol. 20, 229–236.

Smith, P. (2008). Land use change and soil organic carbon dynamics. *Nutr Cycl Agroecosyst,* Vol. 81, 169–178.

Smith, P.; Powlson, D.S.; Glendining, M.J. & Smith, J.U. (1997). Potential for carbon sequestration in European soils: preliminary estimates for five scenarios using results from long-term experiments. *Global Change Biol,* Vol. 3, 67– 79.

Smith, P.; Powlson, D.S.; Smith, J.U.; Falloon, P.D. & Coleman, K. (2000). Meeting Europe's climate change commitments: quantitative estimates of the potential for carbon mitigation by agriculture. *Global Change Biol,* Vol. 6, 525–539.

Stuczynski, T.; Kozyra, J.; Lopatka, A.; Siebielec, G.; Jadczyszyn, J.; Koza, P.; Doroszewski, A.; Wawer, R. & Nowocien, E. (2007). Przyrodnicze uwarunkowania produkcji rolniczej w Polsce *Studia i Raporty IUNG-PIB,* Wyd. IUNG, Pulawy, No 7, 77-115.

Terelak, H.; Motowicka-Terelak, T.; Wroblewska, E.; Gawrysiak, L. & Pietruch, C. (2001). Mapa zawartości substancji organicznej w glebach użytków rolniczych Polski. Wyd. IUNG, Puławy.

VDLUFA (2004): VDLUFA Standpunkt Humusbilanzierung, Methode zur Beurteilung und Bemessung derHumusversorgung von Ackerland, Bonn 2004, http://www.vdlufa.de, (accessed April 2004).

Vleeshouwers, L.M. & Verhagen, A. (2002). Carbon emission and sequestration by agricultural land use: a model study for Europe. *Global Change Biol,* Vol. 8, 519–530.

Volk, T.A.; Verwijst, T.; Tharakan, P.J.; Abrahamson, L.P. & White, E.H. (2004). Growing fuel: a sustainability assessment of willow biomass crops. *Front Ecol. Environ,* Vol. 2, No 8, 411- 418.

West, T.O. & Marlnd, G. (2003). Net carbon flux from agriculture: Carbon emissions, carbon sequestration, crop yield, and land use change. *Biogeochemistry,* Vol. 63, 73-83.

West, T.O. & Post, W.M. (2002). Soil Organic carbon sequestration rates by tillage and crop rotation: a global analysis. *Soil Science Society of America Journal,* Vol. 66, 1930-1946.

10

The Fate and Transport of *Cryptosporidium parvum* Oocysts in the Soil

X. Peng[1,2], S. Macdonald[2], T. M. Murphy[3] and N. M. Holden[2]
[1]*State Key Laboratory of Soil and Sustainable Agriculture, Institute of Soil Science, Chinese Academy of Sciences, Nanjing,*
[2]*UCDSchool of Biosystems Engineering, University College Dublin, Belfield, Dublin*
[3]*Central Veterinary Research Laboratory, Young's Cross, Celbridge, Co. Kildare*
[1]*P.R. China*
[2,3]*Ireland*

1. Introduction

Cryptosporidium parvum (*C. parvum*) is a protozoan pathogen that is commonly present in surface water used for recreation, treated and untreated sewage and even drinking water. The zoonotic protozoan parasite *C. parvum* poses a significant risk to public health and has become a global concern to the water resource management since the 1993 outbreak in Milwaukee, US (Fayer, 2008). Ingestion of a small number of Cryptosporidium oocysts of can lead to potentially fatal consequences for immuno-suppressed individuals and it is thought that infection can be caused by even a single oocyst (Rose, 1997). At present, no specific drug treatment exists for cryptosporidiosis. *C. parvum* oocysts are biologically dormant yet resistance to the levels of chlorine routinely used in potable water treatment thus making waterborne transmission of cryptosporidiosis one of the most prominent public health concerns worldwide (Fayer, 2008).

The main source of *C. parvum* oocysts is from infected calves when agricultural water including runoff, infiltration and subsurface flow from dairies, calving house, silage and grazing lands may be loaded with high concentrations of oocysts. Thus, the fate and transport of *Cryptosporidium* oocysts in the soil is of critical importance to manage and assess risk (Pachepsky et al., 2006). Thus, three key processes are evaluated in this study, including i) oocyst survival in the soil, ii) the attachment of oocysts to soil particles and iii) the transport of oocyst with physical straining in porous soils. Our study may provide an insight into the oocyst life cycle in the soil and help us make a better risk assessment.

2. The fate of *Cryptosporidium* oocysts in the soil

Cryptosporidium oocysts can retain infectivity for months and resist environmental stresses more readily than many other pathogens because of a hard protective wall. As a result, characterisation of the fate of oocysts in the soil has received much attention. In general, the survival of oocysts in the soil slows down exponentially with time. To account for shoulder and tailing effects, a first-order exponential model has usually been used to simulate the die-off curve of oocysts in soils with Eq. 1 (Peng et al., 2008):

$$y_t(t) = y_0 e^{-Kt}, \tag{1}$$

where K (dimensionless) is the die-off rate coefficient, y_0 and y_t are the oocyst number at initial condition and time t (any suitable unit of time), respectively. If normalized by initial oocyst number, Eq. 1 can be rewritten as:

$$y_t'(t) = e^{-Kt} \quad \text{or} \quad \ln y_t'(t) = -Kt, \tag{2}$$

where

$$y_t'(t) = \frac{y_t(t)}{y_0}. \tag{3}$$

In Eq. 2 K is independent of the initial oocyst number and represents a constant die-off rate (K) over the entire incubation period. Alternatively, for a given percentage of inactivated oocysts, K is inversely proportional to the incubation time. Using Eq. 2 it is possible to estimate the infectivity of oocysts at a given time.

2.1 Effect of soil moisture on oocyst survival

Besides temperature, oocyst survival is subjected to stress in soils driven by the interaction of soil moisture and texture. Soil water potential indicates whether oocysts are exposed to wet or dry conditions. Desiccation is probably lethal to oocysts. Walker et al. (2001) reported that decreasing soil water potential by adjusting NaCl solution linearly increased the rate of population degradation. Nasser et al. (2007) also found incubation for 10 days in dry loamy soil at 32°C resulted in a 3-$\log_{10}$ reduction in oocyst infectivity, but a saturated soil caused only 1-$\log_{10}$ reduction at a similar temperature. Jenkins et al. (2003) reported the estimated K values were increased from 0.014 day^{-1} to 0.416 day^{-1} at 25°C when soil water potential was decreased from -0.10 MPa down to -3.2 MPa, but they also pointed out that when soil water potential indicated a water content in excess of field capacity (> -0.10 MPa), then the effect on oocyst infectivity was negligible. Jenkins et al. (2002) did not find any effect on oocyst inactivation of soil water potential in the range -0.033 to -1.5 MPa using a sentinel chamber. This discrepancy compared to other reports perhaps resulted from the one open end of the sentinel chamber being exposed to free water, through which the water potential of the soil in the chamber could be mediated by the moisture of the surrounding soil. Published results suggest that soil water potential or moisture content can influence oocyst die-off rate.

2.2 Effect of soil texture on oocyst survival

Soil texture, a static property of the soil also appears to affect oocyst survival in the soil. Jenkins et al. (2002) reported oocyst survival appeared to be significantly greater in a silt loam soil (50% silt, 16% clay) than in silty clay loam (69% silt, 29% clay) and loamy sand (14% silt, 5% clay). Davies et al. (2005) found oocysts were inactivated faster in loam soil (27% silt, 24% clay) than in clay loam (55% silt, 38% clay). Soil particles probably do not directly affect oocyst survival, but they can modify oocyst metabolic activity through changing other physiochemical and biological soil properties and their attachment properties.

3. The attachment of *Cryptosporidium* oocysts in the soil

As mentioned, soil particles affected the fate of *Cryptosporidium* oocysts. Oocysts attachment to soil particles during transport may have a significant influence on retention in the soil environment and thus longevity of the organism. Attached oocysts seem to be protected from inactivation by ultraviolet radiation, temperature or desiccation (Morita et al., 2002; Hijnen et al., 2006; Peng et al., 2008), but may be a part of the general food web of the soil. Soil might be regarded as a habitat that can promote survival and provide a source of oocysts especially after spreading of animal manure and slurry. The interaction between oocysts and soil particles can increase the oocysts' effective settling velocity and play an important role in regulating the transport and retention in agricultural catchments (Searcy et al., 2005; Searcy et al., 2006). Application of bovine manure was found to enhance the attachment of oocysts to soil particles and this mechanism was partially reversible if manure was diluted (Kuczynska et al., 2005). Searcy et al. (2005) reported that more oocysts were attached to iron oxide and kaolinite particles than illite particles because the latter are more negative charged, but Walker and Montemagno (1999) found that oocysts were readily attached to Al_2O_3 but not to Fe_2O_3, SiO_2 and hydrophobic substrates. Therefore, whether or not oocysts attach to soil particles may depend on the particles physical and chemical properties. Unfortunately obtaining data on soil mineralogy is difficult as it is rarely collected during wide area surveys. This means it would be very useful to establish whether oocyst attachment can be related to more commonly measured soil properties that can be found in may soil surveys and could therefore be used for improving quantitative environmental risk assessments.

Soil code	Sand§	Silt§	Clay§	SOC¶	CEC¶	Total pore	Macropore¶	K_s¶	pH
	------ g kg^{-1} ------				mmol kg^{-1}	-- cm^3 cm^{-3} --		cm day^{-1}	[-]
S02	445	397	159	27	169	0.48	0.11	2.94×10^3	6.08
S01	373	440	187	41	202	0.57	0.06	7.00×10^2	5.72
CG	357	431	212	44	187	0.53	0.10	4.42×10^2	5.32
EG	355	424	221	48	212	0.60	0.04	9.26×10^1	6.57
OG	473	364	162	30	151	0.53	0.04	7.30×10^1	5.60
JSG	668	217	116	16.3	92	0.42	0.01	2.16×10^1	6.33
KG	386	513	102	91	259	ND#	ND#	$<1.0\times10^0$	6.30
RG	394	444	162	38	163	ND#	ND#	$<1.0\times10^0$	6.56
CCG	288	496	217	72	225	ND#	ND#	$<1.0\times10^0$	5.29

§Sand: 0.05-2 mm, Silt, 0.002-0.05 mm, Clay, <0.002 mm.
¶SOC, soil organic carbon, CEC, Cation exchange capacity, Macropore, >60 μm diameter, K_s, saturated hydraulic conductivity of distilled water.
#ND, not determined.

Table 1. Selected soil properties of nine soils present in permanent pastures in Ireland

In the laboratory, we established an attachment experiment to analyse how soil properties influence the interaction between oocysts and soil particles. Nine typical grassland soils from Ireland with a limited range of soil texture but a wide range of soil organic C were used (Table 1). The soils were selected by two reasons: (i) those previously identified as being characteristic of Irish grassland agriculture (CG, EG, OG, and RG soils) (Ryan and Fanning, 1996); and (ii) those found in a specific study catchment used for developing risk assessment associated with *Cryptosporidium* transport (S01, S02 and KG soils). In addition, two other soils (JSG and CCG) were selected to increase the range of soil texture in study. For each soil a composite disturbed sample from five points around a a field was taken, air-dried and ground to <2 mm for determining soil properties. Routine methods of the Institute of Soil Science, Chinese Academy of Sciences (Ru, 2000) were used to analyse basic soil properties: particle-size distribution was determined by the pipette method, particle density was measured by pycnometer method, soil organic carbon was by oxidation with potassium dichromate, cation exchange capacity (CEC) by the ammonium acetate method, and pH by a pH meter, saturated hydraulic conductivity (Ks) by a constant head method. Table 1 presents the measured properties of each soil used.

3.1 Oocysts preparation

C. parvum oocysts were purchased from Creative Science Company, UK. They were purified from manure of experimentally infected calves by sucrose and Percoll gradient centrifugation and water washes, and stored in 0.1% phosphate buffered saline solution at 4 ^{o}C before use. The number of oocysts in the stock solution was about 5.0 x 10^8 oocysts ml^{-1}. Oocysts ranged from 3.9 to 5.9 μm in diameter, and their density was between 1.0 and 1.1 g cm^{-3}.

3.2 The batch attachment experiment

Batch experiments were conducted by spiking ca. 2×10^4 oocysts into 10 ml, pH 7.0 <2 mm soil particles solution in which particle concentration was 2 mg ml^{-1}. A blank experiment (no particles) was run as a reference. Each treatment was run in triplicate. The tubes were rotated overnight in dark room at 15 ^{o}C to let oocysts attach to particles adequately. The mixed suspension was then stood on a vibration-free bench. After 30 min, the top 1 ml (ca. 6.4 mm height) was aspirated either from an oocyst-particle tube or from a blank treatment for direct oocyst staining. The percentage of particle-attached oocysts was estimated by the ratio of the oocysts concentration from the oocyst-particle solution ($Oocyst_{particle}$) to the oocyst concentration from the blank treatment ($Oocyst_{blank}$) as counted in the top 1 ml.

$$Oocyst_{attached}\% = (1 - \frac{Oocyst_{particle}}{Oocyst_{blank}}) \times 100 \quad (4)$$

3.3 Oocyst staining protocol

The oocyst staining followed the instructions of the Dynabeads® anti-Cryptosporidium kit (Invitrogen, Norway). Briefly: (i) each sample was thoroughly vortexed, a 100 μl aliquot was pipetted into a slide well (diameter 12 mm) and air dried at room temperature; (ii) two drops of methanol were added to each well and allowed to air dry; (iii) 50 μl of a combined fluorescein isothiocyanate (FITC) conjugated anti-Cryptosporidium monoclonal antibody was added to each well; (iv) the slide was placed in a humid chamber at room temperature for 60 min; (v) monoclonal antibody was aspirated from the well and washed twice with PBS; and (vi) the slide was covered with a glass slip. The stained slides were enumerated immediately at 200× magnification with an epifluorescence microscope (Olympus, Japan) containing a filter cube with an emission of 530 nm and an excitation wavelength of 490 nm.

3.4 Relation between *Cryptosporidium* oocysts and soil particles

The relationship between oocyst and soil particles for the nine test soils (Figure 1) indicated that the KG soil had a poor interaction with oocysts. In EG and RG soils, oocysts showed a strong attachment to soil particles. For the other six test soils the percentage of particle-free oocysts was between 50% and 65%. The attachment of oocysts depended on fine particle content ($r=0.611$, $P<0.1$). Muirhead et al. (2006) reported that pathogenic *E. coli* appeared to attach preferentially to small particles (<2 μm), which is consistent with the result that the percentage of particle-free oocysts was negatively related to clay content, but not to silt or sand content.

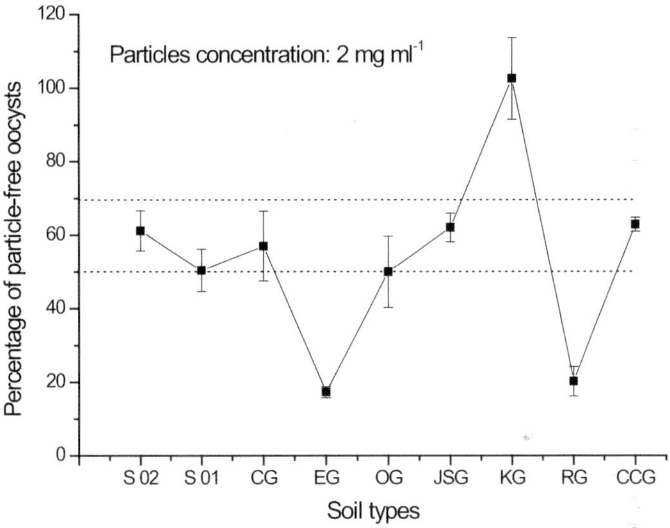

Fig. 1. Oocyst-particle attachment of 9 test soils

Of the 9 test soils, over 35% of oocysts were attached to soil particles with the exception of the high organic carbon KG soil. Negative charged soil organic matter may prevent oocysts, which also have a negative surface charge, from attaching to soil particles. Although there was no significant relationship between attached oocysts and soil organic carbon ($r=-0.492$, $P=0.18$), the sand-free soil organic carbon was observed to be negatively related with attached oocysts ($r=-0.681$, $P<0.05$). This finding is consistent with the results of Searcy et al. (2005) who reported that considerably less oocysts were removed from suspension in the present of more negative charged illite than in the presence of iron oxide or kaolinite. The electrostatic repulsive forces should increase as the suspended particles become more negative. The hydrophobicity of soil organic matter may also contribute to the reduced attachment of oocysts, which is supported by results from Walker and Montemagno (1999) who observed no sorption of oocysts onto hydrophobic substrates.

4. The transport of *Cryptosporidium* oocysts in the soil

The transport and retention behavior of Cryptosporidium oocysts has been examined using homogeneous columns under steady-state, constant flux (CF) conditions controlled by a peristaltic pump (Brush et al., 1999; Harter et al., 2000; Logan, et al., 2001; Bradford and Bettahar, 2005; Hijnen et al., 2005). These studies concentrated on how oocysts moved through homogeneous porous media as affected by the particle size, flux rate or solution chemical properties. Decreasing the median sand size tends to produce lower effluent concentrations and greater oocyst retention near the column inlet (Harter et al., 2000; Logan et al., 2001). High flux rate can produce more oocysts in effluents than low flux rate (Harter et al., 2000; Hijnen, et al., 2005). Hus et al. (2001) and Tufenkji et al. (2005) found that the oocyst removal efficiency depended on ionic strength and solution pH. For the colloid-sized oocysts breakthrough curve (BTC), three physical mechanisms of attachment, detachment and irreversible straining may be involved in the tailing and heterogeneous distribution of oocysts at depth (Bradford et al., 2003; Bradford and Bettahar, 2005). However, the pump used in some of these experiments may force oocysts to pass through smaller pores in order to maintain the CF conditions. In other words, under CF conditions, the pressure applied by the pump will increase with clogging in order to keep a CF. The transport behaviour and breakthrough curve may be different from those under the constant pressure conditions (CP). For the same initial flow rate, we hypothesized that the CF condition may therefore overestimate the oocyst transport potential compared with CP conditions.

In this study, we tested three different porous materials. One coarse sand, one fine sand and soil material <2 mm sieved from a Haplic Cambisol soil from CG site, were uniformly repacked into stainless steel cylinders (20 cm in height, 10 cm in diameter). The median grain size (d_{50}) of coarse sand, fine sand and disturbed soil was 0.12, 0.17 and 0.32 mm, respectively (Table 2). The uniformity index (d_{60}/d_{10}) was 2.4 for coarse sand, 2.3 for fine sand, and 8.2 for disturbed soil. The three homogeneous columns were used to compare oocysts BTC between constant flux (CF) and constant pressure head (CP) systems. For each porous medium, the initial volumetric flow rate between the CF and CP systems was identical.

Porous media	d_{10} (mm)†	d_{50} (mm)	d_{60} (mm)	Uniformity Index (d_{60}/d_{10})	Porosity ($cm^3\ cm^{-3}$)	Ks§ ($cm\ min^{-1}$)
Coarse sand	0.07	0.15	0.17	2.4	0.44	1.7
Fine sand	0.06	0.12	0.14	2.3	0.43	1.3
Disturbed soil	0.05	0.32	0.41	8.2	0.59	0.13

†d_{10}, d_{50}, d_{60} mean 10%, 50%, 60% of the mass finer than the size, respectively
§K_s, saturated hydraulic conductivity

Table 2. Properties of the three homogeneous porous media studied

4.1 The oocysts transport experiment

Prior to BTC investigation, the columns were saturated from the bottom with distilled water for two days. Then about 1-2 pore volume of 0.1 M NaBr solution ran through the column from top to bottom, followed by another 1-2 pore volume of 2.0×10^3 oocysts ml^{-1} solution.

The oocysts reservoir was placed on a stirrer to keep the oocyst homogeneously mixed during the experiment. Br^- and oocysts transport was pushed through the column under a 22 cm CP between the inlet and the outlet or under the CF conditions controlled by a peristaltic pump (Figure 2). Each porous medium was run in triplicate.

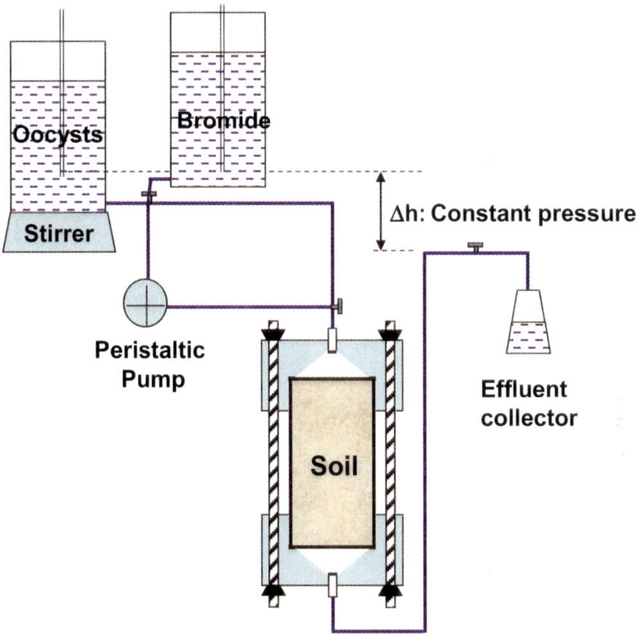

Fig. 2. A schematic diagram of the soil column experiment setup employed in the Br- and oocysts breakthrough study under a constant pressure (CP, Δh=22 cm in this study) or under a constant flux (CF) controlled by a peristaltic pump

4.2 Simulation analysis

The transport of colloid-sized *Cryptosporidium* oocysts can be described by the convective-dispersion equation (CDE), which has been widely used to model solute transport (Jury, 1991). To describe oocysts transport through saturated columns, we employed the one-dimensional CDE:

$$R\frac{\partial C}{\partial t} = D\frac{\partial^2 C}{\partial x^2} - v\frac{\partial C}{\partial x} - \mu C \tag{5}$$

where C is the reduced concentration of the liquid phase (dimensionless), t is time (min), x is vertical distance in the direction of flow (cm), D is the hydrodynamic dispersion coefficient (cm^2 min^{-1}), v is the average water velocity (cm min^{-1}), and μ is the first-order rate decay coefficient describing straining/filtration (min^{-1}). R is the retardation factor, defined as

$$R = 1 + \frac{\rho K_d}{\theta} \tag{6}$$

where ρ is the bulk density of the porous media (g cm^{-3}), θ is the volumetric water content, and K_d is the distribution coefficient describing solute adsorption to the solid phase (cm^3 g^{-1}). Using the computer program CXTFIT for solving nonlinear CDE regression (Toride et al.,1995), we firstly obtained the fitted v and D parameters from the tracer Br^- BTC, with R = 1.0 and μ =0 min^{-1}. Pore water v and D of the oocyst BTC were assumed to equal those of the Br^- trace in each column. With this assumption, we estimated R and μ of the *C. parvum* oocyst BTC.

4.3 Transport of oocysts in the columns

The relative concentration (C_i/C_0) of Br^- and oocysts in effluents as a function of pore volume following step input (Figure 3) reduced the experimental time and the time-consuming enumeration, as compared to using a pulse step method, because the latter requires definition of the rising and falling limbs. For colloid-sized *C. parvum* oocysts transport, three physical mechanisms of straining, attachment and detachment have recently been discussed using results from the pulse step method (Harter et al., 2000; Bradford et al., 2004; Bradford and Bettahar, 2005). Straining was identified as perhaps the most important mechanism. Use of a step input method may highlight the straining mechanism, but any detachment was ignored. Therefore, the retardation parameter R and the straining factor μ of oocyst BTC were estimated from CDE theory (Table 3).

Porous media	Flux system	Bromide			Oocysts		
		v	D	r^2	R	m	r^2
Coarse	CF	3.60±0.13 a†	1.62±0.39 a	1.00±0.00	0.94±0.04 a	0.00±0.00 a	0.91±0.03
sand	CP	3.90±0.12 a	1.76±0.61 a	0.99±0.01	0.79±0.09 a	0.00±0.00 a	0.90±0.05
Fine	CF	3.10±0.04 a	0.89±0.11 b	1.00±0.00	0.92±0.07a	0.01±0.00 b	0.96±0.01
sand	CP	3.10±0.29 a	1.10±0.04 a	1.00±0.00	0.89±0.14 a	0.07±0.01 a	0.96±0.02
Disturbed	CF	0.81±0.04 a	4.30±0.97 a	0.98±0.00	1.40±0.57 a	0.13±0.02 a	0.36±0.19
soil	CP	0.76±0.21 a	12.47±5.4 a	0.98±0.02	5.98±6.25 a	0.21±0.11 a	0.48±0.28

†Different letters indicate significant difference between CF and CP systems at $P<0.05$

Table 3. Fitted parameters of the convective-dispersion equation using the CXTFIT package for the homogeneous porous media. v = pore water velocity (cm min^{-1}), D = hydrodynamic dispersion coefficient (cm^2 min^{-1}), R = retardation factor, μ = straining factor (min^{-1}), and r^2 = squared relation coefficient. CF = constant flux, CP = constant pressure. Mean±standard deviation (n=3)

For the coarse sand column, the colloid-sized oocyst BTC was earlier than the tracer Br^- in both CF and CP systems (Figure 3). This phenomenon of oocysts travelling faster than the tracer Br^- was also observed in fine sand column under the CF conditions. However, a remarkable delay took place in the disturbed soil material. The results were consistent with the fitted retardation factor R values less than 1 for the coarse and fine sand column but

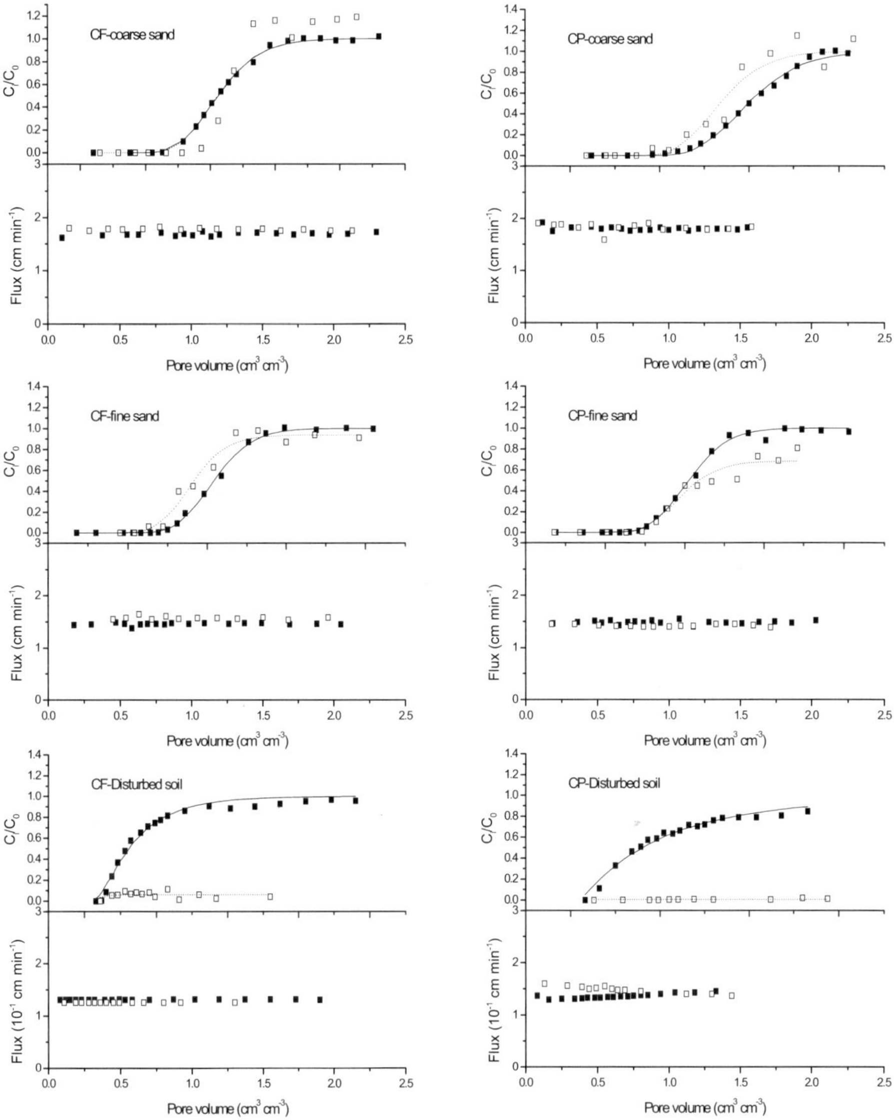

Fig. 3. Comparison of the breakthrough curves of Br^- (solid squares) and oocysts (open squares) and their water flux as a function of pore volume between CF and CP systems for the three homogeneous pore media. The lines of Br^- (solid) and oocysts (dot) are the curves fitted with Equation 1. C_0 is the applied concentration, C_i is the measured concentration

larger than 1 for the disturbed soil (Table 3). The *R* values less than 1 in this study agree with previous reports (Brush et al., 1999; Harter et al., 2000) that oocysts were excluded from the pores finer than their size and from the margins of the pore throats, but this was not true for tracer Br^-. The size exclusion and pore inaccessibility would most likely increase the average pore velocity of colloid-sized oocysts.

Both CP and CF systems displayed an identical steady-state water flux either in ionic Br^- transport or in colloid-sized oocyst transport in the coarse sand column. From the simulation results, there was no significantly difference of the means of the parameters v, D, R and μ. In the coarse sand column, $\mu=0$ indicated that physical straining was negligible. In the fine sand column, oocyst BTC was consistent with the Br^- BTC under the CF conditions, but slower at the later part under the CP conditions, although the water flux remained at a steady state in both Br^- and oocyst transport. The parameters of D and μ were significantly greater under the CP conditions than under the CF conditions, which further demonstrated that more oocysts were retained in the former, even when the water flow rates were identical. The oocysts BTC was overestimated when the consequences of physical straining were overlooked in the CF system. The small values of μ, 0.01 and 0.07 min^{-1} in the CF and CP systems, respectively, showed that the straining of oocysts was not predominant in the fine sand column. The physical straining that occurred in the fine sand meant a steady state of water flow was maintained, but with sufficiently reduce effluent oocyst concentrations. The grain size-dependent oocyst transport was in accordance with other findings (Harter et al., 2000; Bradford et al., 2003; Bradford et al., 2005), which indicate that decreasing grain size tends to produce lower effluent concentration and greater oocyst retention in the column.

In the disturbed soil column, oocyst BTC was much slower than Br^- in both CF and CP systems. Oocysts passing through columns were much less under the CP conditions than under the CF conditions. Under the CP conditions, the water flux during the oocyst breakthrough process always decreased over the entire experiment, whereas the water flux during the Br^- transport increased due to some fine soil particles leaching from the column. The decrease in the water flux resulted from physical straining and trapping of oocysts in the soil pores (Bradford and Bettahar, 2005), or from change of ionic strength (Hsu et al., 2001; Tufenkji et al., 2004). The fitted D, R and μ values in the CP system were greater than in the CF system, although the simulation was poor for the oocyst BTC ($r^2 < 0.5$).

Oocyst concentration decreased exponentially from the inlet to the outlet in coarse and fine sand columns (Figure 4), which is in generally agreement with the shape of the oocyst spatial distribution observed by Bradford and Bettahar (2005) and Harter et al. (2000). Due to a very low recovery for soils, the measurement was not applied to the disturbed soil samples. In coarse sand, a similar mass fraction in effluent (51.5-56.6%) was observed under the CF and the CP conditions. This identical result was for the fraction retained in the columns (29.4-29.9%). In fine sand, however, a significantly greater mass fraction in the effluent was under the CF conditions compared to the CP conditions, although the oocyst fraction retained in the columns was similar.

From the results of the three homogenous columns, the physical straining was more pronounced in disturbed soil than in the sand columns. This may be caused by oocysts preferentially attached to soil particles and by finer pore size distributions in the disturbed soil columns. The consequences of physical straining of oocysts through fine sand were not observed under the CF conditions. Use of a peristaltic pump to keep the flux at a steady state has been widely applied to investigate colloid-sized oocysts transport (Brush et al., 1999; Harter et al., 2000; Tufenkji et al., 2004; Bradford and Bettahar, 2005; Hijnen et al., 2005; Bradford et al., 2006). However, the pump may force oocysts to pass through pores or pore necks which are smaller than their critical size. As a result, the oocyst BTC may be overestimated in the CF system when physical straining takes place. On the other hand, if pore space is large enough for oocyst transport, the physical straining may become a minor mechanism, as occurred in the coarse sand column even in the CP system.

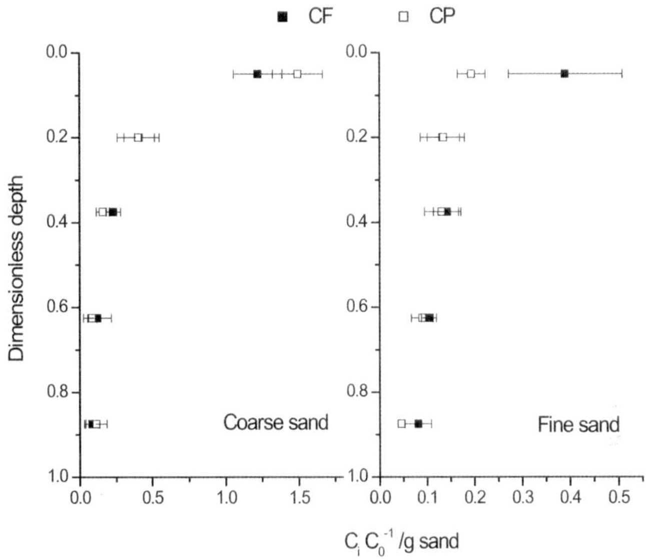

Fig. 4. Comparison of oocysts retention profiles in the coarse and fine sand columns between CF and CP systems. C_0 is the applied concentration, C_i is the measured concentration

Porous media	Flux system	$Mass_{effluent}$	$Mass_{retained}$	Total recovery (%)
Coarse sand	CF	56.6±1.0a	29.4±7.0a	86.6±6.0
	CP	51.5±8.2a	29.9±3.4a	81.4±11.2
Fine sand	CF	46.7±4.9a	14.1±3.3a	60.8±6.9
	CP	30.7±1.1b	10.6±0.7a	41.4±1.8
Disturbed soil	CF	6.8±2.3a	ND	ND
	CP	0.4±0.2b	ND	ND

†Different letters indicate significant difference between CF and CP systems at P<0.05
ND=Not determine

Table 4. Oocyst fractions in effluents and retained in the columns. CF = constant flux, CP = constant pressure. Mean±standard deviation (n=3)

5. Conclusions

For colloid-sized *Cryptosporidium* oocysts the fate and transport processes depend much on the soil physical and chemical properties. The survival of oocysts is affected by soil temperature, moisture and texture. Oocysts preferentially attach to fine particles, but soil organic carbon prevents some oocyst attachment to soil particles. The transport of oocysts under constant pressure head conditions displays a decrease of water flux and a reduction of oocysts concentration in effluents as a result of physical straining, which is perhaps underestimated under constant flux conditions by a peristaltic pump. The understanding of the fate and transport of oocysts in the soil is a basic knowledge requirement in order to properly assess risk associated with *Cryptosporidium* transport through the wider environment.

6. Acknowledgement

This work was funded by the European Commission Framework Programme 6 Marie Curie Transfer of Knowledge project "Cryptonet.ie" (MTKD-CT-2005-029454). The first author gratefully thanks the 'Hundred Talent Program' of the Chinese Academy of Sciences.

7. References

Bradford, S.A., & Bettahar, M. (2005). Straining, attachment, and detachment of *Cryptosporidium* oocysts in saturated porous media. J. Environ. Qual., 34:469-478.

Bradford, S.A., Bettahar, M., Simunek, J. & van Genuchten, M.T. (2004). Straining and attachment of colloids in physically heterogeneous porous media. Vadose Zone J. 3:384-394.

Bradford, S.A., Simunek, J., Bettahar, M., van Genuchten, M.T., & Yates, S.R. (2003). Modeling colloid attachment, straining, and exclusion in saturated porous media. Environ. Sci. Technol. 37:2242-2250.

Bradford, S.A., Simunek, J., & Walker, S.L. (2006). Transport and Straining of *E. coli* O157:H7 in saturated porous media. Water Resour. Res. 42: doi:10.1029/2005WR004805.

Brush, C.F., Ghiorse, W.C., Anguish, L.J., Parlange, J-Y., & Grimes, H.G. (1999). Transport of *Cryptosporidium parvum* oocysts through saturated columns. J. Environ. Qual. 28:809-815.

Davies, C.M., Altavilla, N., Krogh, M., Ferguson, C.M., Deere, D.A., & Ashbolt, N.J. (2005). Environmental inactivation of *Cryptosporidium* oocysts in catchment soils. J. Appl. Microbiol. 98:308-317.

Fayer, R., 2008. Biology. p. 1-41. In R. Fayer and L. Xiao (ed.) *Cryptosporidium* and Cryptosporidiosis, 2nd. CRC Press and IWA Publishing, Boca Raton, FL.

Harter, T., Wagner, S., & Atwill, E.R. (2000). Colloid transport and filtration of *Cryptosporidium parvum* in sandy soils and aquifer sediments. Environ. Sci. Technol. 34: 62-70.

Hijnen, W.A.M., Beerendonk, E.F., & Medema, G.J. (2006). Inactivation credit of UV radiation for viruses, bacteria and protozoan (oo) cysts in water: A review. Water Res. 40: 3-22.

Hijnen, W.A.M., Brouwer-Hanzens, A.J., Charles, K.J., & Medema, G.J. (2005). Transport of MS2 Phage, *Escherichia coli, Clostridium perfringens, Cryptosporidium parvum*, and *Giardia intestinalis* in a gravel and a sandy soil. Environ. Sci. Technol. 39: 7860-7868.

Hsu, B., Huang, C., & Pan, J.R. (2001). Filtration behaviors of *Giardia* and *Cryptosporidium*-ionic strength and pH effects. Water Res. 35: 3777-3782.

Jenkins, M.B., Trout, J. M., Higgins, J., Dorsch, M., Veal, D., & Fayer, R. (2003). Comparison of tests for viable and infectious *Cryptosporidium parvum* oocysts. Parasitol. Res. 89:1-5.

Jenkins, M.B., Bowman, D.D., Fogarty, E.A., & Ghiorse, W. (2002). *Cryptosporidium parvum* oocyst inactivation in three soil types at various temperatures and water potentials. Soil Bio. Biochem. 34:1101-1109.

Jury, W.A., Gardner, W.R., & Gardner, W.H. (1991). Soil Physics, 5th version. John Wiley & Sons, Inc., New York.

Kuczynska, E., Shelton, D.R., & Pachepsky, Y. (2005). Effect of bovine manure on *Cryptosporidium parvum* oocyst attachment on soil. Appl. Environ. Microbiol. 71: 6394-6397.

Logan, A.J., Stevik, T.K., Siegrist, R.L., & Rønn, R.M. (2001). Transport and fate of *Cryptosporidium parvum* oocysts in intermittent sand filters. Water Res. 35: 4359-4369.

Medema, G.J., Schets, F.M., Teunis, P.F.M., & Havelaar, A.H. (1998). Sedimentation of free and attached *Cryptosporidium* oocysts and *Giardia* cysts in water. Appl. Environ. Microbiol. 64: 4460-4466.

Morita, S., Namikoshi, A., Hirata, T., Oguma, K., Katayama, H.., Ohgaki, S.., Motoyama, N., & Fujiwara, M. (2002). Efficacy of UV irradiation in inactivating *Cryptosporidium parvum* oocysts. Appl. Environ. Microbiol. 68: 5387-5393.

Muirhead, R.W., Collins, R.P., & Bremer, P.J. (2006). Interaction of *Escherichia coli* and soil particles in runoff. Appl. Environ. Microbiol. 72: 3406-3411.

Nasser, A.M., Tweto, E., & Nitzan, Y. (2007). Die-off of *Cryptosporidium parvum* in soil and wasterwater effluents. J. Appl. Microbiol. 102:169-176.

Pachepsky, Y.A., Sadeghi, A.M., Bradford, S.A., Shelton, D.R., Guber, A.K., & Dao, T. (2006). Transport and fate of manure-borne pathogens: modeling perspective. Agr. Water Manage. 86: 81-92.

Peng, X., Murphy, T., & Holden, N.M. (2008). Evaluation of temperature on die-off rate of *Cryptosporidium parvum* oocysts in water, soils and faeces. Appl. Environ. Microbiol. 74:7101-7107.

Rose, J.B. (1997). Environmental ecology of *Cryptosporidium* and public health implications. Annual Review of Public Health 18, 135-161.

Ru, R.K. (2000). Methods of Soil Analysis (In Chinese). Chinese Agricultural Publisher, Beijing.

Ryan, M., & Fanning, A. (1996). Effects of fertiliser N and slurry on nitrate leaching - lysimeter studies on five soils. Irish Geography 29: 126-136.

Searcy, K., Packman, A.I., Atwill, E.R., & Harter, T. (2005). Association of *Cryptosporidium parvum* with suspended particles impact on oocyst sedimentation. Appl. Environ. Microbiol. 71: 1072-1078.

Searcy, K., Packman, A.I., Atwill, E.R., & Harter, T. (2006). Deposition of Cryptosporidium oocysts in streambeds. Appl. Environ. Microbiol. 72: 1810-1816.

Smith, J.E., & Perdek. J.M. (2004). Assessment and management of watershed microbial contaminants. Cri. Rev. Env. Sci. Tec. 34:109-139.

Toride, N., Leij, F.J., & van Genuchten, M.T. (1995). The CXTFIT code for estimating transport parameters from laboratory or field tracer experiments. Version 2.0, Research Report NO. 137, U.S. Salinity Laboratory, USDA, ARS, Riverside, CA.

Tufenkji, N., & Elimelech, M. (2005). Spatial distributions of *Cryptosporidium* oocysts in porous media: evidence for dual mode deposition. Environ. Sci. Technol. 39: 3620-3629.

Tufenkji, N., Miller, G.F., Ryan, J.N., Harvey, R.W., & Elimelech, M. (2004). Transport of *Cryptosporidium* oocysts in porous media: role of straining and physicochemical filtration. Environ. Sci. Technol. 38:5932-5938.

Walker, M.J., Leddy, K., & Hagar, E. (2001). Effects of combined water potential and temperature stresses on *Cryptosporidium parvum* oocysts. Appl. Environ. Microbiol. 67:5526-5529.

Walker, M.J., & Montemagno, C.D. (1999). Sorption of *Cryptosporidium parvum* oocysts in aqueous solution to metal oxide and hydrophobic substrates. Environ. Sci. Technol. 33: 3134-3139.

11

Soil Indicators of Hillslope Hydrology

Johan van Tol, Pieter Le Roux and Malcolm Hensley
Department Soil, Crop and Climate Sciences, University of the Free State
South Africa

1. Introduction

The demand for water doubles every 20 years which is more than twice the rate of the world's population growth. New water resources are becoming scarcer and to treat and remediate existing sources more expensive (Clothier et al., 2008). The protection and management of surface and groundwater resources, especially in the highly variable water regime of semi-arid areas, requires accurate analysis of hydrological processes. This involves the identification, definition and quantification of the pathways, connectivities, thresholds and residence times of components of flow making up stream discharge. It is essential that these aspects be efficiently captured in hydrological models for accurate water resource predictions, estimating the hydrologic sensitivity of the land for cultivation, contamination and development, and for quantifying low flow mechanisms (Lorentz et al., 2007; Uhlenbrook et al., 2005; Wenninger et al., 2008). Ideally these hydrological models can best be developed using measurements of the surface and subsurface lateral flow paths, water table fluctuations, connectivity of the various water bodies and the residence flow time of water through the landscape. The landscape unit that is of particular importance is the hillslope (Karvonen et al., 1999; Lin et al., 2006; Ticehurst et al., 2007), hence the accent here on this landscape unit. The measurements named are however expensive and time consuming since these processes are dynamic in nature with strong temporal and spatial variation (McDonnell et al., 2007; Park & Van de Giesen, 2004; Ticehurst et al., 2007).

The need for predictions of the named hydrological processes is becoming increasingly important and led to the launch an International Association of Hydrological Sciences (IAHS) initiative called Predictions in Ungauged Basins or PUB (Sivapalan 2003; Sivapalan et al., 2003) encouraging researchers and modellers to focus their efforts on predicting the hydrological behaviour of catchments based on physical principles without relying on calibrations of hydrological models.

Soils integrate the influences of parent material, topography, vegetation/land use, and climate and can therefore act as a first order control on the partitioning of hydrological flow paths, residence time distributions and water storage (Park et al., 2001; Soulsby et al. 2008). The influence of soil on hydrological processes is due to the ability of soil to transmit, store and react with water (Park et al., 2001). Hydrologists agree that the spatial variation of soil properties significantly influences hydrological processes but that hydrologists lack the skill to gather and interpret soil information (Lilly et al., 1998). The relationship between soil and hydrology is interactive. Water is a primary agent in soil genesis, resulting in the formation of soil properties containing unique signatures of the way they formed. Almost every hydrological process of interest to hydrologists is difficult to observe and measure

(Sivapalan 2003a). Soil properties are not dynamic in nature in the short term, and their spatial variation is not random (Webster, 2000). The correct interpretation of spatially varying soil properties associated with the interactive relationship between soil and hydrology can serve as an indicator of the dominant hydrological processes (Ticehurst et al., 2007; Van Tol et al, 2010a), and improve the understanding of hydrological behaviour of different hillslopes (Lin et al., 2006). The hillslope scale is considered ideal as it is the smallest unit on which most hydrological processes can be observed. The improved understanding can facilitate the development of conceptual, qualitative 2-dimensional descriptions of hillslope hydrological behaviour. These descriptions reflect the physical processes and should be used in the configuration and structure of hydrological models in order to mimic the hydrological behaviour of hillslopes and thereby aid in PUB.

2. Hydropedology

The science of hydrology has been poorly serviced by soil science in the past. Soil science started as a natural science in the late 19th century but changed to an applied science in agriculture in the 20th century. This was mainly due to the quest for food security, heightened by the 1st, 2nd world wars and the 'cold' war which followed. During the post-war years attitudes changed internationally as fears about food shortages decreased while apprehension regarding environmental sustainability grew. Agriculture occupies more than 90% of the land surface in most countries and accounts for around 70% of fresh water utilization in many developed countries. Rapidly expanding industrialization, increased urbanization, intensified agricultural production and population growth have all contributed during the last six decades towards drastically increasing the environmental pollution hazard. A large fraction of this pollution ends up in the soil, making it a vital role player in pollution control and rehabilitation, thereby broadening its contribution to holistic environmental studies.

The term hydropedology was introduced in 1966 by Kutilek (Kutilek & Nielson, 2007), and can be defined as the "...synergistic integration of pedology with hydrology to enhance the holistic study of soil-water interactions and landscape-soil-hydrology relationships across space and time, aiming to understand pedologic controls on hydrologic processes and properties, and hydrologic impacts on soil formation, variability, and functions" (Lin et al., 2008). This field aims to bridge gaps between pedology, soil physics, hydrology and geomorphology and also between micro and macroscopic scales of water interactions. Issues covered by hydropedology include: i) hydrology as factor of soil formation, ii) soil as essential component of hydrological cycle and filter of water, iii) soil morphology as signatures of soil hydrology and iv) landscape-soil-water relationships across scales. For excellent reviews and comprehensive discussions on hydropedology see Lin (2003), Kutilek et al. (2007), Van Huyssteen (2008) and Lin (2010).

2.1 Hydrological flowpaths

Three major flow pathways exist in a typical hillslope: overland flow, subsurface lateral flow and bedrock flow (Karvonen et al. 1999; Ticehurst et al., 2007). Subsurface lateral flow can be divided into: subsurface macropore flow, subsurface lateral flow at A-B horizon interface, return flow at the footslope and toeslope and flow at the soil-bedrock interface (Lin et al., 2006). These flowpaths are not mutually exclusive, and water tends to move between them. Some paths are only connected when the hillslope is wet. The relative importance of the various pathways is determined by soil characteristics, the macropore network and the parent material at the base of the soil (Mosley, 1982). Hydrologic conditioning is influenced by soil depth, pore size and organic matter distribution, tortuosity and the surface and subsurface topography (Sidle et al., 2001).

The role of topography varies with the moisture content of the soil. In drier periods the main controlling factor of movement is soil characteristics. In wetter periods, the topography becomes increasingly important (Lin et al., 2006; McGlynn et al., 2002; Park & van de Giesen, 2004).

2.1.1 Overland flow

Overland flow occurs either as infiltration excess or as saturation excess. In general steeper slopes generate large volumes of overland flow with significant erosive energy. Thinner A horizons usually indicate that the overland flow is dominant, in thicker soils more infiltration due to the greater volume of water needed to saturate the soil is expected. The assumption can be made that thicker soils support more vegetation and this causes a decrease in the overland flow proportion (Ticehurst el al., 2007).

Breaks in slope (normally between midslopes and valley bottoms) reduce the velocity of water and enhance infiltration. The soils in this region are generally thicker, due to deposition of alluvial material and organic matter, which further enhance infiltration. In valley bottoms the runoff rate tends to slow down because of the smaller gradient. These soils are however the wettest in typical hillslopes and the saturated conditions reduce the infiltration rate promoting overland flow.

Where overland flow occurs the water may downslope encounter an area where the soil water deficit has not yet been satisfied, the water then infiltrates. This is called the run-on pathway and is often ignored in rainfall and runoff studies. The water available for infiltration then includes the precipitation as well as water supplied from the upperslope (Nahar et al., 2004).

The amount of overland flow is greatly affected by the texture of the soil, specifically the percentage clay and sand. Sandy soil is generally more permeable and has a greater hydraulic conductivity than clay rich soil, and therefore infiltration excess induced overland flow seldom occurs in sandy soils. In a study by Karnoven et al. (1999) the conductivity of sandy loam soils was 15 times higher than clayey soils.

2.1.2 Subsurface lateral flow

Macropores conduct a considerable amount of water during large storms in forested catchments. Water moves through tree root channels, pores created by organisms (earthworms), as well as cracks. Cracks are usually present in soils with a high 2:1 clay content (vertic soils), especially in drier periods (Lin et al., 2006). There are three factors determining the contribution of subsurface macropore flow of water namely; size of the macropores, the accessibility and continuity of the pores. The continuity of these pores seems to increase with an increase in soil moisture (Nieber et al., 2000). Soil pipes are usually flow pathways parallel with the slope and are formed by soil fauna (moles & mice) as well as dead root channels. They contribute a significant amount of subsurface water to streamflow and are usually quick to respond to rainfall. Pipe flow has a smaller influence in hillslopes with high drainable porosity because water table response is lower due to the high storage potential of the profile (Uchida et al., 2006).

Lateral flow occurs at A/B horizon interfaces due to differences in the structures, densities and hydraulic conductivities of the horizons. Vertical flow will be hindered and water will tend to move laterally if the A horizon is more permeable than the B horizon. In a study by Lin et al. (2006), the lowest water content was recorded below the interface between the A and B horizons due to the great amount of lateral flow. The lower gradient of lower slope terrain units would limit this lateral flow and promote water logging, as well as overland flow due to excess saturation (return flow). Similar mechanisms might result in a flowpath at the bottom of the profile at the interface between the soil and the underlying parent material. Continuous flow after a storm even with little water in the top of the profile suggested that the water moved vertically in the upperslopes and then laterally at or near the soil-bedrock interface in a study by Lin et al, (2006). The permeability, the depth as well as the differentiation between horizons would affect the amount of water moving through this flowpath. Since the clay content of the B horizon in lower slopes usually shows an increase due to luviation, this pathway would generally originate in the upper slopes.

2.1.3 Bedrock flow

Ticehurst et al. (2007) found in their study that the soils from the summit area, which were sandy and shallow, provided and important water intake area for water supply to the bedrock flowpath. The general movement of water in this region is vertical and soils are usually well drained. Due to the age of the soils and the small amount of deposition, little differentiation between horizons is generally present and water drains vertically through the B horizon into the C horizon. The water that doesn't move on top of the bedrock moves through cracks in the bedrock or on solid bedrock within the saprolite. The bedrock flowpath is extremely important for recharge of lower slopes, groundwater levels and generating baseflow in some catchments (Fanning & Fanning 1989; Ticehurst et al., 2007).

2.2 Signatures and control mechanisms

Soil properties, soil horizons, soil profiles and soil patterns are not randomly distributed (Webster, 2000). These soil features are influenced by five soil forming factors i.e. climate, topography, geology, organisms and time. The combination of these factors results in unique soil properties with distinctive vertical and horizontal distributions. These properties and their distribution not only influence hydrological processes but can also serve as verification on the way they were formed. A few examples are presented.

2.2.1 Redox morphology

Soil morphology developed by oxidizing, reducing and redox conditions serves as signatures of flowpaths and storage mechanisms in soils, hillslopes and catchments. Reducing conditions increase down the profile, down slope and with increasing rainfall if all other factors playing a role in the redox process, remains the same. Redox features in soils involve localities where there is depletion in Fe^{3+} and Mn^{2+} concentrations and localities where there is accumulation of Fe^{3+} and Mn^{2+} (Soil Survey Staff, 1992). Depletion in Fe^{3+} and Mn^{2+} is associated with low chroma values (grey colours), and accumulation of Fe^{3+} and Mn^{2+} is associated with high chroma colours (yellow, red and black) in the form of mottles and concretions (Le Roux, 1996).

Micro organisms utilize O_2, NO_3^-, Mn^{2+}, Fe^{3+} and SO_4^{2-} as oxidation agents (electron acceptors) and easily oxidisable organic matter as reduction agent. These reactions occur sequentially from most likely to least likely to be reduced. In the oxidized state Fe^{3+} and $Mn^{2+}$are insoluble, and in the reduced state, very soluble. Reduced localities have high Fe^{2+} and Mn^{+} concentrations in solution. They diffuse to oxidised localities where the concentration in solution is low to be oxidized again (Van Breedeman & Brinkman, 1976). Grey colours of the silicate clay, quartz and feldspar soil minerals are grey and therefore grey colours appear where the Fe coatings are removed (Vepraskas & Bouma, 1976). Yellow, red and black colours occur juxtapositioned where Fe^{3+} and Mn^{2+} accumulate in sequence with an increase in Fe^{3+} and Mn^{2+} concentration (Le Roux, 1996).

Redox features are easily observed in plinthic soils. Plinthic horizons have an accumulation of iron in the form of oxides and hydroxides and are localized in the form of high chroma mottles and concretions. The simple processes leading to the formation of such a horizon are eluviation (removal of constituents), illuviation (accumulation of eluviated material), oxidation and reduction (Fanning & Fanning, 1989). Fe^{3+} is reduced and together with sesquioxides eluviated from the upper lying horizons and Fe^{2+} oxidized and accumulates in the lower horizon. A fluctuating water level is necessary for this to take place.

Plinthite normally occurs in highly weathered soils of the regions with rainfall exceeding 500 mm and where a fluctuating water table is active. High temperatures and a high evaporative demand favour plinthite formation since they influence the fluctuation of water levels. The formation of plinthite on different topographical positions corresponds to the climate. In the drier climates plinthite forms in the lower lying areas. Redoximorphic features occur in soils of semi-arid climates and wetter. The key factor is a ratio of rainfall/evapotranspiration resulting in water flowing to the deep subsoil and impermeable deep subsoil preventing water loss to the fractured rock and result in subsoil saturation. These conditions typically occur in semi-arid climates and wetter.

The relationship between wetness and position in the landscape is reflected through the variation in soil colour. In a typical catena the red Fe rich soils are typically found on the higher lying drier positions of the hillslope, whereas grey gleyed soils can generally be found in the wetter lower-slope terrain positions. Long term data proves that yellow soils are normally better drained than grey soils, but wetter than red soils (Van Huyssteen & Ellis, 1997; Van Huyssteen et al., 2005).

2.2.2 Presence of calcium carbonate ($CaCO_3$)

The dependence of calcareous precipitates on the presence and behaviour of water makes it a good indicator of hillslope hydrology. Calcretes are materials formed by cementation or selective replacement of the soil particles by carbonate. Calcareous layers in soils are controlled by the soil water regime and are typically found in arid to sub-humid regions. Lime precipitates in the soil due to limited leaching which can be brought about by two processes: leaching that can be limited due to low rainfall/high evapotranspiration, or restricting subsoil layers and associated saturated conditions (Driessen & Deckers., 2001; Netterberg, 1978).

In sub-humid to arid regions, calcification is one of the main processes in soils with carbonate rich parent materials. Weathering of the parent material results in the formation of soils with calcium as the major cation on the cation exchange complex. $CaCO_3$, the dominant carbonate in these soils, is pedogenically formed as follows:

$$Ca^{2+} + CO_2 + H_2O \rightarrow CaCO_3 + H_2 \quad (1)$$

Weathered Ca^{2+} dissolves in water leaches towards lower soil horizons and flows downslope, and filling voids and pores. Plant roots extract water and precipitation in the form of $CaCO_3$ occurs due to the presence of CO_2. The CO_2 are present in the soil as a consequence of diffusion from the atmosphere, but CO_2 generated by oxidation of plant roots enhance this process, especially when the natural vegetation consists of grasses and shrubs. This process is the first stage of the formation of a calcic horizon (Fanning et al., 1989; Shankar & Achyuthan, 2007).

According to Netterberg (1978), the presence of calcretes can serve as indicators of previous (fluctuating) ground water levels as well as of preferred flow paths (faults).When the parent material contains small amounts of $CaCO_3$ and the amount of $CaCO_3$ in the profile exceeds the amount that could be released by weathering, the presence of $CaCO_3$ in the soil can be ascribed to the second process namely, deposition of $CaCO_3$ rich dust from coastal shelves (Bockheim & Douglass, 2006).

2.2.3 Soil depth and porosity

Soil depth is the result of the balance between the rate of weathering and the rate of erosion. Under similar climate and hydro-topographical conditions the rate of weathering of the parent material is controlled by the nature of the rock. Porosity (f) is a measure of the total void space in a porous material and is measured, either as a percentage (between 0 and 100%), or as a fraction (between 0 and 1) of the bulk volume. It is defined by the formula:

$$f = 1 - \rho_d \div \rho_s \tag{2}$$

Where ρ_d is the bulk density (Mg. m^{-3}) and ρ_s is the particle density (Mg. m^{-3}, generally taken as 2.65 in soils low in organic matter). Soil depth together with the porosity determines the storage capacity of the soil.

In a South African case study, the storage capacity of two semi-arid catchments was determined (Van Tol et al, 2010a). The average soil depth of catchment B3 was 450 mm and that of B4 & B5 was 190 mm (due to similarities in B4 and B5 they were considered as one catchment). The average porosity of B3 was 301.5 mm compared to 130.6 mm of B4 & B5. Although the area of B3 is smaller (40.7 km^2) than that of B4 & B5 (49.4 km^2), it can store almost twice the volume of water (12.5×10^6 m^3 compared to 6.7×10^6 m^3). This facilitates more water infiltration, greater water holding capacity, a greater volume of water contained at saturation and at drained upper limit. This results in more interflow at the A/B-horizon interface and at the soil/bedrock interface, more water contributing to groundwater bodies and consequently a longer duration of streamflow. More water is available for transpiration resulting in a denser vegetative cover.

The thickness of the soil further influences the residence times of water in catchments. This influence can be more important than that of the slope length, or upslope contribution area (Asano et al., 2002). Shallow soils, with a small storage capacity tend to saturate quickly favouring the generation of overland flow due to saturation excess, resulting in short residence times and high peak flows.

3. Hydrology of soil types

"Theory development will advance if we can develop simple models (which may be caricatures of the basin system but, nevertheless, contain within them the basic properties of the actual basins), provided, importantly, that they can be verified with large-scale patterns extracted from the observed data" (Sivapalan, 2003). In order to develop simple conceptual hydropedological models (and to improve our understanding of the role of hydropedology in both the natural environment and agriculture), it is necessary to understand key hydrological processes, the impact of soil on these processes and the influence of these processes on soil formation.

This relationship between soil and water is however difficult to comprehend at hillslope or catchment scale. For example; water may drain from the soil into the rock and then return to the soil. It may also exit the soil again as return flow. Where a water table occurs in the soil it is often uncertain whether the soil is feeding the rock aquifer or *vice versa*. This interaction between soil and hydrology can be simplified by firstly studying this interaction at a pedon scale. In this section soils are divided into different soil types based on their hydrological behaviour, similar to the Hydrology of Soil Types (HOST) classification system. In HOST the soils of the UK were divided into 29 classes based on their hydrological response (Boorman et al., 1995). In this section we only focus on three main response mechanisms of soils and use six years of soil moisture content measurements to support the classification. Because hydropedology is a rather young and complex subject, with relatively few quantitative measurements worldwide to verify hypotheses, we considered it wise to include only local case studies about which we have sufficient knowledge and as much quantitive information as possible.

3.1 Hydrology of soil types

It is hypothesized that soils can be grouped in three main hydropedological types based on their hydrological response: recharge soils, interflow soils and responsive soils. Data from the Weatherley research catchment (31°06′6″S/28°20′13″E) in South Africa (Van Huyssteen et al., 2005) was used to distinguish between these soil types using the degree of water saturation (s), measured over six years. The degree of water saturation is the volume of water relative to the (f) (Hillel, 1980). Porosity can be calculated using equation 2 and the degree of water saturation as:

$$s = V_W \div V_f \tag{3}$$

Where s is the degree of saturation (as fraction), V_W is the water content (mm^3 mm^{-3}) and V_f is the total pore volume (mm^3 mm^{-3}). Complete saturation (s = 1) is seldom reached since air is usually trapped in pores by water (Hillel, 1980). The drained upper limit (DUL) i.e. the water content below which drainage due to gravity virtually ceases is expected to be around 0.65 in most soils.

The term "annual duration of degree of water saturation above 0.7 of porosity" ($AD_{s>0.7}$) is the first approximated threshold value for the onset of reduction (Van Huyssteen et al., 2005). The degree of saturation before the start of reduction will however differ between areas, soil forms and horizons since numerous factors influence redox conditions in soils. It is because redox reactions of significant extent in soils leave well defined morphological footprints e.g. mottling and/or grey colours, that $AD_{s>0.7}$ is considered to be a useful parameter in hydropedological studies. $AD_{s>0.7}$ was measured in days per year. The mean annual duration in days of events with s>0.7 ($D_{s>0.7}$) was calculated as follows (Van Huyssteen et al., 2005):

$$D_{s>0.7} = AD_{s>0.7} \div F_{s>0.7} \tag{4}$$

Where $F_{s>0.7}$ is the mean annual frequency of events where $s>0.7$ (events year^{-1}).
Soils were classified according to Soil Classification - A taxonomic system for South Africa (Soil Classification Working Group, 1991) although equivalent classification accordance with WRB (IUSS Working Group WRB, 2006).

3.1.1 Recharge soils

Several soils in the Weatherley catchment qualify as recharge soils. The average annual duration of saturation above 0.7 of porosity ($AD_{s>0.7}$), expressed as a % of 365 days, is not significant in these soils (Fig. 1) as conditions near saturation only occur when drainable water accumulates. The short degree of saturation in the subsoil shows that water draining through the soil exits the solum to enter the fractured rock underneath.

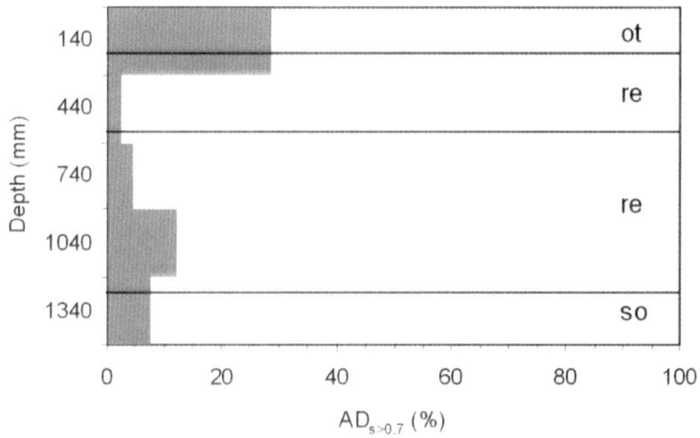

Fig. 1. Mean $AD_{s>0.7}$ (%) values in a typical recharge soil: P221, Hutton 2100 (WRB - Orthidistric Cambisol), Weatherley (after Van Huyssteen et al., 2005)

In recharge soils, the hypothesis is that dominant flow direction is vertical. These soils typically occur on the crest or midslope positions on hillslopes with gentle slopes. Precipitation infiltrates the soil and water flows vertically through the pedon under gravitational forces. The underlying permeable bedrock facilitates infiltration of water. From a hydrological perspective the formation and distribution of recharge soils is therefore dependant to a large extent on the permeability of the underlying material. Depending on the nature of the underlying material the infiltrated water can either recharge regional water tables directly, or in the case of aquicludes or aquitards, move laterally after leaving the soil. This lateral moving water can then recharge the stream through transient or perennial groundwater. Its contribution to transient groundwater may be uncertain. Since these flowpaths through the bedrock are usually the longest, recharge soils are important for generating base flow. Recharge soils show no evidence of saturation in any part of the profile. The annual rainfall and potential evapotranspiration should however be considered when classifying a soil as a recharge soil. In arid areas, precipitation is insufficient for redoximorphic features to form and the soils would be classified as recharge soils based on morphological properties even though they are not *freely drained.*
The contribution of recharge soils to catchment hydrology by implication stops when the soil water balance is negative (i.e. ET>P). This limits its activity to the wet part of the rain

season (Fig. 2). Three phases are clearly visible in the graph namely a wetting up cycle with the start of the rain season, a wet phase during the rainy season and a drying phase in the waning portion of the rain season. The drying phase is only stopped by the start of the wetting up phase of the following rain season or when the water content is lower than the lower limit of plant available water.
The wetting up cycle depends on the precipitation, atmospheric demand (ET) and the size of the reservoir. As the grass vegetation of the Weatherley catchment mainly extracts its water from the upper 900 mm (Zere, 2005) of soil, a relative large volume of soil has to be brought to drained upper limit (DUL) before draining starts. In the majority of years (four out of six) this, cycle is two weeks in duration. In the wet cycle the water content of the recharge soils depends mainly on the distribution of rainfall events. Profile water exceeding DUL drains beyond reach of the grass roots.

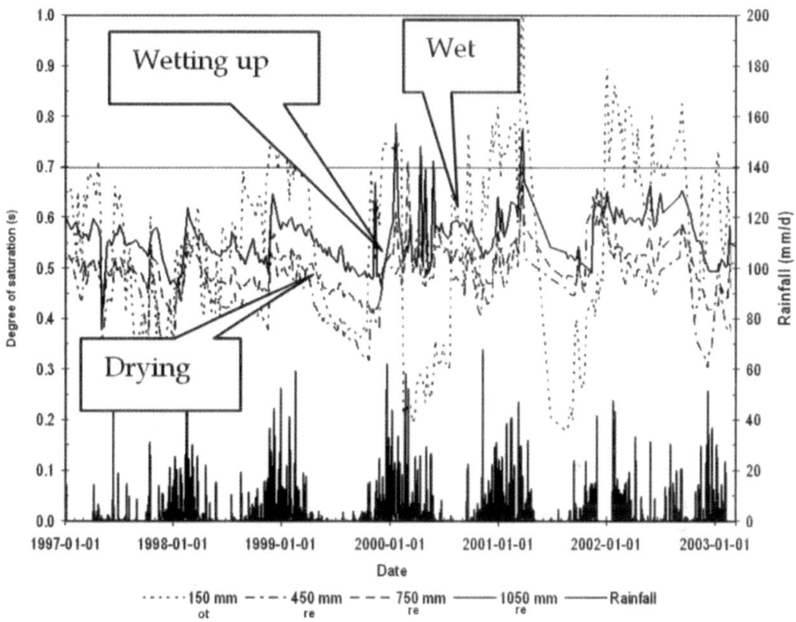

Fig. 2. Degree of saturation vs. rainfall over 6 years of a recharge soil: P221, Hutton 2100 (WRB - Orthidistric Cambisol) in the Weatherley catchment (after Van Huyssteen et al., 2005)

3.1.2 Interflow soils

The ($AD_{s>0.7}$) values in the subsoils of interflow soils is distinctive (Fig. 3). Conditions of water contents near saturation (drainable water) occur in all horizons but typically increase with depth. Interflow soils are associated with subsurface lateral flowpaths. For interflow to occur a layer with lower hydraulic conductivity must be present (B horizon or bedrock with restricted permeability) as well as a slope favouring lateral movement down the slope. Interflow soils are therefore typically found in midslope positions with fairly steep gradients. Water starts moving laterally when infiltrated water encounters a layer with lower hydraulic conductivity (A/B horizon interface; soil - bedrock interface or a saturated layer) or when water, fed from upslope recharge soils, encounters such a layer and may return to the soil.

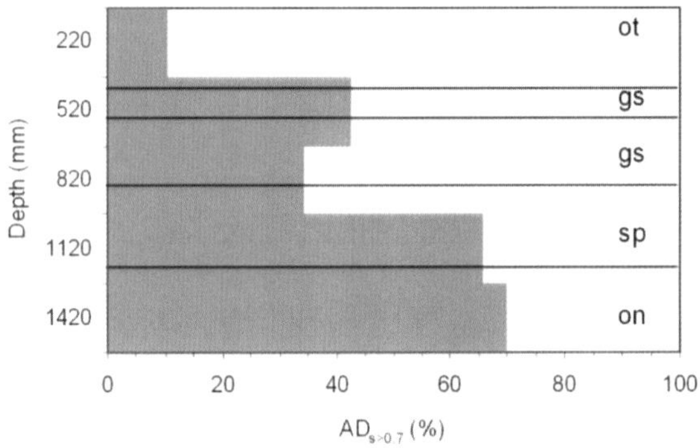

Fig. 3. Mean $AD_{s>0.7}$ (%) values in a typical interflow soil: P225, Longlands 1000 (WRB - Ferric-Endoeutric Albeluvisol) in the Weatherley catchment (Van Huyssteen et al., 2005)

Interflow soils have in contrast with the three phases of recharge soils, a distinctive drainage phase, above DUL (Fig. 4). The duration of ADs>0.7 in the soft plinthic (sp) horizon of ± 67% i.e. 244 days or 8 months (Fig. 3) is an indication that this soil body generally releases water up to the end of August. This implies a 5 month draining phase i.e. stretching from the end of the rain season (early April) to the end of the dry season in August.

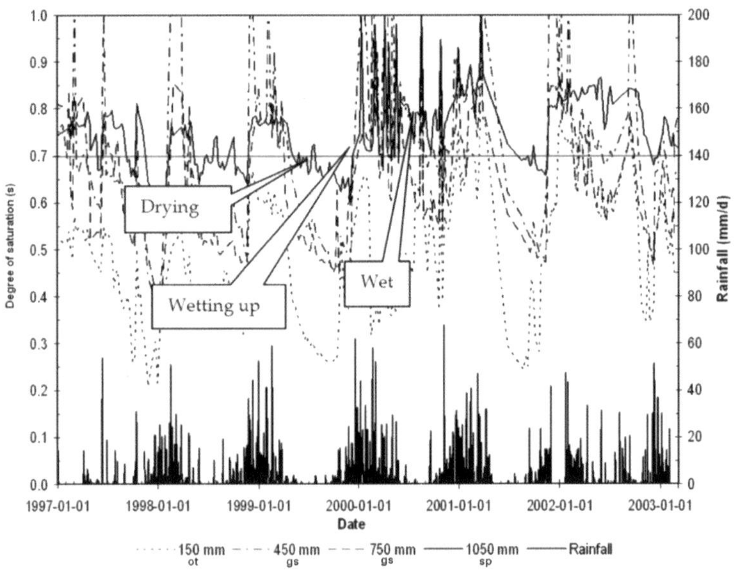

Fig. 4. Degree of saturation vs. rainfall over 6 years of an interflow soil: P225, Longlands 1000 (WRB - Ferric-Endoeutric Albeluvisol) in the Weatherley catchment (after Van Huyssteen et al., 2005)

During the wet phase losses of water by drainage and ET are sometimes slower than additions of water by precipitation, and interflow results in a rise of the transient groundwater into the plinthic, E and A horizons. Such a fluctuating water table is typical of subsoils with plinthic and E horizons (Soil Classification Working Group, 1991). These fluctuations are event driven and can be related to rainfall events. The catchment must first fill up before transient groundwater can occur.

Sub soil flowpaths are associated with a residence time shorter than the bedrock flowpaths and longer than overland flow. Interflow soils would therefore contribute mainly to the shoulder of the hydrograph, and to some extent to baseflow. Interflow soils normally have morphological indications of periodic saturation in the profile. If dominant flow exists on the A/B horizon interface, eluvial horizons form. These horizons show marked removal of colloidal material and organic matter. When interflow occurs at the soil/bedrock interface, the transitional horizon usually show indications of periodic saturation.

3.1.3 Responsive soils

Responsive soils can either be very shallow soils with low infiltration capacity, saturated soils which prohibit water infiltration or soils prone to form crusts resulting in low infiltration rates and generating *Hortonian* overland flow. In the Weatherley catchment responsive soils generate overland flow due to saturation excess. The overland flow component contributes to peak flow as the first part of the peak of the hydrograph. The influence of the water content of the topsoil on the generation of overland flow is illustrated in Fig. 5. The results show that overland flow only becomes significant when the topsoils are close to saturation. Overland flow from responsive soils is therefore expected in the wettest positions in landscapes i.e. valley bottoms and wetlands. In the Weatherley catchment these soils are at or near saturation for long periods (Fig. 6), resulting in conditions called saturation excess overland flow in the rain season.

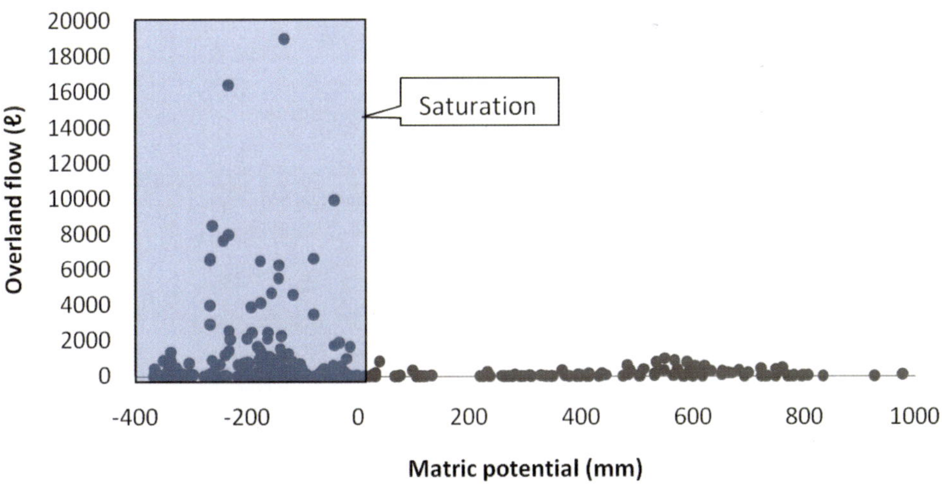

Fig. 5. Volume of overland flow measured at five runoff plots vs. topsoil matric potential in the Weatherley catchment

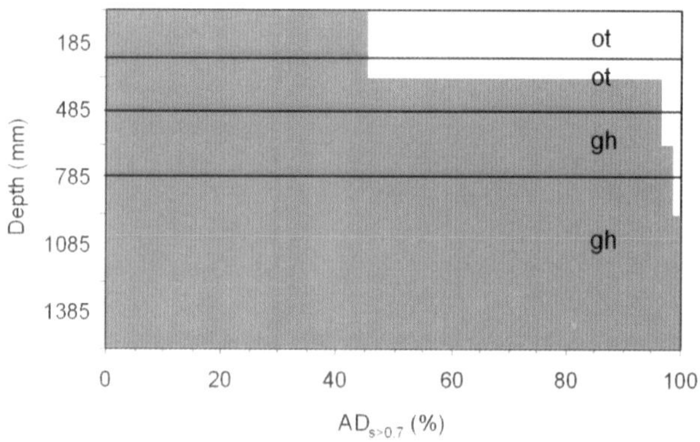

Fig. 6. Mean $AD_{s>0.7}$ (%) values in a typical responsive soil: P235, Katspruit 1000, (WRB - Hyperdistric Gleysol) in the Weatherley (Van Huyssteen et al., 2005)

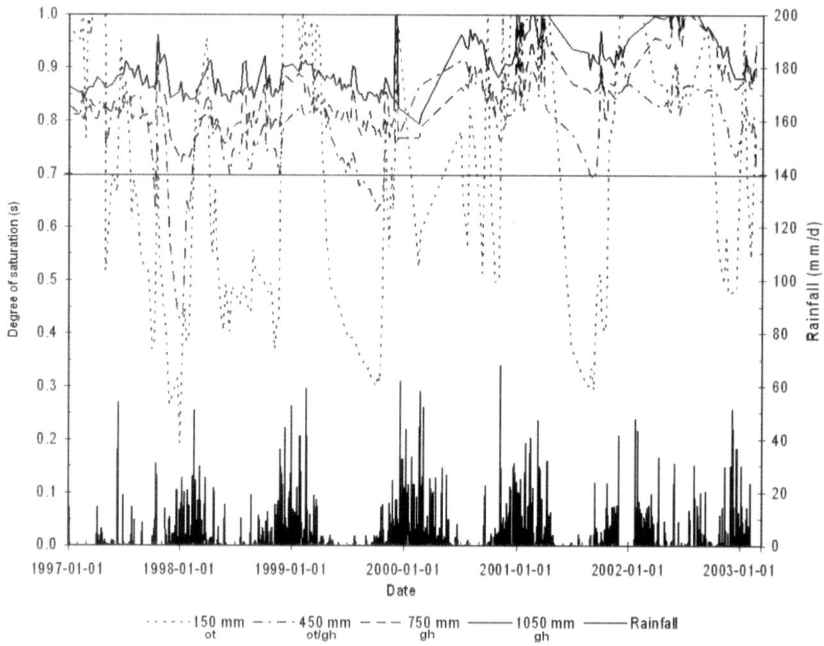

Fig. 7. Degree of saturation vs. rainfall over 6 years of a responsive soil: P235, Katspruit 1000, (WRB - Hyperdistric Gleysol) in the Weatherley catchment (after Van Huyssteen et al., 2005)

Due to long periods of saturation the subsoil (*gh* horizon) lacks an obvious wetting and draining phase since it is saturated or close to saturation throughout the year. Only the topsoil horizon loses water to ET during the dry season (Fig. 7). In order for these subsoils to remain saturated for such long periods under incessant ET demand there needs to be a constant supply of water. It is hypothesized that the recharge soils of the upper slopes supply water to the responsive soils via the bedrock flowpath and to a lesser extent through interflow.

4. Conceptual models of hillslope hydrological behaviour

Deducing the hydrological behaviour of a soil profile without considering its position in the landscape might lead to false interpretations. It is therefore important to obtain a holistic understanding of how hillslopes behave hydrologically. In the case studies reported here soil properties and their spatial distribution (both vertically and horizontally) were interpreted and related to their hydrological response. From these interpretations 2-dimensional descriptions of the hydrological behaviour were formulated. Two examples of these conceptual models are presented in this chapter. For detailed discussions on these examples, study area descriptions and more conceptual examples see Le Roux et al. (2010) and van Tol et al. (2010a).

4.1 Conceptual model of the hydrological behaviour of a hillslope in the Weatherley catchment

A conceptual description of a hillslope in the Weatherley catchment is presented in Fig. 8 (van Tol et al., 2010). The soil forms and their associated properties along with their spatial distribution were interpreted to develop the conceptual model of the hydrological behaviour of the hillslope.

The dominant processes (flowpaths and storage mechanisms) in Fig. 8 are indicated by numbers plus a letter in box inserts related in each case to arrows of different sizes; these inserts are included in the text enclosed in a bracket. A discussion of these processes follows in what can be considered as a hypothesis, based on hydropedology, of the hillslope hydrology.

When it rains infiltration dominates in the upper regions of this hillslope (1a). Gentle slopes as well as dense vegetation impede overland flow and facilitate infiltration. The absence of any signs of wetness in the soil of the upperslope indicates that vertical drainage through the profile is dominant. The texture is non-luvic and the clay content is therefore relatively uniform with depth. No, or very little, lateral flow is expected to occur at the A/B horizon interface. These are considered to be true recharge soils since no signs of wetness were recorded in soil profile 240 up to a depth of 1500 mm, indicating that water does not perch in the pedon within this depth. Water draining through 240 therefore either infiltrates the subsurface layers (2a) or flows at the soil/bedrock interface (3a). The latter was not reached with auger observations down to 2400 mm.

Any water which does infiltrate the fractured rock would then either flow vertically and recharge regional aquifers (2b) or, when it encounters a layer with restricted permeability (aquitard), it would flow laterally (3b) and recharge perennial hillslope groundwater downslope.

The presence of interflow soils located where the rock bedding plain surfaces near Uc8 is an indication that the bedding plane (sandstone rock shelf of the Molteno formation) has restricted vertical permeability promoting considerable flow at the soil/bedrock interface (3a). The greater part of the water draining through the soil of the upper slope is therefore expected to flow laterally at this soil/bedrock interface.

Return flow (ex-filtration) to the soil surface (4) is expected as water flowing at the soil/bedrock interface reaches the protruding Molteno shelf. The amount of water exceeds the storage capacity of the soil and returns to the surface contributing to overland flow. It is expected that the overland flow has a relatively short duration as the water will re-infiltrate when it reaches the soils below the rock outcrop (1b).

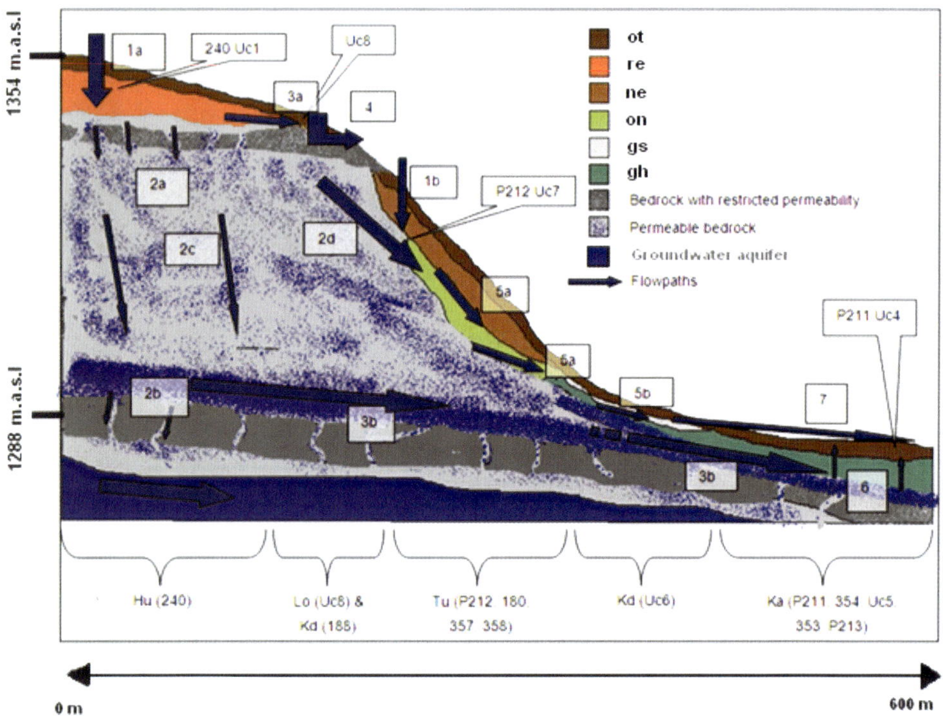

Fig. 8. Conceptual hydrological behaviour of the selected hillslope based on soil interpretations. Various processes are indicated by the numbered arrows

Subsurface lateral flow (5a) in the form of flow at the soil/bedrock interface is indicated by the *on* horizon (WRB - *cambic*) present in the deep subsoil of the Tukulu (WRB - (Orthieutric Cambisol) soil of the midslope at P212. This soil body is situated on the Molteno Formation. Groundwater responsible for the redoximorphic features of the *on* horizon is evidently supplied from the recharge soils as return flow from the bedrock (2d). This return flowpath is expected to result in a fairly constant supply of water during the wet seasons to the *on* horizon, reflecting its association with perennial groundwater.

The *gs* horizon (WRB - *albic*) in the soils of the lower slopes is an indication of the lateral flow of groundwater dominating at the A/B horizon interface (5b). Gleyed soils cover the entire TMU 4 & 5 positions of this hillslope. The gleyed conditions are indications that these profiles are saturated for long periods. The *gh* horizons (WRB - *gley*) have a low hydraulic conductivity that impedes infiltration. Precipitation does not infiltrate into these soils due to the saturated state of the *gh* horizon. The water maintaining saturation in these lower areas must therefore have another origin. It is believed that there is another layer with restricted permeability present in the hillslope (Fig. 8). This layer deflects water which has infiltrated through the recharge soils of the upperslope towards the lower lying areas (3b), resulting in the presence of a perennial aquifer. These very wet soils respond rapidly to precipitation providing overland flow to the stream, the process described as saturation excess overland flow (7). Near surface macropore flow might also play a significant role in this area, as water from the *gh* horizon pushes up into the more porous *ot* horizon (WRB - *ochric*) and then flows laterally. The *ot* horizons of the soils in the lower slope have Fe and Mn mottling, confirming periodic saturated conditions.

Since the *gh* horizons in the lower footslope and toeslope positions (Fig. 8) are saturated for long periods, the dominant flow direction within the pedon is upwards (6). Evapotranspiration will presumably extract more water from the soil than can infiltrate.

4.2 Conceptual model of hillslope hydrological behaviour in the Two Streams catchment

The conceptual model of the hydrological behaviour of two hillslopes in the Two Streams catchment (29°12′23″S/30°39′10″E) is presented in Fig. 9 (Le Roux et al., 2010). The number arrows represent dominant hydrological processes. Note that the slope on the right faces south and that on the left hand side of Fig. 9 faces north. This difference in aspect has resulted in distinct morphological characteristics of the soils in the slopes; north facing slopes in the southern hemisphere are generally drier compared to south facing slopes due to more direct sunlight. The soils on the south facing slope are therefore more reduced and exhibit yellow colours compared to the dominant red colours on the north facing slope.

The dominant process in this catchment is infiltration and vertical drainage of precipitation (1 in Fig. 9). The soils in the upper slopes are considered freely drained with no indications of periodic saturation in the A and B horizons. These soils are considered to be recharge soils. Water exits the solum and drains into the highly weathered saprolite (*so*).

Vertical drainage remains dominant until more impermeable rock is reached, where the flow is deflected in a lateral direction (2). This lateral moving component feeds the stream channel from the side causing prolonged conditions of saturation and gleyed soils (*gh* horizons) next to the stream channel. Some water can however enter cracks and fissures in the relatively impermeable rock and feed deeper lying water bodies.

The vertical sequence of horizons of the upper slopes; orthic A (*ot*), yellow brown apedal B (*ye*) and red apedal B (*re*) (WRB - *ferrilic*), might be an indication of lateral movement close to the surface horizons (3). Near surface macropore flow is often observed in areas with topsoil horizons rich in organic matter. Organic matter increases the macroporosity of soils and favours the generation of preferential flow. The accumulation of clay in the *on* horizon in the midslopes of the south facing slope confirms this lateral flowpath (3). Clay is eluviated from the upperslopes and transported laterally until it accumulates at the break of slope in the *on* horizons.

The importance of studying hydrological processes at hillslope scale is emphasized by the soil distribution pattern of the Two Streams catchment. In both hillslopes the subsoil colour sequence is red then yellow and grey in the valley bottom. Although most of the soils show no indications of lateral flow, the colour sequence suggests that there is significant lateral flow at hillslope scale.

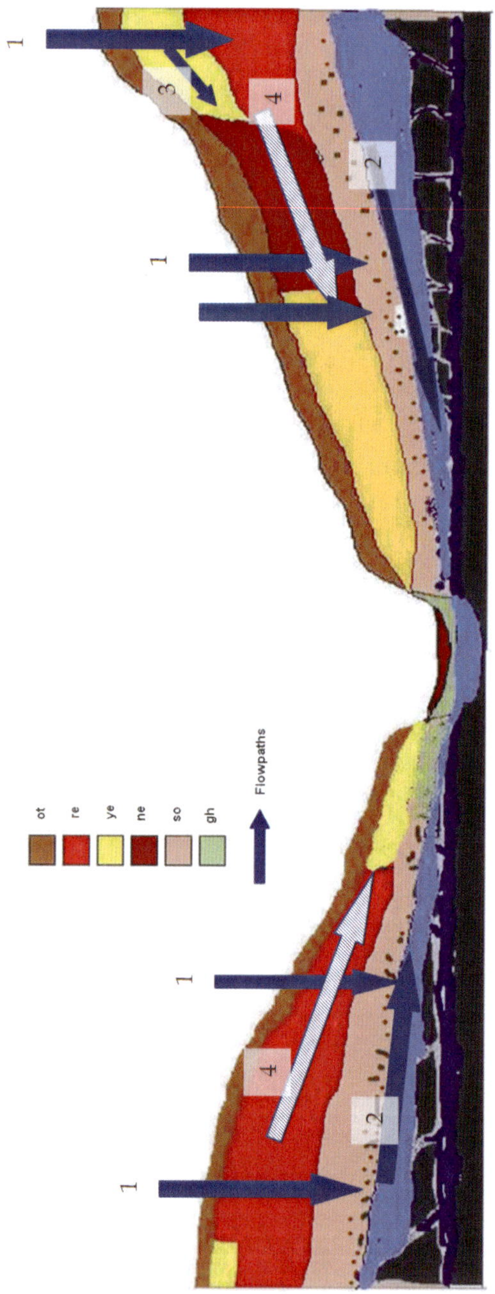

Fig. 9. Conceptual model of hillslope hydrological behaviour in the Two Streams catchment (Le Roux et al., 2010)

5. Hydropedological applications in hydrological modelling

5.1 Introduction

The hydrological response of catchments is dependent on the combined responses of the individual hillslopes within the catchment (Sivapalan, 2003). The hillslope is generally accepted as a fundamental landscape unit (Lin et al., 2006; Weiler & McDonnell, 2004), and is the smallest scale for holistically understanding and simulating hydrological processes (Tromp van Meerveld & Weiler, 2008). It is therefore not surprising that the hillslope forms the basic building block for a number of hydrological models.

The current dominant paradigm in hydrological modelling involves using an *a priori* set of small scale theories and process descriptions (e.g. Darcy and Richard's equations) and splitting the catchment into small enough uniform elements for these theories to work. The models arising from this paradigm emphasize the explicit mapping of landscape heterogeneities and process complexities which, according to McDonnell et al. (2007), are an impossible task in even the most intensively studied catchments. Consequently the models based on current theories rely strongly on calibration, mimicking past data, to account and compensate for the lack of understanding of the actual hydrological processes and heterogeneities in the landscape (McDonnell et al., 2007; Sivapalan 2003). This results in models that 'work' but for the wrong process reasons (Weiler et al., 2004) and models highly over parameterized with many combinations of the parameters resulting in the same final result. This leads to a large degree of modelling uncertainty and models unsuitable for predictions in ungauged basins (Beven, 2001 and McDonnell et al., 2007).

Another current weakness in hydrological modelling is the gap between experimentalists and modellers. For example, an experimentalist may propose a conceptual model of hydrological behaviour of a system based on observations, measurements and experience. The appropriateness of the concept can only be verified when a numerical model is built (Bredehoeft, 2005). Unfortunately modellers usually do not incorporate the experimentalist's knowledge into the model structure (McDonnel et al., 2007; Sivapalan 2003; Tromp-van Meerveld et al., 2008; Weiler et al., 2004). When they do, simulations are followed by calibration exercises, which, together with limited understanding of the complex processed involved lead to further confusion by "correcting" with imperfect data (Dunn et al., 2008). On the other hand experimentalists have focussed on the documentation of the unconventional behaviour of new hillslopes instead of the systematic examination of first order controls of hillslope hydrological behaviour, without intercomparisons to obtain common process behaviours (Weiler et al., 2004). The transference or extrapolation value of these hillslope studies is therefore minimal. Some researchers argue that every hillslope is therefore unique (Beven, 2001). This is possibly true to a certain extent, since after hundreds of experiments we appear to be no further towards a common conceptualization of hillslope hydrology, and experimentalists have not yet expressed what the minimal set of measurements are to characterize even a single hillslope (McDonnel et al., 2007; Tromp-van Meerveld et al., 2008; Weiler et al., 2004)! There is a great need for closer collaboration between experimentalist and modellers (Siebert and McDonnell, 2002). Neither the conceptual model of the experimentalist nor the numerical model of the modeller should be considered untouchable, but the common focus should be to, through iteration of concepts and numbers, represent the physical process numerically.

In this section a modelling exercise using hydropedological data for model parameterization and configuration is described. Although model outputs are presented, the main aim of the exercise was to evaluate the capability of a mechanistic model to utilize hydropedological inputs. Simulations were run over two rainfall seasons in the Weatherley research catchment and simulated streamflow and soil moisture contents were compared to measured values. Only one model run was conducted i.e. there were no calibrations with measured data to improve model outputs.

5.2 The hydrological model and model setup

5.2.1 ACRU-Int

ACRU is an agrohydrological, daily time step, multi-layered soil water budgeting model. The standard version, *ACRU2000*, comprises of two soil layers (A and B- horizon) and a deep groundwater layer. Soil inputs include; the thickness of soil horizons, water contents at the start of simulation (SMAINI and SMBINI), Permanent Wilting Point (PWP), Drained Upper Limit (DUL), saturation (Po), Plant Available Water (PAW), drainage rates (ABRESP and BFRESP) and the erodibility of the soil (*K*-factor). Except for the latter all inputs are required for both soil horizons (Schulze, 2007).

PWP, DUL and Po are largely determined by the soil texture, organic matter and the bulk density. Typical values for these parameters are proposed in chapter 5 of the *ACRU user manual* (Smithers and Schulze, 2004) for different textural classes and clay distribution models, i.e. change in clay content with depth. The clay distribution models and typical texture classes were assigned to the 501 soil series of the binomial soil classification of South African soils (MacVicar et al., 1977). Relatively accurate PWP, DUL and Po values are therefore easily accessible for all South African soils. The PAW is the difference between DUL and PWP and is used to calculate the initial water content, expressed as a percentage of PAW (Smithers *al.*, 2004).

ACRU-Int is a revised version of the standard *ACRU2000* model. In addition to the 2 soil layers (A&B-horizons) an intermediate layer (saprolite) between the soil and deep groundwater levels was introduced (Lorentz et al., 2007). The intermediate layer has a mechanism whereby lateral release of water can be induced when certain threshold positive pressures at the saprolite/bedrock interface is achieved. The lateral releases from the intermediate zone can be routed to any layer of a downslope land segments. This is ideal for imitating flowpaths at hillslope scale.

The model allows redistribution of saturated water (RESP), i.e. between DUL and Po, from the A to the B-horizon (ABRESP) and from the B-horizon to the groundwater (BFRESP). The distribution is expressed as a fraction of the water above DUL draining vertically downwards from the respective horizons on a daily time step. In Schulze (1995) typical RESP values are presented for different textural classes. Low RESP values in a particular horizon will result in the buildup of water above DUL in the soil horizons above it favouring the generation of lateral flow in that horizon. Thus a low ABRESP value will favour lateral flow in the A horizon and a low BFRESP value will favour lateral flow in the B horizon. Reductions in RESP are therefore suggested based on the "Interflow Potential" (IP) of different soil series, high IP = RESP × 0.3 and moderate IP = RESP = 0.6 (Schulze, 1995). The influence of the redistribution fractions on the simulated soil water contents is illustrated in Fig. 10.

Water contents in one land segment were simulated with *ACRU-Int* for six years with actual climatic data from the Weatherley catchment (Fig. 10). Four different RESP fractions were used *a*) ABRESP = 0.01 & BFRESP = 0.01, *b*) ABRESP = 0.99 & BFRESP = 0.99, *c*) ABRESP = 0.99 & BFRESP = 0.01 and *d*) ABRESP = 0.01 & BFRESP = 0.99. Simulation *a* and *b* show similar trends with a buildup of water in the A-horizon, relatively low water contents in the

B-horizon and a general decrease in the water content of the C-horizon due to very little vertical drainage from the A to lower horizons. In simulation *b*, water is allowed to drain freely to the B-horizon and then to the C-horizon, resulting in relatively low water contents in the A and B-horizons but accumulation in the C-horizon. Simulation *c* show water freely draining from the A-horizon but due to the impeding C-horizon (BFRESP = 0.01) buildup in the B-horizon, slowly reducing the water content in the C-horizon.

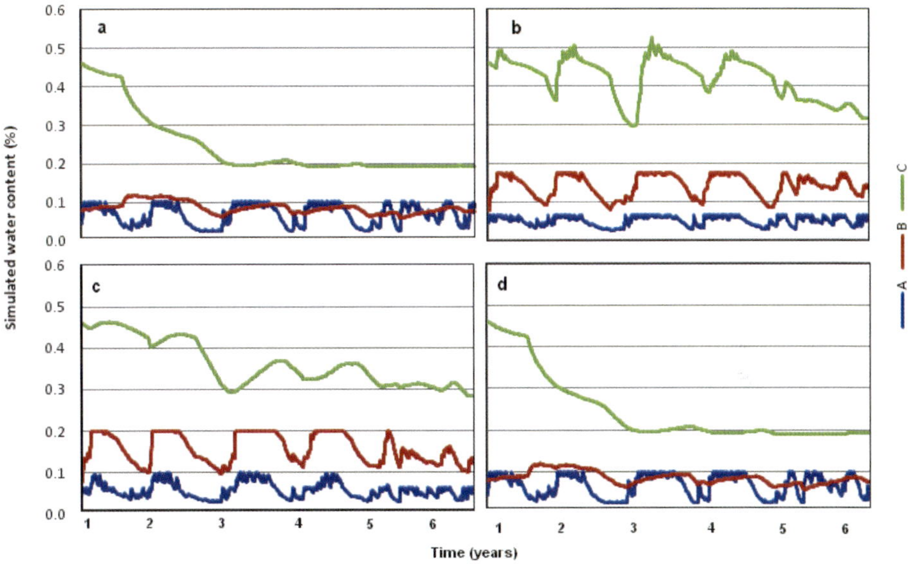

Fig. 10. Water contents (%) for three horizons in the Weatherley catchment with different RESP values simulated over 6 years with *ACRU-Int*

It is clear from Fig. 10 that the RESP values play an integral part in the simulation of soil water contents and consequently on the outflow of different land segments. Exactly how these values were obtained is however not clear. For example; one would expect a direct relationship between the saturated hydraulic conductivity (K_s) and the RESP value. In Schulze (1995) however a siltyloam horizon with K_s of 6.8 mm.h^{-1} has a RESP of 0.45 whereas a sandyclayloam with K_s of 4.3 mm.h^{-1} has a RESP of 0.50, similarly, a loam horizon and a sandyslayloam soil were attributed the same RESP value (0.50) although the K_s of the former is triple that of the latter. The heterogeneous horizonation in terms of the textural distribution is the driving force for lateral flow generation in soils and is the basis for assigning "Interflow Potential" values to different soil series. If texture differences are the main reason for differences in RESP values is it not spurious to reduce the RESP value based on the IP? The volume of water draining vertically in a profile is also related to the slope of the land. Steeper slopes generally favour more lateral flow, and less vertical distribution of saturated water. Also, soils with shallow horizons ought to distribute a greater percentage of water in a particular day compared to soils with deep horizons although the texture and hydraulic conductivities are similar.

Another important parameter in *ACRU-Int*, not considered a soil input but definitely influenced by the soil, is QFRESP. According to definition QFRESP is: *Stormflow response fraction for the catchment/subcatchment, i.e. the fraction of the total stormflow (1.0) that will run off from the catchment/subcatchment on the same day as the rainfall event* (Smithers *et al.*, 2004). QFRESP is inversely correlated with catchment area and will increase with an increase in slope angle, area covered by impervious material, and rainfall intensity. Soils prone to topsoil crusting as well as very shallow or very wet soils should therefore give high QFRESP values. No physical based methods to obtain suitable QFRESP values are however proposed by the modellers, and thus the appropriateness of the values used depends on calibration. *ACRU* can also account for unsaturated flow of water (IUNSAT) and flow through cracks or fissures in swelling soils (ICRACK). The latter is divided into three classes based on the clay content. Both IUNSAT and ICRACK can be excluded from simulations.

5.2.2 Model configuration

For the purpose of the modelling exercise the Weatherley catchment was divided into 7 land segments, each one with distinct hydrological responses (Fig. 11). The division was made derived from several pedological, soil physical, hydrogeological, geophysical and geochemical studies, as well as in-field observations of visible hydrological processes in the selected catchment over the past few years (Lorentz. 2001; Lorentz et al., 2004; Lorentz et al., 2007; van Huysteen et al., 2005; van Tol et al., 2010b). Some soil and landscape attributes, obtained from representative soil profiles of the land segments are presented in Tables 1 and 2. Two representative hillslopes, with diverse hydrological behaviour, were identified based on the properties of the land segments, their sequence from the crest to the river, and the area covered by the individual segments. Conceptual 2-dimensional flow models were then developed and applied to construct flow routings for the two hillslopes (Le Roux et al., 2010). Hillslope 1 includes land segments 1 - 4, and hillslope 2 includes land segments 5 - 7

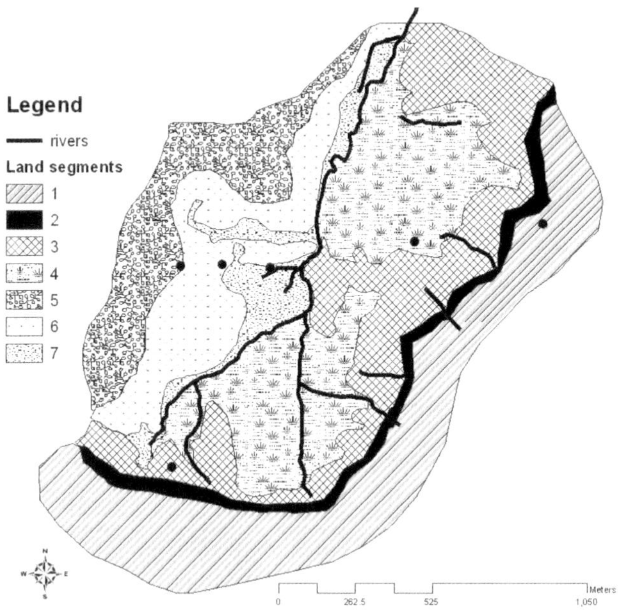

Fig. 11. Different land segments in the Weatherley catchment

(Fig. 12). Land segments 4 and 7 represent the valley bottom or wetland area; they drain to a separate land segment (9) which represents the stream network. Since the deep groundwater levels are always below the stream channel and do not contribute to low flows, all the drainage out of the intermediate zone into the groundwater layer was routed to another land segment (8). Land segment 8 is therefore not linked to any streamflow generation process. The routing between land segments is presented in Fig. 12.

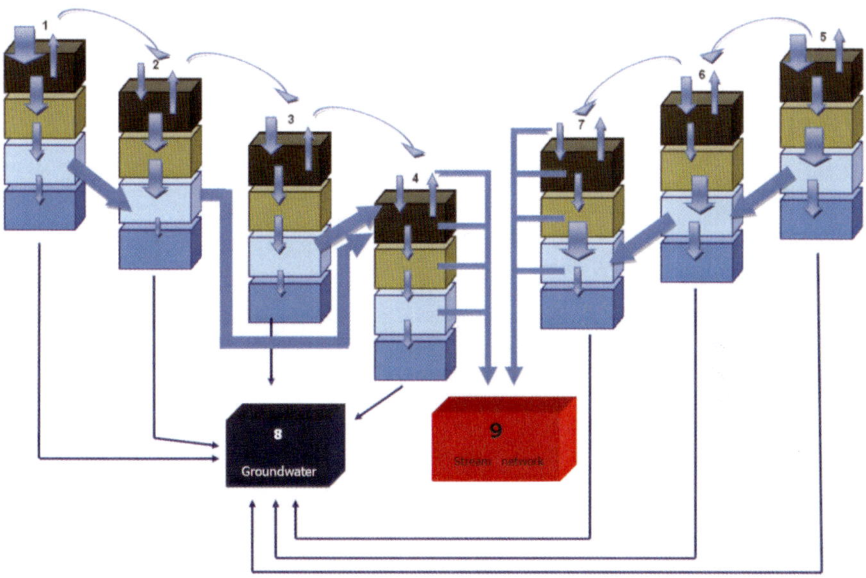

Fig. 12. Flow routing for the simulation, the magnitude of the arrows giving an indication of dominant flow directions in various land segments, the number of each one corresponding to the number in Fig. 11

5.2.3 Model parameterization

Imitating the dominant hydrological processes was one of the major aims of this simulation exercise. According to Sivapalan (2003), parameters without measured values require calibration with gauged data in order to reflect reality. This leads to uncertainty in predictions when dealing with unguaged basins. Parameter values, not directly available from soil profile descriptions, were therefore physically estimated based on the definition of the parameter and the (limited) understanding of the process influenced by the parameter. These calculations include:

- **ABRESP, BFRESP** and **INTRESP:** The difference between K_s of the top and lower horizon gives an indication of the vertical distribution from the former to the latter. K_s values were calculated using ROSSETA (Schaap, 2000) for all the horizons of profiles representing the different land segments using texture class distributions and bulk densities (D_b). The ratio of the K_s of the lower horizon to the K_s of its overlying horizon was used to estimate the particular RESP value. K_s of the R-horizon were measured. Representative texture class, D_b and estimated K_s values are presented in Table 1 and resulting ABRESP, BFRESP and INTRESP values in Table 2.

LandSeg.	Representative soil profiles*1	Horizon*2	Sand (%)	Silt (%)	Clay (%)	Texture class	Db (Mg m-3)	Ks (cm day-1)	s >0.7*3
1	202	A	79.1	10.9	10.0	Sandy Loam	1.68	46.00	1
		B	71.4	12.6	16.0	Sandy Loam	1.61	27.00	
		INT	57.3	16.7	26.0	Sandy Clay Loam	1.72	6.00	
		R						0.10	
2	204	A	78.0	10.0	12.0	Sandy Loam	1.60	52.48	278
		B	78.0	12.0	10.0	Sandy Loam	1.71	36.61	
		INT	75.0	13.4	11.6	Sandy Loam	1.71	26.87	
		R						0.02	
3	212&205	A	74.0	16.0	10.0	Sandy Loam	1.59	45.84	62
		B	70.0	15.0	15.0	Sandy Loam	1.68	19.25	
		INT	50.0	32.0	18.0	Loam	1.74	5.87	
		R						0.20	
4	206	A	50.0	40.0	10.0	Loam	1.52	22.37	622
		B	46.3	37.8	15.9	Loam	1.67	7.79	
		INT	35.1	26.9	38.0	Clay Loam	1.73	2.09	
		R						0.20	
5	210	A	56.0	28.4	15.7	Sandy Loam	1.59	15.36	99
		B	56.4	23.9	19.7	Sandy Loam	1.71	7.60	
		INT	56.2	23.5	20.3	Sandy Clay Loam	1.66	9.18	
		R						2.00	
6	209	A	50.4	33.9	15.7	Loam	1.54	14.75	437
		B	37.7	35.0	27.3	Loam	1.67	3.11	
		INT	5.0	50.8	44.2	Silty Clay	1.73	1.27	
		R						0.20	
7	208	A	45.8	39.8	14.5	Loam	1.47	17.04	541
		B	35.9	43.2	21.0	Loam	1.62	5.12	
		INT	31.2	42.9	26.0	Loam	1.71	2.58	
		R						0.20	

*1 Detailed descriptions together with analytical data for these profiles are presented in Van Huyssteen et al. (2005).

*2 A = A horizon, B = B horizon, INT = C horizon, R = Rock

*3 Number of days during overall period of 662 days with rainfall > 1 mm, when the water content of the A-horizon is >0.7 of Po

Table 1. Some attributes of different land segments used in to calculate model parameters

LandSeg	Area (km²)	Horizon	Depth (m)	Po	PWP	DUL	ABRESP	BFRESP	INTRESP	QFRESP
					$m^3\ m^{-3}$					
1	0.315	A	0.4	0.37	0.09	0.19	0.6	0.2	0.02	0.002
		B	0.42	0.39	0.09	0.19				
		INT	1.6	0.35	0.16	0.25				
2	0.072	A	0.1	0.40	0.09	0.19	0.7	0.7	0.00	0.92
		B	0.3	0.36	0.09	0.19				
		INT	1.2	0.35	0.09	0.19				
3	0.270	A	0.3	0.40	0.09	0.19	0.4	0.3	0.03	0.1
		B	1	0.37	0.09	0.19				
		INT	1	0.35	0.13	0.25				
4	0.375	A	0.5	0.43	0.13	0.25	0.3	0.3	0.10	0.94
		B	0.25	0.37	0.13	0.25				
		INT	0.25	0.35	0.20	0.31				
5	0.183	A	0.16	0.40	0.09	0.19	0.5	1.0	0.22	0.15
		B	0.9	0.36	0.09	0.19				
		INT	0.8	0.37	0.16	0.25				
6	0.225	A	0.45	0.42	0.13	0.25	0.2	1.0	0.16	0.66
		B	0.6	0.37	0.13	0.25				
		INT	0.7	0.35	0.25	0.32				
7	0.091	A	0.35	0.45	0.13	0.25	0.3	1.0	0.08	0.82
		B	0.2	0.39	0.13	0.25				
		INT	1.1	0.36	0.13	0.25				

Table 2. Soil parameters used simulating the Weatherley catchment

- **QFRESP:** Except for the Molteno outcrop, very few areas in the catchment have impervious top layers. Soil crusting and very shallow soils are also not the norm. The only soil related factor influencing the quick flow response is therefore the water content of the topsoil before and during rain events, where saturated A-horizons will impede infiltration, generating overland flow and consequently peak flow. Zere (2007) estimated daily soil water contents, based on weekly neutron probe measurements, over a 6 year period for 28 profiles in the Weatherley catchment. This was used to determine the number of days that A-horizons of the representative profiles were close to saturation (>0.7 of Po) on days with more than 1 mm of rain, during an overall number of 662 days in a 6 year period. The result is presented in Table 1 and resulting QFRESP values in Table 2. Approximately half of land segment 2 is covered by the impervious Molteno rock outcrop and was taken into account for calculation of QFRESP.

The thickness of A and B-horizons was obtained from Van Huysteen et al., 2005 for the profiles representing the different land segments. Where the lower depth of the profile was reached the B2 or C horizon was used as the depth of the intermediate zone, if not, an extra 0.5 m was added to the B2 or C horizon to acquire the intermediate zone depth. PWP, DUL and Po values were estimated based on typical texture class values proposed in Smithers et al., 2004.

The simulation period is from 1st Jan 1998 till 31st August 2001. Simulated results are reported for two rain seasons, starting from the 1st of September 1999 to allow the model to 'settle' and incorrect data regarding initial water contents to even out. Trees were planted on parts of the catchment in 2002 and 31st August 2001 was therefore selected as the end of simulation to avoid dissimilarity between vegetative covers.

Rainfall and, when possible, minimum and maximum temperature data were obtained from the BEEH (2003) database. Other climatic data was obtained from the quaternary catchment database. Streamflow was measured at a Crump weir at the catchment outlet and data regarding the streamflow obtained from BEEH (2003) database. Daily simulated soil water contents were compared to daily water contents calculated from weekly neutron water meter readings (Zere, 2005). Soil water contents are expressed as a fraction of porosity (Po).

5.3 Modelling results

5.3.1 Streamflow

Simulated and measured outflow from catchment for the simulation period is presented in Fig. 13. A R^2 value of 0.64 was attained with a linear line deviating almost 100% from the 1:1 line. The divergence is towards measured flow, i.e. a greater volume of flow was measured than simulated. Figures 15 and 16 accentuate the cause of the deviation from the 1:1 line.

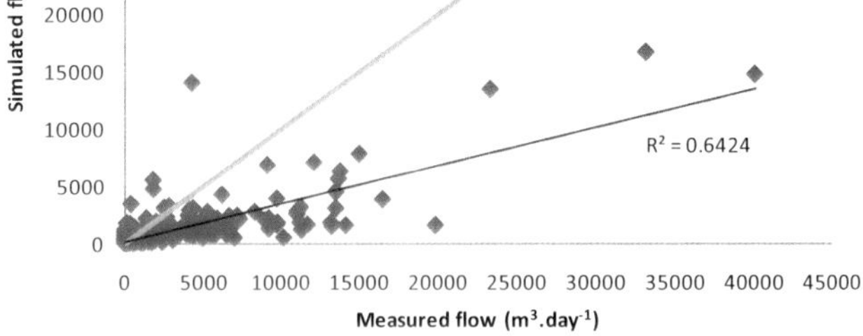

Fig. 13. Simulated vs. measured flow during the selected simulation period

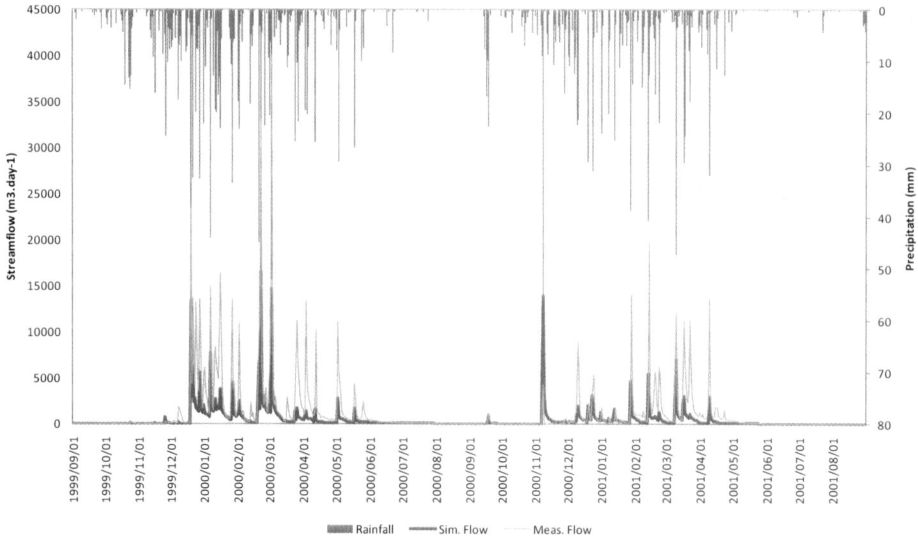

Fig. 14. Simulated vs. measured daily streamflow flow plotted against daily rainfall for the simulation period

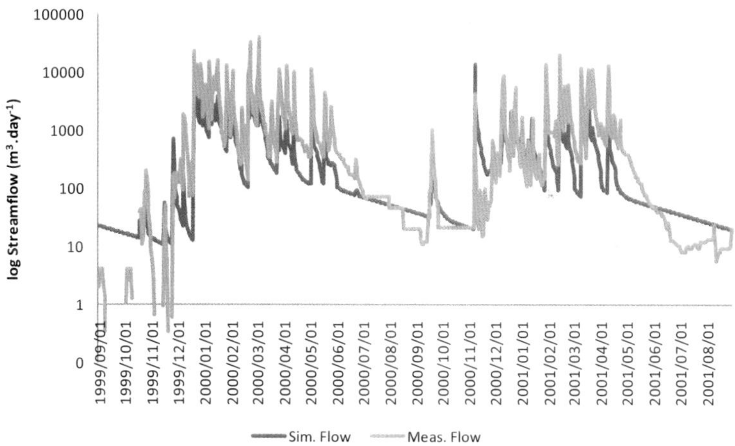

Fig. 15. Log of simulated vs. measured flow over the selected simulation period

Fig. 14 illustrates daily measured flow compared to simulated flow and also the influence of rainfall on flow volumes. It is clear from Fig. 14 that flows are overestimated especially towards the end of the rain seasons. Simulated low flows compared well with measured flows. Rain early in each season resulted in much smaller volumes of streamflow compared to similar size storms just before the end of the rain season. This over estimation of high flows and good representation of low flows is also emphasized in Fig. 15 where comparisons are plotted on a log scale.

Fig. 15 shows high flows being overestimated by an order of magnitude in under some conditions. Low flows and streamflow recession are however simulated reasonable well in drier periods. The average daily difference between simulated and measured results over the simulation period is 854 m^3 day^{-1}. This increases to 1194 m^3 day^{-1} during the rainfall months (October to April) and decreases to 114 m^3 day^{-1} for months normally associated with little or no rainfall. For ten day periods after any recorded rainfall, the difference between simulated and measured streamflows is only 16 m^3 day^{-1}.

5.3.2 Soil water contents

The comparisons between simulated and measured soil water contents of land segment 1 are presented in Fig. 16. The figure is divided into four segments starting with daily rainfall data at the top then simulated vs. measured water contents of the A and B-horizon and then simulated vs. measured water contents of the intermediate zone or C-horizon at the bottom. The water contents are expressed as a fraction of the porosity for the different horizons.
Fig. 16 shows a very good correlation between measured and simulated water contents of the A-horizon. Measured water contents for this horizon is slightly higher for most of the simulation period but drainage is quicker than simulated at the end of the rain season. Measured water contents are far higher than those simulated for the B-horizon (Fig. 16). Seasonal variation is evident in the simulated water contents but not as profound in the measured values. Sharp increases and decreases are noted in measured values, but not in simulated water contents. Measured and simulated water contents of the C-horizon compared well, although the measured contents show less seasonal variation than those simulated (Fig. 16). The response of this horizon to rainfall at the beginning of the simulation period shows a lag time of about 2 months.

5.4 Discussion

Underestimation of quick flows can be attributed to low QFRESP values. The method used to estimate QFRESP awarded high values for the wetland regions i.e. land segments 4 and 7. Van Tol (2010) showed that approximately 92% of the precipitation on the wetland will arrive at the catchment outlet on the same day. It was therefore surprising that QFRESP values of 0.92 and 0.82 for land segments 4 and 7 respectively were insufficient to generate comparable peak flows (Table 2). There are two possible explanations for this: some portions of land segments 3 and 5 form part of the wetland and should therefore be given higher QFRESP values; upslope land segments 1 and 5 make a larger contribution to daily flows than what their QFRESP values suggest. The latter is supported by findings of Lorentz et al. (2007), as well as our later trench and slotted pipe experiments which show a significant volume of water flowing laterally in the A-horizon of upslope land segments. This near 'surface macropore flow' can contribute approximately 16% of event water (Lorentz *et al.*, 2007) and should be taken into consideration when assigning QFRESP values to land segments. The result for the A horizon of land segment 1 presented in Fig. 16 do not however support that suggestion. The reliable simulation of water content for this horizon indicates that the simulated amount of water infiltrating the profile is more or less in agreement with actual infiltration, and that the QFRESP value i.e. 0.002, for this land segment is satisfactory. One could argue that increasing QFRESP and lowering ABRESP would result in similar water contents in the A-horizon, as there would be less recharge of the B horizon and therefore a larger volume build-up in the A horizon. This will however deprive water from the already underestimated B-horizon and will alter the reasonably well simulated water balance of the intermediate zone (Fig. 16). Similarly one cannot simulate less surface water from the A-horizon of land segment 5 by assigning a higher QFRESP value, as water contents of this horizon are underestimated for the greatest part of the simulated period. Lowering the ABRESP would decrease the water content of the already underestimated B-horizon and changing the BFRESP will adversely adjust the relatively well simulated water contents of the intermediate zone.

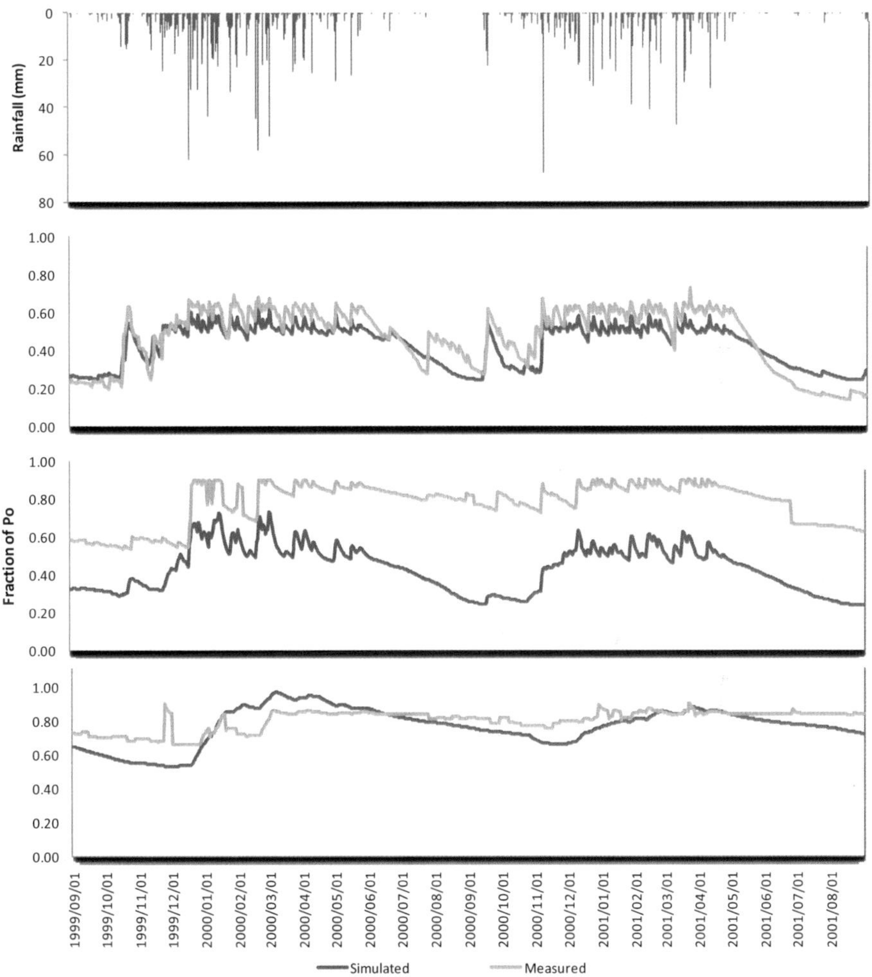

Fig. 16. Rainfall and simulated vs. measured water contents of A, B and C-horizons (top to bottom respectively) of land segment 1

In both land segments 1 and 5 the water contents of the C-horizon are simulated fairly well. Assigning higher QFRESP values to compensate for underestimation of peak flows and then lowering ABRESP, BFRESP and also INTRESP should probably result in comparatively water content simulations and might increase the accuracy of streamflow predictions. That would however be the antithesis of the aim of this study and will definitely not contribute to predictions in ungauged basins.

It is clear that a method for attaining accurate QFRESP values is needed. This method should encompass soil and landscape properties and should preferable be dynamic in nature, as one of the major driving forces for quick flow generation is the antecedent water content of the topsoil. The latter is in accord with the well known 'variable source area' concept. The influence of the antecedent water on streamflow is further emphasized in Fig. 13. In the 1999/2000 season the first significant increase in streamflow was recorded on the 8th of December after 153 mm of precipitation had been recorded from the beginning of September. For the 2000/2001 season the first significant increase in streamflow was recorded on the 7th of November following 110 mm of precipitation from the beginning of September. Before the 8th December 2000 and the 7th of November, 'peak flows' were overestimated by the model as the storage capacity of the catchment had not yet been filled.

Simulated low flows correspond very well with measured flows (Fig. 14). Simulations in Le Roux et al. (2010) using *ACRU2000* (in lumped and distributed mode) and *ACRU-Int* (using 1 and 3 land segments), could simulate high flows moderately well but was unsuccessful in simulating low flows. It is believed that the model configuration used in this study represents the actual processes generating base flow, i.e. drainage from the soil and not from groundwater, and it was therefore encouraging to observe the connection between simulated and measured outflows.

Water contents of the A-horizon were over estimated for the majority of the simulation period in land segments 2, 3, 4, 6 and 7. The reality is that the land segments represent an area, whereas the profiles with the measured data only represent a point in that area. This should be kept in mind. Lateral flow, such as near surface macropore flow, occurs in a downslope direction. When the representative profile is in the lower regions of the land segment, these lateral contributions might have accumulated in the horizons resulting in higher measured compared to simulated water contents. The same applies for the underestimation of water contents in the B-horizons of all land segments except for 3. Routing of A and B-horizons to A, B and C horizons of different land segments might aid in solving this problem and ensure a better imitation of actual hydrological processes.

6. Conclusions

The grand challenge for predictions in ungauged basins (and for hydrological modeling as a whole) requires a holistic approach based on an improved understanding of the complex hydrological system and landscape heterogeneity. Soil acts as first order control in catchment hydrology through governing of water flowpaths and thereby influencing residence times and storage mechanisms. Water controls soil formation to such an extent that the soil carries signatures (morphological properties) of the hydrological processes involved in soil genesis. Based on this interactive relationship, soil scientists can make a valuable contribution to the improved understanding of the hydrological behaviour of the system.

The interactive relationship between soil and hydrology was demonstrated through conceptualizing the dominant hydrological response of soil profiles in the Weatherley catchment with six years of measured soil water contents. Results supported the interpretations and emphasized the different hydrological behaviour of different soils. This led to identification of three hydrological soil types namely recharge, interflow and responsive soil types, based on water movement and storage within these soils. The interpretation and conceptualization of soil distribution patterns of hillslopes in the Weatherley and Two Streams catchments reflected the actual hydrological behaviours of these hillslopes. Soil property distribution at hillslope scale proves to be a vehicle for identifying and studying first order controls of hillslope hydrological behaviour.

There is abundant evidence from international and local sources that soil information can improve efficiency of hydrological models. Because hydropedological studies are still in their infancy in South Africa (and in the world) huge gaps exist between the modelers (hydrologists) and experimentalists (soil scientists). Future studies should aim to close the gap by more interaction with all stakeholders. Only through interdisciplinary research can problems like pollution, land use change, floods and climate change be solved.

There is need to expand these hydropedological studies in order to improve qualitative and quantitative understanding, conceptualization and characterization of flowpaths, connectivity's, thresholds, non-linearity's and residence times at hillslope scale. This can only be achieved through intensive studies of soils in hillslopes over a wide range of climates and geologies.

7. References

Asano, Y., Uchida, T. & Ohte, N., 2002. Residence times and flow paths of water in steep unchannelled catchments, Tanakami, Japan. *Journal of Hydrology* 261, pp. (173-192).

BEEH, 2003. Weatherley Database V1.0. School of Bio-resources Engineering and Environmental Hydrology, University of Natal, Pietermaritzburg.

Beven, K. J., 2001. On fire and rain (or predicting the effects of change). *Hydrological processes* 15, pp. (1397 - 1399).

Bockheim, J. G. & Douglass, D. C., 2006. Origen and significance of calcium carbonate in soils of southwestern Patagonia. *Geoderma* 136, pp. (751 - 762).

Boorman, D. B., Hollis, J. M. & Lilly, A., 1995. Hydrology of soil types: a hydrologically based classification of the soils of the United Kingdom. IH Report No.126. Institute of Hydrology, Oxfordshire, UK.

Clothier, B. E., Green, S. R. & Deurer, M., 2008. Preferential flow and transport in soil: progress and prognosis. *European Journal of Soil Science* 59, pp. (2 - 13).

Driessen, P. & Deckers, J., 2001. Lecture notes on the major soils of the world. http://www.fao.org/DOCREP/003/Y1899E/y1899e09.htm. (Retrieved 26/05/2008).

Dunn, S. M., Freer, J., Weiler, M., Kirkby, M. J., Seibert, J., Quinn, P.F., Lischied, G., Tetzlaff, D. & Soulsby, C., 2008. Conceptualization in catchment modelling: simply learning? *Hydrological processes* 22, pp. (2389 - 2393).

Fanning, D. S. & Fanning M. C. B., 1989. *Soil: Morphology, Genesis and Classification*. Wiley & Sons, New York. pp. (360–369).

Hillel, D., 1980. *Fundamentals of soil physics*. Academic Press, New York.

IUSS Working Group WRB. 2006. *World reference base for soil resources* 2006. 2nd edition. World Soil Resources Reports No. 103. FAO, Rome.

Karvonen, T., Koivusalo, H., Jauhiainen, M., Palko, J. & Weppling, K., 1999. A hydrological model for predicting runoff from different land use areas. Journal of Hydrology. 217, pp. (253-265).

Kutílek, M. & Nielson, D. R., 2007. Interdisciplinary of hydropedology. *Geoderma*, 138, pp. (252 - 260).

Le Roux, P. A. L., 1996, Die aard, verspreiding en genese van geselekteerde redoksmorfe gronde in Suid Afrika. Ph. D. Thesis. University of the Orange Free State, Bloemfontein.

Le Roux, P. A. L., van Tol, J. J., Kuenene, B. T., Hensley, M., Lorentz, S. A., Everson, C. S., van Huyssteen, C. W., Kapangaziwirri, E. & Riddell, E. S., 2010. Hydropedological interpretations of the soils of selected catchments with the aim of improving the efficiency of hydrological models. Report No. K5/1748. Water Research Commission, Pretoria, South Africa.

Lilly, A., Boorman, D.B. & Hollis, J.M., 1998. The development of a hydrological classification of UK soils and the inherent scale changes. *Nutrient Cycling in Agroecosystems* 50, pp. (299 - 302).

Lin, H. S., 2003. Hydropedology: bridging disciplines, scale and data. *Vadose Zone Journal*, 2, pp. (1 - 11).

Lin, H. S., Kogelman, W., Walker, C. & Bruns, M. A., 2006. Soil moisture patterns in a forested catchment: A hydropedological perspective. *Geoderma* 131, pp. (345 - 368).

Lin, H. S., Bouma, J., Owens, P. & Verpraskas, M., 2008. Hydropedology: Fundamental issues and practical applications. *Catena* 73, pp. (151 - 152).

Lin, H. S., 2010. Earth's critical zone and hydropedology: concepts, characteristics and advances. *Hydrol. Earth. Syst. Sci.* 14, pp. (25 - 45).

Lorentz, S. A., 2001. Hydrological systems modelling research programme: hydrological processes. Report No. 637/1/01. Water Research Commission, Pretoria.

Lorentz, S. A., Thorton-Dibb, S., Pretorius, C. & Goba, P., 2004. Hydrological systems modelling research programme: hydrological processes. Report No. 1061 & 1086/1/04. Water Research Commission, Pretoria.

Lorentz, S. A., Bursey, K., Idowu, O., Pretorius, C. & Ngeleka, K., 2007 (a). Definition and upscaling of key hydrological processes for application in models. Report No. K5/1320. Water Research Commission, Pretoria.

MacVicar, C. N., De Villiers, D. E., Loxton, R. F., Verster, E., Lambrechts, J. J. N., Merryweather, R. F., Le Roux, J., Van Rooyen, T. H. & Von Harmse, H. J., 1977. Soil Classification - a binomial system for South Africa. Dept. Agric. Tech. Serv. Pretoria.

McDonnell, J. J., Sivapalan, M., Vaché, K., Dunn, S., Grant, G., Haggerty, R., Hinz, C., Hooper, R., Kirchner, J., Roderick, M. L., Selker, J. & Weiler, M., 2007. Moving beyond heterogeneity and process complexity: A new vision for watershed hydrology. *Water Resources Research* 43, pp. (1-6).

McGlynn, B. L., McDonnell, J. J. & Brammer, D. D., 2002. A review of the evolving perceptual model of hillslope flowpaths at the Maimai catchments, New Zealand. *Journal of Hydrology* 257, pp. (1-26).

Mosley, M. P., 1982. Surface flow velocities through selected forest soils South Island, New Zealand. *Journal of Hydrology* 55, pp. (65-92).

Nahar, N., Govindaraju, R. S., Corradini, C. & Morbidelli, R., 2004. Role of run-on for describing field-scale infiltration and overland flow over spatially variable soils. *Journal of Hydrology*. 286, pp. (36-51).

Netterberg, F., 1978. Dating and correlation of calcretes and other pedocretes. *Trans. Geol. Soc. S. Afr.*, 81, 379 - 391.

Nieber, J. L., Bauters, T. W. J. Steenhuis, T. S. & Parlange, J. Y., 2000 Numerical simulation of experimental gravity-driven unstable flow in water repellent sand. *Journal of Hydrology*, 231, 295-307Park, S. J., Mcsweeney, K. & Lowery, B., 2001. Identification of the spatial distribution of soils using a process-based terrain characterization. Geoderma. 103, pp. (249-272).

Park, S. J. & Van De Giesen, N., 2004. Soil-landscape delineation to define spatial sampling domains for hillslope hydrology. *Journal of Hydrology*. 295, pp. (28 - 46).

Schaap, M. G., 2000. ROSETTA version 1.2. U. S. Salinity Laboratory. ARS-USDA.

Schulze, R. E. 1995. Hydrology and agrohydrology: A text to accompany the ACRU 3.00 agrohydrological modelling system. Water Research Commission, Report No 63/2/84. WRC, Pretoria.

Schulze, R. E., 2007. Soils: Agrohydrological information needs, information sources and decision support. *In* Schulze, R.E (Ed). 2007. South African atlas of climatology and agrohydrology. Report No. 1489/1/06. Water Research Commission, Pretoria.

Seibert, J. & McDonnell. J. J., 2002. On the dialog between experimentalist and modeller in catchment hydrology: Use of soft data for multicriteria model calibration. *Water Resour. Res.* 38 (11), 1241, doi:10.1029/2001WR000978.

Shankar, N. & Achyuthan, H., 2007. Genesis of calcic and petrocalcic horizons from Coimbatore, Tamil Nadu: Micromorphology and geochemical studies. *Quaternary International* 175, pp. (140 - 154).

Siddle, R. C., Noguchi, S., Tsuboyama, Y. & Laursen, K., 2001. A conceptual model of preferential flow systems in forested hillslopes: evidence of seld-organization. Hydrol. Process. 15, pp. (1675-1692).

Sivapalan, M. 2003a. Prediction in ungauged basins: a grand challenge for theoretical hydrology. *Hydrol. Process.*, 17, pp. (3163 - 3170).

Sivapalan, M., Takeuchi, K., Franks, S.W., Gupta, V.K., Karambiri, H., Lakshmi, V., Liang, X., McDonnell, J.J., Mendiono, E. M., O'Connell, P.E., Oki, T., Pomeroy, J.W., Schertzer, D., Uhlenbrook, S., & Zehe, E., 2003. IAHS decade on prediction in ungauged basins (PUB), 2003 - 2012: Shaping an exciting future for the hydrological sciences. *Hydrol. Sci. J.* 48 (6) pp. (857 - 880).

Smithers, J. & Schulze, R. E., 2004. ACRU Agrohydrological modelling system: user manual v4.00. School of Boiresources Engineering and Enviromental Hydrology, University of Natal, Pietermaritzburg.

Soulsby, C. & Tetzlaff, D., 2008. Towards simple approaches for mean residence time estimation in ungauged basins using tracers and soil distributions. *Journal of Hydrology* 363, pp. (60 - 74).

Soil Classification Working Group, 1991. Soil Classification - A taxonomic system for South Africa. *Mem, agric. Nat . Resour. S. Afr.* No. 15. Dept. Agric. Dev., Pretoria.

Soil Survey Staff, 1992. Keys to Soil Taxonomy, 5th edn. Pocahontas Press Inc., Blacksburg, Virginia.

Ticehurst, J. L., Cresswell, H. P., McKenzie, N. J. & Clover, M. R., 2007. Interpreting soil and topographic properties to conceptualise hillslope hydrology. Geoderma 137, pp. (279 - 292).

Tromp-van Meerveld, I. & Weiler, M., 2008. Hillslope dynamics modelled with increasing complexity. *Journal of Hydrology* 361, pp. (24 - 40).

Uchida, T., McDonnell, J. J. & Asano, Y., 2006. Functional intercomparison of hillslope and small catchments by examining water source, flowpath and mean residence time. Journal of Hysrology 327, pp. (627-642).

Uhlenbrook, S., Wenninger, J. & Lorentz, S., 2005. What happens after the catchment caught storm? Hydrological processes at the small, semi-arid Weatherley catchment, South-Africa. *Advances in Geosciences* 2, pp. (237 - 241).

Van Breedeman, N. & Brinkman, R., 1976. *Chemical equilibria and soil formation*. In G. H. Bolt & M. G. M. Bruggrnwert (eds.). Soil chemistry. A. Basic elements. Elsevier, Amsterdam.

Van Huyssteen, C. W. & Ellis, F., 1997. The relationship between subsoil colour and degree of wetness in a suite of soils in the Grabouw district, Western Cape I. Characterization of colour-defined horizons. *S. Afr. J. Plant Soil*, 14, pp. (149 - 153).

Van Huyssteen, C. W., Hensley, M., Le Roux, P. A. L., Zere, T. B. & Du Preez, C. C., 2005. The relationship between soil water regime and soil profile morphology in the Weatherley atchment, an afforestation area in the Eastern Cape. Report no. 1317/1/05. Water Research Commission, Pretoria.

Van Huyssteen, C. W., 2008. A review of the advances in hydropedology for application in South Africa. *S. Afr. J. Plant Soil*, 25, pp. (245 - 254).

Van Tol, J. J., Le Roux, P. A. L., Hensley, M. & Lorentz, S. A., 2010. Soil as indicator of hillslope hydrological behaviour in the Weatherley Catchment, Eastern Cape, South Africa. *Water SA*. 36, pp. (513 - 520).

Van Tol, J. J., Le Roux, P. A. L. & Hensley, M., 2010a. Soil properties as indicators of hillslope hydrology in the Bedford catchments. *S. Afr. J. Plant & Soil* 27, pp. (242 - 251).

Verpraskas, M. J. & Bouma, J., 1976. Model experiments on mottle formation simulating field conditions. *Geoderma* 15, pp. (217-230).

Webster, R. 2000. Is soil variation random? *Geoderma* 97, pp. (149 - 163).

Weiler, M. & McDonnell, J., 2004. Virtual experiments: a new approach for improving process conceptualization in hillslope hydrology. *Journal of Hydrology* 285, pp. (3 - 18).

Wenninger, J., Uhlenbrook, S., Lorentz, S. & Leidbungut, C., 2008. Identification of runoff generation processes using combined hydrometric, tracer and geophysical methods in a headwater catchment in South Africa. *Hydrological Sciences-Journal-des Sciences Hydrologiques* 53, pp. (65 - 80).

Zere, T. B., 2005. The hydropedology of selected soils in the Weatherley catchment in the Eastern Cape of South Africa. Ph.D. Dissertation. University of the Free State, Bloemfontein.

12

Multiscaling Analysis of Soil Drop Roughness

R. García Moreno[1], M.C. Díaz Álvarez[2],
A. Saa Requejo[2] and J.L. Valencia Delfa[3]
[1]*Departamento de la Navegación y de la Tierra, Facultad de Ciencias, Universidad de La Coruña, A Coruña,*
[2]*CEIGRAM (Centre for Studies and Research on Agricultural and Environmental Risk Management). School of Agricultural Engineering, Polytechnic University of Madrid, Madrid,*
[3]*Escuela Universitaria de Estadistica. Universidad Complutense de Madrid, Madrid*
Spain

1. Introduction

Soil surface roughness (SSR), which describes the microvariation in soil elevations across a field resulting primarily from tillage practices and soil texture, is one of the major factors in wind and water erosion (Porta Casanellas et al., 2003). Soil surface roughness and the complementary soil microrelief depression pattern determine water infiltration and drainage network development (Vidal Vázquez et al., 2006). Most studies on SSR have focused on the mathematical description of the variations appearing after rainfall to predict water infiltration and runoff (Linden and Van Doren, 1986; Kamphorst et al., 2000; Darboux and Huang, 2003).

Soil surface roughness is defined as the standard deviation of surface elevation readings. After tillage, soil microtopography exhibits randomly oriented tillage roughness marks of different sizes as well as clods (Allmaras et al., 1966; Zobeck and Onstad, 1987; Huang, 1998). Each specific tillage tool creates its own oriented roughness pattern, which is relatively easy to quantify using a simple geometric model. The challenge consists in quantifying the spatial distribution of randomly oriented SSR (Huang, 1998).

Soil surface roughness, taken on a scale ranging from centimeters to millimeters, plays a very important role in increasing water infiltration and the amount of crop water available and in reducing runoff on cultivated lands (Podmore and Huggins, 1981; Armstrong, 1986; Kamphorst et al., 2000). At the same time, it is an important factor in predicting wind erosion (Zobeck, 1991; Larney et al., 1995), one of the main forms of soil degradation in semiarid and arid climates. The concomitant loss of organic matter and nutrient-rich topsoil occasions a decline in soil productivity (Hagen, 1988; Potter et al., 1990; Larney et al., 1998). Soil surface roughness quantification is therefore crucial to understand soil erosive processes and how soil properties are altered by human action, primarily tillage (Perfect et al., 1990; Saxton, 1995; Murillo et al., 2004).

During the past few years, SSR analysis has focused on developing a unified conceptual framework for describing the geometric complexity of the data with the aid of fractal parameters. A number of methods have been proposed to estimate the fractal dimensions of soil microtopography (Linden and Van Doren, 1986; Malinverno, 1990; Perfect and Kay,

1995; Vidal Vázquez et al., 2005, 2006). The fractal techniques used can be divided into two groups: nonvariational and variational. Nonvariational techniques implicitly assume soil surface self-similarity across a range of scales and aim to characterize soil microrelief features by calculating a single index. Because microrelief fractal behavior is better modeled on the basis of either self-similar or prefractal surfaces, the use of nonvariational techniques has been highly criticized, which has in turn encouraged the use of variational methods (Vivas Miranda, 2000; Vidal Vázquez et al., 2005). The first group includes tortuosity (Bertuzzi et al., 1990) and the Richardson number (Gallart and Pardini, 1996; Pardini and Gallart, 1998). The second group of methods, in turn, is comprised of the semivariogram method (Armstrong, 1986; Huang and Bradford, 1992; Eltz and Norton, 1997; Vivas Miranda, 2000; Vivas Miranda and Paz González, 2002), spectral analysis (Burg, 1967), and the several existing versions of the root mean square or roughness length method (Malinverno, 1990; Gallant et al., 1994; Vivas Miranda, 2000; Vivas Miranda and Paz González, 2002).
Variational techniques are considered to provide a better description of SSR (Vidal Vázquez et al., 2006, 2007). The ones most commonly used to estimate the fractal indices of soil profiles or surfaces are semivariance and local root mean square. Both of these methods are based on the calculation of the Hurst exponent, H, from which the fractal dimension, D, is assessed; moreover, variational methods involve an additional parameter, the so-called crossover length, l. The fractal dimension, D, is a descriptor of horizontal variations in soil roughness, whereas crossover length, l, is related to vertical differences in point elevation data (Vidal Vázquez et al., 2006).
In addition, multifractal models have been used to analyze the scale-invariant properties of objects in very different domains, from turbulent flows to financial data. Scale invariance has been found to be of increasing importance in understanding the complexity of natural phenomena. Multifractal analysis (MFA) has been used intensively in geomorphometry or digital terrain heights (digital elevation models) (Pike, 2000), but only recently to study agricultural soils. Manninen (2003) showed that bare soil exhibits multiscale behavior and Roisin (2007) that MFA can effectively analyze the variability in the inner heterogeneity of tilled soils from soil strength measurements.
To analyze soil surface roughness and related systems, the concepts of scale-invariance and multifractality provide to most productive framework for data analysis, one of such methods, it is called Structure Function, and it focuses on the absolute values of the differences that occur in the data over arbitrarily large or small scales.
To this end, several soil types and tillage tools were selected to study heterogeneity based on soil height readings and the application of multifractal concepts for characterizing soil microrelief (García Moreno, 2006; García Moreno et al., 2008a, 2008b, 2010). Based on Multiscale Analysis, the present study aimed to apply Structure Function analyses and gray differences distribution to original and absolute differences distribution of SSR data. To remark the differences of soil surface roughness the authors applied absolute average of the differences around each point, similar to the low pass filter to the image, called soil drop roughness (SDR). The Structure Function, and associated parameters, was then applied to extract generalised Hurst exponent depending on soil types and tillage tool used.

2. Materials and methods

2.1 Experimental sites

The field experiments were conducted on different soil types at three sites in semiarid central Spain. The first experimental plot was located in the province of Madrid, in fields belonging to the Polytechnic University of Madrid's School of Agricultural Engineering (the Madrid site). The other two were located at La Higueruela (Santa Olalla, province of Toledo), in the Spanish National Research Council's Experimental Station for Environmental Science (La Higueruela site). The main soil characteristics, tested according to ISRIC/FAO (Merrill, 1995) and Soil Science Society of America (Sparks, 1996) methodologies are given in Table 1.

Site	Conductivity (dS/m)	Organic matter (%)	pH	USDA textural analysis (%)			USDA textural class
				Sand	Silt	Clay	
Madrid	1.90 (0.34)	1.8 (0.4)	7.8 (0.2)	57 (1)	17 (2)	26 (1)	Sandy clay loam
La Higueruela	0.21 (0.05)	2.6 (0.1)	6.2 (0.2)	53 (2)	23 (3)	24 (1)	Sandy clay loam
La Higueruela	0.68 (0.55)	1.5 (0.2)	5.7 (0.1)	63 (2)	19 (2)	18 (1)	Sandy loam

Table 1. Properties of the soils studied. (The values in parenthesis are the standard deviation of 12 samples for each type, three per subplot.)

The three types of tools used to till each soil type, namely chisel, tiller, and roller, are the three most common in the central regions of Spain. All measurements were taken immediately after tillage to preclude the effects of other factors. In other words, SSR was analyzed in a total of nine scenarios. Tillage was performed using John Deere equipment: a Model 2810 moldboard plow, a Model 610 integral chisel plow, and a roller level.

The field data were gathered in 2005, one of the driest seasons in Spain in the last 100 yr, with no rainfall recorded in either the spring or the summer. Indeed, while the average annual rainfall in the area is 411 mm, only 125 mm fell in the experimental region between 1 Sept. 2004 and 31 Aug. 2005 (Instituto Nacional de Meteorologia, 2005).

2.2 Soil surface roughness data

Field measurements were obtained with a full-scale pin meter shown in figure 1.

Fig. 1. Pin-meter used to measure soil surface roughness

This instrument consisted of a row of 35-cm-high pins, placed in a frame in which they could slide up or down to conform to surface irregularities. The pin heads were marked with a blue band to better visualize their respective positions when in contact with the soil. The frame, 85 cm high in all, was designed to be able to move the instrument across the soil without disturbing the pin pattern. The instrument was made of lightweight aluminium for ease of handling. With rows containing 50 pins spaced at 20-mm intervals, one full meter could be measured along the x axis with each reading. The y axis readings were taken by sliding the instrument across the plot, on tracks, stopping at 20-mm intervals. As the cells on the resulting grid measured 20 by 20 mm, a total of 2500 readings were taking per 1.0 m^2 of area. An earlier study (García Moreno, 2006) showed this spacing to be sufficient to measure the surface roughness of the three types of soil.

Each corner of the instrument was marked with a red dot and Visual Basic software was developed that would detect these marks as the vertical and horizontal references for shifts in row position.

A Kodak DC 4800 digital camera, set on a tripod, was used to capture pin positions. The lens was focused on a point at the center of the pin meter, i.e., at the average height of the red marks, to ensure the image would not be distorted. After comparing several models, a Silk tripod was found to be best suited to the 40-cm camera height required. The 3.1-megapixel camera was fitted with a 3× (28–84) optical zoom lens.

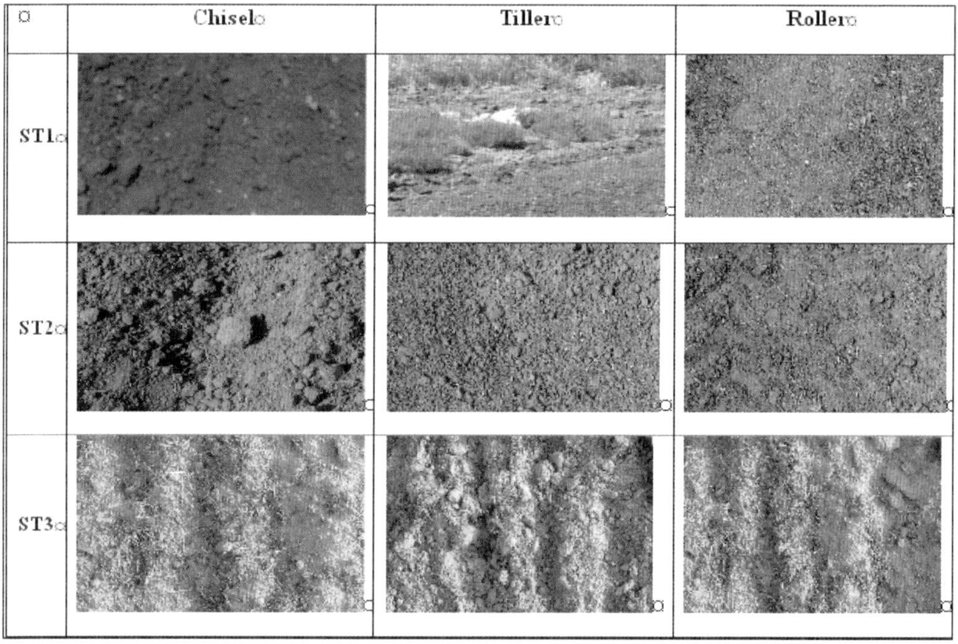

Fig. 2. Images of the plots after passing Chisel (left), Tiller (centre) and Roller (right) of the experimental fields at: ST1, E.T.S.I.A Technical University of Madrid with a Sandy Clay Loam soil; ST2, La Higueruela (C.S.I.C.) with a Sandy Clay Loam; ST3, La Higueruela (C.S.I.C.) with a Sandy Loam soil

Four sampling subplots of 1.0-m^2 were randomly chosen across a tilled area measuring 5 by 10 m^2. These four subplots served to evaluate the effects of each soil type and tillage tool. The data obtained were statistically analyzed to compare the effects of the different tools and soil types studied.
The field procedure consisted in placing the pin meter on the surface of a 1.0-m^2 patch of soil and capturing the initial pin positions and the positions after each 20-mm shift along the y axis. The camera was initially placed at a distance of 2 m from the pin meter. The x axis measurements were the positions of the row of 50 pins. The instrument was moved along the y axis over two rails perforated at 20-mm intervals. It was fitted with a hand brake to halt the process when soil was suspected to be on a light grade.
Consequently, the area measured was 2 by 2 m^2, with a resolution of 20 by 20 mm2. A total of 10,000 elevation readings were taken on each field surface, sufficient to estimate SSR indices based on Multiscale Analysis (Merel and Farres, 1998; Tarquis et al., 2003) and shadow analysis interpretation (Garcia, 2006). Photographs of the nine scenarios studied, after tillage, are shown in figure 2.

2.3 Soil surface roughness data

The original soil surface data was compared to the data submitted to absolute differences distribution to better evaluate the behaviour of SSR related to soil type and tillage tool. After comparing the results, the modified data seems to better define the effects of the resulting soil surface roughness after passing the tillage tools to the different soils, and give a clearer image of the influence of soil type. For that reason after applying a gray scale distribution and Multifractal Analysis to both types of data, only the modified was evaluate using the Structure function. The resulting modified data was named by authors Soil Drop Roughness (SDR), since represent the highest differences on soil surface roughness values.
The gray scale distribution was applied giving a value of gray intensity (0 to 250 intensities) to the heights related to the soil surface roughness.

2.4 Multiscale analysis through structure function

Standard Gaussian-type statistics are not really adequate to describe Soil Surface Roughness. Means and standard deviations depend on the scale of space sampling. For such systems, the concepts of scale-invariance and multifractality provide to most productive framework for data analysis. One method for doing this is called "Structure Functions" focuses on the absolute values of the differences that occur in the data over arbitrarily large or small scales. For nonstationary processes x(t), with stationary increments, the Structure Function of order q is defined as the q-th moment of the increments of x(t) by the following equation:

$$S_i(q,s) = \left\langle |x_i(t+s) - x_i(t)|^q \right\rangle \qquad (1)$$

Where i value represents the ith data point, and $\langle\ \rangle$ denotes the average.
Structure Functions are generalized correlation functions, which are particularly evident from equation 1 for the case where q=2. This second order is equivalent to the power spectrum. In general, q may be any real number not just integers, and can even be negative. However, there are divergence problems inherent to the negative-order exponent so that computations are best restricted to positive real number (Davis, 1994). If the process x(t) is

scale-invariant and self similar or self-affine over some range of time lags $s_{\min} \leq s \leq s_{\max}$, then the qth-order structure function is expected to scale as:

$$S_i(q,s) \approx s^{\zeta(q)} \tag{2}$$

$\zeta(q)$, scaling exponent is monotonically non-decreasing function of q. The behaviour described by equations 1 and 2 is called "multiscaling" because each statistical moment is scaling but with a different exponent. Therefore, a hierarchy of exponents can be defined using $\zeta(q)$ for obtain the generalized Hurst exponent H(q) introduced by Marshak et al. (1994)

$$H(q) = \frac{\zeta(q)}{q} \tag{3}$$

The first moment, characterizing the scaling of the average absolute fluctuations, corresponds to the scaling exponent H= $\zeta(q=1)$. H characterizes the difference from the conserved pure multiplicative process; it is the degree of non conservation of the process (scale by scale), a quantifier of nonstationarity in the observed variability (Marshak et al 1994). The second moment, H(q=2) is the Hurst exponent (HI). Processes with a linear $\zeta(q)$ are monofractal, H (q) is constant. For multifractal process, this exponent is non-linear and concave. We can obtain an estimator of $\zeta(q)$ by log-log plot the slope for different q values.
We use $\zeta(q)$ for q=2, 3 and 6 for obtain two parameters. The first one reflects the deviation from a linear structure function as:

$$\lambda_{desv} = 3\zeta(q=1) - \zeta(q=3) \tag{4}$$

Where λ_{desv} higher we move away from the monofractality. For that is a multifractality measure.
We define the second parameter, a measure of the intermittency when the scaling is in space (Mahrt, 1989) as

$$\mu_{Int} = 2 - \zeta(q=6). \tag{5}$$

3. Results and discussion

3.1 Gray scale interpretation

The original soil surface roughness resulting from each tillage tool for the different types of soils was represented as gray scale distribution, figure 3. Since the results were not very significant, the original data was submitted to the absolutes differences distribution and again represented as gray image interpretation, figure 4, index called SDR.
The results obtained when the gray scale interpretation is applied to the SSR and SDR data are proportional, for original and modified data, to each the treatment, tillage-tool. The images resulting for each soil and treatment for the original an absolute differences data is quiet similar. The SDR data eliminates the interferences found from original data and express the fingerprint of the resulting SSR for each type of soil and tillage tool, for that reason the differences in gray tones are more extreme. The resulting images for the original data present more contrast and tonalities or gray.

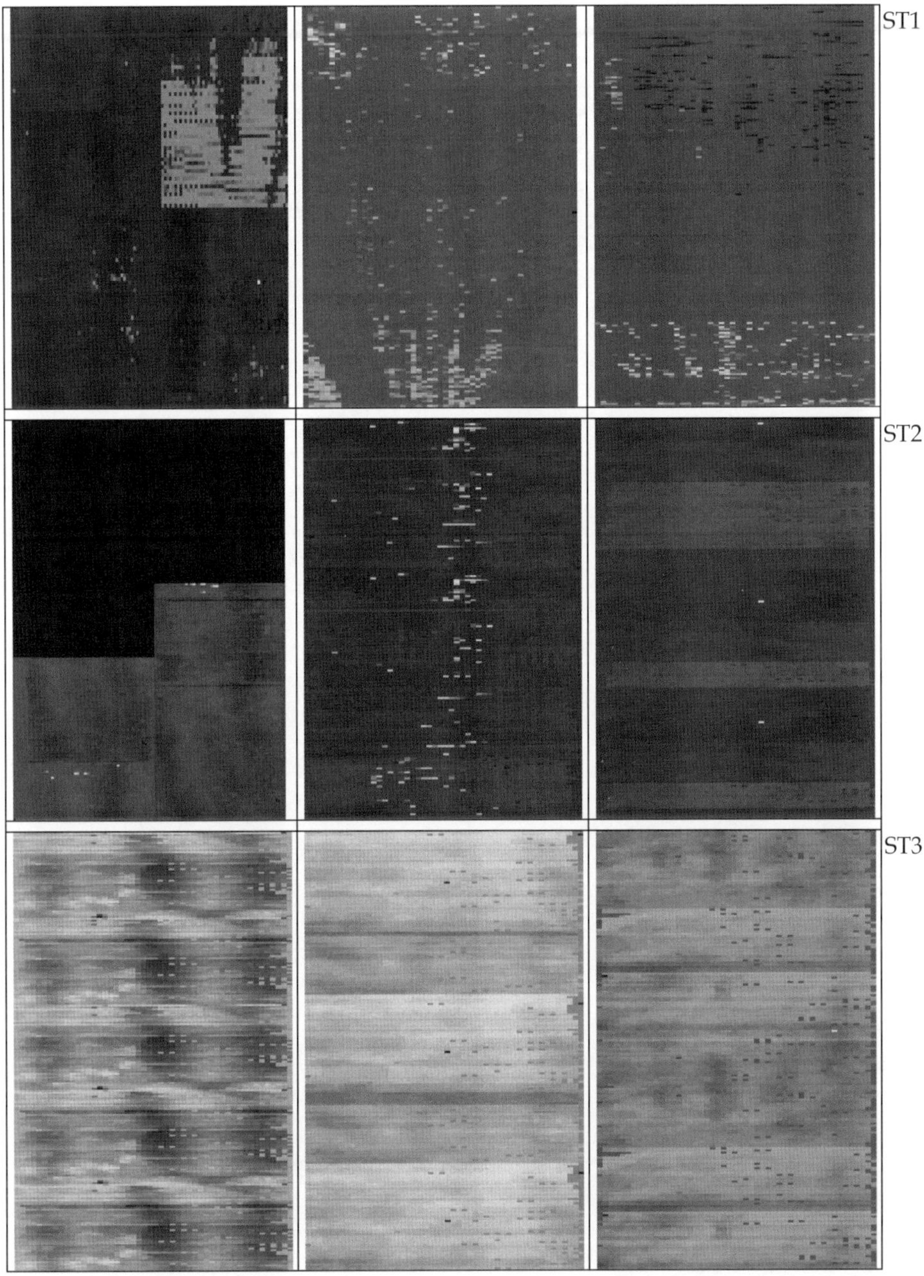

Fig. 3. Gray scale interpretation of soil surface roughness (SSR) distribution after applying chisel, roller and tiller (left to right) from E.T.S.I.A-U.PM. Soil (ST1), Sandy Clay Loam soil at La Higueruela (ST2) and Sandy Loam soil at La Higueruela (ST3) after passing chisel, tiller and roller (left to right)

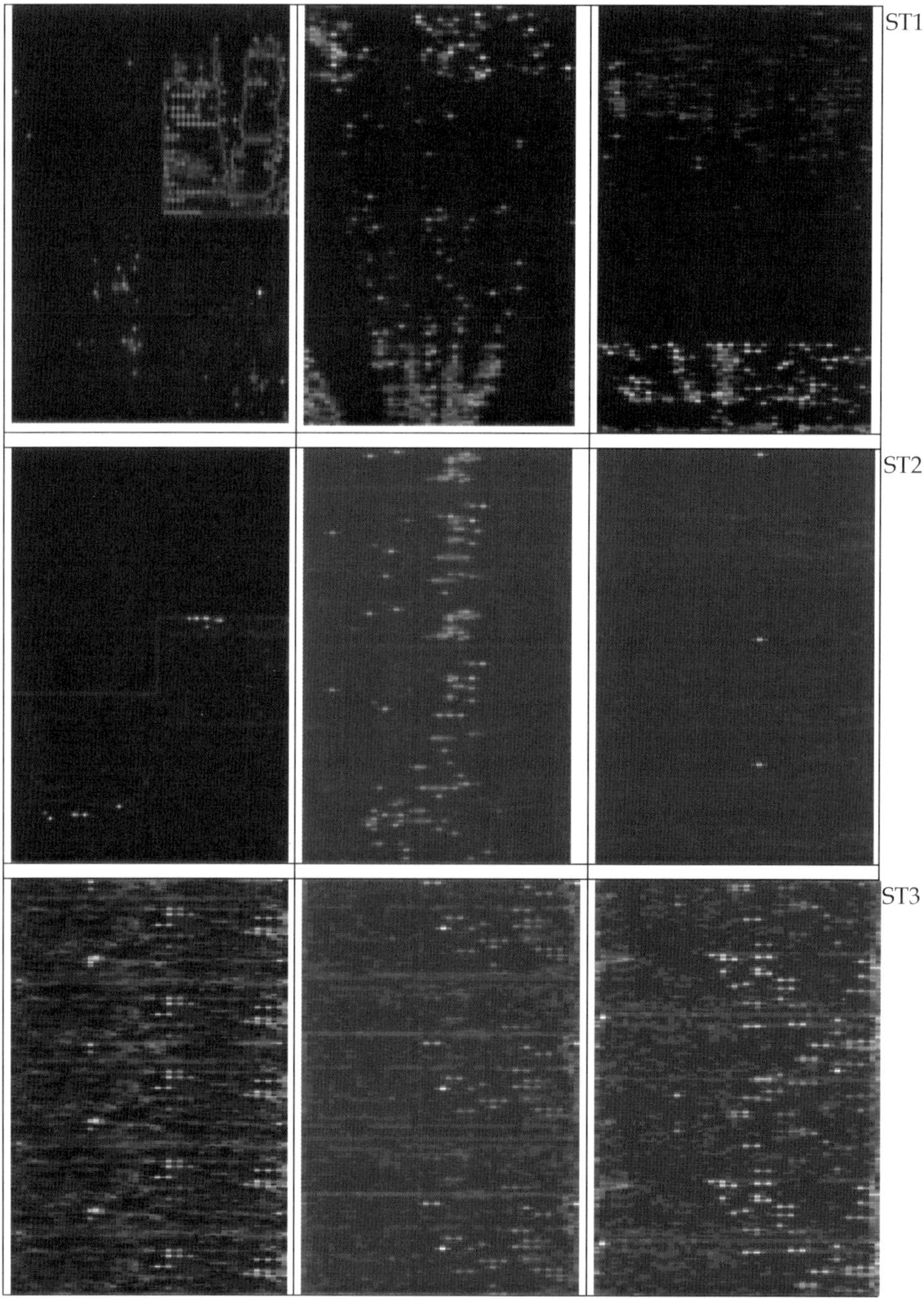

Fig. 4. Gray scale interpretation of soil surface roughness absolute differences distribution at E.T.S.I.A-U.PM. (ST1), Sandy Clay Loam soil at La Higueruela (ST2) and Sandy Loam soil at La Higueruela (ST3) after applying chisel, tiller and roller (left to right)

The results obtained are consistent with the photographs in figure 2. In general, the images for the distribution for absolute differences data represent more exactly the expecting results for the conditions applied fort each soil.
Differences amongst treatment reveal that we find more extreme differences in height for chisel followed by tiller and roller, also the soil of ETSIA , the soil type prevailing at the Madrid site, gives more extreme results than the soils from La Higueruela. These fields, characterized by a high incidence of clods and fragmented stones, have been tilled conventionally for many years. The pedogenic influence is greater in the more highly developed La Higueruela site soils, where conservation tillage has been in place for the last 20 yr.
An exception to this rule arose in the sandy clay loam at La Higueruela, where the soil was so dry that chisel was nearly ineffectual. For that reason the results form the sandy loam soil shows more differences on SSR AND SDR data.
From the gray interpretation values the results form the ETSIA and sandy Clay Loam are the most extreme values (figure 3), corroborate with the absolute differences images (figure 4). However the results form the sandy loam of La Higueruela gives less drastic differences but more continuous microelevations. These differences on the gray scaling, mainly based on SDR index, show that in the case of the ETSIA and the sandy clay loam the SSR is originated by tillage tool, while in the case of the sandy loam is due to the SSR of the soil itself.

3.2 Multiscale analysis through structure function

The results obtained from the ζ (q) curves, as can be observed in figure 5, corroborate the former results (Garcia Moreno et al., 2008a, 2008b), previous published on these data on multifractal nature of soil surface roughness analysed through spectrum evaluation. The soil drop roughness (SDR) results give greater variability for the ETSIA, followed by the sandy clay loam and the sandy loam of La Higueruela. In this sense, the tiller gives the greater variability based on the tillage tool comparison followed by chisel and roller for all the cases except for the sandy clay loam. The main difference with the spectrum results is that SDR is greater for the tiller, in the case of SSR the variability is greater for chisel because of the geometry of the resulting roughness based on the tool. In both cases, the soil type does not influence the resulting geometry of the fingerprint.
The data resulting from both soils of La Higueruela for figure 5 are similar, comparing to the Madrid site SDR results, where we observe differences.
The multifractality index, table 2, resulting from the different soil-tillage tool treatments gives a complementary information of the rest of parameters expressed in tables 2, 3, 4 and 5.
In general, observing the tendency of the ζ (q) curves the differences found in the HI and H are the smallest (tables 2 and 3 respectively). Lower q means decreasing differences, as it is shown in figure 5.
On the hand, multifractal index, λ, and intermittency, μ, require ζ (q) values for q equal 3 and 6 respectively, and therefore the variation is the higher than HI and H, (tables 4 and 5).
The resulting multifractal parameters seem to evaluate that the soil drop roughness resulting from the different treatments have different origin from soil and tillage tool casuistic. In this sense, and according to the gray scaling interpretation the soil drop roughness from the Madrid site seems to be originated more by tillage tool, while the cases from La Higueruela seem that the soil drop roughness are due to the microelevation in the land itself. The variability of the SDR based on the ζ (q) of the sandy clay loam of La Higueruela seems to be intermediate that means that the soil drop roughness is due in part to the tillage tool and in part to the land. In this case it is normal since as mentioned the soil was too dried to pass the chisel and the tiller in the sandy clay loam soil.

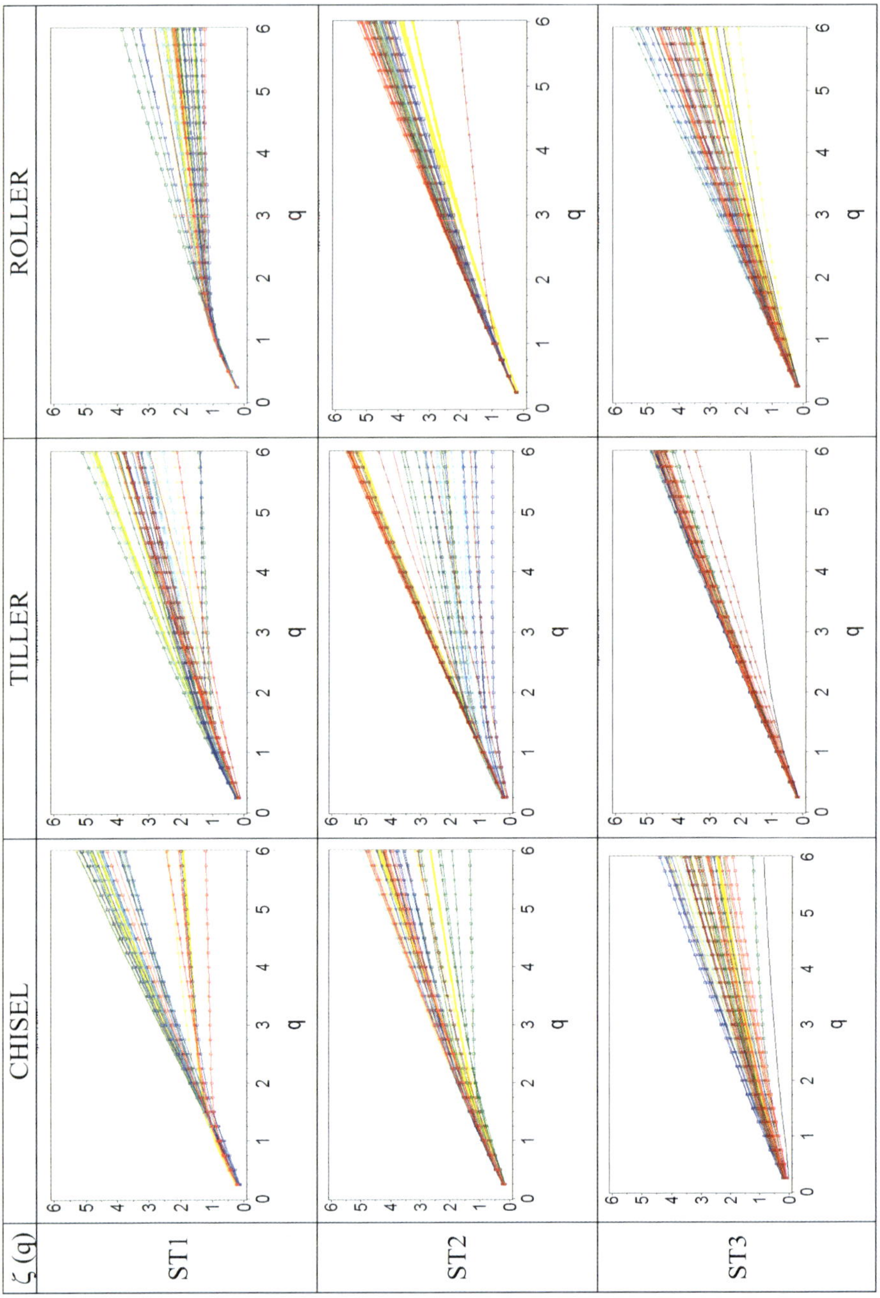

Fig. 5. ζ (q) curves obtained by SF for SSR related to soil type and tillage tool at E.T.S.I.A-U.PM. (ST1), Sandy Clay Loam soil at La Higueruela (ST2) and Sandy Loam soil at La Higueruela (ST3) after applying chisel, tiller and roller

Soil	TILLAGE TOOL	HI_{min}	HI_{max}	HI_{av}	HI_{std}	HI_k	HI_s
ST1	CHISEL	0.12	0.86	0.60	0.11	4.85	-1.67
	TILLER	0.36	0.99	0.74	0.18	-1.08	-0.45
	ROLLER	0.55	0.90	0.70	0.08	-0.44	0.22
ST2	CHISEL	0.36	0.85	0.62	0.11	-0.09	-0.37
	TILLER	0.19	0.84	0.56	0.14	0.16	-0.32
	ROLLER	0.10	0.85	0.58	0.18	-0.20	-0.71
ST3	CHISEL	0.12	0.74	0.40	0.15	-0.81	0.07
	TILLER	0.16	0.66	0.44	0.11	0.06	-0.48
	ROLLER	0.27	0.79	0.61	0.11	0.10	-0.81

Table 2. Hurst Index (HI) derived from Multiscale analysis (MA) of soil drop roughness for three tillage treatments and the following soils: ST1, sandy clay loam at Madrid; ST2, sandy clay loam at La Higueruela; ST3, sandy loam at La Higueruela. Parameters of the HI distribution are: minimum value (HImin), maximum value (HImax), average (HIav), standard deviation (HIstd), kurtosis (HIk) and skesness (HIs)

Soil	Tillage Tool	H_{min}	H_{max}	H_{av}	H_{std}	H_k	H_s
ST1	CHISEL	0.11	0.94	0.72	0.14	3.96	-1.71
	TILLER	0.42	1.02	0.83	0.16	0.21	-0.98
	ROLLER	0.78	0.97	0.87	0.04	-0.39	0.44
ST2	CHISEL	0.46	0.92	0.72	0.09	0.20	-0.09
	TILLER	0.25	0.97	0.68	0.16	-0.16	-0.51
	ROLLER	0.27	0.91	0.67	0.18	-0.64	-0.66
ST3	CHISEL	0.14	0.80	0.45	0.17	-0.76	0.03
	TILLER	0.21	0.78	0.50	0.13	-0.32	0.10
	ROLLER	0.38	0.85	0.69	0.11	-0.04	-0.75

Table 3. Value of H derived from Multiscale Analysis (MA) of soil drop roughness for three tillage treatments and the following soils: ST1, sandy clay loam at Madrid; ST2, sandy clay loam at La Higueruela; ST3, sandy loam at La Higueruela. Parameters of the H distribution are: minimum value (Hmin), maximum value (Hmax), average (Hav), standard deviation (Hstd), kurtosis (Hk) and skewness (Hs)

Soil	TILLAGE TOOL	λ_{min}	λ_{max}	λ_{av}	λ_{std}	λ_k	λ_s
ST1	CHISEL	-0.26	1.49	0.66	0.39	-0.77	-0.18
	TILLER	0.11	1.30	0.52	0.30	0.09	0.90
	ROLLER	0.17	1.37	0.89	0.25	0.31	-0.38
ST2	CHISEL	-0.11	1.78	0.59	0.44	0.28	1.11
	TILLER	0.07	1.54	0.69	0.38	-0.89	0.20
	ROLLER	0.10	1.51	0.50	0.31	2.21	1.32
ST3	CHISEL	-0.06	0.73	0.26	0.19	-0.34	0.42
	TILLER	-0.09	1.40	0.36	0.29	2.65	1.41
	ROLLER	0.08	0.94	0.46	0.25	-0.90	0.38

Table 4. Multifractality index (λ) derived from Multiscale analysis (MA) of soil drop roughness for three tillage treatments and the following soils: ST1, sandy clay loam at Madrid; ST2, sandy clay loam at La Higueruela; ST3, sandy loam at La Higueruela. Parameters of the λ distribution are: minimum value (λ_{min}), maximum value (λ_{max}), average (λ_{av}), standard deviation (λ_{std}), kurtosis (λ_k) and skewness (λ_s)

Soil	TILLAGE TOOL	μ_{min}	μ_{max}	μ_{av}	μ_{std}	μ_k	μ_s
ST1	CHISEL	-2.07	2.71	-0.15	0.91	0.42	0.50
	TILLER	-3.46	1.43	-1.25	1.39	-1.20	0.30
	ROLLER	-2.75	1.13	-0.51	0.85	-0.16	-0.26
ST2	CHISEL	-2.69	2.56	-0.29	1.11	0.06	0.49
	TILLER	-2.64	2.95	0.13	1.12	0.31	-0.05
	ROLLER	-2.19	3.63	-0.35	1.17	1.47	1.01
ST3	CHISEL	-1.70	2.11	0.31	0.92	-0.62	-0.06
	TILLER	-1.22	2.30	0.27	0.82	0.04	0.49
	ROLLER	-2.07	1.82	-0.45	0.93	-0.07	0.36

Table 5. Intermittency index (μ) derived from Multiscale analysis (MA) of soil drop roughness for three tillage treatments and the following soils: ST1, sandy clay loam at Madrid; ST2, sandy clay loam at La Higueruela; ST3, sandy loam at La Higueruela. Parameters of the μ distribution are: minimum value (μmin), maximum value (μmax), average (μav), standard deviation (μstd), kurtosis (μk) and skewness (μs)

4. Conclusions

Multiscaling analyses of soil surface roughness data are useful descriptors of soil drop roughness structure and complexity. Also these results were valuable supplements to image interpretations of gray scales for microrrelief distribution.
As regards variations due to soil type, in roller and tiller tillage, the images interpretation of gray scales were substantially extreme than for roller tool, whereas in chisel tillage, the resulting images exhibited the greatest variation. The latter finding is an indication of the scant complexity of the structure generated by the tool in the case of the Madrid site soil. However, the method shows that for the soils of La Higueruela the differences of the microrrelief are due to the soil itself more than the tillage tools. Since extremely dry soil may affect the results in such regions, however, field problems must be taken into account when interpreting the resulting soil drop roughness data.
Structure Function is highly sensitive to roughness of soil height measurements. It appears to be better suited than conventional indexes to comparing differently managed plots. The methodology highlights the steepest roughness that means the border values amongst the smoothest surface.
Structure Function of the absolute differences of the soil drop roughness of the sandy loam at La Higueruela are less multifractal, that means that the results of SDR are due mainly to the soil, and have less influence by tillage tools. While the case of the Madrid site is the opposite and the multifractality of results are originated mainly by the different tillage tool. The values obtained with different tillage tools are higher in clayey soils, where the presence of clods and rock fragments adds to heterogeneity, raising the associated indexes, case of Madrid site. In the studied cases the tiller results in greatest variations in all the three soil types, whereas the other tillage tools are more homogeneous.
The Structure Function seems to be a good tool to explain the variability and origin of the soil drop roughness depending on the tillage tool and soil type, mainly in extreme cases. Even this methodology is coming from the turbulence field analyses, where the amount of data is much higher than those who have in this project, the concept of generalised Hurst exponent improves to differentiate dissimilar cases.

5. Acknowledgment

The authors wish to thank Dr. Carlos Lacasta, director of the Spanish National Research Council's Experimental Centre for Environmental Science at La Higueruela Santa Olalla, Toledo, and his team, for their assistance in the field experiments. This study was possible by the economic support of Conselleria de Economia e Industria, Xunta de Galicia, through the Angeles Alvariño Program and the INCITE Human Resources Program (Cofinanced by the European Social Fund).

6. References

Agterberg, F.P., Cheng, Q., Brown, A., Good, D., 1996. Multifractal modelling of fractures in Lac Du Bonnet bathlolith, Manitoba. *Computers & Geosciences*, 22(5), 497-507.

Ahammer, H. and DeVaney, T.T.J.. 2005. The influence of noise on the generalized dimensions. *Chaos, Solutions and Fractals*, 26(3), 707-717.

Allain, C., and M. Cloitre. 1991. Characterizing the lacunarity of random and deterministic fractal sets. *Physical Review A*, 44, 3552-3558.

Allmaras, R.R., R.E Burwell., W.E Larson, and R.F. Holt. 1966. Total porosity and random roughness of the inter row zone as influenced by tillage. *USDA Conserv. Res. Rep.* 7, 1-22.

Armstrong, A.C., 1986. On the fractal dimensions of some transient soil properties. *J. Soil Sci.*, 37, 641-652.

Bertuzzi, P., G. Raws, G., and D. Couralt, D.,. 1990. Testing Roughness indices to estimate soil surface changes du to simulated rainfall. *Soil and Tillage Research* 17:87-99.

Bird, N., M.C. Díaz, A. Saa and AM Tarquis, 2006. A Review of Fractal and Multifractal Analysis Of Soil Pore-Scale Images. *J. of Hydrology*, 322, 211-219.

Buczhowski, S.; Hildgen, P., Cartilier, L., 1998. Measurements of fractal dimension by box-counting: a critical analysis of data scatter. *Physica A*, 252, 23-24.

Burg, J., 1967. Maximum entropy spectral analysis. In: *Proceedings of the 37th Meeting Society of Exploration Geophysicist*, Oklahoma City.

Burrough, P.A. 1983. Multiscale source of spatial variation in soil: I. The application of fractal concepts to nested levels of soil variation. *J. Soil Sci.* 34:577–597

Cheng, Q. and F.P. Agterberg, 1996. Multifractal modelling and spatial statistics. *Mathematical Geology*, 28(1), 1-16.

Cheng, Q., 1997a. Discrete multifractals. *Mathematical Geology*, 29(2), 245-266.

Cheng, Q., 1997b. Multifractal modeling and lacunarity analysis. *Mathematical Geology*, 29(7), 919-932.

Cheng, Q., 1999. The gliding box method for multifractal modeling. *Computer&Geoscience*, 25, 1073-1079.

Currence, H.D., Lovely, W.G., 1970. The analysis of soil surface roughness. *Trans. of the ASAE* 13, 710-714.

Davis, A., Marshak, A., Wiscombe, W. and Cahalan, R. 1994. Multifractal Characterisations of Simulated. *J. Geophysics* Research, 99, 8055-8072.

Eltz, F.L.F., Norton, L.D., 1997. Surface roughness changes as affected by rainfall erosivity, tillage and canopy cover. *Soil Sci. Soc. Am. J.* 61: 1746-1755.

Evertsz, C. J. G., Mandelbrot, B.B., 1992. Multifractal Measures. In Peitgen, H., Juergens, H., Saupe, D. (Editors): *Chaos and Fractals*. Springer, Berlin, pp 921-953.

Feder, J., 1989. *Fractals*, Plenum Press: New York.

Gallant, J.C., Moore, I.D., Hutchinson, M.F., Gessler, P., 1994. Estimating fractal dimension of profiles: a comparison of methods. *Math. Geol.* 26, 455-481.

Gallart, F., Pardini, G., 1996. Perfilru: un programa para el análisis de perfiles microtopograficos mediante el estudio de la geometría fractal. *Cuadernos Lab. Xeolóxico de Laxe*, 21, 169-178 (in spanish).

Huang, C.-h. 1998. Quantification of soil microtopography and surface roughness. *Fractals in soil science*. Ed. By Philippe Baveye, Jean Yves Parlange and Bobby A. Stewart. Advances in Soil Science. CRC.

García Moreno, R. (2006). Desarrollo de una metodología para la medición de la rugosidad del suelo. Ph. D. Dissertation, Polytechnic University of Madrid (UPM), pp. 120.

García Moreno, R., Díaz Álvarez, M. C., Tarquis Alfonso, A., Barrington, S. and Saa Requejo, A. 2008a. Tillage and soil type effects on soil surface roughness at semiarid climatic conditions, Soil and Tillage Research 98 (1), 35-44.

García Moreno, R., Saa Requejo, A., Tarquis, A.M., Barrington, S., and Díaz Álvarez, M.C., 2008b. A shadow analysis method to measure soil surface roughness. Geoderma 146 (2008) 201–208.

García Moreno, R., Díaz Álvarez, M.C., Tarquis, A. M., Paz González, A., and Saa Requejo, A Shadow analysis of soil surface roughness compared to the chain set method and direct measurement of micro-relief, *Biogeosciences*, 7: 2477-2487, 2010.

Grau, J, Médez, V.; Tarquis, A.M., Saa, A. and M.C. Díaz., 2006. Comparison of gliding box and box-counting methods in soil image analysis. *Geoderma*, 134, 349-359.

Hagen, L. J. 1988. New wind erosion model developments in the USDA. In 1988 *Wind Erosion Conference Proceedings*. Texas Tech. University, Lubbock, 11-13 April.

Huang, C.H. 1998. Quantification of soil microtopography and surface roughness. *Fractals in soil science*. Ed. By Philippe Baveye, Jean Yves Parlange and Bobby A. Stewart. Advances in Soil Science. CRC.

Huang, C.H., Bradford, J.M., 1992. Applications of a laser scanner to quantify soil microtopography. *Soil Sci. Soc. Am. J.* 56:14-21.

Instituto Nacional de Meteorologia, 2005. Ministerio de Medio Ambiente. http://www.inm.es/web/izq/noticias/meteonoti/

Kamphorst, E.C., Jetten, V, Guérif, J. Pitkänen, J. Iversen, B.V., Douglas, J.T., and A. Paz, A. 2000. Predicting depressional storage from soil surface roughness. *Soil Sci, Soc. Am. J.* 64:1749-1758.

Larney, F.J., M.S. Bullock, H.H. Janzen, B.H. Ellert, and E.C.S. Olson.1998. Wind erosion effects on nutrient redistribution and soil productivity. *J. Soil Water Conserv.* 53:133–140.

Larney, F.J., M.S. Bullock, S.M. McGinn, and D., Black, A. L.,W. Fryrear, D. W., Saleh, A., Zobeck, T.M., Halvorson, A. D., Tanaka, D. L.,. 1995. Quantifying wind erosion on summer fallow in southern Alberta. *J. Soil Water Conserv.* 50:91–95.

Linden, D.R., Van Doren, J.D.M., 1986. Parameters for characterizing tillage-induced soil surface roughness. *Soil Sci. Soc. Am J.* 50: 1550-1565.

Lovejoy, S. D. Schertzer, 2007: Scaling and multifractal fields in the solid earth and topography, *Nonlin. Processes Geophys.*, 14 , 465-502.

Mahrt, L. 1989. Intermittency of atmospheric turbulence. *Journal of the Atmospheric Sciences* 46, 79-95.

Malinverno, A., 1990. A single method to estimate the fractal dimension of self-affine series. *Geophys. Res. Lett.* 17 (11), 1953.

Manninen A. T., 2003. Multiscale surface roughness description for scattering modelling of bare soil. *Physica A*, Volume 319, 1 march 2003, pp. 535-551.

Marshak A., Davis, R., Cahalan R and Wiscombe W. 1994. Bounded cascade models as nonstationary multifractals. *Physical Review E*. Vol 49, nº1, 55-69.

Merel, A.P., Farres, P.J., 1998. The monitoring of soil surface development using analytical photogrammetry. *Photogrametric record*. 16(92): 331-345.

Merrill, S., 1995. Procedures for soil analysis. ISRIC, FAO, Holanda.

Murillo, J M., Moreno, E., Girón, I.F., and Oblitas, M. I., 2004. Conservation tillage: long term effect on soil and crops under rained conditions in south-west Spain (Western Andalusia). *Spanish Journal of Agricultural Research* (2004) 2 (1), 35-43.

Pardini, G., Gallart, F.A., 1998. Combination of laser technology and fractals to analyze soil surface roughness. European *J. Of Soil Sci.* 49:197-202.

Perfect, E., P.H. Groenevelt, B.D. Kay, and C.D. Grant. 1990. Spatial variability of soil penetrometer measurements at the mesoscopic scale. *Soil Tillage Res.* 16:257–271.

Perfect, E., Kay, B.D., 1995. Applications of fractals in soil and tillage research: a review. Soil Tillage Res. 36, 1-20.

Perrier, E., Tarquis, A.M. and Annette Dathe, 2006. A Program for Fractal and Multifractal Analysis of Two-Dimensional Binary Images . Computer Algorithms versus Mathematical Theory. *Geoderma,* 134, 284-294.

Pike, R.J., 2000. Geomorphometry -diversity in quantitative surface analysis. *Physical Geography,* 24(1), 1-20.

Podmore, T.H., and L.F. Huggins. 1981. An automated profile meter for surface roughness measurements. *Trans. ASAE* 24:663–665, 669.[

Porta Casanellas, J., López-Acevedo Reguerín, M., Roquero de Laburu, C. 2003. *Edafología para la agricultura y el medio ambiente.* Ed. Mundi-Prensa

Potter, K.N., T.M. Zobeck and L.J. Hagen. 1990. A microrrelief index to estimate soil erodibility by wind. *Trasns. ASAE* 33:151-155.

Roisin, C.J.C., 2007. A multifractal approach for assessing the structural state of tilled soils. *Soil Sci. Soc. Am. J.* 71:15-25.

Saxton, K.E. 1995. Wind erosion and its impact on off-site air quality in the Columbia plateau—An integrated research plan. *Trans. ASAE* 38:1031–1038.

Schertzer, D. and Lovejoy, S., 1991. *Nonlinear Variability in Geophysics.* Kluwer Academic Publishers, Dordrecht, The Netherlands, 318 pp.

Soil Science Society of America. 1996. *Methods of Soil Analysis. Part 3 – Chemical Methods.* SSSA Book Series:5. Ed. D.L. Sparks.

Tarquis, A.M., Giménez, D., Saa, A., Diaz, M.C., Gascó, J.M. 2003. Scaling and multiscaling of soil pore systems determined by image analysis. In *Scaling methods in soil physics,* Pachepsky, Radcliffe and Selim, Eds., CRC Press, chap. 2.

Vidal Vázquez, E., Vivas Miranda, J.G., Paz Gónzalez, A., 2005. Characterizing anisotropy and heterogeneity of soil surface microtopography using fractal models. *Ecological Modelling,* 182, 337-353.

Vidal Vázquez, E., Miranda, J.G.V., Alves, M.C. and A. Paz Gónzalez, 2006. Effect of tillage on fractal indices describing soil surface microrelief of a Brazilian Alfisol. *Geoderma,* 134, 428-439.

Vidal Vázquez, E., Miranda, J. G. V. and A. Paz González. 2007. Describing soil surface microrelief by crossover length and fractal dimension. *Nonlin. Processes Geophys.,* 14, 223-235.

Vivas Miranda, J.G., 2000. Análisis fractal del microrrelieve del suelo (in spanish). Ph. D. Dissertation, University of Coruña, p. 313.

Vivas Miranda, J.G., Paz González, A., 2002. Fractal models for the description of soil surface roughness. *Proceedings of the third International Congress Man and Soil at the Third Millennium.* J.L. Rubio, R,P.C. Morgan S. Asins and V. Andreu, eds. Geoforma Ediciones. 2099- 2112

Zobeck, T.M. and Onstad, C.A., 1987. Tillage and rainfall effect on random roughness: A review. *Soil Tillage Res.* 9(1):1-20.

13

Soil-Landscape Modelling – Reference Soil Group Probability Prediction in Southern Ecuador

Mareike Ließ[1], Bruno Glaser[2] and Bernd Huwe[1]
[1]*University of Bayreuth, Department of Geosciences/ Soil Physics*
[2]*Martin-Luther University Halle Wittenberg, Soil Biogeochemistry*
Germany

1. Introduction

Since long, soils are understood as a function of their genetic factors: parent material, relief, climate, organisms and time, a concept first described by Dokutschajew (1883) and better known from Jenny (1941). The complex interaction of these factors activates particular soil forming processes, which in dependence of their intensity and duration, lead to characteristic soil properties. The resulting profile reflects the balance of these processes in its properties (Grunwald, 2006). Soil-landscape modelling uses the knowledge about soil genesis to predict soil distribution in a landscape based on continuously available environmental parameters by statistical models. The early conceptual models have resulted into quantitative soil-landscape models, which do not only make the spatial prediction of continuous soil properties possible, but include model uncertainty. Being at first an unwelcome nuisance that reduced map reliability, gradually soil variation and its unpredictability was seen as a key soil attribute by itself (Burrough et al., 1994).
Soil research within the area of the scientific research station San Francisco in the southern Ecuadorian Andes has been carried out for many years. A first preliminary soil map was provided in 2009 (Liess et al., 2009). This soil map, based on Reference Soil Groups (RSGs) from the World Reference Base for Soil Resources (WRB) (FAO, IUSS Working Group WRB, 2007) does not include prediction uncertainty and neglects Cambisols. However, Cambisols are part of the dataset and have also been described by Yasin (2001) and Wilcke et al. (2002, 2003). Other soils that occur under natural vegetation within the area are Histosols, Stagnosols, Umbrisols and Regosols (FAO, IUSS Working Group WRB, 2007). Histosols were described by Yasin (2001) and Schrumpf et al. (2001) as Haplosaprists according to Soil Taxonomy classification (Soil Survey Staff, 2006). Yasin (2001) investigated forest soils only between 1900 - 2240 m a.s.l., whereas Schrumpf et al. (2001) explored soils along an altitudinal gradient from 1850 - 3050 m a.s.l. Thus, Histosols were found on slope angles varying from 10 - 50° at 1850 - 2700 m a.s.l.; Stagnosols were described between 2080 - 2850 m a.s.l. (Yasin, 2001; Schrumpf et al., 2001; Liess et al., 2009). Umbrisols were assigned by Schrumpf et al. (2001) and Liess et al. (2009).
Prediction of soil types from terrain factors by statistical models is a standard approach within the field of soil-landscape modelling. Lagacherie & Holmes (1997) as well as Moran & Bui (2002) assigned soil classes by CTs based on parameters calculated from a digital

elevation model (DEM). Skidmore et al. (1996), Thomas et al. (1999) and Dobos et al. (2000) spatially predicted soil types from terrain analysis. Furthermore, Gessler et al. (1995), Moore et al. (1993) and Odeh et al. (1994) predicted soil attributes from terrain parameters. Bourennane et al. (2000) and Hengl et al. (2004) regionalised soil horizon and topsoil thickness from a DEM. Several statistical models are available to relate soils to environmental predictors. Bishop & Minasny (2006) compared some of them: Linear, generalized linear (GLM) and generalized additive (GAM) models, classification and regression trees (CART) and artificial neural networks (ANN). Among the considered model types, only ANN were assigned a better predictive power than CART, but lack the ease of use, parsimony, interpretability and computational efficiency that applies for CART.

By extending the dataset of Liess et al. (2009) and constructing various classification trees (CT), we expect to develop a more precise RSG map and include prediction uncertainty by displaying the RSG probability. The investigated soils will be related to terrain parameters by a CT (Breimann et al., 1984) that organises the dataset according to the respective RSG. The tree model can then be used to assign the RSG probabilities to the whole area covered by a DEM.

2. Material and methods

2.1 Research area

The research area is situated between the provincial capitals Loja and Zamora (Figure 1) in the southern Ecuadorian Andes from 1670 to 3160 m a.s.l. It extends in UTM-Zone 17M from west to east between 710500 and 716000, and from north to south between 9561500 and 9557000 (Figure 1). The San Francisco River divides the area into two parts: The north-west facing slopes south of the river are covered by montane rain forest and subpáramo vegetation above the tree line. Within this area, Homeier et al. (2002) differentiated various

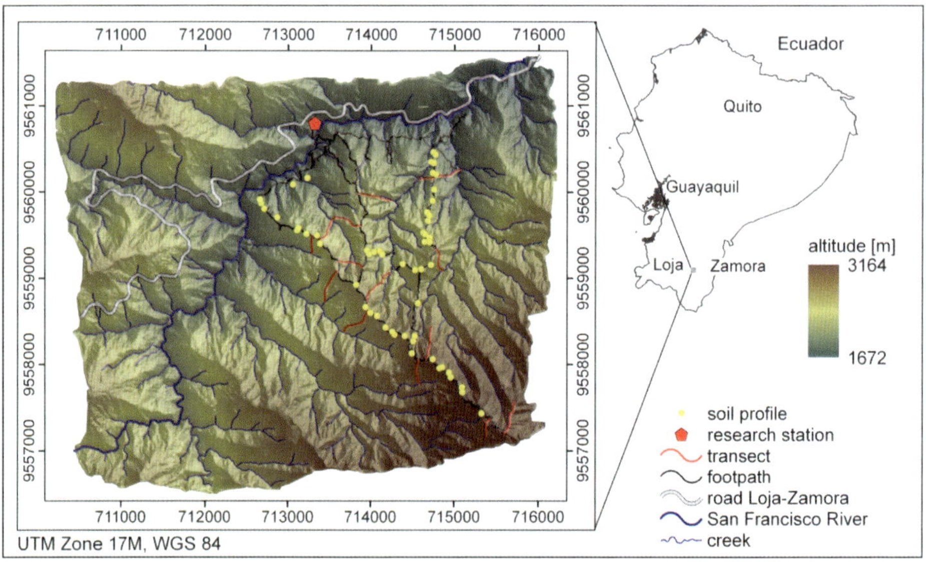

Fig. 1. Research area. Overlaid hill shading with light source from north-east (adapted from Liess et al., 2009)

forest types according to their altitude and position on the ridge or in the valley. The south-eastern facing slopes north of the river are mainly covered by pastures and succession vegetation after fire clearance when sites were left unused. For soil model development, only sites under natural vegetation were considered.

As part of the Chiguinda unit, the research area is lithologically covered by metasiltstones, siltstones and quartzites which are intermixed with layers of phyllite and clay schists (Litherland et al., 1994). Furthermore, it is influenced by the regular occurrence of landslides. Average total annual rainfall increases from 2050 mm at an altitude of 1960 m a.s.l. to approximately 4400 mm at 3100 m a.s.l. (Rollenbeck, 2006). Average air temperature decreases with increasing altitude from 19.4 to 9.4 °C (Fries et al., 2009).

2.2 Classification trees

Classification trees (CTs), a method first described by Breimann et al. (1984), were used to relate the RSGs to terrain parameters. It was conducted with the rpart library of the R-Project for Statistical Computing (Therneau & Atkinson, 2003).

In CTs subdivision is based on a categorical response variable, i. e. RSG. The final subsets, also called end nodes, should be as pure as possible. This is done by trying to assign them to only one category in the response variable, e.g. to Histosol. The Gini criterion (Equation 1) is applied as a measure of purity (Breiman et al., 1984). It serves as a decision criterion, to determine which terrain parameter best separates the dataset continuously into always two subsets to create the purest end nodes.

$$gini(t) = 1 - \sum_{i=1}^{k} P_i^2 \qquad (1)$$

The Gini-Index (Equation 1) reaches its maximum in a particular node t if all categories k within this node are equally represented. On the other hand, when the probability P_i is equal to zero for all but one category within any node, the Gini-Index reaches its minimal value. The categorical value accounting for the majority within each end node is then assigned to the corresponding parameter values, indicating the typical position within the landscape (e.g. Liess et al., 2009). However, another option is to assign the percentage of each categorical value within an end node as occurrence probability to the corresponding landscape position. This is of course only justified if the applied sampling scheme guarantees sampling of all landscape positions to an equal extent.

The CT is pruned to avoid overfitting and obviate random variation. To assess model performance, the cross validation error (CV) is calculated. The dataset is subdivided into 10 subsets, and the process is repeated 10 times with 9 parts for model training and the 10th part as the evaluation dataset. Eventually, among all trees considered for the final model, the tree with the lowest cross validated error rate is chosen. CV and model pseudo R^2 are calculated. Pseudo stability indices are constructed to satisfy the different interpretations, e.g. explained variance or square of correlation. They are similar to R^2 in that they also range between 0 and 1 and a higher value represents a better adaptation to the data.

2.3 Dataset and GIS methodology to gain terrain data

Topographic data for the research area is available on a continuous landscape level. The DEM used to obtain terrain parameters for the establishment of a prediction model of RSG occurrence has 2 m cell size (Liess et al., 2009). For model application, this accuracy was reduced to 10 m to decrease calculation time. The used terrain parameters include altitude a.s.l., aspect, slope angle, terrain curvature, upslope contributing catchment area and overland flow distance to the channel network (OFD).

Slope angle, aspect and curvature were computed with a 2nd degree polynomial fit from Zevenbergen and Thorne (Zevenbergen & Thorne, 1987; Cimmery, 2007). The contributing area was calculated with two methods; (1) based on the Kinematic Routing Algorithm (KRA CA) (Lea, 1992) and (2) based on the Braunschweiger Digital Relief Model (BS CA) (Bauer et al., 1985). In addition to the OFD, the horizontal (HOFD) and vertical (VOFD) overland flow distances were also calculated. The channel network itself was assessed applying the Strahler stream order ≥ 5 as initiation threshold (Strahler, 1957). Terrain curvature was computed using directly adjacent cells. Finally, the terrain parameters were calculated and the RSGs were predicted for each individual raster grid cell. The free and open source GIS software, SAGA, was used (Böhner et al., 2006).

The research area was sampled at 367 sites, including 311 auger points and 56 soil profiles. Soil sampling covered 24 sampling classes produced by an overlay of four altitudinal, three slope angle and two aspect classes to guarantee representative area coverage. Transects for auger sampling (Figure 1) were laid according to the catena concept (Milne, 1935) from hilltop to valley bottom. For more detailed information on the applied sampling design, see Liess et al. (2009).

Two methods were used to assign terrain parameters to the soil dataset. On the one hand, the nearest neighbour (n. n.) value was allocated to each soil profile or auger point. On the other hand, a buffer representing the radius of GPS accuracy was placed around the sampled location, and the calculated mean value of the corresponding area was assigned. This assignment was completed for each of the described parameters apart from the slope angle and aspect. These were directly measured in the field. The slope angle and aspect which were computed from the DEM were solely used for model application.

2.4 Probability calculation

The probability of each RSG was predicted via a CT which grouped the soil sampling points regarding the existence or absence of that RSG. Thus, the percentage of sampling points assigned to the corresponding RSG in each end node of the tree was used to predict the probability of that RSG. Thereby, the diagnostic properties necessary for assigning the particular RSG were used, whereas the necessary absence of other properties was neglected. This was done in particular to establish a good prediction scheme for Stagnosols. It was decided that the occurrence and thickness of a sufficient stagnic colour pattern and/ or albic horizon is more important than the limitation in organic layer thickness. As a consequence, soils with a 40 cm organic layer displaying also a thick stagnic horizon were classified as Histosols and Stagnosols. Any other proceeding would have made the development of a Stagnosol prediction scheme incomplete and complex.

To sum the individual probabilities and standardize them by relating each RSG to the total probability sum, is one option. This option neglects WRB (FAO, IUSS Working Group WRB, 2007) hierarchy, because all RSGs are competing on an equal level and no soil process is given dominance over another. As a consequence, the probabilities refer to the probability of the diagnostic property necessary for RSG assignation. Later we will refer to these as WRB independent probabilities.

Figure 2 shows the probability calculation scheme based on WRB (FAO, IUSS Working Group WRB, 2007) hierarchy.

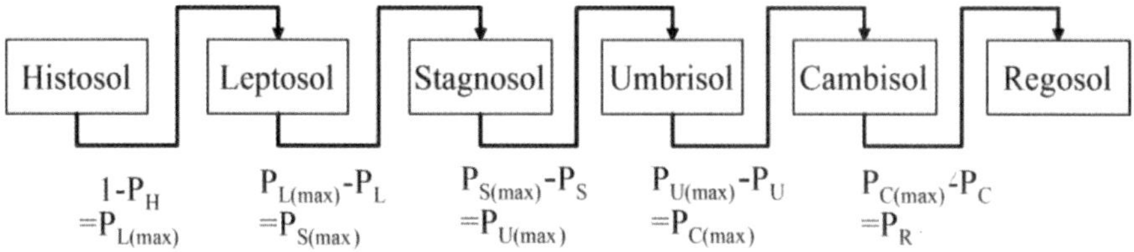

Fig. 2. Hierarchical calculation scheme for the maximum possible probability of each RSG according to WRB hierarchy. P_X is the actual probability of the respective RSG: H Histosol, L Leptosol, S Stagnosol, U Umbrisol, C Cambisol, R Regosol. $P_{X(max)}$ is the maximum possible probability of the RSG

It is used to calculate the maximal possible probability for each RSG from the probability predicted by the CTs. Maximal Leptosol probability is left after subtracting Histosol probability from 1. Maximal Stagnosol probability is left after also subtracting the actual Leptosol probability and so on. Equation 2 shows the calculation of the actual probability, P_X, according to the CT probability, $P_{X(tree)}$, and the maximal possible probability, $P_{X(max)}$.

$$P_{x(\max)} \cdot P_{x(tree)} = P_x \tag{2}$$

3. Results and discussion

3.1 Classification tree models and digital soil maps

Figure 3 presents the CT models to predict Histosol, Leptosol and Stagnosol occurrence probability from nearest neighbour (n. n.) and mean terrain values.

The RSG Histosol is assigned to soils with an organic layer ≥ 40 cm (FAO, IUSS Working Group WRB, 2007). Its probability within the research area was found to depend on two hydrological parameters (Figures 3a and 3d): KRA CA and VOFD. Probability is predicted with at least 0.2 (Figure 3a, d) throughout the research area. The highest probability (0.87) as predicted by n. n. relief values (Figure 3a) was obtained for small catchments (KRA CA < 258 m^2) within a distance of 14 – 23 m from the channel network. Though, probabilities are also high, 0.65, for small catchments (KRA CA < 258 m^2) within a VOFD of 54 – 176 m. The latter is a more conservative prediction, since it is based on 206 sampled sites and not only 15 as for the first differentiation criteria (Figure 3a). Sites seem to coincide in some parts with upper slope areas and ridges (Figure 4a).

Prediction by mean terrain values (Figure 3d) again shows high probabilities in similar landscape positions, i.e. for small catchments < 254 m^2 from 54 – 175 m VOFD (0.65) and < 26 m VOFD (0.70). The former is the safest prediction similar to the Histosol prediction from n. n. terrain values (188 sampled sites). Areas likely to be covered by Histosols with this 0.65 probability are again found along ridges. In contrast to the CT from n. n. terrain values, the highest probabilities, 0.85, by mean relief values (Figure 3d) are assigned to large catchments (≥ 254 m^2) with a VOFD from 103 – 145 m, dominating in dark colours as broad belts at 103 m distance around the creeks (Figure 4b). This also accounts for the major difference between the two models (Figure 4c). But since the corresponding end node in the tree model (Figure 3d) is only supported by 13 sampled sites, this finding is not representative for the research area.

Leptosols refer to soils limited to 25 cm depth by continuous rock (FAO, IUSS Working Group WRB, 2007). During soil sampling continuous rock was rarely attained, and refusal typically occured at the C horizon. This made the establishment of a model predicting soil depth to continuous rock impossible. Therefore, to calculate Leptosol occurrence probability expert knowledge was applied in addition to the CT methodology. From field work and data review it was known that Leptosols are found on steep slopes ≥ 50° and close to the creeks at approximately < 20 m HOFD. Other soils, which occurred at the same landscape positons with even higher probability, were excluded for model development. Afterwards, they were included again to calculate the probabilities of the tree end nodes. This explains the rather untypical appearance of the Leptosol CTs (Figures 3b and 3e). Usually, for any final subdivision into two end nodes, one of them would always display a probability > 0.5 and the other < 0.5. However, for the reason of adding more datasets after tree development this is not the case. This procedure was necessary in order to develop a reasonable model and account for true probabilities. Leptosol CTs established with n. n. and mean terrain values are very similar. In the already mentioned positions, Leptosol probability was assumed 0.30 – 0.36 (Figure 3b and 3e).

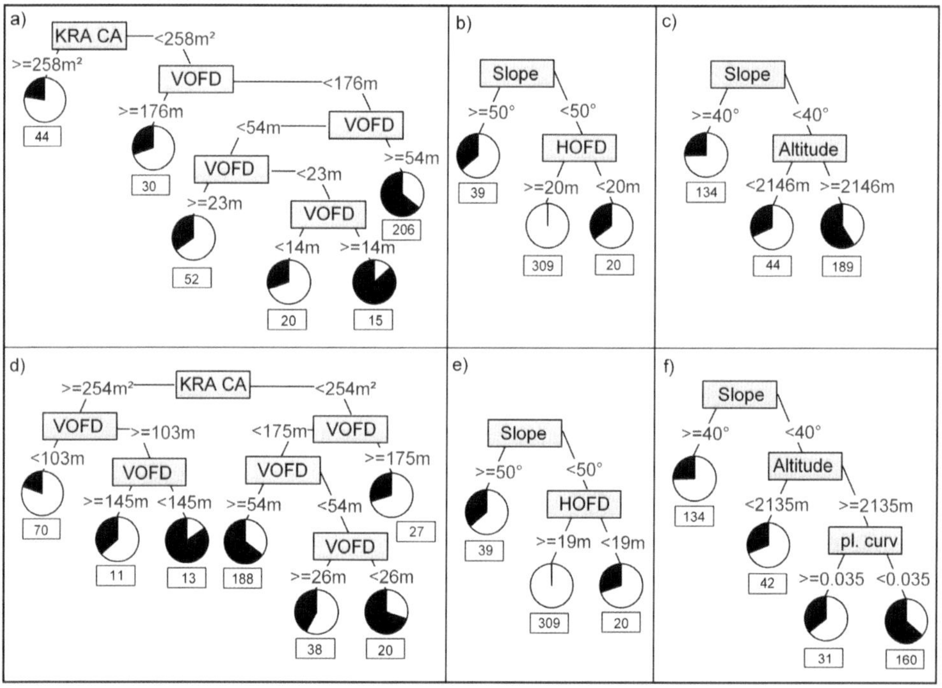

Fig. 3. Classification trees predicting RSG probability. The pie charts' black parts represent the occurrence probability in the corresponding landscape positions. The numbers in the boxes underneath the charts refer to the number of sampling sites used for the probability prediction in each end node. Prediction by n. n. terrain values: a) Histosol probability, b) Leptosol probability and c) Stagnosol probability. Prediction by mean terrain values: d) Histosol probability, e) Leptosol probability and f) Stagnosol probability. (KRA CA = upslope contributing catchment area according to the Kinematic Routing Algorithm, VOFD = vertical overland flow distance, HOFD = horizontal overland flow distance, pl. curv = plan curvature)

The lighter colours in Figure 4b compared to Figure 4a are due to the fact that a probability of 0.20 (< 103 m VOFD, Figure 3d) falls into a smaller mapping class than 0.23 (Figure 3a) in the map layout. The similarity between the two models for the mentioned sites is indicated by yellow colours in Figure 4c. The sites mapped in red colours refer to a 0.1 - 0.3 higher probability as predicted by mean relief values. Comparison of the two tree models (Figures 3a and 3d) shows that differences are not higher than 0.13. The models differ only by a probability of 0.03 - 0.13, neglecting the mentioned 13 sites.

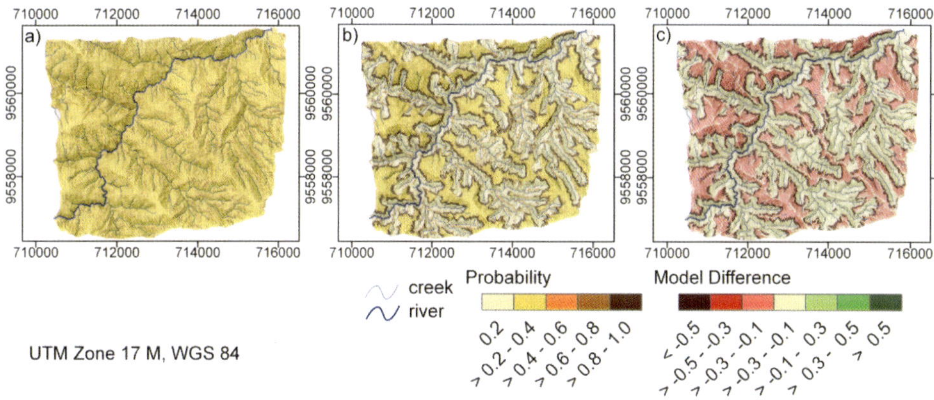

Fig. 4. Maps of Histosol occurrence probability (Overlaid hill shading with light source from north-east): a) Prediction by n. n. terrain values, b) prediction by mean terrain values and c) model difference

Figures 5a and 5b show the Leptosol probability distribution within the research area after model application. With the inclusion of WRB (FAO, IUSS Working Group WRB, 2007) hierarchy, Leptosol probability also depends on Histosol probability. But since Histosol probability close to the creeks (< 103 m VOFD) is predicted with only 0.2, model from mean relief values (Figure 3d), and 0.3, model from n. n. relief values (Figure 3a), it does not influence Leptosol probability much for those sites. Model difference regarding prediction by n. n. and mean terrain values (Figure 5c) is always ≤ ± 0.1 (0.05); including WRB hierarchy (FAO, IUSS Working Group WRB, 2007), model difference (Figure 5f) is increasing (hardly recognisable in the map). Model difference regarding probability predicted directly by the CTs and probability being calculated based on WRB hierarchy (Figures 5g and 5h) shows a similar picture. The difference between the WRB independent and dependent prediction by n. n. values (Figure 5g) is ≤ ± 0.1, but higher regarding the prediction difference by mean terrain values (Figure 5h).

Stagnosols are "soils exhibiting hydromorphic features for some time during the year in some part within 50 cm of the mineral soil surface and show a stagnic colour pattern and/ or an albic horizon in half or more of the soil volume" (FAO, IUSS Working Group WRB, 2007). Planosols are classified by similar diagnostic properties, but in addition display an abrupt textural change, which could not be confirmed for the investigated soils. Stagnosol probability is predicted throughout the research area with at least 0.25 (Figures 3c and 3f). The probability in both models depends on slope angle and altitude. It is higher on slopes < 40°. Above 2146 m a.s.l. for the prediction by n. n. and above 2135 m a.s.l. by mean terrain values, the probability increases even further. While curvature is of no importance for Stagnosol probability prediction by n. n. relief values, mean terrain values assign an even higher probability for concave plan curvature with 0.64. Landscape positions < 2146 m a.s.l. for prediction by n. n. and < 2135 m a.s.l. by mean terrain values, and high slope angles account for the lowest probability of Stagnosols.

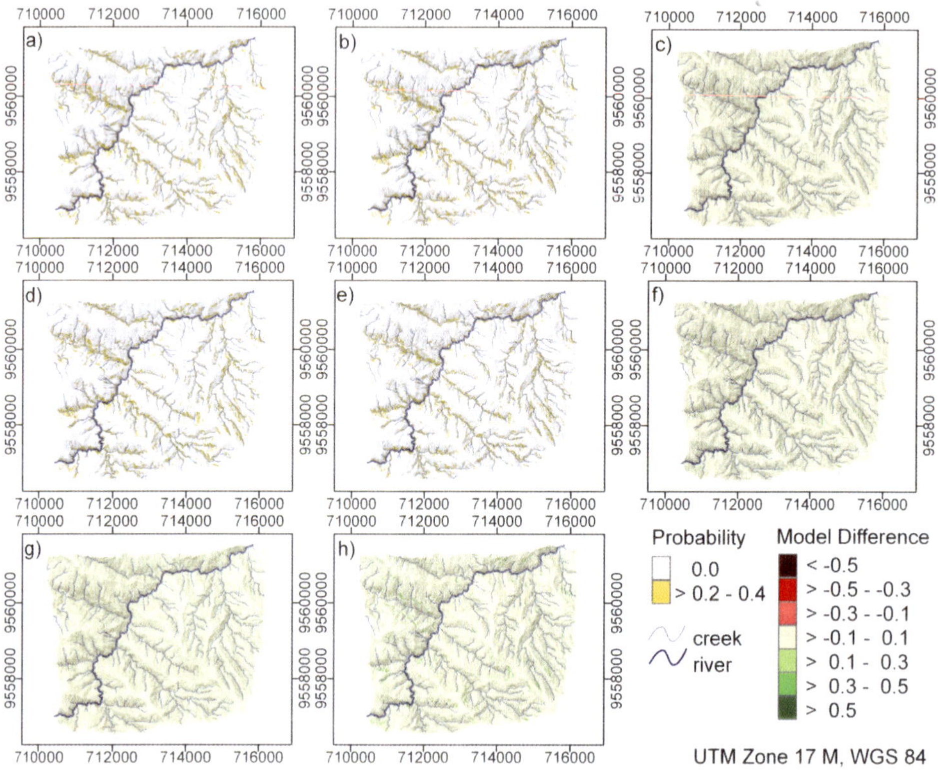

Fig. 5. Maps of Leptosol occurrence probability (Overlaid hill shading with light source from north-east). Independent on WRB hierarchy: a) prediction by n. n. terrain values, b) prediction by mean terrain values and c) model difference. Dependent on WRB hierarchy: d) prediction by n. n. terrain values, e) prediction by mean terrain values and f) model difference. Difference between independent and WRB hierarchy dependent prediction: g) n. n. terrain values and h) mean terrain values

Model application to the research area is shown in Figure 6. Stagnosols reach higher probabilities by the mean terrain values model (Figure 6b) compared to the prediction from n. n. terrain values (Figure 6a). Figures 3c and 3f show that the difference between the probability prediction by n. n. and mean relief values (Figure 6c), + 0.1 – 0.3, is not due to this higher Stagnosol probability on high altitudes as predicted by mean terrain values on concave sites. This difference accounts for only 0.05. However, it is due to the reduced probability assigned to convex sites ≥ 2135 m a.s.l. (0.24 difference). As a conclusion to this, the two models are quite similar, mainly differing by the dependence on curvature, which is not included in the model from n. n. relief values. Including WRB (FAO, IUSS Working Group WRB, 2007) hierarchy in the probability prediction, a site classified as Histosol or

Leptosol cannot be classified as Stagnosol. Accordingly, Histosol probability reduces Stagnosol probability to a perceptible extent (Figures 6d and 6e). Figures 6g and 6h show that these differences account for 0.1 to 0.3 for most of the research area with the prediction by n. n. terrain values (Figure 6g) still yielding less differences in the lower altitudes compared to the prediction by mean terrain values (Figure 6h). Differences between the two models are extended while including WRB hierarchy (Figure 6f), compared to that being independent of WRB hierarchy (Figure 6c).

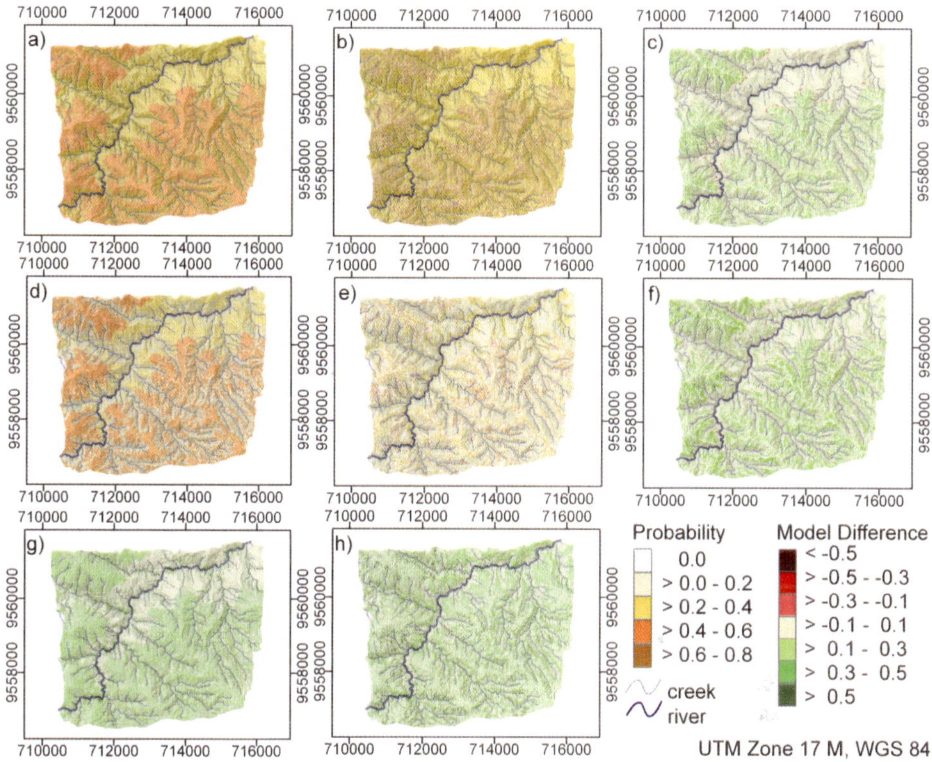

Fig. 6. Maps of Stagnosol occurrence probability (Overlaid hill shading with light source from north-east). Independent on WRB hierarchy: a) prediction by n. n. terrain values, b) prediction by mean terrain values and c) model difference. Dependent on WRB hierarchy: d) prediction by n. n. terrain values, e) prediction by mean terrain values and f) model difference. Difference between independent and WRB hierarchy dependent prediction: g) n. n. terrain values and h) mean terrain values

CTs for Umbrisols, Cambisols and Regosols cannot be provided. Umbrisol prediction was impossible, since the used dataset contains only 7 Umbrisols among 367 sampled sites and is not enough to gain a clear prediction scheme. Furthermore, not all but some of the determined Umbrisols are situated within the accumulation zone of former landslides so that an additional variable to predict their occurrence would be necessary. Cambisols and Regosols, on the other hand, are rather unspecific RSGs which makes their prediction difficult. Cambisols need a cambic horizon, but apart from that they are rather determined by the absence of diagnostic criteria that would classify the soil for another RSG. Regosols are even worse, since they do not have any characteristic on their own, but refer to all soils that do not classify as another RSG.

3.2 Model performance and uncertainty

Overall CT model performance is limited (Table 1). Terrain attributes can likely only explain RSG distribution to a limited extent within this mountainous tropical landscape. Unfortunately, no information is available about parent material distribution, but rapid bedrock changes were discovered during field work. The profound influence of landslides causes shifts in soil material and mixes it with rock material, leading to quite different soil properties. Although there has been a landslide inventory based on visible landslide scars on a time series of aerial photographs from 1962 to 1998 (Stoyan, 2000), most former landslides remain hidden under the regrown dense forest cover as was experienced during field work.

RSG	Terrain Parameters	Model Pseudo R^2	CV Pseudo R^2
Histosol	nearest neighbour	0.34	0.22
	mean	0.35	0.21
Stagnosol	nearest neighbour	0.22	0.19
	mean	0.28	0.13

RSG= Reference Soil Group, CV = cross validation

Table 1. Model quality of classification trees to predict Histosol and Stagnosol probability

CTs in general have certain disadvantages: (1) They are very dependent on the dataset used, i.e. some sample points more or less may lead to rather different models and (2) they predict abrupt values due to the grouping into end nodes. A continuous probability distribution of the RSGs in reality therefore is replaced by some probability classes according to Figure 3.

What makes WRB RSG prediction in general problematic is the character of the WRB itself. Assignment of some RSGs requires exceeding an absolute (Histosols) and for others a relative (Stagnosols) thickness value of a diagnostic horizon. If a soil has an organic layer ≥ 40 cm, it is classified as Histosol independent of its mineral properties. If the organic layer is 1 cm less, these mineral properties abruptly become important. Relating the extent of the stagnic horizon to soil depth obviously is not characteristic enough to allow for a good model relating the Stagnosol occurrence pattern to terrain parameters. This is probably the reason why model accuracy is limited. As a consequence the low R^2 are not considered as a problem, but as a natural phenomenon in predicting complex entities such as RSGs.

Furthermore, the calculated CT R^2 refers to a one value prediction. As was described earlier, a CT model usually assigns the category which forms the majority within each end node to the respective landscape position. It does not consider other categories assigned to that end node as classification possibility, but neglects them. Any soil map has a certain degree of uncertainty. Usually boundaries between soil units are drawn according to expert knowledge or GIS interpolations. However, the degree of uncertainty which is a logical phenomenon in any below ground investigation usually is not included within the soil map. The new generation of digital soil maps provides a new development in this area. Accordingly, our digital soil maps include this model uncertainty through assigning RSG occurrence probabilities instead of unique values. Other authors mainly used fuzzy-logic to include this uncertainty, e.g. McBratney & De Gruiter (1992), Hannemann (2010).

Another aspect to be considered, is that generally soil maps are gained on a much larger scale. Lagacherie & Holmes (1997) use a spatial resolution of 50 m, Moran & Bui (2002) use 250 m. Therefore, the small scale, 10 m resolution, in our soil maps might be another reason for the low R^2. The soils within the research area change within a few meters radius as typical for tropical soils. Accordingly, the highest possible resolution was used. This way low scale soil variability is included within the models, which would be neglected while working on a larger scale. To conclude, the size of the applied dataset is not enough to represent the investigated soil-landscape at this high precision.

3.3 Comparison with earlier soil map

A RSG probability prediction is also possible from a single CT which predicts all RSGs at once. Liess et al. (2009) established such a CT for the research area (Figure 7), but did not predict probabilities from it. The percentage of the RSGs within each end node of this tree was interpreted as occurrence probability for the RSGs according to the related landscape position and compared it to the findings from the various CTs of this study. The difference between RSG probability by the tree model from Liess et al. (2009) and our predictions is displayed in Figure 8. The first column maps the RSG probabilities according to Liess et al. (2009), the second column presents the differences between the latter and our prediction from n. n. relief values (WRB dependent), and the third column shows the differences regarding the prediction from mean relief values.

The model from Liess et al. (2009) (Figure 7) assigned a very high Histosol probability with 0.6 – 0.8 to about half of the research area. For some sites the predicted probability was even higher. In our new model, Histosol probability was less, 0.2 – 0.4 for most of the area (Figure 4b), but continuous on all sites with at least 0.2 (Figure 3a and 3d). It was shown that Histosol probability is high within some landscape positions and for a VOFD from 54 – 175 m this is supported by a high number of sampled sites. In contrast to this, the end nodes in the tree model from Liess et al. (2009) mostly contain only a very limited number of sampling sites, e.g. the end nodes that predict particularly high Histosol probabilities (≥ 0.8) only contain 12 – 15 sampled sites. The end node with the most sites predicting Histosol probability with 0.78, refers to landscape positions in small catchments < 214 m HOFD, similar to our findings. The importance of the catchment size as first subdividing variable for model development was confirmed. For smaller catchment sizes, i.e. sites through which a smaller area discharges, Histosol occurrence is more likely.

Leptosols were predicted with low probability on steep slopes and close to the creeks (< 20, < 19 m HOFD). The latter is confirmed by Liess et al. (2009) who predicted Leptosols < 21 m HOFD, but with a high probability of 0.71 (Figure 8d). 0.71 of 7 sampled sites that are contained within the respective end node no. 7 (Figure 7) are 5 sampled sites. To use only five Leptosol sites to predict such a high probability seems unreasonable. On steep slopes, especially in an area influenced by landslides, soils have less chance to develop. Hence, it is no surprise to find Leptosols in these landscape positions. Close to the creeks soil material is probably removed downslope within the channel system during times of high rainfall; through these sites a high amount of water discharges due to a high contributing catchment area. On many sites the organic layer directly overlies continuous rock.

Stagnosols were predicted with a higher probability by n. n. relief values compared to the model from Liess et al. (2009) (Figure 8h). This is due to the fact that Stagnosols were predicted as all soils that display sufficient stagnic properties, but it was neglected that some of them carry a sufficiently thick organic layer to qualify as Histosols. Stagnic properties and

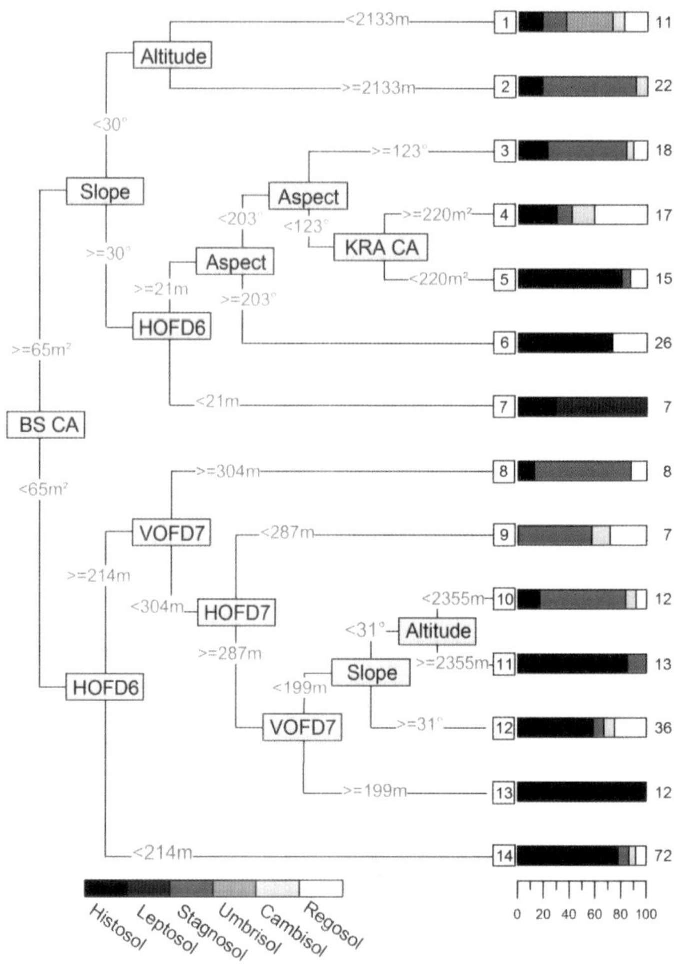

Fig. 7. Classification tree model to predict RSG probability within the research area. Numbers before the boxplots indicate the node number, numbers behind the boxplots indicate the number of sample sites per end node. BS CA and KRA CA upslope contributing catchment area according to the Braunschweiger relief model and kinematic routing algorithm, HOFD horizontal and VOFD vertical overland flow distance, 6, 7 refer to different precision in channel network (adapted from Liess et al., 2009)

thick organic layers occur at the same landscape position: The WRB (FAO, IUSS Working Group WRB, 2007) describes Histosols as soils in "poorly drained basins and depressions" and "highland areas with a high precipitation–evapotranspiration ratio". Nevertheless, these two properties are seen as competing if it comes to soil classification by WRB. Two soils showing both a thick organic layer and stagnic properties are assigned to different RSGs even if they are different only by 1 cm in organic layer thickness. Prediction from mean relief values shows more similarities in Stagnosol probability to Liess et al. (2009) (Figure 8i) than prediction by n. n. terrain values. This is because Liess et al. (2009), who used a subset of our dataset, predicted the RSGs by mean relief values, too. Stagnosol probability increases above an altitude of 2146 m a.s.l. on slope angles < 40°. An increase in Stagnosol abundance with increasing altitude and decreasing slope angle was also described by Liess et al. (2009). Schrumpf et al. (2001) stated an increase in hydromorphic properties with increasing altitude and designated soils as Humaquepts (Soil Survey Staff, 2006). The increase with altitude can be attributed to the increasing rainfall (Rollenbeck, 2006). Lesser steep slope angles account for a slower discharge.

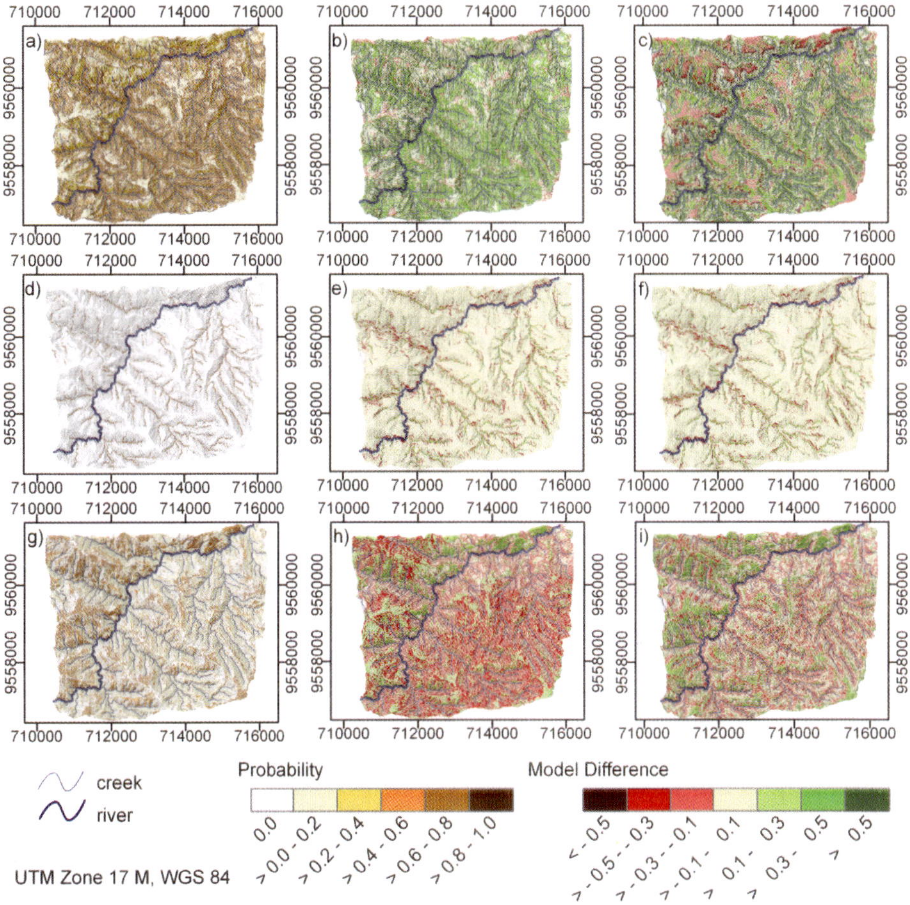

Fig. 8. RSG probability prediction by the simple tree model from Liess et al. (2009) (column 1: a, d and g) and calculated difference in probability prediction between that model and the WRB dependent model from n. n. (column 2: b, e and h) and mean terrain values (column 3: c, f and i). Histosol (1st row), Leptosol (2nd row) and Stagnosol (3rd row). Overlaid hill shading with light source from north-east

We assume the RSG probability predicted by various CTs, to better represent soil reality within the research area, since the dataset does not consist of all RSGs to an equal extent so that some are preferred over others during the tree subdivision process. Furthermore, the multiple CTs rather predict probabilities of soil diagnostic properties, which can occur simultaneously at one site within the soil profile. Accordingly, the model from Liess et al. (2009) overestimated Histosol probability for most sites as can be seen by the mainly green colours in Figure 8b and c. However, at the same time it underestimated Stagnosols in most of the area as can be deduced from the prevailing red colours in Figures 8h and i. In a similar way, Leptosols are overestimated by the model from Liess et al. (2009).

4. Conclusions

Models adapted for n. n. compared to those adapted for mean terrain values showed only minor differences. We conclude that predicting all RSGs at once is not as good as predicting each RSG on its own by a CT. The dataset does not consist of all RSGs to an equal extent, so some RSGs are preferred over others during the tree subdivision process.

Model performance might be improved by choosing a lower resolution to exclude small scale diversity, reducing model dependence on the dataset, applying a different statistical model or predicting soil properties instead of the complex RSG entities. However, further research is needed to prove these assumptions.

Model uncertainty in the digital soil maps is represented by the occurrence probabilities of the RSGs. Probabilities of various RSGs at the same landscape position can be understood as competing RSGs. But the probabilities of the various RSGs can also be interpreted as a soil composed of the various RSGs, i.e. various diagnostic horizons or various soil processes running simultaneously or successively as has been part of soil genesis theory for a long time (Simonson, 1959; Schelling, 1970). Thereby, this provides a good means to acknowledge inter-relations between the RSGs. An even better chance to acknowledge this would be the prediction of the diagnostic properties necessary for WRB classification by themselves. In accordance with McBratney & De Gruiter (1992), who thought to improve the existing soil classification systems via fuzzy sets, we would like to contribute the above-mentioned ideas to the development of a continuous soil systematisation system.

5. Acknowledgements

The authors are indebted to the German Research Foundation (DFG) for funding the study in the framework of the Research Unit FOR816. Logistic support of the foundation Nature and Culture International (NCI, San Diego – Loja) is gratefully acknowledged. Furthermore, we would like to thank Christopher L. Shope for English language revision.

6. References

Bauer, J., Rohdenburg, H., & Bork, H.-R. (1985). Ein digitales Reliefmodell als Vorraussetzung für ein deterministisches Modell der Wasser- und Stoff-Flüsse. In: *Landschaftsgenese und Landschaftsökologie H. 10, Parameteraufbereitung für deterministische Gebiets-Wassermodelle, Grundlagenarbeiten zur Analyse von Agrar-Ökosystemen*, Bork, H.- R., Rohdenburg, H., pp. (1 – 15). Selbstverlag Abteilung für Physische Geographie und Landschaftsökologie der Technischen Universität Braunschweig, ISSN 0170-7299, Braunschweig

Bishop, T. F. A., Minasny, B., 2006. Digital Soil-Terrain Modeling: The Predictive Potential and Uncertainty. In: Grunwald, S. (Ed.) *Environmental Soil-Landscape Modeling*. ISBN 0-8247-2389-9, CRC Press, Boca Raton.

Böhner, J., McCloy; K.R., & Strobl, J. (2006). *Göttinger Geographische Abhandlungen 115: SAGA – Analysis and Modelling Application*, Geographisches Institut der Universität Göttingen, Verlag Erich Goltzt GmbH & Co. KG, Göttingen

Bourennane, H., King, D., Couturier, A., 2000. Comparison of kriging with external drift and simple linear regression for predicting soil horizon thickness with different sample densities. *Geoderma* 97, pp.(255– 271).

Breimann, L., Friedmann, J.H., Olshen, R.A., & Stone, C.J. (1984). *Classification and regression trees*, CRC press, Wadsworth.

Burrough, P. A., Bouma, J., Yates, S. R., 1994. The state of the art in pedometrics. *Geoderma*, Vol 62, pp.(311 – 326), ISSN 0016-7061

Cimmery, V. (2007). *User guide for SAGA, version 2.0*, Available from: http://sourceforge.net/ projects/saga-gis/ files/(access: 25.11.2009)

Dobos, E., Micheli, E., Baumgardner, M.F., Biehl, L., & Helt, T. (2000). Use of combined digital elevation model and satellite radiometric data for regional soil mapping. *Geoderma*, Vol. 97, pp. (367–391), ISSN 0016-7061

Dokutschajew, W. W., 1883. Russkij Cernozem, St. Petersburg.

FAO, IUSS Working Group WRB (2007). World Reference Base for Soil Resources, ISRIC, Rome

Fries, A., Rollenbeck, R., Göttlicher, D., Nauss, T., Homeier, J., Peters, T., & Bendix, J. (2009). Thermal structure of a megadiverse Andean mountain ecosystem in southern Ecuador, and its regionalization. *Erdkunde*, Vol. 63, pp. (321–335), ISSN 0014-0015

Gessler, P., Moore, I., McKenzie, N., & Ryan, P. (1995). Soil-landscape modelling and spatial prediction of soil attributes. *International Journal of Geographical Information Systems*, Vol. 9, 4, pp. (421– 432), ISSN 1365-8816

Grunwald, S., 2006. What Do We Really Know about the Space-Time Continuum of Soil Landacapes? In: Grunwald, S. (Ed.), *Environmental Soil-Landscape Modeling*, pp. (3 – 36), ISBN 0-8247-2389-9, CRC Press, Boca Raton.

Hannemann, J. (2010). *Die Berücksichtigung inhaltlicher und räumlicher Unschärfe bei der GIS-gestützten Erstellung von Bodenkarten.* Dissertation Universität Bayreuth, Geowissenschaft. Shaker Verlag, Aachen

Hengl, T., Heuvelink, G.B.M., & Stein, A. (2004). A generic framework for spatial prediction of soil variables based on regression-kriging. *Geoderma*, Vol. 120, No. 1-2, pp. (75–93), ISSN 0016-7061

Homeier, J., Dalitz, H., & Breckle, S.-W. (2002). Waldstruktur und Baumartendiversität im montanen Regenwald der Estacón Cientíca San Franscisco. Südecuador. *Ber. d. Reinh. Tüxen-Ges*, Vol. 14, pp. (109–118), ISSN 0940-418X

Jenny, H. (1941). *Factors of soil formation. A system of quantitative pedology*, McGraw-Hill, New York.

Lagacherie, P. & Holmes, S. (1997). Addressing geographical data errors in a classification tree soil unit predicton. *International Journal of Geographic Information Science*, Vol. 11, pp. (183–198), ISSN 1365-8816

Lea, N. L. (1992). An aspect driven kinematic routing algorithm. In: *Overland Flow Hydraulics and Erosion Mechanics. Parsons.* A. J. and Abrahams, A. D., pp. (393 – 407), London

Liess, M., Glaser, B., & Huwe, B. (2009). Digital Soil Mapping in Southern Ecuador. *Erdkunde*, Vol. 63, 4, pp. (309–319), ISSN 0014-0015

Litherland, M., Aspen, J.A., & Jemielita, R.A., 1994. The metamorphic belts of Ecuador. *Overseas Memoirs - British Geological Survey*, Vol. 11, pp. (1–147), ISSN 0951-6646

McBratney, A.B. & DeGruiter, J.J. (1992). A continuum approach to soil classification by modified fuzzy k-means with extragrades. *Journal of Soil Science*, Vol. 43, pp. (159–175), ISSN 0038-075X

Milne, G. (1935). Some Suggested Units of Classification and Mapping, Particularly for East African Soils. *Soil Research*, Vol. 4, pp. (183–198), ISSN 0309-133

Moran, C.J. & Bui, E.N. (2002). Spatial data mining for enhanced soil map modelling. *International Journal of Geographic Information Science*, Vol. 16, pp. (533–549), ISSN 1365-8816

Moore, I., Gessler, P., Nielsen, G., Peterson, G., 1993. Soil attribute prediction using terrain analysis. In: *Soil Science Society of America Journal*, 57/ 2, pp.(443- 452).

Odeh, I., McBratney, A., Chittleborough, D., 1994. Spatial prediction of soil properties from landform attributes derived from a digital elevation model. *Geoderma*, 63/3-4, pp.(197- 214).

Rollenbeck, R. (2006). Variability of precipitation in the Reserva Biólogica San Francisco / Southern Ecuador. *Lyonia, A Journal of Ecology and Application*, Vol. 9, No. 1, pp. (43–51), ISSN 0888-9619

Schelling, J. (1970). Soil genesis, soil classification and soil survey. *Geoderma*, 4, 3, pp. (165–193), ISSN 0016-7061

Schrumpf, M., Guggenberger, G., Valarezo, C., & Zech, W., 2001. Tropical montane rainforest soils. Development and nutrient status along an altitudinal gradient in the South Ecuadorian Andes. *Die Erde*, Vol. 132, pp. (43–59), ISSN 0013-9998

Simonson, W.R. (1959). Outline of a Generalized Theory of Soil Genesis. *Soil Science Society of America Journal*, Vol. 23, pp. (152–156), ISSN 0361-5995

Skidmore, A.K., Watford, F., Luckananurug, P., & Ryan, P.J. (1996). An operational GIS expert system for mapping forest soils from a geographical information system. *International Journal of Geographic Information Science*, Vol. 5, pp. (431–445), ISSN 1365-8816

Soil Survey Staff (2006). Keys to Soil Taxonomy. 10th ed. United States Department of Agriculture, Natural Resources Conservation Service, Available from: http://soils.usda.gov/technical/classification/ taxonomy/ (access: 22/10/2007)

Stoyan, R., 2000. *Aktivität, Ursachen und Klassifikation der Rutschungen in San Francisco/ Südecuador*. Diplomarbeit Universität Erlangen.

Strahler, A.N. (1957). Quantitative analysis of watershed geomorphology. *Transactions of the American Geophysical Union*, Vol. 38, No. 6, pp. (913–920), ISSN 0002-8606

Therneau, T.M. & Atkinson, B. (2003). The rpart Package, Available from: http://cran.r-project.org/web/ packages/ rpart/ rpart.pdf (access: 28/02/2008)

Thomas, A.L., King, D., Dambrine, E., Couturies, A., & Roque, A. (1999): Predicting soil classes with parameters derived from relief geologic materials in a sandstone region of the Vosges mountains (northeastern France). *Geoderma*, Vol. 90, pp. (291–205), ISSN 0016-7061

Wilcke, W. ,Yasin, S., Abramowski, U., Valarezo, C., & Zech, W. (2002). Nutrient storage and turnover in organic layers under tropical montane rain forest in Ecuador. *European Journal of Soil Science*, Vol. 53, pp. (15–27), ISSN 1351-0754

Wilcke, W., Valladarez, H., Stoyan, R., Yasin, S., Valarez, C., & Zech, W. (2003). Soil properties on a chronosequence of landslides in montane rain forest, Ecuador. *Catena*, Vol. 53, pp. (79–95), ISSN 0341-8162

Yasin, S. (2001). *Water and Nutrient Dynamics in Microcatchments under Montane Forest in the South Ecuadorian Andes*. Bayreuther Bodenkundliche Berichte, Band 73.

Zevenbergen L.W. & Thorne C.R. (1987). Quantitative Analysis of Land Surface Topography. *Earth Surface Processes Landforms*, Vol. 12, pp. (47 – 56), ISSN 0197-9337

14

Spatial Sampling Design and Soil Science

Gunter Spöck
University of Klagenfurt
Austria

1. Introduction

Spatial statistics is the science of the analysis of geo-referenced data and loosely speaking may be divided into the three sub-areas analysis of point processes, analysis of areal data and geostatistics. Point processes naturally arise for example in geophysics, when the locations of earthquakes are noticed, in epidemiology, where new illness cases of certain epidemies are mapped geographically or in biology, where cell centers of a certain tissue are mapped under the microscope. Areal data are data that are attached to areas like number of illness cases in certain medical districts, percent of grassland in a certain county or number of votes in certain political districts. The topic of this article is but geostatistics, the science of continuous stochastic processes or so-called random fields that are defined over some region in 2- or 3-dimensional geographic space $\mathbf{X}$ or in space-time. A random field $\{Y(x) : x \in \mathbf{X}\}$ is a set of random variables or so-called regionalized variables $Y(x)$ that are attached to every location $x \in \mathbf{X}$. The probability law of the random field is uniquelly determined by its so-called projective family, the set of all finite dimensional distributions of any finite set of $Y(x)$ obeying symmetrie and consistency with marginal distributions although we will see next that geostatistics most often deals only with the first and second order characteristics of random fields, the trend or mean function $m(x) = \mathrm{E}(Y(x))$ and the covariance function $C(x, y) = \mathrm{Cov}(Y(x), Y(y))$. Most often a linear trend function $m(x) = \mathbf{f}(x)^T\beta$, where $\mathbf{f}(x)$ is a fixed vector-valued function and β is a regression parameter vector to be estimated, is sufficient for modelling purposes. For $x \in \Re^2$, $\mathbf{f}(x)$ could be for example a vector of polynomials in the coordinates $x = (x_1, x_2)$. The covariance function $C(.,.)$ must be positive semidefinite, meaning that it must give any linear combination of $Y(x_1), Y(x_2), \ldots, Y(x_n)$ positive variance. Most often an additional assumption of second-order stationarity must be met, meaning that $C(x, x + h) = C(h)$ is dependent only on lag h and not on the locations x and $x + h$ themselves. The next stronger assumption is the assumption of isotropy, meaning $C(x, x + h) = C(||h||_2)$, where $||h||_2$ is the Euclidean length of h. Both the assumption of second order stationarity and of isotropy are met in order to make the covariance function estimable from only single realized variables or data $y(x_1), y(x_2), \ldots, y(x_n)$. The task of geostatistics is to produce a prediction map of all $y(x_0), x_0 \in \mathbf{X}$ based on the available data $y(x_1), y(x_2), \ldots, y(x_n)$ and to report on the accuracy of these predictions. The best known methodology for this task of interpolation or map drawing is kriging, also known as best linear unbiased prediction. The so-called universal kriging predictor is dependent on both the data $\mathbf{y} = (y(x_1), y(x_2), \ldots, y(x_n))^T$, the covariance matrix $\mathbf{K}$ of the corresponding random variables and the covariance vector $\mathbf{c}_0$ between these random variables and the random variable $Y(x_0)$ to be predicted. It can be written

$$\hat{Y}(x_0) = \mathbf{f}(x_0)^T\hat{\beta} + \mathbf{c}_0^T\mathbf{K}^{-1}(\mathbf{y} - \mathbf{F}\hat{\beta}), \tag{1}$$

where

$$\hat{\beta} = (\mathbf{F}^T\mathbf{K}^{-1}\mathbf{F})^{-1}\mathbf{F}^T\mathbf{K}^{-1}\mathbf{y} \tag{2}$$

is the generalized least squares estimate of the regression parameter vector β and $\mathbf{F}$ is the design matrix corresponding to $\mathbf{f}(.)$ and the locations $x_1, x_2, \ldots, x_n$. The mean squared error (MSEP) of this unbiased predictor is given by

$$\mathbf{E}(Y(x_0) - \hat{Y}(x_0))^2 = \sigma^2(x_0) = C(\mathbf{0}) - \mathbf{c}_0^T\mathbf{K}^{-1}\mathbf{c}_0 + \mathbf{g}(x_0)^T(\mathbf{F}^T\mathbf{K}^{-1}\mathbf{F})^{-1}\mathbf{g}(x_0), \tag{3}$$

where

$$\mathbf{g}(x_0) = (\mathbf{f}(x_0) - \mathbf{F}^T\mathbf{K}^{-1}\mathbf{c}_0). \tag{4}$$

Obviously, the mean squared error $\sigma^2(x_0)$ is dependent on the arrangement of the sampling locations $x_1, x_2, \ldots, x_n$ via the covariance matrix $\mathbf{K}$, the covariance vector $\mathbf{c}_0$ and the design matrix $\mathbf{F}$.

Like with any statistical prediction method one aim of the statistician in kriging is to make best use of the available data, maybe under certain budget constraints or constraints on the number of locations that can be sampled. Thus, the aim is to get as good predictions as possible for the complete area of investigation $\mathbf{X}$. One possibility to formalize this aim is to try to select the sampling locations $x_1, x_2, \ldots, x_n$ in such a way that the sum of all kriging MSEP's

$$\int_{\mathbf{X}} \sigma^2(x_0)dx_0 \tag{5}$$

becomes a minimum over the area of investigation $\mathbf{X}$. Notably, this is a very complicated optimization problem and becomes still more complicated by the fact that the covariance matrix $\mathbf{K}$ enters the kriging MSEP in its inverse form $\mathbf{K}^{-1}$. It has been tried to solve this sampling design problem besides other criteria that also measure the accuracy of kriging predictions in a number of recent research papers. The aim of the next section is to discuss some of these papers.

In soil science especially the remediation of contaminated waste soils is of importance. Not only the area of contamination must be best detected but the contamination concentrations must be optimally predicted. Most often then soil with contamination level above a certain threshold must be remediated. Sampling design for contamination detection in soil science most often works in three phases:

1. The area of contaminated soil must be detected: Here most often samples on a coarse regular rectangular grid are taken to detect the contaminated area.
2. The spatial variability or the covariance function must be best estimated: Here the design from stage 1.) is updated by some nearby locations in order to get also some good estimates of the nugget effect and the small scale variability (the appearance of the covariance function near the origin).
3. Contamination concentrations must be best predicted and their uncertainties be documented.

This work especially deals with phase 2.) and phase 3.). We consider as design criterion for 3.) first the integrated kriging variance averaged over the area of contamination. Next as a combined design criterion for 2.) and 3.) the averaged expected lengths of predictive intervals are considered. Finally it is shown how these design criteria can be generalized to spatial variables having skewed distributions. In this context a criterion for spatial sampling design with Box-Cox transformed spatial variables is proposed.
Mathematically speaking we will approximate the investigated spatial random field by means of a large regression model consisting of cosine-sine Bessel surface harmonics with random amplitudes (Sections 3, 4). This approximating regression model is a direct result of the polar spectral representation theorem of isotropic random fields (Section 3). We show that kriging prediction in the original random field model is equivalent to trend prediction in this approximating Bayesian linear regression model (Section 4). Thus, standard convex experimental design theory for Bayesian linear regression may be used to find spatial sampling designs. Based on this theory we propose two algorithms for the calculation of exact designs (Section 7.3):

- An algorithm for removing redundant design locations.
- An algorithm for adding sampling locations to an existing network.

In Section 5 we generalize our approach to the case that the covariance function is estimated by restricted maximumum likelihood (REML) and thus is uncertain. The result is that the actual predictive intervals are wider than when not taking into account this uncertainty. The Smith and Zhu design criterion adjusts to this uncertainty by considering the average of the expected lengths of predictive intervals as a design criterion. Section 6 again is a generalization, because here we assume the data distribution to be skewed and to be transformable to a Gaussian distribution by means of a Box-Cox transformation. Once again the average of the expected lengths of predictive intervals is considered as a design criterion. The proposed methodologies are illustrated by a true data example from soil science (Section 8). Finally, one from the author freely available MATLAB and Octave toolbox is described which can perform all the tasks of spatial sampling design and spatial interpolation (Section 7).

2. A review on spatial sampling design

The importance of (optimal) spatial sampling design considerations for environmental applications and soil science has been demonstrated in quite a few papers and monographs (Brus and de Gruijter, 1997; Brus and Heuvelink, 2007; Caeiro et al., 2003; Cox Jr., 1999; Delmelle and Goovaerts, 2009; Diggle and Lophaven, 2006; Dobbie et al., 2008; Groenigen et al., 1999; Hengl et al., 2003; Lark, 2002; US-EPA, 2002). The papers on spatial sampling design may be divided into several categories of which some are overlapping. First of all we may differentiate between design criteria for spatial prediction and for estimation of the covariance function and between combined criteria for both desires at once. Works falling into the category of criteria for prediction are (Brus and Heuvelink, 2007; Fedorov and Flanagan, 1997; Müller and Pazman, 1998; 1999; Müller, 2005; Pazman and Müller, 2001). Criteria for the estimation of the covariance function are considered by (Lark, 2002; Müller and Zimmerman, 1999; Zimmerman and Homer, 1991). Combined criteria we can find in the article (Zhu and Stein, 2006), who condsider the minimization of the average expected length of predictive intervals. Further papers falling into this category of combined criteria are Bayesian articles specifying a priori distributions over covariance functions like (Brown et al., 1994); Diggle and Lophaven, 2006; Fuentes et al.,

2007; Müller et al., 2004). Actually (Brown et al., 1994)) and (Fuentes et al., 2007) consider the covariance function to be non-stationary and deal with an entropy based design criterion according to which the determinant of the covariance matrix between locations to be added to the design must be maximized. Both make use of simulated annealing algorithms to find optimal designs obeying their criteria.

At this stage we are at a second distinguishing feature of optimal design algorithms. We can distinguish beween stochastic search algorithms like simulated annealing (Aarts and Korst, 1989) or evolutionary genetic algorithms and deterministic algorithms for optimizing the investigated design criteria. With the exception of the works of (Fedorov and Flanagan, 1997; Müller and Pazman, 1998; 1999; Müller, 2005; Pazman and Müller, 2001; Spöck and Pilz, 2010) almost all algorithms for spatial sampling design optimization use stochastic search algorithms for the finding of optimal configurations of sampling locations $x_1, x_2, \ldots, x_n$. The term spatial simulated annealing (SSA) finds its first manifestation in the work of (Groenigen et al., 1999). (Trujillo-Ventura and Ellis, 1991) consider multiobjective sampling design optimization.

Available free software for spatial sampling design optimization is quite rare. Up to the knowledge of the author there are only 6 sampling design toolboxes freely available including an own one: (Gramacy, 2007) has implemented sampling design for treed Gaussian random fields in the R-package tgp. In treed Gaussian random fields the area **X** of investigation is partitioned by means of classification trees into rectangular sub-areas with sides parallel to the coordinate axes. This software is especially useful for the design of computer simulation experiments, where parameters guiding the computer simulation output are identified as spatial coordinates. Another software especially useful for computer simulation experiments and sequential design is the DAKOTA package (http://dakota.sandia.gov). Further papers falling into the category of computer simulation experiments are (Chen et al., 2006; Gramacy and Lee, 2010; Kleijnen et van Beers, 2004; Lim et al., 2002; Mitchell and Morris, 1992; Morris et al., 1993; Schonlau, 1997). Software freely available upon request for research purposes and monitoring network optimization is (Le and Zidek, 2009). This software implements the entropy based design criterion mentioned above. (Gebhardt, 2003) implements a branch and bound algorithm for designing with the criterion (5). (Baume et al. , 2011) compare different greedy algorithms for spatial design. Maybe this small list of software for spatial sampling design is not complete and hopefully more software can be obtained from the different authors of research papers upon request. Since for the practioneer there is a strong need for spatial sampling design and almost no software is freely available this was an impetus for the author of this chapter to program a toolbox for spatial sampling design. The spatDesign Matlab toolbox of the author is freely available online at

- http://wwwu.uni-klu.ac.at/guspoeck/spatDesignMatlab.zip

An Octave version not implementing as many design criteria as the Matlab toolbox is available at

- http://wwwu.uni-klu.ac.at/guspoeck/spatDesignOctave.zip

3. The spatial mixed linear model

This section starts our own approach to spatial sampling design. No stochastic search algorithms like simulated annealing to optimize the design criterion are needed in this approach because we make use of the mathematical structure of the investigated design criteria.

We consider a mean square continuous (m.s.c.) and isotropic random field $\{Y(x) : x \in \mathbf{X} \subseteq \Re^2\}$ such that

$$Y(x) = \mathbf{f}(x)^T\beta + \varepsilon(x), \quad \mathrm{E}\varepsilon(x) = 0, \tag{6}$$

where $\mathbf{f}(x)$ is a known vector of regression functions, $\beta \in R^r$ a vector of unknown regression parameters and

$$\mathrm{Cov}(Y(x), Y(y)) = C(||x - y||); \quad x, y \in \mathbf{X}. \tag{7}$$

Then, according to (Yaglom, 1987), the covariance function can be represented in the form

$$C(t) = \int_0^\infty J_0(t\omega)dG(\omega), \; t \geq 0, \tag{8}$$

where $J_0(.)$ is the Bessel function of the first kind and order 0, $t = ||x - y||$ is the Euclidean distance between x and y, and $G(.)$ is the so-called (polar) spectral distribution function associated with $C(.)$. As such $G(.)$ is positive, monotonically increasing and bounded from above. On the other hand, knowing $C(.)$ its spectral distribution can be obtained from the inversion formula

$$\frac{G(\omega^+) + G(\omega^-)}{2} = \int_0^\infty J_1(t\omega)\omega C(t)dt, \tag{9}$$

where $G(\omega^+)$ and$G(\omega^-)$ denote the right- and left-hand side limits at ω and J_1 denotes the Bessel function of first kind and order 1. Approximating $G(.)$ by means of a step function with positive jumps $a_i^2 = G(\omega_{i+1}) - G(\omega_i)$ at preselected points ω_i, $i = 0, 1, , \ldots, n-1$, and changing to polar coordinates $(t, \varphi) = (radius, angle)$ the polar spectral representation theorem for m.s.c. isotropic random fields tells us that the error process may be approximated as

$$\varepsilon(t, \varphi) \approx \sum_{m=0}^{\infty} \{\cos(m\varphi) \sum_{i=1}^{n} J_m(\omega_i t) U_{m,i}\} + \sum_{m=1}^{\infty} \{\sin(m\varphi) \sum_{i=1}^{n} J_m(\omega_i t) V_{m,i}\}, \tag{10}$$

where all the random variables $U_{m,i}$ and $V_{m,i}$ are uncorrelated, have mean zero, and their variances are $\mathrm{var}(U_{m,i}) = \mathrm{var}(V_{m,i}) = d_m a_i^2$; and $d_m = 1$ for $m = 0$ and $d_m = 2$ for $m \geq 1$. By truncating the above series at a sufficiently large $m = M$, we get an approximation of our random field in form of a mixed linear model

$$Y(x) \approx \mathbf{f}(x)^T\beta + \mathbf{g}(x)^T\alpha + \varepsilon_0(x) \tag{11}$$

From above it becomes clear that the components of the additional regression vector $\mathbf{g}(\cdot)$ are made up of the following radial basis functions (cosine-sine-Bessel-harmonics)

$$g_{m,i}(t, \varphi) = cos(m\varphi) J_m(\omega_i t); \tag{12}$$

$$m = 0, \ldots, M; i = 1, \ldots, n$$

$$g_{m,i}(t, \varphi) = sin((m - M)\varphi) J_{m-M}(\omega_i t);$$

$$m = M + 1, \ldots, 2M; i = 1, \ldots, n.$$

4. Classical Bayesian experimental design problem

Starting from our spatial mixed linear model (11) we may gain further flexibility with a Bayesian approach incorporating prior knowledge on the trend. To this we assume that the regression parameter vector β is random with

$$\mathrm{E}(\beta) = \mu \in R^r, \quad \mathrm{Cov}(\beta) = \mathbf{\Phi}. \tag{13}$$

This is exactly in the spirit of (Omre, 1987) who introduced Bayesian kriging this way. He used physical process knowledge to arrive at "qualified guesses" for the first and second order moments, μ and $\mathbf{\Phi}$. On the other hand, the state of prior ignorance or non-informativity can be modelled by setting $\mu = \mathbf{0}$ and letting $\mathbf{\Phi}^{-1}$ tend to the matrix of zeroes, thus passing the "Bayesian bridge" to universal kriging, see (Omre and Halvorsen, 1989).
Now, combining (11) and (13), we arrive at the **Bayesian spatial linear model** (BSLM)

$$Y(x) = \mathbf{h}(x)^T\gamma + \varepsilon_0(x), \tag{14}$$

where

$$\mathbf{h}(x) = \begin{pmatrix} \mathbf{f}(x) \\ \mathbf{g}(x) \end{pmatrix}, \gamma = \begin{pmatrix} \beta \\ \alpha \end{pmatrix}, \mathrm{E}\gamma = \begin{pmatrix} \mu \\ 0 \end{pmatrix} =: \gamma_0, \mathrm{Cov}(\gamma) = \begin{pmatrix} \mathbf{\Phi} & 0 \\ 0 & \mathbf{A} \end{pmatrix} =: \mathbf{\Gamma}.$$

Here $\varepsilon_0(x)$ is white-noise with variance σ_0^2 and $\mathbf{A}$ denotes the covariance matrix of α, resulting after the polar spectral approximation of the random field. (Spöck and Pilz, 2010) demonstrate that Bayesian linear trend estimation in the above BSLM actually approximates Bayesian linear kriging in the original model abitrarily closely. The same is true for the total mean squared error (TMSEP) of the trend prediction and the TMSEP of Bayesian kriging.
Thus taking the TMSEP of the trend prediction in the approximating model as a substitute for the Bayes kriging TMSEP we arrive at the following classical experimental design problem for so-called I-optimality:

$$\int_{\mathbf{X}} \mathbf{h}(x_0)^T(\mathbf{H}^T(d_n)\mathbf{H}(d_n) + \sigma_0^2\mathbf{\Gamma}^{-1})^{-1}\mathbf{h}(x_0)dx_0 \to \min_{d_n}. \tag{15}$$

Here $d_n = \{x_1, x_2, \ldots, x_n\}$ collects either the design points to be added to the monitoring network or in the case of reducing the network the design points remaining in the monitoring network. $\mathbf{H}(d_n)$ expresses the dependence of the design matrix $\mathbf{H} = (\mathbf{h}(x_i)^T)_{i=1,2,\ldots,n}$ on the design points in the set d_n.
At this point we advise the reader not familiar with Bayesian experimental design theory to read the Appendix of (Spöck and Pilz, 2010). The key point in this theory is that the above so-called concrete design problem seemingly showing no mathematical structure may be expanded to a so-called continuous design problem that has the nice feature to be a convex optimization problem. Thus, the whole apparatus of convex optimization theory is available to approximately solve the above design problem for I-optimality. In particular, directional derivatives may be calculated and optimal continuous designs may be found by steepest descent algorithms. Continuous designs are just probability measures ξ on $\mathbf{X}$ and may be rounded to exact designs d_n. Defining the so-called continuous Bayesian information matrix

$$\mathbf{M}_B(\xi) = \int_{\mathbf{X}} \mathbf{h}(x)\mathbf{h}(x)^T\xi(dx) + \frac{\sigma_0^2}{n}\mathbf{\Gamma}^{-1} \tag{16}$$

and

$$\mathbf{U} = \int_{\mathbf{X}} \mathbf{h}(x_0)\mathbf{h}(x_0)^T dx_0, \tag{17}$$

it may be shown that the set of all such information matrices is convex and compact and that the extended design functional

$$\Psi(\mathbf{M}_B(\xi)) = \text{tr}(\mathbf{U}\mathbf{M}_B(\xi)^{-1}) \tag{18}$$

is convex and continuous in $\mathbf{M}_B(\xi)$. The above design functional $\Psi(.)$ thus attains its minimum at a design $\xi^* \in \Xi$, where Ξ is the set of all probability measures defined on the compact design region $\mathbf{X}$, see (Pilz, 1991). The closeness of exact designs d_n to the optimal continuous design ξ^* may be judged by means of a well-known efficiency formula, see Appendix of (Spöck and Pilz, 2010).

5. The Smith and Zhu design criterion

In real world applications the isotropic covariance function $C_\theta(t)$ is always uncertain and estimated. The kriging predictor used is then based on this estimated covariance function $C_{\hat{\theta}}(t)$. Thus, the kriging predictor is always a plug-in predictor and the reported (plug-in) kriging variance underestimates the true variance of this plug-in predictor.
(Smith and Zhu, 2004) consider spatial sampling design by means of minimizing the average of the expected lengths of $1-\alpha$ predictive intervals:

$$\int_{\mathbf{X}} \text{E}(\text{length of predictive interval at } x_0) dx_0. \tag{19}$$

Their predictors of the $\alpha/2$ and $1-\alpha/2$ quantiles of the predictive distributions are selected in such a way that the corresponding predictive intervals have coverage probability bias 0. The predictors of the mentioned quantiles are essentially the plug-in kriging predictors based on restricted maximum likelihood (REML) estimation of the covariance function plus/minus a scaled plug-in kriging standard error term that is corrected to take account of REML estimation. Based on Laplace approximation they show that this design criterion up to order $O(n^{-2})$, where n is the number of data, is equivalent to:

$$\int_{\mathbf{X}} [\sigma_\theta^2(x_0) + \text{tr}(\kappa_\theta^{-1}\{\frac{\partial\lambda_\theta(x_0)}{\partial\theta^T}\}^T \mathbf{K}_\theta \frac{\partial\lambda_\theta(x_0)}{\partial\theta^T}) + z_{1-\alpha/2}^2\{\frac{\partial\sigma_\theta(x_0)}{\partial\theta}\}^T \kappa_\theta^{-1} \frac{\partial\sigma_\theta(x_0)}{\partial\theta}] dx_0 \tag{20}$$

$$\rightarrow \underset{d_n=\{x_1,\ldots,x_n\}}{Min}$$

Here

$$\kappa_{\theta,i,j} = \text{tr}(\mathbf{W}_\theta \frac{\partial\mathbf{K}_\theta}{\partial\theta_i} \mathbf{W}_\theta \frac{\partial\mathbf{K}_\theta}{\partial\theta_j}) \tag{21}$$

is the Fisher information matrix for REML,

$$\mathbf{W}_\theta = \mathbf{K}_\theta^{-1} - \mathbf{K}_\theta^{-1}\mathbf{F}(\mathbf{F}^T\mathbf{K}_\theta^{-1}\mathbf{F})^{-1}\mathbf{F}^T\mathbf{K}_\theta^{-1}, \tag{22}$$

$z_{1-\alpha/2}$ is the $1-\alpha/2$-quantile of the standard normal distribution, $\sigma_\theta^2(x_0)$ is the universal kriging variance at x_0 and $\lambda_\theta(x_0)$ is the universal kriging weights vector for prediction at x_0.

This design criterion takes both prediction accuracy and covariance uncertainty into account. Sections 3 and 4 have demonstrated that by using the BSLM (14) as approximation to the true isotropic random field the I-optimality design criterion can be completely expressed in terms of the Bayesian information matrix

$$\mathbf{M}_B = \mathbf{H}^T\mathbf{H} + \sigma_0^2\mathbf{\Gamma}^{-1}. \tag{23}$$

Going from this information matrix to its continuous version

$$\mathbf{M}_B(\xi) = \int_{\mathbf{X}} \mathbf{h}(x)\mathbf{h}(x)^T\xi(dx) + \frac{\sigma_0^2}{n}\mathbf{\Gamma}^{-1}, \tag{24}$$

where ξ is a probability measure on the design space **X**, the extended design functional

$$\Psi(\mathbf{M}_B(\xi)) = \mathrm{tr}(\mathbf{U}\mathbf{M}_B(\xi)^{-1}) \tag{25}$$

becomes continuous and convex on the compact and convex set of all such information matrices $\mathbf{M}_B(\xi)$. This was the reason why classical convex experimental design algorithms could be used to find optimal spatial sampling designs minimizing the criterion (18).
In (Spöck et al., 2011) it is shown that also the Smith and Zhu design criterion has some favourable properties, so that classical convex experimental design theory can be applied to this design criterion, too:

- Expression (20) can be expressed completely in terms of the Bayesian information matrix $\mathbf{M}_B$.
- The design functional is continuous on the convex and compact set of all $\mathbf{M}_B(\xi)$ and has some advantageous properties according to which classical experimental design algorithms may be used in order to find spatial sampling designs.

Assuming the BSLM (14) the covariance function actually is parametrized in the diagonal matrix **A** and the nugget variance σ_0^2. Since the Smith and Zhu design criterion assumes the covariance parameters to be estimated by restricted maximum likelihood we actually estimate this diagonal matrix **A** and σ_0^2 by this methodology. The a priori covariance matrix $\mathbf{\Phi} = \mathrm{cov}(\beta)$ must be given almost infinite diagonal values because the (Smith and Zhu, 2004) approach assumes the trend parameter vector β to be estimated by generalized least squares and $\mathbf{\Phi} \to \infty$ bridges the gap from Bayesian linear to generalized least squares trend estimation. The a priori mean $\mu = \mathrm{E}(\beta)$ can be set to **0** then.
According to the polar spectral representation (10) several values in the diagonal matrix **A** are identical:

$$\mathbf{A} = \mathrm{diag}(\{d_m a_i^2\}_{m=0,\ldots,M;i=1,\ldots,n;k=1,2}), \tag{26}$$

where the definitions of d_m and a_i^2 and the indexing derive from the polar spectral representation (10). For restricted maximum likelihood estimation of **A** we have two possibilities:

- We can leave the a_i's unspecified: This approach is almost nonparametric because a lot of a_i's and corresponding frequencies w_i are needed to get the isotropic random field properly approximated and corresponds to a semiparametric estimation of the spectral distribution function via a step function.

- We can specify a parametric model for the a_i^2's: The polar spectral density function for an isotropic random field over $\Re^2$ possesing for example an exponential covariance function $B(h) = C\exp(-\frac{3h}{\alpha})$ is given by

$$g(w) = \frac{C\frac{3}{\alpha}w}{((\frac{3}{\alpha})^2 + w^2)^{3/2}}. \tag{27}$$

The polar spectral density function is defined just as the first derivative of the polar spectral distribution function $G(w)$. A possible parametrization for the a_i^2's then is

$$a_i^2 = \frac{g(w_i) + g(w_{i-1})}{2}(w_i - w_{i-1}), i = 1, 2, \ldots, n, \tag{28}$$

where $0 = w_0 < w_1 < \ldots, w_n$ are fixed frequencies.

For the optimization of the Smith and Zhu design criterion we make use of the same exchange design algorithms as described in Section 7.3 . We must only replace $\Psi(\mathbf{M}_B(\xi))$ by the Smith an Zhu design functional given in (Spöck et al., 2011).

6. Spatial sampling design for trans-Gaussian kriging

In trans-Gaussian kriging the originally positive valued data $Z(x_i), i = 1, 2, \ldots, n$ are transformed to Gaussianity by means of the Box-Cox transformation

$$g_\lambda(z) = \begin{cases} \frac{z^\lambda - 1}{\lambda} & : \lambda \neq 0 \\ \log(z) & : \lambda = 0 \end{cases}. \tag{29}$$

Let $\mathbf{Z} = (Z(x_1), Z(x_2), \ldots, Z(x_n))^T$ be the vector of original data and

$$\mathbf{Y} = (g_\lambda(Z(x_1)), g_\lambda(Z(x_2)), \ldots, g_\lambda(Z(x_n)))^T \tag{30}$$

be the vector of transformed data. The predictive density for trans-Gaussian kriging at a location x_0 then may be written:

$$\Phi(g_\lambda(z); \hat{Y}_{OK}(x_0), \sigma^2_{OK}(x_0)) * z^{\lambda-1}, \tag{31}$$

where $\Phi(.; \hat{Y}_{OK}(x_0), \sigma^2_{OK}(x_0))$ is the Gaussian density with mean the ordinary kriging predictor $\hat{Y}_{OK}(x_0)$ at x_o and based on the transformed variables $\mathbf{Y}$, and variance the ordinary kriging variance $\sigma^2_{OK}(x_0)$. $z^{\lambda-1}$ is the Jacobian of the Box-Cox transformation.
There is a nice relationship between the α-quantiles z_α of the predictive density (31) and the Gaussian density $\Phi(.; \hat{Y}_{OK}(x_0), \sigma^2_{OK}(x_0))$ which makes spatial sampling design an easy task: It can be shown that the transformed α-quantiles $g_\lambda(z_\alpha)$ are actually equivalent to the α-quantiles y_α of the Gaussian density $\Phi(.; \hat{Y}_{BK}(x_0), \sigma^2(x_0))$ that is truncated at 0 and where we only take the positive interval $[0, \infty)$ for calculating these y_α-quantiles.
For spatial sampling design we can consider again the average expected length of $1 - \alpha$-predictive intervals. Once we have calculated the Gaussian density $\Phi(.; \hat{Y}_{OK}(x_0), \sigma^2_{OK}(x_0))$, truncated it at 0, and have determined its $\alpha/2$- and $1 - \alpha/2$-quantiles $y_{\alpha/2}$ and $y_{1-\alpha/2}$ the corresponding quantiles $z_{\alpha/2}$ and $z_{1-\alpha/2}$ can be calculated easily by means of applying the inverse Box-Cox transformation to $y_{\alpha/2}$ and $y_{1-\alpha/2}$. In order to make the expected length of predictive intervals also dependent on REML-estimation of the covariance function, we can

consider instead of the Gaussian density $\Phi(.;\hat{Y}_{OK}(x_0),\sigma^2_{OK}(x_0))$ that unique Gaussian density $\hat{\Phi}$ whose 0.025- and 0.975-quantiles are given by the (Smith and Zhu, 2004) 95% predictive interval

$$\begin{aligned}&\hat{Y}_{OK}(x_0) \pm \\ &1.96 \quad \sigma_\theta(x_0)\{1+\frac{1}{2\sigma^2_\theta(x_0)}[\mathrm{tr}(\kappa_\theta^{-1}\{\frac{\partial\lambda_\theta(x_0)}{\partial\theta^T}\}^T\mathbf{K}_\theta\frac{\partial\lambda_\theta(x_0)}{\partial\theta^T})+ \\ &1.96^2 \quad \{\frac{\partial\sigma_\theta(x_0)}{\partial\theta}\}^T\kappa_\theta^{-1}\frac{\partial\sigma_\theta(x_0)}{\partial\theta}]\}. \end{aligned} \tag{32}$$

Last but not least to get expected predictive intervals we must replace in the statistic $t(\mathbf{Y}) = \hat{Y}_{OK}(x_0)$ every variable $Y(x_i)$ for which we do not have data by its ordinary kriging predictor based on the available data. Furthermore, we note that in the above approach we have not taken into account the fact that the transformation parameter λ itself is estimated too, i.e. by maximum likelihood, and then is plugged-into the ordinary kriging predictor. A future paper will take account of also this additional uncertainty.

7. The spatDesign toolbox

The spatial sampling design and geostatistics toolbox spatDesign has been developed since 2003. It can be run in both MATLAB and Octave and can be downloaded from:

- http://wwwu.uni-klu.ac.at/guspoeck/spatDesignMatlab.zip
- http://wwwu.uni-klu.ac.at/guspoeck/spatDesignOctave.zip

The toolbox underlies the GNU Public Licence Version 3 or higher and thus is freely available. In MATLAB (www.mathworks.com) the toolbox is fully functional but assumes that also the MATLAB Optimization and Statistics toolboxes are installed. In Octave functions like weighted-least-squares that make in MATLAB use of the Optimization Toolbox and especially of the function FMINCON do not work. But in near future also an Octave workaround for the function FMINCON will become available from the author. The spatial sampling design functions corresponding to the Smith and Zhu design criterion need on a standard PC a lot of computation time. For this reason this part of the toolbox has been parallelized to work with NVIDIA GPU's and the freely available MATLAB parallelization package GPUmat (www.gp-you.org). If you have a CUDA (www.nvidia.com) compatible graphics card installed then this will be automatically detected and the parallelized algorithms for the Smith and Zhu design criterion will be used. Unfortunately, almost no efforts have been undertaken to date to parallelize also Octave, so the Smith and Zhu part of the Octave toolbox will not work unless you have a lot of time to wait for the results in the unparallelized version. Wolfgang Nowak from the Institute of Hydraulic Engineering (IWS), University of Stuttgart, has added his FFT-Kriging Toolbox to the software package. This is really fast MATLAB and Octave code for Bayesian linear kriging with external drift and works in both 2- and 3-dimensional Euclidean space. The original kriging code of the spatDesign toolbox is not so fast as this code and implements only 2D interpolation. The reason is that the original code works with local kriging neighborhoods and as a consequence for every new prediction an often high dimensional covariance matrix $\mathbf{K}$ between given local observations must be inverted.

The spatDesign Toolbox provides code for all three areas of geostatistics,

- covariance estimation and variography

- spatial interpolation and kriging
- spatial sampling design and planning of monitoring networks.

But the largest emphasis of the toolbox surely is on spatial sampling design and the planning of monitoring networks.

7.1 Covariance estimation and variography software

For covariance estimation and variography the empirical semivariogram estimator of (Matheron, 1962)

$$\hat{\gamma}(h) = \frac{1}{2|N(h)|} \sum_{(i,j)\in N(h)} (Y(x_i) - Y(x_j))^2, \tag{33}$$

where

$$N(h) = \begin{cases} \{(i,j) : x_i - x_j \approx h\} : stationarity \\ \{(i,j) : ||x_i - x_j||_2 \approx h\} : isotropy \end{cases} \tag{34}$$

is implemented in both forms the isotropic one and the general stationary one. The corresponding MATLAB and Octave functions are EMPVARIOGRAM.m and EMPVARIOGRAMANISO.m.

Having calculated the empirical semivariogram a theoretical semivariogram model can be fitted to the empirical semivariogram by means of weighted-least-squares. As theoretical semivariogram model a nested model or convex combination of an exponential and a Gaussian semivariogram model with different ranges and nugget is implemented. The corresponding MATLAB and Octave functions are WEIGHTEDLEASTSQUARES.m for the isotropic case and WEIGHTEDLEASTSQUARESANISO.m for the geometrically anisotropic case.

The toolbox implements also transformed-Gaussian kriging. For this reason a maximum likelihood estimation procedure has been implemented that estimates both the Box-Cox transformation parameter λ and the eventually geometrically anisotropic covariance function at once by means of a profile likelihood approach. Iteratively any kind of parameters are fixed and the likelihood function is maximized in the rest of the parameters until convergence. This function is quite slow since the likelihood function is iteratively maximized either in λ, the covariance parameters or the anisotropy parameters. The function performing these tasks is called ESTIMATE_TRANSFO_COV_ML.m.

In the up to here mentioned functions for covariance estimation the trend of the random field is assumed to be constant. For the case that the trend is of the form

$$m(x) = \mathbf{f}(x)^T \beta \tag{35}$$

it must be eleminated before the covariance function or the semivariogram can be estimated. For this case the function CALCULATE_RESIDUALS_TREND_COVARIANCE.m has been implemented. Starting with a least squares estimate of β and the corresponding residuals iteratively empirical semivariograms of residuals, corresponding weighted least squares fits to the empirical semivariogram, generalized least squares estimates of β and again residuals and semivariograms.....are calculated until convergence.

7.2 Spatial interpolation and kriging software

As simplest interpolation routine Voronoi interpolation is implemented in the function VORONOIPOLYGONALINTERPOLATIONONGRID.m. Ungauged locations are given the same value as the closest datum location. Besides Voronoi interpolation also Bayesian linear kriging with external drift and transformed-Gaussian kriging based on the Box-Cox transformation are implemented.

The corresponding functions are named KRIGELINEARBAYESONGRID.m and TRANSGAUSSIANKRIGINGONGRID.m. The output of the first function are the Bayesian kriging predictions and corresponding TMSEPs on a grid. The output of the second function are the skew predictive distributions from transformed-Gaussian kriging on a grid. Because the complete predictive distributions are calculated several statistics may be derived from them: The function VISUALIZEPOSTQUANTILE.m allows to calculate and visualize several quantiles of the predictive distributions as well as the means, medians and modal values of these distributions. The function VISUALIZEPROBGREATER.m allows to calculate maps that give the probabilities that certain thresholds are exceeded. Finally the functions CROSSVALIDATION.m and VISUALIZECROSSVALIDATION.m calculate crossvalidation statistics like mean absolute errors, percents of actual data below the quantiles of the predictive distribution and percent actual data above threshold vs. expected percent data above threshold.

7.3 Spatial sampling design software

spatDesign implements three design criteria for Bayesian linear kriging, where the first two are criteria for prediction only with the covariance function assumed to be certain. The third criterion is the (Smith and Zhu, 2004) criterion taking also account of the fact that the covariance function is estimated. Concerning trans-Gaussian kriging also the Smith and Zhu design criterion is implemented. The actual version of the toolbox is spatDesign V.2.2.0.

The implemented criteria for prediction only are:

- I-optimality:

$$\Psi(\mathbf{M}_B(d_n)) = \mathrm{tr}(\mathbf{U}\mathbf{M}_B(d_n)^{-1}) \to \min_{d_n} \tag{36}$$

$$\mathbf{U} = \sum_{i,j=1}^{m} \mathbf{h}(x_{i,j})\mathbf{h}(x_{i,j})^T; \tag{37}$$

  where the integral in (17) has been replaced by the sum over a fine grid of locations $x_{i,j} \in \mathbf{X}$.

- D-optimality:

$$\Psi(\mathbf{M}_B(d_n)) = |(\mathbf{H}^T(d_n)\mathbf{H}(d_n) + \sigma_0^2\mathbf{\Gamma}^{-1})^{-1}| \to \min_{d_n} \tag{38}$$

The Smith and Zhu design criterion can be expressed according to Section 6 also as

$$\Psi(\mathbf{M}_B(d_n)) \to \min_{d_n}, \tag{39}$$

although because of place restrictions we cannot give this expression in detail here. The interested reader is referred to (Spöck et al., 2011).

The basic algorithm for calculating spatial sampling designs is an exchange algorithm from experimental design theory for calculating exact designs and going back to (Fedorov, 1972). Contrary to the construction of optimal discrete designs, here we cannot prove convergence of the exact designs to the functional value $\Psi(d^*)$ of an optimal exact design d^*; we can only guarantee stepwise improvement of a given exact starting design, i.e. the sequence of functional values $\Psi(d_{n,s})$ decreases monotonically with increasing iteration index s. The algorithm is an exchange algorithm improving n-point designs and starting from an initial design.

7.3.1 Exchange algorithm

Step 1. Use some initial design $d_{n,1} = \{x_{1,1}, \ldots, x_{n,1}\} \in \mathbf{X}^n$ of size n.
Step 2. Beginning with $s = 1$ form the design $d_{n+1,s} = d_{n,s} + (x_{n+1,s})$ by adding the point

$$x_{n+1,s} = \arg\min_{x \in \mathbf{X}} \Psi(\mathbf{M}_B(d_{n,s} + (x)))$$

to $d_{n,s}$.
Then form $d^j_{n,s} = d_{n+1,s} - (x_{j,s}), j = 1, 2, \ldots, n+1$ and delete that point $x_{j^*,s}$ from $d_{n+1,s}$ for which

$$\Psi(\mathbf{M}_B(d^{j^*}_{n,s})) = \min_{j \in \{1,\ldots,n+1\}} \Psi(\mathbf{M}_B(d^j_{n,s}))).$$

Step 3. Repeat Step 2 until the point to be deleted is equivalent to the point to be added.

For the design functional (18) Step 2 is determined as follows:

$$x_{n+1,s} = \arg\max_{x \in \mathbf{X}} \frac{\mathbf{h}(x)^T \mathbf{M}_B(d_{n,s})^{-1} \mathbf{U} \mathbf{M}_B(d_{n,s})^{-1} \mathbf{h}(x)}{n + \mathbf{h}(x)^T \mathbf{M}_B(d_{n,s})^{-1} \mathbf{h}(x)}$$

$$j^* = \arg\min_{1 \le j \le n+1} \frac{\mathbf{h}(x_{j,s})^T \mathbf{Q}_B(d_{n+1,s}) \mathbf{h}(x_{j,s})}{n + 1 - \mathbf{h}(x_{j,s})^T \mathbf{M}_B(d_{n+1,s})^{-1} \mathbf{h}(x_{j,s})},$$

where

$$\mathbf{Q}_B(d_{n+1,s}) = \mathbf{M}_B(d_{n+1,s})^{-1} \mathbf{U} \mathbf{M}_B(d_{n+1,s})^{-1}.$$

For the Smith and Zhu design criterion no such simplification exists and the complete design functional $\Psi(.)$ must be recalculated in every step.

7.3.2 Generation of an initial design

The initial design is a one-point design which minimizes the design functional among all designs of size $n = 1$. Note that such a design exists since the Bayesian information matrix is positive definite even for designs of size $n = 1$.

Step 1. Choose $x_1 \in \mathbf{X}$ such that
$x_1 = \arg\min_{x \in \mathbf{X}} \Psi(\mathbf{M}_B((x)))$, and set $d_1 = (x_1)$.
Step 2. Beginning with $i = 1$, find x_{i+1} such that $x_{i+1} = \arg\min_{x \in \mathbf{X}} \Psi(\mathbf{M}_B(d_i + (x)))$ and form $d_{i+1} = d_i + (x_{i+1})$. Continue with i replaced by $i+1$ until $i+1 = n$.
Step 3. If $i + 1 = n$ then stop and take
$d_{n,1} = \{x_1, \ldots, x_n\}$ as an initial design.

7.3.3 Combination of the algorithms 7.3.1 and 7.3.2

It is a good idea to combine the initial design algorithm 7.3.2 and the exchange algorithm 7.3.1 in the following way:

Step 1. Start with the initial design algorithm and find a design with one first design point.
Step 2. Having found a design with $m \geq 1$ design points apply the exchange algorithm to this design to improve it.
Step 3. Add to the design from Step 2 one further design point by means of the initial design algorithm to get $m+1$ design points.
Step 4. Go back to Step 2 and iterate Step 2 and Step 3 until you have found n desired design points.

7.3.4 Reduction of experimental designs

Often it is desired to reduce a given experimental design $d = \{x_1, x_2, \ldots, x_n\}$ to one including only $m < n$ design points from d:

Step 1. Delete that design point x_{j^*} from d for which
$x_{j^*} = \arg\min_{x_j \in v} \Psi(\mathbf{M}_B(d - (x_j)))$, and set
$d := d - (x_{j^*})$.
Step 2. Iterate Step 1 until the design d contains only m design points.

Also this algorithm may be combined with an improvement step similar to the exchange algorithm 7.3.1. In algorithm 7.3.1 merely the calculation of $x_{n+1,s}$ has to be replaced by

$$x_{n+1,s} = \arg \min_{x \in d - d_{n,s}} \Psi(\mathbf{M}_B(d_{n,s} + (x))),$$

where d is the inital design that has to be reduced. This improved algorithm has the advantage that design points once deleted can reenter the design in the exchange step.

7.3.5 Inverse of the information matrix

The calculation of exact designs requires in every step the calculation of the inverses of the information matrices $\mathbf{M}_B(d_{n,s})$ or $\mathbf{M}_B(d_{n+1,s})$. In the next Sections we will see that these information matrices can have a quite high dimension of about 3000×3000. So, how can one invert such large matrices in affordable time? There is computationally no need to make explicit use of numerical matrix inversion algorithms, when one considers the update formulas (13.26) and (13.28) in (Pilz, 1991):

$$\mathbf{M}_B(d_{n,s} + (x))^{-1} = \frac{n+1}{n}\{\mathbf{M}_B(d_{n,s})^{-1} - \frac{\mathbf{M}_B(d_{n,s})^{-1}\mathbf{h}(x)\mathbf{h}(x)^T\mathbf{M}_B(d_{n,s})^{-1}}{n + \mathbf{h}(x)^T\mathbf{M}_B(d_{n,s})^{-1}\mathbf{h}(x)}\},$$

$$\mathbf{M}_B(d^j_{n,s})^{-1} = \frac{n}{n+1}\{\mathbf{M}_B(d_{n+1,s})^{-1} + \frac{\mathbf{M}_B(d_{n+1,s})^{-1}\mathbf{h}(x_{j,s})\mathbf{h}(x_{j,s})^T\mathbf{M}_B(d_{n+1,s})^{-1}}{n+1-\mathbf{h}(x_{j,s})^T\mathbf{M}_B(d_{n+1,s})^{-1}\mathbf{h}(x_{j,s})}\}$$

Obviously only matrix- and vector multiplications are needed in these update formulae.

7.3.6 Computation

The I-optimality and D-optimality criteria have the additional advantage that the optimizations in algorithms 7.3.1 - 7.3.4 can be computationally further simplified. What is more, the averaging over the design area **X** corresponding to eq. (17) has to be done only once before actual sampling design by means of computing the matrix **U**. The Smith and Zhu design criterion and the design criterion for trans-Gaussian kriging no longer have this advantage and the averaging over the design area **X** corresponding to eq. (20) and (32) must be done whenever $\Psi(\mathbf{M}_B(d))$ is calculated. Both eq. (17) and (20) have been simplified by means of replacing the integrals by discrete sums over a fine grid of locations $x_{i,j} \in \mathbf{X}$. Also the calculation of the minima in algorithms 7.3.1 - 7.3.4 has been simplified by means of searching for the minimum over a fine grid of locations. When the minimum is found over the discrete grid we further iterate with a line search algorithm with this minimum as a starting value to the actual global minimum. For all that, the averaging for calculating the integral in the Smith and Zhu design criterion and the design criterion for trans-Gaussian kriging is computationally too intensive to be calculated on a standard PC.
Recently CUDA technology (www.nvidia.com) has been developed for NVIDIA graphic cards. This allows to make use of the parallel performance of these graphic cards and to put intensive floating point operations to these GPU's. Thus, we have invested in such a graphics card and now do the averaging operations correponding to expression (20) and (32) in parallel on a multiprocessor NVIDIA GTX 580 GPU. To this, we have installed GPUmat (www.gp-you.org) a free software for MATLAB (www.mathworks.com) that automatically can thread operations to the NVIDIA GPU. The usage of this software is quite easy. We just have to specify MATLAB objects that we want to calculate on the GPU as GPUdouble or GPUsingle. Performance is immense. Processing on the NVIDIA GTX 580 is between 100 to 200 times faster than on a standard Pentium 8 Core 3.06 Ghz CPU.

7.3.7 Basic sampling design functions

The basic spatial sampling design functions are:

- OPTIMALLY_DELETE_N_LOCATIONS_FROM_POOLDELETE.m
- OPTIMALLY_ADD_N_LOCATIONS_FROM_POOLCOMPLETE.m
- OPTIMALLY_ADD_N_LOCATIONS_FROM_POOLADD.m
- OPTIMALLY_IMPROVE_POOLDELETE_FROM_POOLCOMPLETE.m
- OPTIMALLY_IMPROVE_POOLDELETE_FROM_POOLADD.m

The names of these functions are self-explanatory: "Pooldelete" is the discrete pool of locations that are allowed to be deleted from the design. "Poolcomplete" is the complete compact area of investigation **X** of points allowed to be added to the design. "Pooladd" is the discrete pool of locations that are allowed to be added to the design. "Improve" means the exchange algorithm, where locations from "Pooldelete" may either be exchanged to locations from "Poolcomplete" or from "Pooladd" and the total number of sampling locations remains constant.

8. An example session

The purpose of this section is to demonstrate some capabilities of the spatDesign toolbox. The most important Matlab function calls related to sampling design for trans-Gaussian kriging are given. The data set considered is the so-called Gomel data set. It comprises 591 sampling

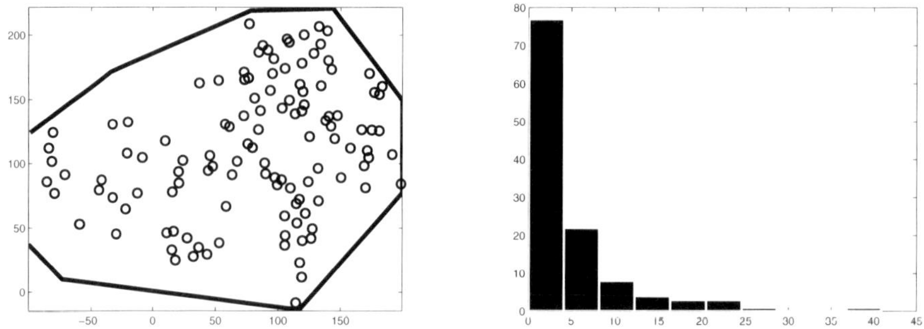

Fig. 1. Left: The 119 sampling locations of the Gomel data set. Right: The histogram of the Cs137 concentrations.

locations in the region of Gomel, Belarus, where ten years after the Chernobyl accident the concentration of Cs137 in soil has been measured. For demonstration purposes we do not use the full data set but only every fifth location, so that we have to deal with only 119 data. The following Matlab code plots the data locations and calculates a histogram of the Cs137 concentrations (see Fig. 1.).

- x=Gomel.x(1:5:end);
- y=Gomel.y(1:5:end);
- z=Gomel.zaver(1:5:end);
- figure()
- plot(x,y,'o')
- hold on
- plot(polyBoxGomeliso.x,polyBoxGomeliso.y)
- figure()
- hist(z)

In Fig. 1. we see that there are obviously areas in the design region that look very empty having no sampling locations. To the East the sampling grid is more dense than to the West. From the histogram of Fig. 1. it becomes obvious that the distribution of Cs137 is very skewed to the right. Thus, this data set is very suitable for testing the trans-Gaussian kriging spatial sampling design routines of the spatDesign toolbox. In Fig. 2. a Voronoi polygonal interpolation of the Cs137 concentrations is given. There are obviously 3 regions having very high Cs137 concentration, but the rest of the Gomel region has average to low concentrations. The 3 regions having high concentrations are regions that had heavy rainfall during the Chernobyl accident so that the radioactive cloud was washed out into the soil there.

- grid=generategrid2(-100,200,-50,250,150,150,polyBoxGomeliso);
- prediction=voronoipolygonalinterpolationongrid(x,y,z,grid);
- a=reshape(prediction.prediction,150,150);

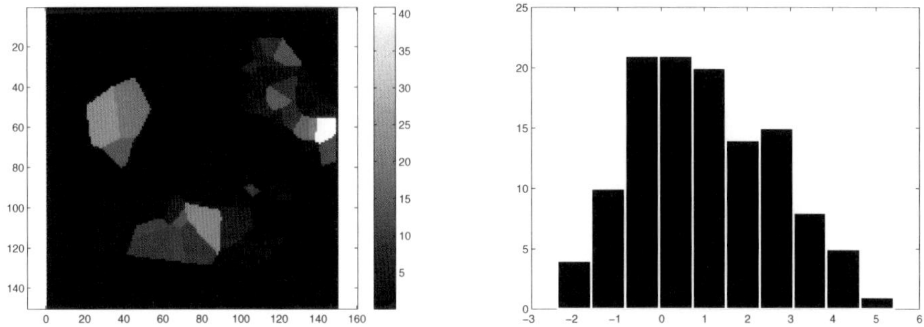

Fig. 2. Left: Voronoi interpolation of the Cs137 values. Right: Histogram of the Box-Cox transformed Cs137 concentrations.

- figure()
- imagesc(a(end:-1:1,:))
- colorbar

Next, let us estimate the Box-Cox transformation parameter lambda0 and an exponential isotropic covariance function for the corresponding Gaussian random field. delta0 comprises the parameters of this estimated exponential covariance function.

- options=optimset('LargeScale','on','MaxFunEvals',200,'TolFun',1e-3,'MaxIter',100);
- [lambda0,delta0]=estimate_transfo_cov_ml(x,y,z,0,[1,0;0,1],[0,2.4,80,0,0],0:15:150,10, -20,20,0.01,1,[1,1,1,0,0],0,0.001,0.001,options);

The estimated parameters are lambda0=0.1919 and delta0=[0.1644, 3.3601, 149.3302]=[nugget, sill, range]. To check whether lambda0 truely transforms the Cs137 concentrations to Gaussianity we calculate the histogram of the transformed data (Fig. 2.). . As a next step

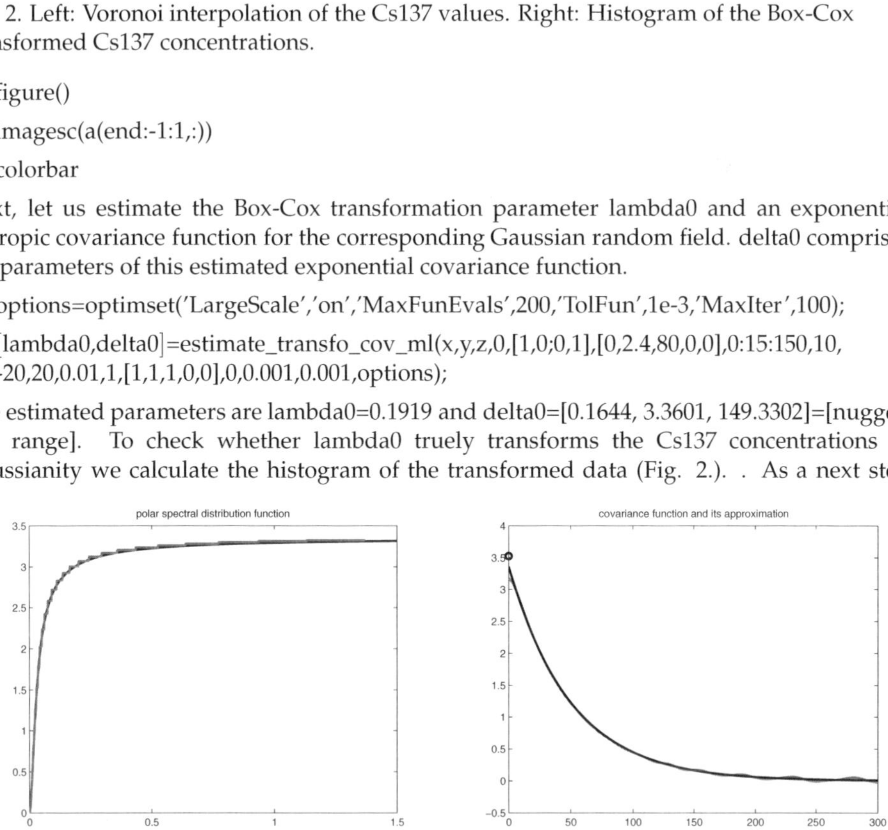

Fig. 3. Left: Polar spectral distribution function and its approximation (grey). Right: True covariance function (black) and its worst approximation (grey).

we calculate the polar spectral distribution function (Fig. 3.).

- plotspectraldist(0:0.01:1.5,delta0);

Obviously this spectral distribution function almost attains its maximum of 3.36 at $w_0 = 1.5$. We now select the frequencies $w_i, i = 1, 2, \ldots, n$, calculate an approximation to the spectral distribution function via a step function (the steps are the a_i^2) and check whether with this selected approximation to the spectral distribution function the original covariance function becomes well approximated (Fig. 3.).

- load wscaled
- w=wscaled*1.5;
- [w,deltastep]=expstep(w,delta0(3),delta0(2));
- approxspectraldist(w,deltastep)
- plotcovarianceapprox(w,45,deltastep,delta0,-100:3:200,250,-100,-100,200,-50,250);

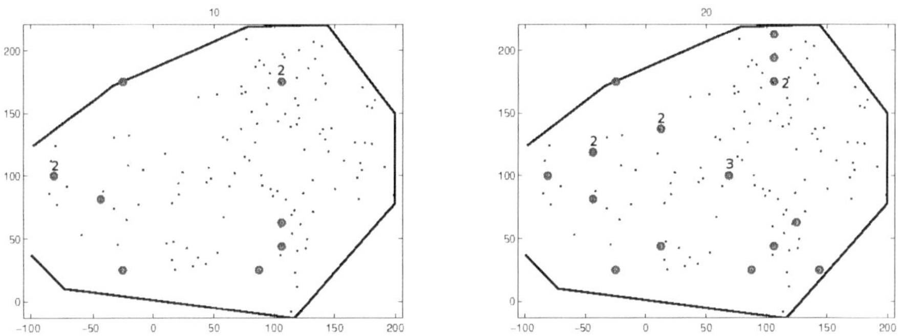

Fig. 4. Left: 10 locations added to design (circles) and their multiplicities. Right: 20 locations added to design (circles) and their multiplicities.

A look at the approximating covariance function in Fig. 3. shows that at the origin the difference between the true covariance function and the approximating covariance function is 0.17. This is small scale variation that the approximating covariance function does not take into account. Later in spatial sampling design we will add this value of 0.17 to the nugget effect 0.1644 of the true covariance function. Thus, 0.17+0.1644 is the variance of the uncorrelated error process $\epsilon_0(x.)$ in our BSLM (14).
We now have everything that we need in order to do spatial sampling design for trans-Gaussian kriging. We consider to add 20 additional sampling locations from the complete design region **X** to the available grid of 119 sampling locations (see Fig. 4.).

- [xadd,yadd,zadd,avglengthpredint]=optimally_add_n_locations_from_poolcomplete(
  {}, % no external drift
  [], % no need to specify the matrix **U**
  x,y,z, % the Gomel data
  0.17+delta0(1), % the variance of the uncorrelated error process $\epsilon_0(x)$
  $[1000000, 0, 0; 0, 0.0000001, 0; 0, 0, 0.0000001]$, % the a priori variance of the constant trend must be given almost infinite variance, no linear drift
  w, % the frequencies of the Bessel harmonics
  45, % the largest angular frequency
  deltastep, % the a_i^2

```
delta0, % the parameters of the exponential covariance function
lambda0, % the Box-Cox transformation parameter
20, % we want to add 20 samples
-100,200,-50,250, % the size of the design region
100, % maximally iterate 100 times in the exchangement step
polyBoxGomeliso, % the polygonal design region
17,17, % discretization in x an y when considering new samples
'z' % we apply the (Smith and Zhu, 2004) design criterion to trans-Gaussian kriging
);
```

Fig. 5. shows the expected lenghts of the 95% predictive intervals for the optimal 10 points and 20 points design, respectively. Fig. 6. plots the decrease of the average of the expected lenghts of the 95% predictive intervals when adding up to 20 design locations. The calculation of

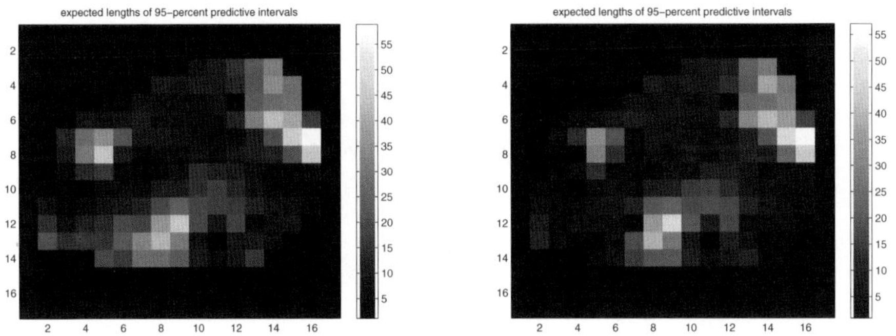

Fig. 5. Left: Expected lengths of 95% predictive intervals. 10 locations added to design. Right: Expected lengths of 95% predictive intervals. 20 locations added to design.

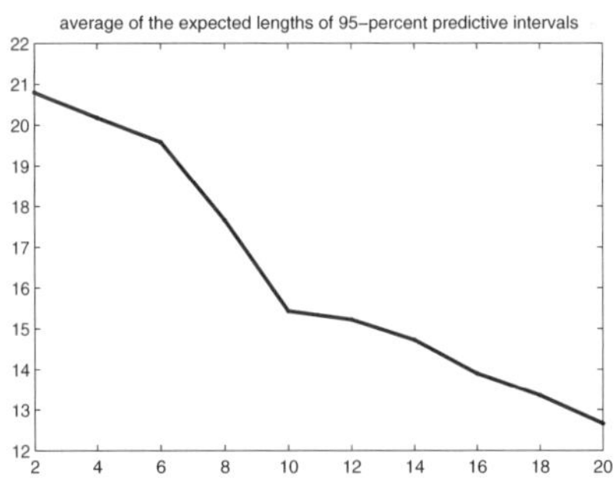

Fig. 6. Average of the expected lengths of 95% predictive intervals.

the optimal 20 point design takes on an Intel i7 8 Core CPU and a NVIDIA 580 GPU with 1.6 Gb RAM 2 weeks. Similar calculations for the simpler design functional (36) take without NVIDIA Cuda support 2 days. From Fig. 4. it is obvious that the first design locations are added to the margin of areas that look empty. In order to get also the covariance function (especially close to its origin) well estimated design locations are selected close to existing locations or they are selected with multiplicities greater than 1. There is a fundamental difference between designing for Gaussian random fields and our designs for trans-Gaussian kriging. Contrary to the designs for Gaussian random fields designs for trans-Gaussian kriging are dependent also on the data by means of the ordinary kriging predictor in eq. (32).

9. Conclusions

Our approach to spatial sampling design is motivated by the fact that up to date most sampling design implementations use stochastic search algorithms like simulated annealing. The approach to spatial sampling design proposed in this Chapter makes use of the mathematical structure of the investigated design functionals. Based on the polar spectral approximation of the isotropic random field we are able to use classical experimental design theory for the calculation of spatial sampling designs. One only missing link for the Smith and Zhu design criterion and for the design criterion of trans-Gaussian kriging is the fact that up to date we were not able to show convexity properties of the corresponding design functionals. This is surely a topic for future resarch. The data example giving radioactivity measurements in soil shows that trans-Gaussian kriging and sampling design for it are relevant also to soil science.

A topic for future research is to extend the proposed approach also to non-stationary random fields. Some work towards this direction has already been done in (Spöck, 2008). The approach discussed therein locally maintains isotropy but globally is non-stationary and is also based on the approximation of a so-called locally isotropic random field by means of a cosine-sine-Bessel surface-harmonics regression model with random amplitudes.

10. References

Aarts, E.H. and Korst, J. (1989). *Simulated Annealing and Boltzman Machines*, Wiley, New York, 284 p..

Baume, O.P., Gebhardt, A., Gebhardt, C., Heuvelink, G.B.M., Pilz, J. (2011). Network optimization algorithms and scenarios in the context of automatic mapping, *Computers & Geosciences*, 37, (3), pp. 289-294.

Brown, P.J., Le, N.D. and Zidek, J.V. (1994). Multivariate spatial interpolation and exposure to air pollutants, *The Canadian Journal of Statistics*, 2, pp. 489-509.

Brus, D.J. and de Gruijter, J.J. (1997). Random sampling or geostatistical modeling? Chosing between design-based and model-based sampling strategies for soil (with discussion), *Geoderma*, 80, pp. 1-44.

Brus, D.J. and Heuvelink, G.B.M. (2007). Optimization of sample patterns for universal kriging of environmental variables, *Geoderma*, 138, pp. 86-95.

Caeiro, S., Painho, M., Goovaerts, P., Costa, H. and Sousa, S. (2003). Spatial sampling design for sediment quality assessment in estuaries, *Environmental Modelling and Software*, 18, pp. 853-859.

Chen, V.C.P., Tsui, K.L., Barton, R.R. and Meckesheimer, M. (2006). A review on design, modeling and applications of computer experiments, *IIE Transactions*, 38, pp. 273-291.

Cox Jr., L.A. (1999). Adaptive spatial sampling of contaminated soils, *Risk Analysis*, 19, pp. 1059-1069.

Delmelle E.M. and Goovaerts, P. (2009). Second-phase sampling designs for non-stationary spatial variables, *Geoderma*, doi:10.1016/j.geoderma.2009.08.007.

Diggle, P. and Lophaven, S. (2006). Bayesian Geostatistical Design, *Scandinavian Journal of Statistics*, 33, pp. 53-64.
Dobbie, M.J., Henderson, B.L. and Stevens Jr., D.L. (2008). Sparse sampling: Spatial design for monitoring stream networks, *Statistics Surveys*, 2, pp. 113-153.
Fedorov, V.V. (1972). *Theory of optimal experiments*, transl. and ed. by W.J. Studden and E.M. Klimko, Academic Press, New York, (Russian Original: Nauka, Moscow 1971), 292 p..
Fedorov, V.V. and Flanagan, D. (1997). Optimal monitoring network design based on Mercer's expansion of the covariance kernel, *Journal of Combinatorics, Information and System Sciences*, 23, pp. 237-250.
Fuentes, M., Chaudhuri, A. and Holland, D.M. (2007). Bayesian entropy for spatial sampling design of environmental data, *Journal of Environmental and Ecological Statistics*, 14, pp. 323-340.
Gebhardt, C. (2003). *Bayesian Methods for Geostatistical Design*, Ph.D. Thesis, University of Klagenfurt, Austria, 139 p..
Gramacy, R.B. (2007). tgp: An R package for Bayesian nonstationary semiparametric nonlinear regression and design by treed Gaussian Process models, *Journal of Statistical Software*, 19, Issue 9., 47 p..
Gramacy, R.B. and Lee, H.K.H. (2010). Optimization under unknown constraints, *Bayesian Statistics* 9, (ed.) Bernardo, JM. et al., Oxford University Press, 18 p..
Groenigen, J.W. van, Siderius, W. and Stein, A. (1999). Constrained optimisation of soil sampling for minimisation of the kriging variance, *Geoderma*, 87, pp. 239-259.
Hengl, T., Rossiter, D.G. and Stein, A. (2003). Soil sampling strategies for spatial prediction by correlation with auxiliary maps, *Australian Journal of Soil Research*, 41, pp. 1403-1422.
Kleijnen, J.P.C. and Beers, W.C.M. van (2004). Application driven sequential design for simulation experiments: Kriging metamodeling, *Journal of the Operational Research Society*, 55, pp. 876-893.
Lark. R.M. (2002). Optimized spatial sampling of soil for estimation of the variogram by maximum likelihood, *Geoderma*, 105, pp. 49-80.
Le, N.D. and Zidek, J.V. (2009). *Statistical Analysis of Environmental Space-Time Processes*, Springer, New York, 356 p..
Lim, Y.B., Sacks, J. and Studden, W.J. (2002). Design and analysis of computer experiments when the output is highly correlated over the input space, *Canadian Journal of Statistics*, 30, pp. 109-126.
Matheron, G. (1962). *Traite de geostatistique appliquee*, Tome I. Memoires du bureau de recherches geologiques et minieres principles, Paris, Editions Technip, 14, 333 p..
Mitchell, T.J. and Morris, M.D. (1992). Bayesian design and analysis of computer experiments: Two examples, *Statistica Sinica*, 2, pp. 359-379.
Morris, M.D., Mitchell, T.J. and Ylvisaker, D. (1993). Bayesian design and analysis of computer experiments: Use of derivatives in surface prediction, *Technometrics*, 35, pp. 243-255.
Müller, P., Sanso, B. and De Iorio, M. (2004). Optimal Bayesian Design by Inhomogeneous Markov Chain Simulation, *Journal of the American Statistical Association*, 99, pp. 788-798.
Müller, W.G. and Pazman, A. (1998). Design measures and approximate information matrices for experiments without replications, *Journal of Statistical Planning and Inference 71*, pp. 349-362.

Müller, W.G and Pazman, A. (1999). An algorithm for the computation of optimum designs under a given covariance structure, *Journal of Computational Statistics*, 14, pp. 197-211.

Müller, W.G. (2005). A comparison of spatial design methods for correlated observations, *Environmetrics*, 16, pp. 495-505.

Müller, W.G. and Zimmerman, D.L. (1999). Optimal designs for variogram estimation. *Environmetrics*, 10, pp. 23-37.

Omre, H. (1987). Bayesian kriging - merging observations and qualified guess in kriging, *Mathematical Geology*, 19, pp. 25-39.

Omre, H. and Halvorsen, K. (1989). The Bayesian bridge between simple and universal kriging, *Mathematical Geology*, 21, pp. 767-786.

Pazman, A. and Müller, W.G. (2001). Optimal design of experiments subject to correlated errors, *Statistics and Probability Letters*, 52, pp. 29-34.

Pilz, J. (1991). *Bayesian Estimation and Experimental Design in Linear Regression Models*, Wiley, New York, 306 p..

Schonlau, M. (1997). *Computer experiments and global optimization*, Dissertation, University of Waterloo, Canada 130 p..

Smith, R.L. and Zhu, Z. (2004). Asymptotic theory for kriging with estimated parameters and its application to network design, www.stat.unc.edu/postscript/rs/supp5.pdf, 21 p..

Spöck, G. (2008). Non-stationary spatial modeling using harmonic analysis, in J.M. Ortiz and X. Emery (ed.), *Proceedings of the Eight International Geostatistics Congress*, Gecamin, Chile, pp. 389-398.

Spöck, G. and Pilz, J. (2010). Spatial sampling design and covariance-robust minimax prediction based on convex design ideas, *Stochastic Environmental Research and Risk Assessment*,24, pp. 463-482.

Spöck, G., Zhu, Z. and Pilz, J. (2011). Simplifying objective functions and avoiding stochastic search algorithms in spatial sampling design, accepted by *Stochastic Environmental Research and Risk Assessment*, 20 p.

Trujillo-Ventura, A. and Ellis, J.H. (1991). Multiobjective air pollution monitoring network design, *Atmospheric Environment*, 25, pp. 469-479.

US-EPA (2002). *Guidance on choosing a sampling design for environmental data collection*, EPA QA/G-5S, United States Environmental Protection Agency, 70 p..

Yaglom, A.M. (1987). *Correlation Theory of Stationary and Related Random Functions I*, Springer-Verlag, New York, 526 p..

Zhu, Z. and Stein, M.L. (2006). Spatial sampling design for prediction with estimated parameters, *Journal of Agricultural, Biological and Environmental Statistics*, 11, pp. 24-49.

Zimmerman, D.L. and Homer, K.E. (1991). A network design criterion for estimating selected attributes of the semivariogram, *Environmetrics*, 2(4), pp. 425-441.

15

Statistical Methods for the Analysis of Soil Spatial and Temporal Variability

Ahmed Douaik[1], Marc van Meirvenne[2] and Tibor Tóth[3,4]
[1]*Research Unit on Environment and Conservation of Natural Resources, National Institute of Agricultural Research (INRA), Rabat,*
[2]*Department of Soil Management, Ghent University, Ghent,*
[3]*Department of Soil Science, Research Institute for Soil Science and Agricultural Chemistry, Budapest,*
[4]*Rural, Water and Ecosystem Resources Unit, Institute for Environment and Sustainability, European Commission – Joint Research Center, Ispra*
[1]*Morocco*
[2]*Belgium*
[3]*Hungary*
[4]*Italy*

1. Introduction

Soils vary in space and time due to the combined effects of physical, chemical, and biological processes that operate at different scales and with different intensities. This variability is caused by soil forming factors, i.e., climate, parent material, time, topography, and vegetation (Jenny, 1980), farming practices (fertilization, irrigation, tillage, etc.), erosion, etc. Each cause may operate independently or in combination with other factors and over a wide range of spatial and/or temporal scales. Spatial scales can manifest at microscale to watershed and more, while temporal scale can span from seconds to decades and longer. These processes and causes create a pattern of nested variability or soil heterogeneity. This scale-dependency means that soil properties may display spatial and/or temporal patterns only over certain distances and not over others.

Characterizing spatial variability of soil properties and crop parameters by inventorying them is needed to evaluate the effectiveness of their management. Also, characterizing their temporal variability by monitoring them is important in order to compare different management systems for sustainability and environmental quality. In addition, an understanding of the spatial and/or temporal variability can provide a framework for developing effective sampling schemes for future site management and efficient experimental designs for research approaches.

Due to the spatial and/or temporal variability of soil properties, numerous samples need to be taken and the measurements need to be repeated as conditions change or to determine if they are changing. Soil properties are traditionally determined in laboratory from samples taken from field. However, this approach, although more precise, is more time consuming and expensive. Alternatively, soil properties can be easily, quickly, cheaply, and more intensively measured, even indirectly, using terrain attributes derived from a DEM and using emerging soil-landscape mapping technologies like GPS, geophysical techniques (EMI, GPR, TDR, etc.), soil spectroscopy and remote sensors, etc. Even with these auxiliary or proxy data sources, a full inventory of soil resources and properties is non-realistic and impossible. Therefore, soil can be measured at only limited space locations and time instants. Soil properties at non-sampled locations and/or times should be predicted. These predictions and related uncertainties can be obtained by using statistical models for soil spatial and/or temporal variability. In this way, soil variability can be considered as the combination of a systematic (deterministic) and a random (stochastic) components. The former represents the gradual or distinct changes (trends) while the latter involves differences not related to a known cause and those not discerned by the nature or the scale of investigation. The choice of the adequate model depends on the behaviour of the phenomenon of interest (a priori knowledge), the objective of the analysis, the nature of the available data, and, from a practical point of view, the existence of good statistical methods and adequate software packages.

Excellent review papers on modelling spatial and temporal variability were published a decade ago (Heuvelink & Webster, 2001; Kyriakidis & Journel, 1999). The latter review focused only on geostatistical methods for modelling particularly the joint space-time variability while the former discussed methods for modelling purely spatial, purely temporal, and joint space-time variability.

Our chapter is basically a review of statistical methods that have been used in order to describe and model the spatial and temporal variability of soil resources and properties. Once this variability is characterized and modelled, it can be advantageously used for different practical applications like optimal sampling of soil resources, efficient designing of experiments, interpolation and mapping of soil properties, etc. Therefore, in this chapter, we will present the main statistical methods with a short description of each of them, discuss eventually the possible links between some of them, identify their assumptions, merits and limitations, stress some of their practical applications, and list some relevant bibliographic references. The chapter fills the gap from the Heuvelink & Webster (2001) paper by considering more statistical methods like temporal stability (Douaik et al, 2006; Vachaud et al, 1985) and dynamic spatial variation (Douaik et al, 2007; Lesch et al, 1998), updates the presentation of the methods from the Kyriakidis & Journel (1999) paper by taking into account the research work done, and especially the last developments in pedometrics, during the past decade and adds some examples of practical applications like mapping using the Bayesian maximum entropy (Christakos, 2000; Douaik et al, 2005; 2009).

After presenting some classical statistical methods (Section 2), we provide an overview of the geostatistical methods including structural analysis (Section 3) and prediction (Section 4). Then, we present the Bayesian maximum entropy, another method of prediction (Section 5). Finally, we discuss the different methods and conclude with some suggestions for choosing among them (Section 6).

2. Classical statistical methods

Classical statistical methods are basically aspatial and atemporal methods where the spatial and temporal coordinates are ignored. The different approaches are: coefficient of variation (2.1), variance component analysis (2.2), temporal stability (2.3), and dynamic spatial variation (2.4).

2.1 Coefficient of variation

It is assumed that the observed value for a soil attribute (z) at any location ($\mathbf{s}$) at a given time instant (t) is:

$$E\left[z(s,t)\right] = \mu + \varepsilon(s,t) \quad (1)$$

with μ the population mean and $\varepsilon(\mathbf{s},t)$ random, spatially and temporally uncorrelated error assumed normally distributed with zero mean and variance σ^2.

The population mean is estimated from a sample:

$$\hat{\mu} = \bar{z} = \frac{1}{n}\sum_{i=1}^{n} z_i \quad (2)$$

with n: the sample size and the variance, σ^2, is estimated by:

$$\hat{\sigma}^2 = \frac{1}{n-1}\sum_{i=1}^{n}\left(z_i - \bar{z}\right)^2 \quad (3)$$

Its root square, $\hat{\sigma}$, is the standard deviation. The coefficient of variation is defined as:

$$CV = 100 * \frac{\hat{\sigma}}{\bar{z}} \quad (4)$$

The coefficient of variation can be calculated for the temporal probability distribution function (PDF) at each space location as well as for the spatial PDF at each time instant. As an example, Van Wesenbeeck & Kachanoski (1988) studied the spatial distribution (from 100 locations) of the surface soil water content throughout the growing season (65 measurement dates). They concluded that there exists a temporal dependence of the spatial variability and a spatial dependence of the temporal variability. They also suggested that random sampling of the locations should be avoided. James et al. (2003) defined the temporal variability of soil moisture content as the coefficient of variation at each site across the whole year (52 time instants) and used an analysis of variance (ANOVA) to test for differences between habitats (three sites) in the temporal variability of soil moisture content. They found that soil moisture varied significantly with time but not between habitats while the interaction between time and habitat was significant which reflects differences between habitats in temporal pattern of soil moisture. The coefficients of variation were significantly different between habitats. Other examples of the use of the coefficient of variation for the assessment of the spatial and temporal variabilities can be found in Ehrenfeld et al. (1997) and Guo et al. (2002).

The standard deviation and, hence, the coefficient of variation does not incorporate spatial and/or temporal information and therefore doesn't provide a representation of the nature of the spatial and/or temporal behaviour of a given soil property. It is considered as an indicator of global variability as opposed to local variability (Carter & Pearen, 1985), the latter is identified using correlogram, covariance function, or variogram which will be presented later in this chapter (sections 3.3 to 3.6).

2.2 Variance component analysis

In the variance component analysis approach, the spatial and temporal contributions are considered as factors with random effects in the ANOVA model. In this way the different components of the total variance can be determined. Campbell et al. (1989) adopted the following model, for the log concentrations of a given solute (log c) in subsample l (3) from sampling area k (2) at site j (2) and at time i (6 dates over 20 days):

$$\log c_{ijkl} = \mu + T_i + S_j + A_k + (A.U)_{kl} + \varepsilon_{ijkl} \tag{5}$$

with μ: the grand mean; T: the time effect; S: the site effect; A: the sampling area effect; $(A.U)$: the subsampling within sampling area effect, and ε: an error term. The temporal and spatial variances were compared to the sampling area variance while the latter was compared to the subsampling within sampling area variance. They found that the temporal variation was far less significant than the variation between the two sites. Also they found that the variation between areas a few meters apart was greater than that between the two sites even though the texture and the management of the sites were quiet different. Other examples of the evaluation of the temporal and spatial variabilities based on the concept of variance component analysis can be found in Van Es (1993) and Van Es et al. (1999).
The variance component analysis is directly related to the coefficient of variation presented in the precedent section. The variance component estimates for each variability source are computed, and then the square root of these estimates is divided by the mean value. The output is exactly the definition of the coefficient of variation. Thus, the variance component analysis is suffering from the same drawbacks as the coefficient of variation approach.

2.3 Temporal stability

The concept of temporal stability was first introduced by Vachaud et al. (1985). It is defined as the time invariant association between spatial location and statistical parameters of soil properties. They distinguished mainly two approaches: relative differences and Spearman rank correlation.

2.3.1 Relative differences

Let z_{ij} be the observed value of a soil property at location i (i=1, …, n) and time j (j=1, …, m). The relative differences δ_{ij} are defined as:

$$\delta_{ij} = \frac{z_{ij} - \bar{z}_j}{\bar{z}_j} \tag{6}$$

with $\bar{z}_j$: the spatial average for the time instant j. The temporal average of δ_{ij} is:

$$\bar{\delta}_i = \frac{1}{m}\sum_{j=1}^{m} \delta_{ij} \tag{7}$$

and its corresponding temporal standard deviation is defined as:

$$\sigma(\bar{\delta}_i) = \sqrt{\frac{1}{m-1}\sum_{j=1}^{m} (\delta_{ij} - \bar{\delta}_i)^2} \tag{8}$$

A zero value for $\bar{\delta}_i$ indicates that the temporal average $\bar{z}_j$ represents the average value over the whole study area at any time. The field average value is overestimated if $\bar{\delta}_i > 0$ while it is underestimated when $\bar{\delta}_i < 0$. A more time stable location will be indicated by a small value of $\sigma(\bar{\delta}i)$ whereas a high value of the latter is an indication of a less time stable location. The $\bar{\delta}_i$ values can be plotted graphically against their rank with the corresponding temporal standard deviations.

Based on the relative differences, Kachanoski & De Jong (1988) showed that a useful test for temporal stability is the Pearson correlation coefficient between soil property values measured at two consecutive time instants. They also showed that temporal stability exists if the relative differences remain constant between two time instants: $\delta_{i2} = \delta_{i1}$. This equation implies that:

$$z_{i2} = \frac{\bar{z}_2}{\bar{z}_1} z_{i1} \qquad (9)$$

$\bar{z}_1$ and $\bar{z}_2$ being spatial averages for the time instants j = 1 and 2, respectively. This equation establishes a linear relationship between the soil property at two different times with an intercept equal to zero and a slope equal to the ratio of the mean value observed at the second time frame to the mean value observed at the first time frame. Consequently, a good test for the temporal stability is the correlation between soil property values measured at consecutive time frames (Kachanoski & De Jong, 1988).

Furthermore, the former equation implies that the regression between relative differences from two consecutive time frames should have a zero intercept and a unity slope. This is another way to check the existence of temporal stability of the spatial pattern of a variable. However, the regression between z_{i2} and z_{i1} is, in general, of the following form:

$$z_{i2} = a * z_{i1} + b \qquad (10)$$

Using equations (9) and (10) we expect that, if temporal stability exists, the following conditions are verified:

$$a = \frac{\bar{z}_2}{\bar{z}_1} \text{ and } b = 0 \qquad (11)$$

So, another way to check the temporal stability is to test for a zero intercept and a unity slope of the regression between two consecutive time instants.

If there is a constant increase or decrease in the soil property, w, in all the locations between two time frames, we expect from Eq. (10) that (Kachanoski & De Jong, 1988):

$$a = 1 \text{ and } b = w \qquad (12)$$

Equation (12) indicates that if the slope of the regression between values at two consecutive time instants is not significantly different from one, then the regression intercept represents the constant change that has occurred between the two time instants.

Based on whether the intercept (b) is significantly different from 0, the slope (a) is significantly different from 1, or both, four possible scenarios can be distinguished (Douaik et al, 2007; Grant et al., 2004):

- *Scenario* 1 ($a = 1$ *and* $b = 0$): there is no net change between the two sampling dates
- *Scenario* 2 ($a \neq 1$ *and* $b = 0$): the mean soil property level has changed and there is a static (uniform) change in the soil property spatial pattern
- *Scenario* 3 ($a = 1$ *and* $b \neq 0$): the mean soil property level has not changed and the change in the soil property spatial pattern is dynamic (non-uniform) but, when averaged, this change is not significant

- *Scenario* 4 ($a \neq 1$ *and* $b \neq 0$): the mean soil property level has changed and the change in the soil property spatial pattern is dynamic (non-uniform).

These refinements of the concept of temporal stability were used by, among others, Da Silva et al. (2001), Douaik (2005), Douaik et al (2006; 2007), Ehrenfeldt et al. (1997), Goovaerts & Chiang (1993), Petrone et al. (2004), Van Wesenbeeck et al. (1988).

The concept of relative differences is based on the same statistical parameters (mean and standard deviation) as the two precedent approaches. However, it uses the standardized data (relative differences) for each time instant instead of the original data and the mean and standard deviation are computed for each space location. This concept is based, like the two precedent approaches, on the Gaussian distribution of data and the absence of autocorrelation. Both assumptions must be checked before the use of relative differences in the study of temporal stability. The normality can be checked using the standard statistical tests and the independence can be verified using either a test for autocorrelation like the Moran test or by computing the variogram and determining its range. Normality can be gained via data transformation and independence by the selection of sample locations separated by distances larger than the autocorrelation range. For non normally distributed data, but still non autocorrelated, the Spearman rank correlation can be used. The latter is presented in the next section.

2.3.2 Spearman rank correlation

It refers to the tendency of a soil property, measured at different locations in space, to maintain their relative ranking over time. It is defined as:

$$r_s = 1 - \frac{6\sum_{i=1}^{n}\left(R_{ij} - R_{ik}\right)^2}{n\left(n^2 - 1\right)} \tag{13}$$

with R_{ij} and R_{ik} the ranks of z_{ij} observed at location i on time instants j and k, respectively. A value of this coefficient equal to one indicates a perfect temporal stability between time instants j and k, and thus identity of ranks for any location whereas a lack of temporal stability implies that $r_s = 0$. Many researchers used this concept to evaluate the temporal stability of different soil properties, among them, Campbell et al. (2001), Douaik (2005), Farley & Fitter (1999), Guo et al. (2002), Heathman et al. (2009), Hu et al (2011), Reichardt et al. (1993), Si (2002), and Wendroth et al. (1999).

The Spearman rank correlation coefficient is a non parametric statistics therefore it can be used even for non normally distributed data. However, in the case of autocorrelation, its use is not recommended and it will be cautious to use statistical methods which consider the spatial and temporal nature of the given soil property. In addition, even for independent data, the Spearman rank correlation inform us only about the relative ranking of space locations between two time instants and do not provide a quantification of the temporal change. This quantification can be obtained using either the paired-*t* test for comparing mean values or the dynamic spatial variation approach. The former is presented in what follows while the latter is discussed in section 2.5.

2.4 The Paired-*t* test

Equation (9) just indicates if there is a constant change or not between two time instants, but we still need to check if the mean values of a soil property between these two times are significantly different or not. To answer this question, the paired-t test (McClave & Sincich, 2006) is used. The test statistic t is computed as:

$$t = \frac{\bar{d}}{s_d / \sqrt{n}} \tag{14}$$

with $\bar{d}$ the mean of the differences, d_i, between the observations for the second and the first time frames:

$$\bar{d} = \frac{1}{n}\sum_{i=1}^{n} d_i = \frac{1}{n}\sum_{i=1}^{n}(z_{i2} - z_{i1}) \tag{15}$$

s_d is the standard deviation of the differences:

$$s_d = \sqrt{\frac{1}{n-1}\sum_{i=1}^{n}(d_i - \bar{d})^2} \tag{16}$$

and n the number of the differences, i.e. the number of locations.
The test statistic t (equation 14) has (n-1) degrees of freedom. The calculated statistic is compared to a tabulated t value with the same degrees of freedom and a probability of error of type I (taken generally equal to 5%). Equivalently, the corresponding probability to the test statistic t is compared to 5% to check if the difference of mean values is significant or not. As an example of application of the paired-t test, Kenny et al. (2002) checked the temporal trend of the mean thickness of the *Ap* horizon.

2.5 Dynamic Spatial variation

The concept of temporal stability handles each soil property separately and, as described before (section 2.3), it can quantify the temporal change even without testing if this change is significant or not. Lesch et al. (1998) discussed another approach that can test for change in the mean value as well as for a dynamic spatial variation between two time instants. Dynamic spatial variation refers to the spatially variable change of a soil property between two time instants, i.e., the change is not the same between two time instants for each location.
To test for the existence of dynamic spatial variation, Lesch et al. (1998) identified different steps. First of all, the pairs of soil data obtained at a first time instant (y and z), from n space locations, are used to establish a calibration equation based on the regression model:

$$z = b_1 * y + b_0 \tag{17}$$

The parameters of this regression equation (b_0 and b_1) are estimated using the ordinary least squares method. In a second step, the residuals of the model are tested for spatial independence, normality, homoscedasticity and the linearity of the relationship. The three last assumptions can be verified using classical regression tools. The assumption of spatial independence can be checked using the Moran test (Lesch et al., 1995). In the next step, a mixed ANOVA model is set:

$$z_{ij} = \mu + t_i + b_j + \varepsilon_{ij} \tag{18}$$

i=1, 2 are the two time instants and j=1, … , n are the locations. This model is equivalent to:

$$z_{2j} - z_{1j} = d_j = d_0 + \eta_j + \left(\varepsilon_{2j} - \varepsilon_{1j}\right) \tag{19}$$

with $d_0 = t_2 - t_1$ is the difference between mean values for the second and first time instants and η_j independently and identically normally distributed with zero mean and variance equal to $\sigma_\theta^{\ 2} : \eta_j \approx N\left(0, \theta^2\right)$. If z_{1j} represents the observed values of the first soil property at the first time instant and at locations j (j=1, …, n) and z_{2k} represents the same property observed at the second time instant at the locations k (k=1, …, m), the difference between the two values is given by Eq. (19) with d_0 representing the shift in the average value between the two time instants and η representing the dynamic spatial variation. These two quantities can be statistically tested: the first if it is different from zero and the variance of the second if it is higher than zero.

Based on whether the first parameter (the mean difference, d_0) is significantly different from zero, the second parameter (spatial shift, θ^2) is significantly higher than zero, or both, four possible scenarios can be distinguished (Douaik et al., 2007):

- *Scenario* 1 ($d_0 = 0$ *and* $\theta^2 = 0$): there is no net change with time in the soil property
- *Scenario* 2 ($d_0 \neq 0$ *and* $\theta^2 = 0$): the mean soil property level has changed and there is a static (uniform) change in the soil property spatial pattern
- *Scenario* 3 ($d_0 = 0$ *and* $\theta^2 > 0$): the mean soil property level has not changed and the change in the soil property spatial pattern is dynamic (non-uniform); again, when averaged, this change is not significant
- *Scenario* 4 ($d_0 \neq 0$ *and* $\theta^2 > 0$): the mean soil property level has changed and the change in the spatial pattern of the soil salinity is dynamic (non-uniform).

An application of the approach of Lesch et al (1998) can be found in Douaik (2005), Douaik et al (2007) and Tóth et al (2002).

The approach of Lesch et al (1998) is based on the determination of a linear regression between two soil properties; hence it assumes that the residuals from the regression model are normally distributed with equality of variances and spatially not autocorrelated. All these assumptions must be checked. If spatial autocorrelation is present in the residuals, the estimates of the regression parameters and their variances are biased and the statistical tests are no longer valid. In this special case, spatial regression and/or geostatistical approaches (sections 3 and 4) are recommended.

3. Geostatistics: Structural analysis

Structural analysis is concerned with the description and the modelling of the spatial and/or temporal variability using geostatistical tools like the covariance function, the correlogram, or the variogram. The analysis of space-time data requires the definition of a space-time continuum with a coordinate system and a measure of space-time distance and models and techniques which make possible the link between spatio-temporally distributed data (Christakos et al., 2002). The spatial coordinates are defined, in general, in two dimensions $\boldsymbol{s} = (s_1, s_2)$ and the temporal coordinate t along the temporal axis such that the space-time coordinates are $\boldsymbol{p} = (\boldsymbol{s}, t)$.

3.1 Space-Time Random Field model

Christakos et al. (2002) defined space-time random field (STRF) model as 'a mathematical construction that rigorously represents the distribution of natural phenomena across space and time. The STRF model provides scientifically meaningful representation, explanation, and prediction of these phenomena in uncertain environments. These uncertainties may be due to measurement errors, heterogeneous databases, erratic fluctuations in the space-time variation of the underlying process, and insufficient knowledge'. An STRF, $X(\boldsymbol{p})$, is a collection of realizations χ of the space-time distribution of a natural variable. It can be viewed as a collection of correlated random variables $\boldsymbol{x}=[x_1, \ldots, x_n]'$ at the space-time points $\boldsymbol{p}=[p_1, \ldots, p_n]'$ and a realization of the STRF at these points is $\chi=[\chi_1, \ldots, \chi_n]'$. It assigns a probability that a realization χ in n dimensions will occur following the multivariate probability distribution function (PDF) of $X(\boldsymbol{p})$ or, equivalently, its cumulative distribution function (CDF). The PDF and CDF represent a complete characterization of the STRF. A partial, but sufficient, characterization of the STRF is provided by its space-time statistical moments. The first order moment, which is the mean function, expresses trends or systematic structures in space-time. The moment of second order, the covariance function, expresses correlations and dependencies between two different points $\boldsymbol{p}$ and $\boldsymbol{p}'$. In addition to these two moments, an STRF can be also characterized by its variogram.

3.2 Conceptual approaches

Christakos and Raghu (1996) reported that some forms of averaging over time or space are done before the analysis of the spatial or temporal random fields. In doing so, the joint space-time variability is not accounted for and there is some loss of information. Initially to consider both space and time variability, some authors resorted to some simplifications. For example, Bilonick (1985) and Egbert & Lettenmaeir (1986) decomposed their STRF in a purely spatial and a purely temporal components whereas Rodriguéz-Iturbe & Eagelson (1987) considered separable models where spatial variations were assumed to be independent of temporal variations.

Three main approaches can be distinguished for the analysis of the geostatistical space-time data (Kyriakidis & Journel, 1999; Stein & Sterk, 1999; Stein et al., 1998):

- Methods based on vectors of spatial random fields (SRF) for the case of many space locations and few time instants (Bogaert & Christakos, 1997; Papritz & Fluhler, 1994). It does not include the temporal dependence existing between observations and can predict only at the observed time instants. The spatial variability is modelled either through a separable spatial variogram for each time instant, or by a single spatial variogram considering time instants as replicates;
- Methods based on vectors of time series (temporal random fields, TRF) for the case of long time series with few space locations (Rouhani & Wackernagel, 1990; Solow & Gorelick, 1986). This approach doesn't take into account the spatial dependence and it predicts only at the observed locations. In a similar way as above, independent TRF or TRF as replicates in space can be considered.
- Methods using an STRF, thus considering the joint space-time variability;

For the first approach, the focus of the analysis is on smooth interpolated soil attribute maps over specific time instants. The intention is to capture single instantaneous snapshots (a static picture) of the natural process. The objective can be also the comparison of the various maps or the detection of the temporal persistence or change in the spatial patterns (Goovaerts & Chiang, 1993; Van Meirvenne et al., 1996). The intention of the second approach is to capture a sequence of successive snapshots at single space locations, which give temporal profiles. Only the third approach includes both the spatial and temporal dependencies so the interpolation is more precise and can be done for unsampled time instants at unsampled space locations. The focus is here on video sequence of successive spatial pictures like a movie.

As a fully satisfactory stochastic model should involve explicitly both spatial and temporal aspects, we present in the following sections the STRF concept and, subsequently, some special cases. The spatial and/or temporal variability can be described and modelled using either the covariance function or the variogram; although the latter is far more used in the geostatistical literature than the former and, therefore, only the variogram forms will be presented here. Also, in the case of two different soil properties or for the same soil property but observed at two different spatial locations or time instants, the joint variability can be characterized using the cross covariance function or the cross variogram. We present in section 3.3 the joint STRF while the spatial random field (SRF) is presented in section 3.4 and the temporal random field (TRF) in section 3.5. Finally, the space-time, the spatial, and the temporal cross variograms are discussed in section 3.6.

3.3 Joint Space-Time Random Field model

The space time variogram is defined as:

$$\gamma_X(h,\tau)=\frac{1}{2}\mathrm{var}\left[X(s+\boldsymbol{h},\boldsymbol{t}+\tau)-X(s,\boldsymbol{t})\right]=\frac{1}{2}E\left\{\left[X(s+\boldsymbol{h},\boldsymbol{t}+\tau)-X(s,\boldsymbol{t})\right]^2\right\} \quad (20)$$

It is estimated, from the observed data, by (Stein et al., 1998):

$$\hat{\gamma}_X(\boldsymbol{h},\tau)=\frac{1}{2N(\boldsymbol{h},\tau)}\sum_{i,j=1}^{N(h,\tau)}\left[x(s_i,\boldsymbol{t}_j)-x(s_i+\boldsymbol{h},\boldsymbol{t}_j+\tau)\right]^2 \quad (21)$$

with $x(s_i,\boldsymbol{t}_j)$ and $x(s_i+\boldsymbol{h},\boldsymbol{t}_j+\tau)$ are pairs of observations with a spatial distance equal h at the time instant t_j, the total number of such pairs is $N_j(h)$, and a temporal distance equal τ, the number of such pairs is $N(\tau)$. Bilonick (1985) and Stein et al. (1998) are examples of application of the space-time variogram while examples of application of the space-time covariance function include Bennett (1975) for the analysis of population diffusion, Bell (1987) for the study of rainfall, Douaik (2005) and Douaik et al. (2004a) for the analysis of soil salinity.

3.4 Separate Spatial Random Field model

To each of the n_t time instants corresponds a spatial random field (SRF) $X(\mathbf{s}, t_j)$, j=1, … , n_t. The spatial variogram, corresponding to each of the above SRF, is defined as:

$$\gamma_j(\boldsymbol{h})=\frac{1}{2}\mathrm{var}\left[X(\boldsymbol{s},\boldsymbol{t}_j)-X(\boldsymbol{s}+\boldsymbol{h},\boldsymbol{t}_j)\right]=\frac{1}{2}E\left\{\left[X(\boldsymbol{s},\boldsymbol{t}_j)-X(\boldsymbol{s}+\boldsymbol{h},\boldsymbol{t}_j)\right]^2\right\} \quad (22)$$

It is estimated by:

$$\hat{\gamma}_j(\boldsymbol{h}) = \frac{1}{2\boldsymbol{N}_j(\boldsymbol{h})} \sum_{i=1}^{N_j(h)} \left[x(\boldsymbol{s}_i, \boldsymbol{t}_j) - x(\boldsymbol{s}_i + \boldsymbol{h}, \boldsymbol{t}_j) \right]^2 \tag{23}$$

with $\boldsymbol{N}_j(\boldsymbol{h})$ is the number of pairs $[x(\boldsymbol{s}_i, \boldsymbol{t}_j),\ x(s_i + h,\ t_j)]$ of data separated by a spatial distance h, at the time instant t_j.
The spatial variogram (equation 22) is a special case of the space-time variogram (equation 20), i.e. when the temporal lag is zero meaning the same time instant:

$$\gamma_j(\mathbf{h}) = \gamma_X(\mathbf{h}, 0) \tag{24}$$

The spatial variograms include the spatial dependence but the dependence in time is not accounted for. It is useful in the case of rich data in space and scarce data in time. The spatial variograms can be one, two or three dimensional.
The above procedure was used, for example, by Ettema et al. (1998) and Sterk & Stein (1997), among others. Many applications of separate spatial variograms, one for each time instant, can be found in the soil science literature. Among others, soil carbon (Chevallier et al., 2000; Van Meirvenne et al., 1996), soil salinity and sodicity (Cetin & Kirda, 2003; Mostafa & Yomota, 1998), soil water (Schume et al., 2003; Wendroth et al., 1999), nitrates (Bruckler et al, 1997), soil biological activity (Gorres, 1997), soil strength (Castrignanò et al., 2002), and soil fertility (Delcourt et al, 1996; Shi et al., 2002) were analyzed for their spatial and temporal change.

3.5 Separate Temporal Random Field model

In a similar way to SRF, for a fixed space location $\boldsymbol{s}_i$, i=1, …, n_s, corresponds the independent temporal random fields $X(\boldsymbol{s}_i, t)$.
The temporal variogram at the location $\boldsymbol{s}_i$, i=1, …, n_s, is:

$$\gamma_i(\tau) = \frac{1}{2}\mathrm{var}[X(s_i, t) - X(s_i, t+\tau)] = \frac{1}{2}E\{[X(s_i, t) - X(s_i, t+\tau)]^2\} \tag{25}$$

It is estimated as:

$$\hat{\gamma}_i(\tau) = \frac{1}{2N_i(\tau)} \sum_{j=1}^{N_i(\tau)} \left[x(s_i, t_j) - x(s_i, t_j + \tau) \right]^2 \tag{26}$$

with $N_i(\tau)$ is the number of pairs $[x(s_i, t_j),\ x(s_i, t_j + \tau)]$ of data separated by a temporal distance (time period) τ, at the space location $\boldsymbol{s}_i$.
Similarly to the spatial variogram, the temporal variogram (equation 25) is a special case of the space-time variogram (equation 20) when the spatial lag is zero meaning the same spatial location:

$$\gamma_i(\tau) = \gamma_X(\mathbf{0}, \tau) \tag{27}$$

The temporal dependence is included whereas the spatial dependence is not considered. Thus, it is useful if many data are collected in time and few in space. The temporal variograms are one dimensional. Application of temporal variograms in the soil literature is scarce and Cichota et al (2006) and Franzluebbers et al (2002) are examples of such application.

3.6 Space-time, spatial, and temporal cross variograms

The space-time cross variogram is defined, for two soil properties X and Y, as:

$$\gamma_{XY}(h,\tau)=\frac{1}{2}E\{[Y(s+h,t+\tau)-Y(s,\tau)][X(s+h,t+\tau)-X(s,\tau)]\} \tag{28}$$

It is estimated by:

$$\hat{\gamma}_{XY}(h,\tau)=\frac{1}{2N(h,\tau)}\sum_{i,j=1}^{N(h,\tau)}\left[x(s_i,t_j)-y(s_i+h,t_j+\tau)\right]^2 \tag{29}$$

In the case of two SRF observed at two time instants t_j and $t_{j'}$, the spatial cross variogram is defined by:

$$\gamma_{jj'}(h)=\frac{1}{2}E\{\left[X(s,t_j)-X(s+h,t_j)\right]\left[X(s,t_{j'})-X(s+h,t_{j'})\right]\} \tag{30}$$

It is estimated by:

$$\hat{\gamma}_{jj'}(h)=\frac{1}{2N_\tau(h)}\sum_{i=1}^{N_\tau(h)}\left[x(s_i,t_j)-x(s_i+h,t_j)\right]\left[x(s_i,t_{j'})-x(s_i+h,t_{j'})\right] \tag{31}$$

with $N_\tau(h)$ is defined as $N(h,\tau)$ for a particular fixed temporal lag τ.
By analogy with the spatial cross variogram, the temporal cross variogram for two TRF observed at two spatial locations s_i and $s_{i'}$ is defined as:

$$\gamma_{ii'}(\tau)=\frac{1}{2}E\{[X(s_i,t)-X(s_i,t+\tau)][X(s_{i'},t)-X(s_{i'},t+\tau)]\} \tag{32}$$

It is estimated by:

$$\hat{\gamma}_{ii'}(\tau)=\frac{1}{2N_h(\tau)}\sum_{j=1}^{N_h(\tau)}\left[x(s_i,t_j)-x(s_i,t_j+\tau)\right]\left[x(s_{i'},t_j)-x(s_{i'},t_j+\tau)\right] \tag{33}$$

with $N_h(\tau)$ is defined as $N(h,\tau)$ for a particular fixed spatial lag h.
Bishop & Lark (2006), Maas et al. (2010), and Motaghian & Mohammadi (2011) are examples of application of cross-variogram in the soil science. The definition of the space-time, the spatial, and the temporal cross-variograms requires that data are observed at the same space locations and time instants, at the same space locations for two different time instants or at the same time instants for two different space locations, respectively. If it is not the case, the space-time, the spatial or the temporal pseudo cross-variogram (Papritz et al., 1993) needs to be used. Vanderlinden et al. (2006) and Zhang et al. (1999) are examples of application of pseudo cross-variogram and cokriging in soil science literature.

4. Geostatistics: Prediction

When a theoretical model is fitted to the experimental (observed) space-time covariance function or variogram, it becomes possible to tackle the problem of predicting attributes at unsampled space locations and/or time instants. Consider the problem of predicting the value of a continuous attribute x at any unsampled space location $\mathbf{s}_0$ and time instant t_0, $x(\mathbf{s}_0, t_0)$, using only the x-data available over the space region S and time domain T, say, n data $\{x(s_{ij}, t_j), j = 1,...,n_t; i = 1,...,n_j\}$.

Kriging is a family of generalized least squares regression algorithms that allows to predict $x(\mathbf{s}_0, t_0)$. The latter is considered, in geostatistics, as a realization of the STRF $X(\mathbf{s}_0, t_0)$. The problem of predicting $x(\mathbf{s}_0, t_0)$ is defined as (Rouhani & Myers, 1990):

$$\hat{X}(s_0, t_0) = \sum_j \sum_i \lambda_{ij} X(s_{ij}, t_j) \tag{34}$$

with λ_{ij}, the kriging weights for the time instant t_j and the space location s_{ij}.
This is the general form and there are some simplifications which are related to the kind of STRF considered. We present first the simple cases where we consider only SRF or TRF, and then we present the more general case for the single STRF.

4.1 Spatial or temporal kriging

In case of spatial or temporal random field (sections 3.4 and 3.5) we saw that the dependence in data is modelled via a spatial or temporal covariance function or variogram. For example, in the case of an SRF, we are dealing in the space domain and the different kriging algorithms defined for SRF can be applied. Let $X(\mathbf{s}, t_0)$ denotes the SRF at the time instant t_0. The predictor $X(\mathbf{s}_0, t_0)$, at the space location $\mathbf{s}_0$, is a linear combination of the n_s observations $x_1, \ldots, x_{ns}$ available at the time instant t_0:

$$\hat{X}(s_0, t_0) = \sum_{i=1}^{n_s} \lambda_i X(s_i, t_0) \tag{35}$$

The n_s weights are calculated such that $X(\mathbf{s}_0, t_0)$ is unbiased and that the variance of the prediction error is minimal. A detailed description of the computation procedure can be found in geostatistical handbooks (Cressie, 1993; Goovaerts, 1997).
Three kriging algorithms can be distinguished (Goovaerts, 1997):

- Simple kriging: the mean is known and constant thorough the space region;
- Ordinary kriging: the mean is unknown but constant in the sub-domains of the space region;
- Kriging with trend model or universal kriging: the mean is unknown, is local and varies smoothly in the sub-domains of the space region.

In a similar way, the predictor $X(\mathbf{s}_0, t_0)$, at the time instant t_0, is a linear combination of the n_t observations $x_1, \ldots, x_{nt}$ available at the space location $\mathbf{s}_0$.
The prediction is done for each time instant (space location) separately and independently from the other time instants (space locations). Many applications can be found in the soil science literature: ordinary kriging using variograms converted from a basic pooled spatial variogram on the sampling time instants (Sterk & Stein, 1997; Stein & Sterk, 1999); ordinary

kriging and universal kriging (Ettema et al., 1998; Figueira et al., 2001; Liu et al, 2009); and ordinary log-normal kriging (Ettema et al., 2000).

4.2 Spatial or temporal cokriging

For the same cases as before (spatial or temporal random fields) more accurate estimates, based on cokriging, can be obtained by using the information available at a precedent time instant as well as the contemporaneous data for spatial prediction (D'Agostino et al., 1998) or information from two different TRFs observed at two different space locations. Cokriging can also be used to estimate the temporal change for a spatially correlated soil property between two time instants using the pseudo cross-variogram (Lark, 2002; Papritz & Fluhler, 1994).

Bogaert (1996) showed that the simple cokriging system between n_t coregionalized space variables (one for each time instant) is equivalent to the simple space-time kriging. This equivalence was shown for a more general form of the covariance functions (Kyriakidis & Journel, 1999). Bogaert (1996) demonstrated also that space ordinary cokriging system with one unbiasedness condition is equivalent to the space-time ordinary kriging system when it is expressed in terms of covariance functions. Pseudo cross-variograms should replace the cross-variograms if the cokriging system has to be written in terms of variograms. However, the use of the unique unbiasedness constraint is restricted to the second order stationary situations. The author noted also that space-time ordinary kriging is preferable to space ordinary cokriging whenever it is possible because, although the differences in the prediction variances are negligible when data are abundant, it is not the case when some time instants involve limited data.

We recall that for SRF, the prediction can be done at unsampled locations only for one or more of the observed time instants. As examples of application, we cite Bogaert & Christakos (1997) and D'Agostino et al. (1998). In the case of TRF, cokriging allows forecasting and hindcasting, the latter is useful in the estimation of missing values. Examples of application can be found in Rouhani et al. (1992a, b). Cokriging was used also to estimate a missing observation at a given time instant by considering the observations at this time instant as the variable of interest and the observations at other time instants as covariables (Stein, 1998).

Let a given soil property be sampled at n_1 space locations during the first time instant, $X_1(s_i)$, $i = 1, \ldots, n_1$, and at n_2 (less than n_1) space locations later during the second time instant, $X_2(s_j)$, $j = 1, \ldots, n_2$. The kriging estimator of X_2 at an unsampled location s_0 is (Goovaerts, 1997):

$$\hat{X}_2(s_0) = \sum_{i=1}^{n_1} \lambda 1_i X_1(s_i) + \sum_{j=1}^{n_2} \lambda 2_j X_2(s_j) \tag{36}$$

with $\lambda 1_i$ and $\lambda 2_j$ are the kriging weights relative to the variables X_1 and X_2, respectively. The weights must satisfy the following conditions:

$$\sum_{i=1}^{n_1} \lambda 1_i = 0 \quad \text{and} \quad \sum_{j=1}^{n_2} \lambda 2_j = 1 \tag{37}$$

4.3 Space-time kriging

For the more general single STRF, two kriging algorithms can be distinguished, i.e. two step space-time kriging and anisotropic space-time kriging. They are presented in the next sections followed by other forms of space-time kriging.

4.3.1 Two-step space-time kriging

The prediction of the value of $\mathbf{X}_0 = \mathbf{X}(\mathbf{s}_0, t_0)$ is a linear combination of n_t predictors in space. In the first step, at each of the n_t time instants t_j, $j = 1, \ldots, n_t$, the predictor of the value of $\mathbf{X}(\mathbf{s}_0, t_j)$ is determined as (Stein et al., 1994):

$$\hat{X}(s_0, t_j) = \sum_{i=1}^{n_j} \lambda_{ij} X(s_{ij}, t_j) \tag{38}$$

The kriging weights are determined in the standard manner for kriging in the space domain (Goovaerts, 1997). In the second step $\mathbf{X}(\mathbf{s}_0, t_0)$ is predicted, using a linear combination of the predictors determined at the first step. This predictor is (Stein et al., 1994):

$$\hat{X}(s_0, t_0) = \sum_{j=1}^{n_t} \theta_j \hat{X}(s_0, t_j) = \sum_{j=1}^{n_t} \sum_{i=1}^{n_j} \theta_j \lambda_{ij} X(s_{ij}, t_j) \tag{39}$$

with θ_j are the kriging weights from the second step. The predictor $\hat{X}(s_0, t_0)$ explicitly uses spatial variograms that change over time. This algorithm was applied in different situations (Douaik, 2005; Stein et al., 1994, 1998).

4.3.2 Anisotropic space-time kriging

For this algorithm, the prediction is done in one and unique step capitalizing on a space-time variogram model which incorporates the spatial and temporal dependencies with a space-time anisotropy ratio (Bechini et al., 2000; Rouhani & Myers, 1990; Stein et al., 1994, 1998). The advantage of the anisotropic over the two step space-time kriging is that the former allows predictions to be made at every point in space and time. Also it avoids the uncertainties related to the predictors determined at the first step for the latter.

4.4 Space-time cokriging

De Iaco et al. (2004) extended the Linear Model of Coregionalization (LMC) to the space-time domain using generalized product sum models for the variograms. In addition, they proposed an extension of ordinary cokriging to the space-time domain. Although De Iaco et al. (2004) discussed the more general case of several secondary variables; we present here only the case of one variable of interest and one covariable. A linear space-time predictor of $\mathbf{X}(\mathbf{s}, t)$ at an unsampled point $(\mathbf{s}_0, t_0)$ in the space-time domain is:

$$\hat{X}(s_0, t_0) = \sum_{i=1}^{n} \Lambda_i(s_0, t_0) X(s_i, t_i) \tag{40}$$

where $\Lambda_i(\mathbf{s}_0, t_0)$ are the (2x2) matrices of kriging weights whose elements $\lambda_i^{\alpha\beta}(s,t)$ are the weights assigned to the value of the βth variable, $\beta = 1, 2$, at the ith sampled point to predict the αth variable, $\alpha = 1, 2$, at the unsampled point ($\mathbf{s}_0$, t_0). The predicted value includes two components:

$$\hat{X}(s_0,t_0) = \left[\hat{X}_1(s_0,t_0), \hat{X}_2(s_0,t_0)\right] \tag{41}$$

such that

$$\hat{X}_\alpha(s_0,t_0) = \sum_{i=1}^{n}\sum_{\beta=1}^{2}\lambda_i^{\alpha\beta}(s_0,t_0)X_\beta(s_i,t_i) \text{ , } \alpha = 1, 2 \tag{42}$$

This corresponds to an extension of ordinary cokriging to the space-time domain. More detail on how to determine the kriging weights $\lambda_i^{\alpha\beta}(s,t)$ can be found in De Iaco et al. (2004). In their work, they compared space-time ordinary kriging to space-time ordinary cokriging. They found that the correlation coefficients between observed and predicted values were higher when the latter algorithm was used. Also, the cokriging error variances were lower than those from kriging.

4.5 Other forms of space-time kriging

Additional forms of space-time kriging exist in the literature. Among these variants, indicator kriging (Bilonick, 1988), kriging with external drift (Snepvangers et al., 2003), and factorial kriging (De Iaco et al., 2003; Rouhani et al., 1992a, b; Van Meirvenne & Goovaerts, 2002) were used.

Now after that the classical geostatistical techniques of space-time interpolation have been laid out, we present a versatile and more general method of interpolation, i.e., the Bayesian Maximum Entropy (BME).

5. Bayesian maximum entropy

For most of the classical geostatistical methods of interpolation, i.e. kriging, the prediction is based solely on the hard data, accurate observed or measured data. However, other data sources can be useful, mostly, for space locations and/or time instants where hard data are missing and they can improve the accuracy of the predictions. Such data can be different types of soft data (interval, probabilistic, etc.), physical laws, moments of higher order, etc.

Bayesian Maximum Entropy (BME) (Christakos, 2000; Christakos & Li, 1998) is a recent approach developed for the spatio-temporal mapping of natural processes using uncertain information. It offers the flexibility to incorporate various sources of physical knowledge. This incorporation of physical knowledge bases enables global prediction features and the adoption of probability distributions without a need for any assumptions like for example to be Gaussian.

5.1 BME knowledge bases

Christakos (2000) defined a knowledge base as 'a collection of knowledge sources relevant to the problem at hand to be involved by a reasoning process aiming at the solution of the problem'. In the BME framework, the total knowledge (K) available regarding a natural process is considered to be formed from two main bases: the general knowledge (G) and the specificatory knowledge (S). The general knowledge represents the knowledge that one has about the distribution of the natural variable to be mapped before any specific data. It encompasses physical laws, statistical moments of any order (including the mean and variogram or covariance function), multipoint statistics, etc. It is said to be general because it does not depend on the specific random field realization at hand. The specificatory knowledge includes the data specific to a given experiment or situation. It refers to a particular occurrence of the natural variable at a particular space location and a particular time instant. It is divided into two main categories depending on the accuracy of the data. The first category is called hard data which are exact measurements of the natural process and considered to be error-free. It encompasses accurate measurements obtained from real-time observation devices, computational algorithms, simulation processes, etc. The second category is called soft data. It involves uncertain observations, empirical charts, assessments by experts, etc. They are incomplete or qualitative data linked to opinions, intuition, etc. Different kinds of soft data are available, often in the form of interval domain data or of a probabilistic nature.

To illustrate these notions of hard and soft data, we give here an example. Soil salinity can be assessed by measuring the electrical conductivity of the soil. It is measured either in the laboratory (*ECe*) or in the field (*ECa*). The measurement in the laboratory is done with high precision in standard conditions and is directly related to soil salinity. Thus, it is considered as an accurate measurement of the soil salinity and represents the hard data. The apparent electrical conductivity measured in the field (*ECa*) is only an indirect measurement of soil salinity and still needs to be calibrated with *ECe* values. Thus, based on a calibration model (frequently a regression model), *ECa* values are converted into 'predicted' *ECe* values. As these values are only predictions from a stochastic model, they are entailed with some uncertainty, and are consequently considered as soft data. The predicted *ECe* values and their corresponding standard deviations allow us to define the two kinds of soft data: the confidence intervals are built which give the interval soft data while the two statistic parameters are used to describe Gaussian PDFs which represent our probabilistic soft data.

5.2 The three steps of BME analysis

The BME has a double goal: informativeness (prior information maximization given the general knowledge) and cogency (posterior probability maximization given specificatory knowledge). The BME analysis is done in three main steps of knowledge acquisition, integration and processing: the structural or prior step, the meta-prior step, and the integration or posterior step.

The goal of the prior step is the maximization of the information content considering only the general knowledge before any use of the data, which corresponds to the first goal of the BME. This step assumes an inverse relation between information and probability: the more informative an evaluation of a mapping situation is, the less probable it is to occur. This means that if a theory is general and vague, it includes more alternatives so it is more probable however it is less informative. A random field is completely defined by its multivariate probability distribution function (PDF), which forms the prior PDF. The latter should be derived by means of an estimation process that takes into consideration physical constraints under the form of prior information or knowledge. This information is measured, in the context of BME, using the Shannon's entropy function (Shannon, 1948), thus the E in BME. The expected information is given by the entropy function which needs to be maximized subject to the physical constraints provided by prior information or knowledge *G*, which justifies the M in the BME acronym:

$$f_G(\chi_{map}) = Z^{-1}\exp\left[\sum_{\alpha=1}^{N_c}\mu_\alpha g_\alpha(\chi_{map})\right] \tag{43}$$

where χ_{map} represents the available data and the one to be predicted, $f_G(\chi_{map})$ is the unknown multivariate PDF associated with the general knowledge *G*, $Z = \exp(-\mu_0)$ is a normalization constant, and μ_α are Lagrange multipliers.

During the meta-prior step, the specificatory knowledge is collected and organized in appropriate quantitative forms that can be easily incorporated in the BME framework. The available data can be divided into two main types: the hard data and the soft data. The hard data are considered exact and accurate measurements of the natural process while the soft data are indirect and inaccurate measures of the variable of interest. The soft data may be in the form of intervals, probability distribution functions, etc.

In the posterior step, the two knowledge bases (*G* and *S*) are integrated. The goal is the maximization of the posterior PDF given the total knowledge *K*, which is the second goal of BME. The prior PDF is updated, by considering the available site-specific knowledge (the data). This updating is performed using a Bayesian conditionalization, thus the B in BME. For interval soft data, the posterior PDF $f_K(\chi_k)$ is:

$$f_K(\chi_k) = (AZ)^{-1}\int_I \exp\left[\sum_{\alpha=1}^{N_c}\mu_\alpha g_\alpha(\chi_{map})\right] d\chi_{soft} \tag{44}$$

and for probabilistic soft data, the posterior PDF $f_K(\chi_k)$ is:

$$f_K(\chi_k) = (AZ)^{-1}\int_I \exp\left[\sum_{\alpha=1}^{N_c}\mu_\alpha g_\alpha(\chi_{map})\right] f_S(\chi_{soft}) d\chi_{soft} \tag{45}$$

with χ_{soft} the soft data and *A* is a normalization constant.

5.3 Some BME estimates

The posterior PDF (equations 44 and 45), which is not limited to the Gaussian type, describes fully the random field at the estimation point. It provides a complete picture of the mapping situation as well as different estimators and their associated estimation uncertainty. The prediction points lie, in general, on a regular grid and the predictions are used to create space-time maps. The different estimates presented in the following sections are specific to a general knowledge involving the first two statistical moments and a specificatory knowledge encompassing hard data as well as interval and probabilistic soft data. From the posterior PDF, different statistical parameters can be derived like the mode, the conditional mean, and the variance of the prediction error. The conditional mean is:

$$\hat{\chi}_{k,mean} = \int f_K(\chi_k)\chi_k d\chi_k \tag{46}$$

while the variance of the prediction error is:

$$\hat{\sigma}_k^{\,2} = \int \left(\chi_k - \hat{\chi}_{k,mean}\right)^2 f_K(\chi_k) d\chi_k \qquad (47)$$

5.4 Kriging as a special case of BME

When the general knowledge is limited to the mean and covariance/variogram functions and the specificatory knowledge is restricted to the only hard data, the BME posterior PDF is Gaussian (the mean and the mode are equal) and the BME mean estimate is equivalent to the kriging (best minimum mean squared error, MMSE) estimate, which is the conditional mean. When the space-time random field is Gaussian, the kriging estimate becomes linear and optimal among all MMSE estimators. Consequently, BME is a more general interpolation approach and kriging is a special case in limiting situations.

5.5 Examples of BME application

The BME approach has been applied in different research fields. Examples of such applications can be found in agricultural sciences (Bogaert & D'Or, 2002; D'Or & Bogaert, 2003; D'Or et al., 2001; Douaik, 2005; Douaik et al., 2004b, 2005, 2009), environmental sciences (Christakos & Serre, 2000; Serre et al., 2001), etc.

5.6 Other space-time interpolation methods

The BME is one of the space-time interpolation methods that allow the incorporation of data with different information support and of different degrees of certainty and the knowledge in the form of physical laws. The data assimilation approach (Swinbank et al., 2003) is another method for such situations. It has two main variants: the Kalman filtering (Kalman, 1960) and the Bayesian data fusion (Fasbender et al., 2009). However, their application in the soil science literature is limited (Bierkens, 2001; Heuvelink et al., 2006), mostly for the latter since it was very recently developed.

6. Discussion and conclusions

This chapter has the main objective of reviewing the mostly used statistical methods for appraising the spatial and temporal variability of soil properties. There is a large panoply of techniques and, as stated in the introduction, the choice of the adequate model will depend on the behaviour of the phenomenon of interest (a priori knowledge), the objective of the analysis, and the nature of the available data.

As a first step, the classical statistical method of coefficient of variation, which ignores the spatial and temporal coordinates, can be used as an exploratory data analysis to get an idea about the variability of data either for a given location, a given time instant or both. The coefficient of variation is considered as an indicator of global variability as opposed to local variability; the latter is identified using, for example, the variogram, a geostatistical tool.

Another classical statistical method, the variance component analysis, in which the spatial and temporal contributions are considered as factors with random effects and the joint space-time contribution as interaction in the ANOVA model, can be used to determine their relative contribution to the total variance. Since this approach is directly related to the coefficient of variation, it is suffering from the same drawbacks as the later.

Still using classical statistical methods, there are opportunities to consider the spatial and temporal dimensions of data using, for example, the concept of temporal stability. It is defined as the time invariant association between spatial location and statistical parameters of soil properties and can be assessed using either the relative differences or the Spearman rank correlation. A good test for the temporal stability is the correlation/regression between soil property values measured at consecutive time frames. A way to check the temporal stability of the spatial pattern is to test for a zero intercept and a unity slope of the regression between two consecutive time instants. If the slope of the regression between values at two consecutive time instants is not significantly different from one, then the regression intercept represents the constant change that has occurred between the two time instants. Based on the significance test of these two regression parameters, four possible scenarios can be distinguished: no net change between the two sampling dates; the mean soil property level has changed and there is a static (uniform) change in the soil property spatial pattern; the mean soil property level has not changed and the change in the soil property spatial pattern is dynamic (non-uniform) but, when averaged, this change is not significant; and the mean soil property level has changed and the change in the soil property spatial pattern is dynamic (non-uniform). The concept of relative differences is based on the same statistical parameters (mean and standard deviation) as the two precedent approaches (coefficient of variation and variance component analysis). This concept is based, like the two precedent approaches, on the assumptions of Gaussian distribution of data and the absence of autocorrelation. Both assumptions must be checked before the use of relative differences in the study of temporal stability. The normality can be checked using the standard statistical tests and the independence can be verified using either a test for autocorrelation like the Moran test or by computing the variogram and determining its range. Normality can be gained via data transformation and independence by the selection of sample locations separated by distances larger than the autocorrelation range. For non normally distributed data, but still non autocorrelated, the Spearman rank correlation can be used. It refers to the tendency of a soil property, measured at different locations in space, to maintain their relative ranking over time. In the existence of autocorrelation, its use is not recommended and it will be cautious to use statistical methods which consider the spatial and temporal nature of the given soil property. In addition, even for independent data, the Spearman rank correlation inform us only about the relative ranking of space locations between two time instants and do not provide a quantification of the temporal change. This quantification can be obtained using the paired-*t* test for comparing mean values. Thus, the paired-*t* test, still a classical statistical method, can be used following a temporal stability analysis based on the concept of relative differences.

The concept of temporal stability handles each soil property separately and quantifies the temporal change even without testing if this change is significant or not. Dynamic spatial variation approach can test for change in the mean value (mean shift) as well as for uniform/non-uniform spatial variation between two time instants (spatial shift). Dynamic spatial variation refers to the spatially variable change of a soil property between two time instants, i.e., the change is not the same between two time instants for each location. Again, like with the relative differences concept, based on whether the mean difference is significantly different from zero, the spatial shift parameter is significantly higher than zero, or both, four possible scenarios can be distinguished: no net change with time in the soil property; the mean soil property level has changed and there is a static (uniform) change in the soil property spatial pattern; the mean soil property level has not changed and the change in the soil property spatial pattern is dynamic (non-uniform), again, when averaged, this change is not significant; and finally, the mean soil property level has changed and the change in the spatial pattern of the soil property is dynamic (non-uniform).

The dynamic spatial variation approach is based on the determination of linear regression between two soil properties; hence, it assumes that the residuals from the regression model are normally distributed with equality of variances and spatially non autocorrelated. All these assumptions must be checked. If spatial autocorrelation is present in the residuals, the estimates of the regression parameters and their variances are biased and the statistical tests are no longer valid. In this special case, spatial regression and/or geostatistical approaches are recommended.

The concept of temporal stability requires a moderate number of measurements in space and time, but once it is checked, the number of future measurements can be reduced to locations with measurements representing important statistics such as the average value. Likewise, based on the dynamic spatial variation test, if it was found that there was a static (uniform) spatial variation between two dates, the sampling effort could be limited to the first soil property survey (an easy to measure and cheap variable) for the first date, while the second soil property (a hard and expensive variable) could be done for two dates but at a reduced number of locations. So the joint use of the concept of temporal stability and the temporal mean shift and spatial shift tests could result in a drastically reduced sampling effort. As illustration, Douaik et al. (2006) found that it was sufficient to measure soil salinity at just two locations and nine dates, from initially 20 locations and 19 time instants, to characterize reliably the average field soil salinity of 25 km^2 study area. The statistical method to use will depend on the data availability and the aim of the study. If data are available only for one of the two soil properties, there is only one possibility for checking the spatial pattern: the concept of temporal stability. For checking the temporal change in the average soil property, there is a choice between the paired-*t* test and the concept of temporal stability. If, on the other hand, data are available for both soil properties, there are more options: the paired-*t* test and the concept of temporal stability for checking the temporal change in the average level separately for both soil properties or the temporal mean shift test for checking the temporal change in the average level combining both soil properties. Regarding the check of the temporal stability of the spatial patterns, the same choices are available except the paired-*t* test.

For an implicit and formal assessment of the spatial and temporal variability of soil properties, structural analysis is used with the goal of describing and modelling this variability using geostatistical tools like the covariance function, the correlogram, or the variogram. The models of spatial and temporal variability can be used, later, for interpolating between space locations and time instants for mapping soil properties. Basically two interpolating methods can be used for space-time data: the geostatistical method of kriging and the Bayesian maximum entropy.

Three main approaches for the analysis of space-time data can be distinguished:

- Methods based on vectors of spatial random fields (SRF) for the case of many space locations and few time instants. It does not include the temporal dependence existing between observations and can predict only at the observed time instants. The spatial variability is modelled either through a separable spatial variogram for each time instant, or by a single spatial variogram considering time instants as replicates;

- Methods based on vectors of time series (temporal random fields, TRF) for the case of long time series with few space locations. This approach doesn't take into account the spatial dependence and it predicts only at the observed locations. In a similar way as above, independent TRF or TRF as replicates in space can be considered.
- Methods using an STRF, thus considering the joint space-time variability.

For the first approach, the focus of the analysis is on smooth interpolated soil attribute maps over specific time instants. The intention is to capture single instantaneous snapshots (a static picture) of the natural process. The objective can be also the comparison of the various maps or the detection of the temporal persistence or change in the spatial patterns. The intention of the second approach is to capture a sequence of successive snapshots at single space locations, which give temporal profiles. Only the third approach includes both the spatial and temporal dependencies so the interpolation is more precise and can be done for unsampled time instants at unsampled space locations. The focus is here on video sequence of successive spatial pictures like a movie.

When a theoretical model is fitted to the experimental (observed) space-time variogram, it becomes possible to tackle the problem of predicting attributes at unsampled space locations and time instants. Kriging, a family of generalized least squares regression algorithms, is one of the methods allowing these predictions. In case of spatial (temporal) random field, the dependence in data is modelled via a spatial (temporal) variogram. The prediction is done for each time instant (space location) separately and independently from the other time instants (space locations). More accurate estimates, based on cokriging, can be obtained by using the information available at a precedent time instant as well as the contemporaneous data for spatial prediction or information from two different TRFs observed at two different space locations. Cokriging can also be used to estimate the temporal change for a spatially correlated soil property between two time instants using the pseudo cross variogram. Space-time kriging is preferable to spatial cokriging whenever it is possible because, although the differences in the prediction variances are negligible when data are abundant, it is not the case when some time instants involve limited data. For the more general single STRF, two kriging algorithms can be distinguished: two-step space-time kriging and anisotropic space-time kriging. For the latter algorithm, the prediction is done in one and unique step capitalizing on a space-time variogram model which incorporates the spatial and temporal dependencies with a space-time anisotropy ratio. The advantage of the anisotropic over the two-step space-time kriging is that the former allows predictions to be made at every point in space and time. Also it avoids the uncertainties related to the predictors determined at the first step for the latter. In the case of the availability of space-time data for more than one soil property, space-time cokriging can be used for improving the interpolation of the soil property of interest by considering its spatial and temporal dependence with one or more auxiliary soil properties. Additional forms of space-time kriging exist in the literature. Among these variants, indicator kriging, kriging with external drift, and factorial kriging were used in soil science.

The prediction for most of the geostatistical methods of interpolation, i.e. kriging, is based solely on the accurate observed or measured data called hard data. However, other data sources can be useful, mostly, for space locations and/or time instants where hard data are missing and they can improve the accuracy of the predictions. Such data can be different types of uncertain data (soft data), physical laws, statistical moments of higher order, etc. Bayesian Maximum Entropy (BME) is a recent approach developed for the spatio-temporal mapping of natural processes using uncertain information in addition to accurate measured data. It offers the flexibility to incorporate various sources of physical knowledge enabling global prediction features and the adoption of probability distributions without a need for any assumptions like for example to be Gaussian. The BME has a double goal of informativeness (prior information maximization given the general knowledge) and cogency (posterior probability maximization given specificatory knowledge or the data at hand). The BME probability distribution function describes fully the random field at the prediction point. It provides a complete picture of the mapping situation and different statistical parameters can be derived like the mode, the conditional mean, and the variance of the prediction error. The prediction points lie, in general, on a regular grid and the predictions are used to create space-time maps. In limiting situations (when the general knowledge is limited to the mean and variogram and the specificatory knowledge is restricted to the only hard data), the BME probability distribution function is Gaussian and the BME mean estimate is equivalent to the kriging estimate. Consequently, BME is a versatile and more general method of interpolation and kriging is a special case in limiting situations.

Besides BME, there are other space-time interpolation methods allowing the incorporation of data with different information support and of different degrees of certainty and the knowledge in the form of physical laws. The general data assimilation approach is one of these methods, with its two main variants: the Kalman filtering and the Bayesian data fusion. However, their application in the soil science literature is limited, mostly for the latter since it was very recently developed.

In conclusion, different statistical methods are presently available for dealing with spatial and temporal variability of soil properties and others may be developed in the future. The behaviour of the phenomenon of interest (a priori knowledge), the objective of the research work, and the nature of the available data should help the practitioner in the choice of the adequate method(s).

7. Acknowledgements

Authors acknowledge the support of the European Commission Joint Research Center and the Hungarian National Science Foundation (OTKA) Grant No NN79835.

8. References

Bechini, L.; Ducco, G.; Donatelli, M. & Stein, A. (2000). Modelling, interpolation, and stochastic simulation in space and time of global solar radiation. *Agriculture, Ecosystems and Environment*, Vol. 81, No. 1, pp. 29–42, ISSN 0167-8809.

Bell, T.L. (1987). A space-time stochastic model of rainfall for satellite remote-sensing studies. *Journal of Geophysical Research*, Vol. 92, No. D8, pp. 9631–9643, ISSN 0148–0227.

Bennett, R.J. (1975). The representation and identification of spatio-temporal systems: an example of population diffusion in North-West England. *Transactions of the Institute of the British Geographers*, Vol. 66, No. 1, pp. 73–94, ISSN 0020-2754.

Bierkens, M.F.P. (2001). Spatio-temporal modelling of the soil water balance using a stochastic model and soil profile descriptions. *Geoderma*, Vol. 103, No. 1, pp. 27–50, ISSN 0016-7061.

Bilonick, R.A. (1985). The space-time distribution of sulphate deposition in the Northern US. *Atmospheric Environment*, Vol. 19, No. 11, pp. 1829–1845, ISSN 0004-6981.

Bilonick, R.A. (1988). Monthly hydrogen ion deposition maps for the North Eastern US from July 1982 to September 1984. *Atmospheric Environment*, Vol. 22, No. 9, pp. 1909–1924, ISSN 0004-6981.

Bishop, T.F.A. & Lark, R.M. (2006). The geostatistical analysis of experiments at the landscape-scale. *Geoderma*, Vol. 133, No. 1-2, pp. 87–106, ISSN 0016-7061.

Bogaert, P. (1996). Comparison of kriging in a space-time context. *Mathematical Geology*, Vol. 28, No. 1, pp. 73–86, ISSN 0882-8121.

Bogaert, P. & Christakos, G. (1997). Stochastic analysis of spatio-temporal solute content measurements using a regression model. *Stochastic Hydrology and Hydraulics*, Vol. 11, No. 4, pp. 267–295, ISSN 0931-1955.

Bogaert, P. & D'Or, D. (2002). Estimating soil properties from thematic soil maps: the Bayesian Maximum Entropy. *Soil Science Society of America Journal*, Vol. 66, No. 5, pp. 1492–1500, ISSN 0361-5995.

Bruckler, L.; De Cockborne, A.M.; Renault, P. & Claudot, P. (1997). Spatial and temporal variability of nitrate in irrigated salad crops. *Irrigation Science*, Vol. 17, No. 2, pp. 53–61, ISSN 0342-7188.

Campbell, C.G.; Ghodrati, M. & Garrido, F. (2001). Temporal consistency of solute transport in a heterogeneous field plot. *Soil Science*, Vol. 166, No. 8, pp. 491–506, ISSN 0038-075X.

Campbell, D.J.; Kinnigurgh, D.G. & Beckett, P.H.T. (1989). The soil solution chemistry of some Oxfordshire soils: temporal and spatial variability. *Journal of Soil Science*, Vol. 40, No. 2, pp. 321–339, ISSN 0022-4588.

Carter, M.R. & Pearen, J.R. (1985). General and spatial variability of Solonetzic soils in north central Alberta. *Canadian Journal of Soil Science*, Vol. 65, No. 1, pp. 157–167, ISSN 0008-4271.

Castrignanò, A.; Maiorana, M.; Fornaro, F. & Lopez, N. (2002). 3D spatial variability of soil strength and its change over time in a durum wheat field in southern Italy. *Soil and Tillage Research*, Vol. 65, No. 1, pp. 95–108, ISSN 0167-1987.

Cetin, M. & Kirda, C. (2003). Spatial and temporal changes of soil salinity in a cotton field irrigated with low-quality water. *Journal of Hydrology*, Vol. 272, No. 1-4, pp. 238–249, ISSN 0022-1694.

Chevallier, T. ; Voltz, M. ; Blanchart, E. ; Chotte, J.L. ; Eschenbrenner, V. ; Mahieu, M. & Albrecht, A. (2000). Spatial and temporal changes of soil carbon after establishment of a pasture on a long-term cultivated vertisol (Martinique). *Geoderma*, Vol. 94, No. 1, pp. 43–58, ISSN 0016-7061.

Christakos, G. (2000). *Modern spatiotemporal geostatistics*, Oxford University Press, ISBN 978-019-5138-95-5, New York, USA.

Christakos, G.; Bogaert, P. & Serre, M.L. (2002). *Temporal GIS: advanced functions for field-based applications*, Springer-Verlag, ISBN 978-354-0414-76-6, New York, USA.

Christakos, G. & Li, X. (1998). Bayesian maximum entropy analysis and mapping: a farewell to kriging estimators. *Mathematical Geology*, Vol. 30, No. 4, pp. 435–462, ISSN 0882-8121.

Christakos, G. & Raghu, V.R. (1996). Dynamic stochastic estimation of physical variables. *Mathematical Geology*, Vol. 28, No. 3, pp. 341–365, ISSN 0882-8121.

Christakos, G. & Serre, M.L. (2000). BME analysis of spatiotemporal particulate matter distributions in North Carolina. *Atmospheric Environment*, Vol. 34, No. 20, pp. 3393–3406, ISSN 0004-6981.

Cichota, R.; Hortado, A.L.B. & Van Lier, Q.D.J. (2006). Spatio-temporal variability of soil water tension in a tropical soil in Brazil. *Geoderma*, Vol. 133, No. 3-4, pp. 231–243, ISSN 0016-7061

Cressie, N. (1993). *Statistics for spatial data*, Wiley, ISBN 978-047-1002-55-0, New York, USA.

D'Agostino, V.; Greene, E.A.; Passarella, G. & Vurro, M. (1998). Spatial and temporal study of nitrate concentration in groundwater by means of coregionalisation. *Environmental Geology*, Vol. 36, No. 3-4, pp. 285–295, ISSN 0943-0105.

D'Or, D.; Bogaert, P. & Christakos, G. (2001). Application of the BME approach to soil texture mapping. *Stochastic Environmental Research and Risk Assessment*, Vol. 15, No. 1, pp. 87–100, ISSN 1436-3240.

D'Or, D. & Bogaert, P. (2003). Continuous-valued map reconstruction with the Bayesian Maximum Entropy. *Geoderma*, Vol. 112, No. 3-4, pp. 169–178, ISSN 0016-7061.

Da Silva, A.P.; Nadler, A. & Kay, B.D. (2001). Factors contributing to temporal stability in spatial patterns of water content in the tillage zone. *Soil and Tillage Research*, Vol. 58, No. 3-4, pp. 207–218, ISSN 0167-1987.

De Iaco, S.; Myers, D.E. & Posa, D. (2003). The linear coregionalization model and the product sum space-time variogram. *Mathematical Geology*, Vol. 35, No. 1, pp. 25–38, ISSN 0882-8121.

De Iaco, S.; Palma, M. & Posa, D. (2004). Modelling and prediction of multivariate space-time random fields. *Computational Statistics and Data Analysis*, Vol. 48, No. 3, pp. 525–547, ISSN 0167-9473.

Delcourt, H.; Darius, P.L. & De Baerdemaker, J. (1996). The spatial variability of some aspects of topsoil fertility in two Belgian fields. *Computers and Electronics in Agriculture*, Vol. 14, No. 2-3, pp. 179–196, ISSN 0168-1699.

Douaik, A. (2005). *Evaluation of the space-time variability of soil salinity by statistical, geostatistical and Bayesian maximum entropy methods*. PhD Dissertation, Ghent University, Ghent, Belgium.

Douaik, A.; Van Meirvenne, M. & Tóth, T. (2004a). Spatio-temporal kriging of soil salinity rescaled from bulk soil electrical conductivity, In: *GeoEnv IV: Geostatistics for Environmental Applications*, Sanchez-Vila, X.; Carrera, J. & Gomez-Hernandez, J. (Eds.), pp. 413-424, Kluwer Academic Publisher, ISBN 978-1-4020-2007-0, Dordrecht, The Netherlands.

Douaik, A.; Van Meirvenne, M.; Tóth, T. & Serre, M. (2004b). Space-time mapping of soil salinity using probabilistic Bayesian maximum entropy. *Stochastic Environmental Research and Risk Assessment*, Vol. 18, No. 4, pp. 219–227, ISSN 1436-3240.

Douaik, A.; Van Meirvenne, M. & Tóth, T. (2005). Soil salinity mapping using spatio-temporal kriging and Bayesian maximum entropy with interval soft data. *Geoderma*, Vol. 128, No. 3-4, pp. 234–248, ISSN 0016-7061.

Douaik, A.; Van Meirvenne, M. & Tóth, T. (2006). Temporal stability of spatial patterns of soil salinity determined from laboratory and field electrolytic conductivity. *Arid Land Research and Management*, Vol. 20, No. 1, pp. 1–13, ISSN 1532-4982.

Douaik, A.; Van Meirvenne, M. & Tóth, T. (2007). Detecting and monitoring temporal changes of spatial patterns of soil salinity using electrode probes. *Soil Science Society of America Journal*, Vol. 71, No. 5, pp. 1629–1635, ISSN 0361-5995.

Douaik, A.; Van Meirvenne, M. & Tóth, T. (2009). Stochastic approaches for space–time modeling and interpolation of soil salinity. In: Metternicht, G. & Zinck, J.A. (eds). *Remote sensing of soil salinization: impact on land management*. CRC Press, Boca Raton, FL. pp: 273 - 290.

Egbert, G.D. & Lettenmaeir D.P. (1986). Stochastic modelling of the space-time structure of atmospheric chemical deposition. *Water Resources Research*, Vol. 22, No. 2, pp. 165–179, ISSN 0043–1397.

Ehrenfeld, J.G.; Han, X.; Parsons, W.F.J. & Zhu, W. (1997). On the nature of environmental gradients: temporal and spatial variability of soils and vegetation in the New Jersey pinelands. *Journal of Ecology*, Vol. 85, No. 6, pp. 785–798, ISSN 0022-0477.

Ettema, C.H.; Coleman, D.C.; Vellidis, G.; Lowrance, R. & Rathbun S.L. (1998). Spatiotemporal distributions of bacterivorous nematodes and soil resources in a restored riparian wetland. *Ecology*, Vol. 79, No. 8, pp. 2721–2734, ISSN 0012-9658.

Farley, R.A. & Fitter A.H. (1999). Temporal and spatial variation in soil resources in a deciduous woodland. *Journal of Ecology*, Vol. 87, No. 4, pp. 688–696, ISSN 0022-0477.

Fasbender, D.; Brasseur, O. & Bogaert, P. (2009). Bayesian data fusion for space-time prediction of air pollutants: the case of NO_2 in Belgium. *Atmospheric Environment*, Vol. 43, No. 30, pp. 4632–4645, ISSN 0004-6981.

Figueira, R.; Sousa, A.J.; Pacheco, A.M.G. & Catarino, F. (2001). Use of secondary information in space-time statistics for biomonitoring studies of saline deposition. *Environmetrics*, Vol. 12, No. 3, pp. 203–217, ISSN 1180-4009.

Franzluebbers, K.; Franzluebbers, A.J. & Jawson, M.D. (2002). Environmental controls on soil and whole-ecosystem respiration from a tallgrass prairie. *Soil Science Society of America Journal*, Vol. 66, No. 1, pp. 254–262, ISSN 0361-5995

Goovaerts, P. (1997). *Geostatistics for natural resources evaluation*, Oxford University Press, ISBN 978-019-5115-38-3, New York, USA.

Goovaerts, P. & Chiang C.N. (1993). Temporal persistence of spatial patterns for mineralizable nitrogen and selected soil properties. *Soil Science Society of America Journal*, Vol. 57, No. 2, pp. 372–381, ISSN 0361-5995.

Gorres, J.H.; Dichiaro, M.J.; Lyons, J.B. & Amador, J.A. (1997). Spatial and temporal patterns of soil biological activity in a forest and an old field. *Soil Biology and Biochemistry*, Vol. 30, No. 2, pp. 219–230, ISSN 0038-0717.

Grant, L.; Seyfried, M. & McNamara, J. 2004. Spatial variation and temporal stability of soil water in a snow-dominated, mountain catchment. *Hydrological Processes*, Vol. 18, No. 18, pp. 3493–3511, ISSN 0885-6087.

Guo, D.; Mou, P.; Jones, R.H. & Mitchell, R.J. (2002). Temporal changes in spatial patterns of soil moisture following disturbance: an experimental approach. *Journal of Ecology*, Vol. 90, No. 2, pp. 338–347, ISSN 0022-0477.

Heathman, G.C.; Larose, M.; Cosh, M.H. & Bindlish, R. (2009). Surface and profile soil moisture spatio-temporal analysis during an excessive rainfall period in the Southern Great Plains, USA. *Catena*, Vol. 78, No. 2, pp. 159–169, ISSN 0341-8162.

Heuvelink, G.B.M. & Webster, R. (2001). Modelling soil variation: past, present, and future. *Geoderma*, Vol. 100, No. 3-4, pp. 269–301, ISSN 0016-7061.

Heuvelink, G.B.M.; Schoorl, J.M.; Veldkamp, A. & Pennock, D.J. (2006). Space-time Kalman filtering of soil redistribution. *Geoderma*, Vol. 133, No. 1-2, pp. 124–137, ISSN 0016-7061.

Hu, W.; Shao, M.; Han, F. & Reichardt, K. (2011). Spatio-temporal variability behavior of land surface soil water content in shrub- and grass-land. *Geoderma*, Vol. 162, No. 3-4, pp. 260–272, ISSN 0016-7061.

James, S.E.; Partel, M.; Wilson, S.D. & Peltzer, D.A. (2003). Temporal heterogeneity of soil moisture in grassland and forest. *Journal of Ecology*, Vol. 91, No. 2, pp. 234–239, ISSN 0022-0477.

Jenny, H. (1980). *The soil resource: origin and behavior* (4th ed.), Springer-Verlag, ISBN 978-038-7905-43-3, New York, USA.

Kachanoski, R.G. & De Jong, E. (1988). Scale dependence and the temporal persistence of spatial patterns of soil water storage. *Water Resources Research*, Vol. 24, No. 1, pp. 85–91, ISSN 0043–1397.

Kalman, R.E. (1960). A new approach to linear filtering and prediction problems. *Transactions of the American Society of Mechanical Engineers: Journal of Basic Engineering*, Vol. 82, No. D1, pp. 35–45, ISSN 0021-9223.

Kenny, E.A.; Hall, J.W. & Wang, C. 2002. Temporal trends in soil physical properties at a soil quality benchmark site in British Columbia, Canada. *Proceedings of the 17th World Congress of Soil Science*, [CD-ROM], Bangkok, Thailand, August 2002.

Kyriakidis, P.C. & Journel, A.G. (1999). Geostatistical space-time models: a review. *Mathematical Geology*, Vol. 31, No. 6, pp. 651–684, ISSN 0882-8121.

Lark, R.M. (2002). Robust estimation of the pseudo-cross variogram for cokriging soil properties. *European Journal of Soil Science*, Vol. 53, No. 2, pp. 253–270, ISSN 1351-0754.

Lesch, S.M.; Herrero, J. & Rhoades, J.D. (1998). Monitoring for temporal changes in soil salinity using electromagnetic induction techniques. *Soil Science Society of America Journal*, Vol. 62, No. 1, pp. 232–242, ISSN 0361-5995.

Lesch, S.M.; Strauss, D.J. & Rhoades, J.D. (1995). Spatial prediction of soil salinity using electromagnetic induction techniques.1 Statistical prediction models: a comparison of multiple linear regression and cokriging. *Water Resources Research*, Vol. 31, No. 2, pp. 373–386, ISSN 0043–1397.

Liu, X.; Zhang, W.; Zhang, M.; Ficklin, D. & Wang, F. (2009). Spatio-temporal variations of soil nutrients influenced by an altered land tenure system in China. *Geoderma*, Vol. 152, No. 1-2, pp. 23–34, ISSN 0016-7061.

Maas, S.; Scheifler, R.; Benslama, M.; Crini, N.; Lucot, E.; Brahmia, Z.; Benyacoub, S. & Giraudoux P. (2010). Spatial distribution of heavy metal concentrations in urban, suburban and agricultural soils in a Mediterranean city of Algeria. *Environmental Pollution*, Vol. 158, No. 6, pp. 2294–2301, ISSN 0269-7491.

McClave, J.T. & Sincich, T. (2006). *Statistics* (10th ed.), Prentice Hall, ISBN 978-013-1497-55-9, Upper Saddle River, USA.

Mostafa, M.M. & Yomota, A. (1998). Use of a covariance variogram to investigate influence of subsurface drainage on spatial variability of soil-water properties. *Agricultural Water Management*, Vol. 37, No. 1, pp. 1–19, ISSN 0378-3774.

Motaghian, H.R. & Mohammadi, J. (2011). Spatial estimation of saturated hydraulic conductivity from terrain attributes using regression, kriging, and artificial neural networks. *Pedosphere*, Vol. 21, No. 2, pp. 170–177, ISSN 1002-0160.

Papritz, A. & Fluhler, H. (1994). Temporal change of spatially autocorrelated soil properties: optimal estimation by kriging. *Geoderma*, Vol. 62, No. 1-3, pp. 29–43, ISSN 0016-7061.

Papritz, A.; Kunsch, H.R. & Webster, R. (1993). On the pseudo cross-variogram. *Mathematical Geology*, Vol. 25, No. 8, pp. 1015–1026, ISSN 0882-8121.

Petrone, R.M.; Price, J.S.; Carey, S.K. & Waddington, J.M. (2004). Statistical characterization of the spatial variability of soil moisture in a cutover peatland. *Hydrological Processes*, Vol. 18, No. 1, pp. 41–52, ISSN 0885-6087.

Reichardt, K.; Bacchi, O.O.S.; Villagra, M.D.L.M.; Turrati, A.L. & Pedrosa, Z.O. (1993). Hydraulic variability in space and time in a dark red latosol of the tropics. *Geoderma*, Vol. 60, No. 1-4, pp. 159–168, ISSN 0016-7061.

Rodriguez-Iturbe, I. & Eagelson, P.S. (1987). Mathematical models of rainstorm events in space and time. *Water Resources Research*, Vol. 23, No. 1, pp. 181–190, ISSN 0043–1397.

Rouhani, S.; Ebrahimpour, R.M.; Yaqub, I. & Gianella, E. (1992a). Multivariate geostatistical trend detection and network evaluation of space-time acid deposition data. – I. Methodology. *Atmospheric Environment*, Vol. 26, No. 14, pp. 2603–2614, ISSN 0004-6981.

Rouhani, S.; Ebrahimpour, R.M.; Yaqub, I. & Gianella, E. (1992b). Multivariate geostatistical trend detection and network evaluation of space-time acid deposition data. –II. Application to NADP/NTN data. *Atmospheric Environment*, Vol. 26, no. 14, pp. 2615–2626, ISSN 0004-6981.

Rouhani, S. & Myers, D.E. (1990). Problems in space-time kriging of geohydrological data. *Mathematical Geology*, Vol. 22, No. 5, pp. 611–623, ISSN 0882-8121.

Rouhani, S. & Wackernagel, H. (1990). Multivariate geostatistical approach to space-time data analysis. *Water Resources Research*, Vol. 26, No. 4, pp. 585–591, ISSN 0043–1397.

Schume, H.; Jost, G. & Katzensteiner, K. (2003). Spatiotemporal analysis of the soil water content in a mixed Norway spruce (Picea abies (L) Karst.)-European beech (Fagus sylvatica L.) stand. *Geoderma*, Vol. 112, No. 3-4, pp. 273–287, ISSN 0016-7061.

Serre, M.L.; Christakos, G.; Howes, J. & Abdel-Rehiem, A.G. (2001). Powering an Egyptian air quality information system with the BME space-time analysis toolbox: results from the Cairo baseline year study. In: *GeoEnvIII: Geostatistics for Enironmental Applications*, Monestiez, P.; Allard, D. & Froidevaux, R. (eds), pp. 91-101, Kluwer Academic Publisher, ISBN 978-079-2371-06-9, Dordrecht, The Netherlands.

Shannon, C.E. (1948). A mathematical theory of communication. *Bell System Technical Journal*, Vol. 27, No. 2, pp. 379–423, ISSN 0005-8580.

Shi, Z.; Wang, K.; Bailey, J.S.; Jordan, C. & Higgins, A.H. (2002). Temporal changes in the spatial distributions of some soil properties on a temperate grassland site. *Soil Use and Management*, Vol. 18, No. 4, pp. 353–362, ISSN 1475-2743.

Si, B.C. (2002). Spatial and statistical similarities of local soil water fluxes. *Soil Science Society of America Journal*, Vol. 66, No. 3, pp. 753–759, ISSN 0361-5995.

Snepvangers, J.J.J.C.; Heuvelink, G.B.M. & Huisman, J.A. (2003). Soil water content interpolation using spatiotemporal kriging with external drift. *Geoderma*, Vol. 112, No. 3-4, pp. 253–271, ISSN 0016-7061.

Solow, A.R. & Gorelick, S.M. (1986). Estimating monthly streamflow values by cokriging. *Mathematical Geology,* Vol. 18, No. 8, pp. 785–810, ISSN 0882-8121.

Stein, A. (1998). Analysis of space-time variability in agriculture and the environment with geostatistics. *Statistica Neerlandica,* Vol. 52, No. 1, pp. 18–41, ISSN 1467-9574.

Stein, A.; Kocks, C.G.; Zadocks, J.C.; Frinking, H.D.; Ruissen, M.A. & Myers, D.E. (1994). A geostatistical analysis of the spatiotemporal development of downy mildew epidemics in cabbage. *Phytopathology,* Vol. 84, No. 10, pp. 1227–1239, ISSN 0031-949X.

Stein, A. & Sterk. G. (1999). Modelling space and time dependence in environmental studies. *International Journal of Applied Earth Observation and Geoinformation,* Vol. 1, No. 2, pp. 109–121, ISSN 0303-2434.

Stein, A.; Van Groenigen, J.W.; Jeger, M.J. & Hoosbeek, M.R. (1998). Space-time statistics for environmental and agricultural related phenomena. *Environmental and Ecological Statistics,* Vol. 5, No. 2, pp. 155–172, ISSN 1352-8505.

Sterk, G. & Stein, A. (1997). Mapping wind-blown mass transport by modelling variability in space and time. *Soil Science Society of America Journal,* Vol. 61, No. 1, pp. 232–239, ISSN 0361-5995.

Swinbank, R.; Shutyaev, V. & Lahoz, W.A. (Eds.). (2003). *Data Assimilation for the Earth System,* Kluwer Academic Publisher, ISBN 978-140-2015-93-9, Dordrecht, The Netherlands.

Tóth, T.; Kuti, L.; Fórizs, I.; Kabos, S. & Douaik, A. (2002). Spatial and temporal aspects of soil salinization in a sodic grassland. In: *Proceedings of the Interntional Symposium on Sustainable use and management of soils in arid and semiarid regions,* Fez Cano, A.; Ortiz Silla, R. & Mermut, A.R. (eds.), vol. 1, pp. 276–288, Cartagena, Spain.

Vachaud, G.; Passerat De Silans, A. ; Balabanis, P. & Vauclin, M. (1985). Temporal stability of spatially measured soil water probability density function. *Soil Science Society of America Journal,* Vol. 49, No. 4, pp. 822–828, ISSN 0361-5995.

Van Es, H.M. (1993). Evaluation of temporal, spatial, and tillage-induced variability for parametrization of soil infiltration. *Geoderma,* Vol. 60, No. 1-4, pp. 187–199, ISSN 0016-7061.

Van Es, H.M.; Ogden, C.B.; Hill, R.L.; Schindelbeck, R.R. & Tsegaye, T. (1999). Integrated assessment of space, time, and management-related variability of soil hydraulic properties. *Soil Science Society of America Journal,* Vol. 63, No. 6, pp. 1599–1608, ISSN 0361-5995.

Van Meirvenne, M. & Goovaerts, P. (2002). Accounting for spatial dependence in the processing of multi-temporal SAR images using factorial kriging. *International Journal of Remote Sensing,* Vol. 23, No. 2, pp. 371–387, ISSN 0143-1161.

Van Meirvenne, M.; Pannier, J.; Hofman, G. & Louwagie, G. (1996). Regional characterization of the long-term change in soil organic carbon under intensive agriculture. *Soil Use and Management,* Vol. 12, No. 2, pp. 86–94, ISSN 1475-2743.

Van Wesenbeeck, I.J. & Kachanoski, R.G. (1988). Spatial and temporal distribution of soil water in the tilled layer under a corn crop. *Soil Science Society of America Journal,* Vol. 52, No. 2, pp. 363–368, ISSN 0361-5995.

Van Wesenbeeck, I.J.; Kachanoski, R.G. & Rolston, D.E. (1988). Temporal persistence of spatial patterns of soil water content in the tilled layer under a corn crop. *Soil Science Society of America Journal,* Vol. 52, No. 4, pp. 934–941, ISSN 0361-5995.

Vanderlinden, K.; Ordonez, R.; Polo, M.J. & Giraldez, J.V. (2006). Mapping residual pyrite after a mine spill using non co-located spatiotemporal observations. *Journal of Environmental Quality*, Vol. 35, No. 1, pp. 21–36, ISSN 0047-2425.

Wendroth, O.; Pohl, W.; Koszinski, S.; Rogasik, H.; Ritsema, C.J. & Nielsen, D.R. (1999). Spatiotemporal patterns and covariance structures of soil water status in two north-east German field sites. *Journal of Hydrology*, Vol. 215, No. 1-4, pp. 38–58, ISSN 0022-1694.

Zhang, R.; Shouse, P. & Yates, S. (1999). Estimates of soil nitrate distributions using cokriging with pseudo cross-variograms. *Journal of Environmental Quality*, Vol. 28, No. 2, pp. 424–428, ISSN 0047-2425.

16

Mineral Nitrogen as a Universal Soil Test to Predict Plant N Requirements and Ground Water Pollution – Case Study for Poland

Agnieszka Rutkowska and Mariusz Fotyma
Institute of Soil Science and Plant Cultivation – National Research Institute in Puławy Poland

1. Introduction

Nearly whole (99,7%) territory of Poland, covering about 313 thousand square kilometers is situated in the Baltic Sea basin. This territory is drained by two big rivers Vistula and Odra and seven small rivers flowing directly to this sea. The load of about 145 Gg mineral nitrogen, originating mainly from the agricultural land is discharged yearly by polish rivers to Baltic Sea. Among all countries situated in this catchment Poland contributes in almost 26% to the pollution of Baltic Sea with biogenic substance, nitrogen. In the bulk, this is a serious contribution although calculated for one inhabitant (3,7 kg N per capita) and/or one area unit (4,6 kg N per ha) it belongs to the smallest among other Baltic countries. The big share of agriculture in this load results from the fact that agricultural land covers over 60% of the whole polish territory and constitutes almost 50% of this land in the entire Baltic catchment's. However, no wonder that Helsinki Commission and UE exert the pressure on our country for limiting the outflow of nitrogen, including the one from agriculture.

Natural farming conditions in Poland are poor, due to prevalence of light, sand- derived soils (60% very light and light soils) and unfavorable climate. The agriculture landscape originates from the period of glaciations. Due to the high diversity of the parent rocks and different geological and pedological processes, a substantial number of soil types and sub-types have developed in Poland. According to the Polish classification system there are 35 different soil types and 78 sub-types. The most common are brown soils, grey brown podsolic soils, rusty soils and podsolis soils (Terelak et al. 2000). A much smaller part of land is covered by chernozem soils, rendzina soils, black soils and alluvial soils. Unfertile and acid soils account for 50 to 65 percent of arable land. Soil acidity is linked with the low content of available phosphorus and magnesium and hence constitutes of one of the most limiting factors for soil productivity.

Rainfall mostly occurs during the summer months, and the average annual precipitation varies from 500 mm to 900 mm for the regions of Poland, depending on height above sea level and distance from the Baltic Sea. Average temperature ranges from 7,0 to 9,5 degrees of Celsius. The length of growing season is on average 220 days and is comparable to that in Scandinavian countries. From May to September evapotranspiration in Poland exceeds rainfall, resulting in a continuous water deficit, especially on light soil. Nevertheless, in the early spring the outflow of water from the soil profile is a common phenomena.

Overall, the natural conditions for crop production in Poland are by 30-40% worse in comparison for those in Western Europe. The efficiency of agricultural production is rather low (about 3,5 $t \cdot ha^{-1}$ cereal grain) but the consumption of nitrogen in mineral (about 65 kg $N \cdot ha^{-1}$) and organic (about 30 kg $N \cdot ha^{-1}$) fertilizers are comparable with most of the EU countries. As a consequence the balance of nitrogen is positive (about 50 kg $N \cdot ha^{-1}$) and the nutrient utilization efficiency is rather low (about 57%). To deal with the problems of nitrogen surplus in the country and bad management of nitrogen fertilizers, Ministry of Agriculture and Rural Development launched in late 90ties the program of soil monitoring for mineral nitrogen.

2. Mineral nitrogen test – the purposes and the needs

Mineral nitrogen, widely understood as the sum of NO_3^- and NH_4^+ ions, consists of a very small part in the total pool of soil nitrogen. Over 98% of all nitrogen present in soil is in organic forms, not available for plant roots until it is mineralized by microbes into inorganic forms. Mineral forms of nitrogen are very labile and prone to be lost through volatilization and leaching to the ground water. Thus, in the last years the content of mineral nitrogen (N_{min}) is commonly recognized as a soil test of nitrogen accessibility for crops and as an indicator for potential pollution of ground water with nitrogen compounds.
Van der Paauw (1963), researching the effects of residual nitrogen was the forerunner for investigations on inorganic N in the soil profile. Later, research in various countries led to N fertilization recommendations based on the linear relationship between N_{min} in the rooting zone of the crop at the beginning of vegetation and the optimum N fertilizer rate for the crop.
In the Netherlands, N fertilizer recommendations are the function of soil type, whereby "a" and "b" represent a coefficient of the linear relationship between recommendation and N_{min} accordingly to the equation (1):

$$N_{rec} = a - b \times N_{min} \tag{1}$$

(as cited in Van Cleemput et al. 2008). In Belgium, an N index method has been proposed. However, this index includes numerous factors, up to 18, as N mineralization, soil compaction, possible N leaching and is calculated as follows (2):

$$N\ index = X_1 + X_2 + X_3 + \ldots + X_{16} + X_{17} + X_{18} \tag{2}$$

The optimum N fertilization recommendation is calculated accordingly to the equation (3):

$$N_{rec} = a - b \times N\ index, \tag{3}$$

where a and b depend on the cultivar and the destination of the harvested products (Vandendriessche et al., 1992). In France and United States of America, the N balance sheet method was developed. The theoretically recommended N fertilizer doses are based on the balance between the crop requirement minus the amount of N_{min} present before sowing plus the mineralization. The residual N_{min} in the soil profile at harvest to the mean rooting depth consists of the amount of mineral N that remains in the rooting zone after optimum N fertilization. The practical N fertilization recommendation can further be adjusted according to expected losses. The potential losses are estimated at between 5 and 20%, depending mostly on soil texture (Carter et al., 1974). This method has been applied also in China, with the following approach (4):

$$W_{input} = W_{output} - \Delta W - (W_n - W_{n+m}), \quad (4)$$

where W_{input} is the N requirement, W_{output} – N requirement of the target yield, ΔW – volatilized N (N mineralized + subsoil mineral N + dry deposition N + wet deposition N, W_n - N available before planting, W_{n+m} – N available after harvest (as cited in Van Cleemput et al., 2008).

In Western Europe, investigations on N_{min} as a soil test for fertilizers recommendations have started at the beginning of 1970s (Nommik 1966, Scharpf and Wehrman 1975) and in Poland in the beginning of 1990s (Fotyma 1995, 1996). Of course, for fertilizer recommendations, N_{min} test needs a calibration, e.g. critical value for crop nutrition has to be estimated. In pioneer German studies (Scharpf, 1977) by calibrating this test for fertilizer recommendations the simple principle "kilogram for kilogram" has been proposed. It means that each kilogram of mineral nitrogen in the soil profile reveals the fertilizer value equal to one kilogram of nitrogen applied in mineral fertilizers. In Polish approach (Fotyma, 1995) the effect of soil mineral nitrogen has been evaluated by using the N_{min} fertilizer replacement value (N_{min} RFV). RFV is a multiplying factor for the content of N_{min} to a given depth in the soil profile in order to obtain the equivalent amount of nitrogen in mineral fertilizers. When RFV value is equal to 1 then the principle "kilogram for kilogram" comes practically into force. This value depends on crop as well as on soil and climatic conditions and changes in the range 0,6 – 1,1 (Fotyma et al., 2007).

Similarly as in the case of all methods above, evaluation of RFV values is laborious and numerous data from experiments including at least 5-6 nitrogen rates are needed for the process. Moreover, the values reveal a strong variability depending on soil and climatic conditions thus results can be generalized only roughly. The lack of universality of RVF poses a serious drawback in direct application of the N_{min} soil test and since late 80ties, another approach gains in popularity. Allowing for the difficulty of evaluating RFV values for fertilizer recommendation, determination of these values has been substituted by establishing the classes of N_{min} content in the soil profile, like classes of available phosphorus and potassium. For such calibration of N_{min} test the results of soil monitoring for the content of mineral nitrogen offer a good solution.

In the late of 90s, N_{min} test starts to be used as a tool for predicting a risk of soil and water pollution by nitrates (Verhagen & Bouma 1997, Enckevort et al. 2002). Nitrates are easily leached down the soil profile to the ground water thus, for environment protection, the content and distribution of nitrates in the soil profile in autumn are of paramount importance. The EU Nitrate Directive (1991) precisely established its content in at 50 mg dm^{-3} NO_3 or 11,3 mg NO_3-N dm^{-3} of drinking water. This limit is a reference value for the soil ground water as well, though the ground water with an exception of shallow wells is not directly used for this purpose in general. However direct estimation of nitrate concentration in a ground water is a cumbersome and costly procedure. Therefore, it seems interesting to calculate this concentration based on nitrates content in the bulk of soil.

This paper is focused on the parameterization of N_{min} soil test both for fertilizer recommendation and environment protection, using numerous data from the soil monitoring program.

3. Materials and methods

The monitoring program for the content of mineral nitrogen has been launched by the Polish Ministry of Agriculture and Rural Development in 1997. Soil samples have been collected in fixed sites all over the country, twice a year – in spring and autumn to the depth of 90 cm in 0-30, 30-60 and 60-90 cm layers. Soil samples were frozen, and the content of N_{min} was determined in each layer by the colorimetric method in 1% extract of K_2SO_4. The content of N_{min} was expressed in mg NO_3-N and NH_4- N·kg^{-1} soil dry matter and recalculated into kilogram of N_{min} in a given soil layer per ha, called further N_{min} amount. Additionally, at the beginning of this project in all monitoring sites, soil texture has been determined. The technical part of the program, including soil sampling and analysis was entrusted to 17 agrochemical laboratories supervised by the Ministry. The results of analysis were collected in the central database at the Institute of Soil Science and Plant Cultivation – National Research Institute in Puławy, processed statistically and reported in due course.

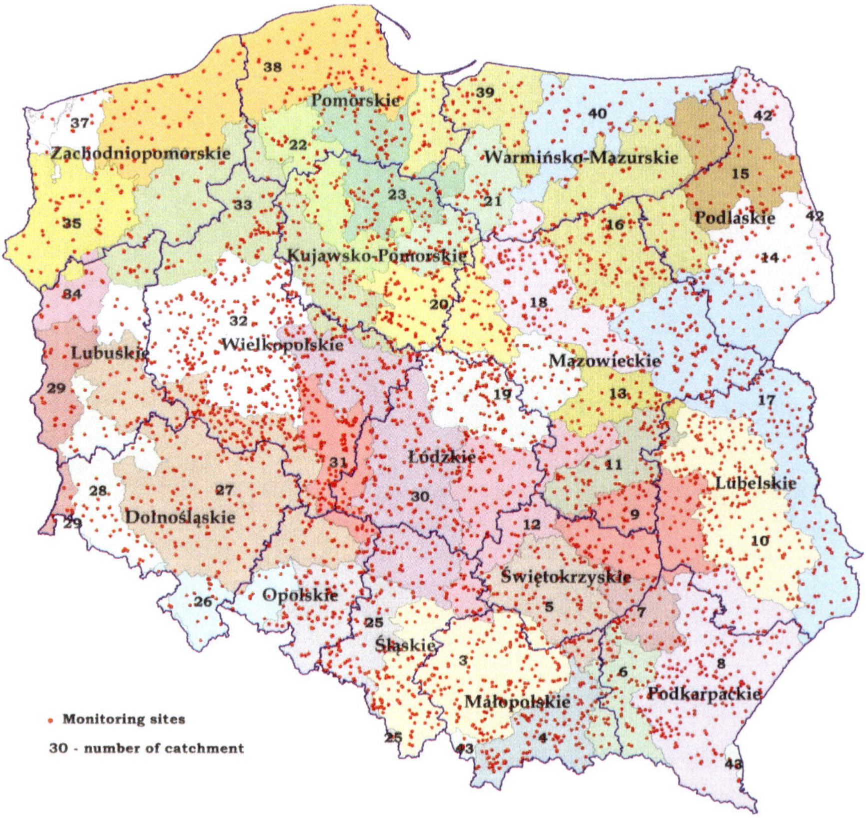

Fig. 1. Localization of the monitoring sites in Poland

In this paper, the results for soil samples, collected from about 5100 sites, in the years 1997-2006 are presented as the first stage of the monitoring program. The total number of samples (records) in the data base is about 95000. Each record includes data for three soil layers and two forms of mineral nitrogen NH_4-N and NO_3-N. The monitoring sites were established on arable land, randomly distributed all over Poland (Fig 1). The distribution of obtained data concerning concentration and content of N_{min} was far from normality. Therefore, the median value was used as a characteristic of a mean and percentile as a measure of date distribution. The preliminary statistical analysis shows the significant dependence of N_{min} content on the soil texture. For this reason, the normative of mineral nitrogen were calculated for each of four textural soil groups: very light (up to 10% of particles below 0,02 mm, light (11 - 20% of particles below 0,02 mm, medium (21-35% of particles below 0,02 mm and heavy (above 35% of particles below 0,02 mm). Similarly, to five classes of soil reaction and P, K, and Mg content officially used in Poland, five intervals of N_{min} content were proposed. Each following interval contains 20% of ordered observations and thus five classes of N_{min} were distinguished - very low, low, medium, high and very high.

4. Results

4.1 Content of mineral nitrogen forms in soil profile

The content of mineral forms of nitrogen has been presented in table 1. The quantity of nitrates and ammonia strongly depends on soil texture, the soil layer and a period of soil sampling. The content of nitrates in soil are higher in autumn than in spring and increases from the very light to heavy soils. The content of nitrate in the upper soil layer is higher in autumn than in spring while the reverse is true for the deepest soil horizon. Ammonia content does not practically differ between the sampling periods and soil textural groups and NH_4 – N is much more evenly distributed in the soil profile irrespective of the season and soil texture. Generally, nitrates content is higher than the content of ammonia.

N form	Soil layer cm	Spring				Autumn			
		Soil texture				Soil texture			
		very light	light	medium	heavy	very light	light	medium	heavy
N_{min}	0 - 30	6,4	7,6	8,4	9,0	8,7	10	10,6	11
	30 -60	3,8	5,5	6,1	7,1	4,7	5,4	6,1	6,6
	60 -90	3,4	5,1	4,8	5,4	3,3	4,2	4,0	4,3
NO_3 - N	0 - 30	3,3	4,4	5,5	6,5	5,6	7,1	7,8	8,3
	30 -60	2,0	3,5	4,0	5,2	2,9	3,7	4,1	4,9
	60 -90	1,9	3,4	3,2	3,8	2,0	2,7	2,5	2,8
NH_4 - N	0 - 30	2,7	2,5	2,2	1,8	2,4	2,2	2,0	1,6
	30 -60	1,4	1,4	1,5	1,3	1,4	1,2	1,4	1,1
	60 -90	1,1	1,3	1,1	1,2	1,0	1,1	1,1	1,0

Table 1. Median of mineral nitrogen forms in soil profile depending on soil layer, soil textural group and sampling period in mg N· kg^{-1} soil

4.2 Total amount of mineral nitrogen in soil

For practical purposes, the content of mineral nitrogen, expressed in mg N $\cdot kg^{-1}$ soil has been recalculated into the nitrogen amount expressed in kg N $\cdot ha^{-1}$ using standardized value of soil density for selected textural soil groups. Generalized values of soil density for each soil textural group as well as a conversion factor used for recalculation of N content into N amount are presented in table 2.

Soil texture	Soil density [kg·m^{-3}]	Conversion factor for 30 cm soil layer* (rounded)
Very light	1533	4,6
Light	1500	4,5
Medium	1433	4,3
Heavy	1300	3,9

* kg N·30cm = mg N·kg^{-1} soil x W

Table 2. Soil density depending on soil textural classes

The tables 3 – 5 shows general statistics for the central tendency and the scatter of the data, in the form of percentiles for amounts of N_{min} and amounts of NO_3 – N. Nitrate nitrogen is the form mostly preferred by plants and simultaneously very prone to leaching to the ground and open waters. Three soil layers, or rather profiles 0-30 cm, 0-60 cm and 0-90 cm have been distinguished. Fertilizer recommendations are usually based on the mineral nitrogen amount in these profiles with the preference for 0-60 cm and/or 0-90 cm profiles. The most interesting is the soil profile 0-90 cm which is explored by the roots of common crops. The difference in the amounts of nitrates in this profile between autumn and spring periods of soil sampling was further considered as nitrogen losses to the underground water. In the soil profile 0 - 90 cm (table 5) the median quantity of N_{min} amounted to 65 – 95 kg N ha^{-1} in spring and 79-96 kg N ha^{-1} in autumn. The quantity of NO_3 – N ranged from 35 – 69 kg N ha^{-1} in spring and 52 -72 kg N ha^{-1} in autumn. The amount of nitrates and total mineral nitrogen increased from the very light soils towards heavy ones. The difference between the N_{min} and NO_3 –N amounts in spring and autumn periods of soil sampling diminished in the same direction.

Characteristics	Spring				Autumn			
	Soil texture				Soil texture			
	Very light	Light	Medium	Heavy	Very light	Light	Medium	Heavy
Number of samples	10784	19585	8718	8710	10821	19584	8792	8016
Median	29,0 (14,7)	33,3 (19,8)	34,8 (22,9)	35,1 (25,3)	39,1 (26,2)	44,5 (32,4)	45,0 (33,5)	43,7 (33,5)
20% - percentile	18,8 (7,36)	22,0 (10,3)	22,5 (12,3)	22,6 (14,0)	23,5 (12,4)	27,4 (16,3)	26,8 (17,0)	25,7 (17,2)
40% - percentile	25,3 (11,9)	29,7 (16,2)	30,6 (19,1)	30,4 (21,4)	34,0 (21,1)	38,7 (26,5)	38,6 (27,6)	37,0 (27,1)
60% - percentile	33,1 (17,9)	37,8 (23,8)	40,3 (27,2)	40,2 (29,6)	45,1 (31,7)	51,7 (38,7)	52,7 (40,8)	50,3 (40,6)
80% - percentile	45,5 (28,0)	52,2 (35,5)	55,6 (39,9)	55,8 (42,5)	63,9 (48,7)	72,4 (58,0)	74,8 (59,9)	72,9 (62,0)
100% - percentile	105 (74,0)	108 (79,2)	112 (82,4)	111 (83,5)	143 (119)	145 (126)	146 (126)	145 (128)

Table 3. Statistical characteristics of N_{min} and N-NO_3 (in parenthesis) amounts in the soil profile 0-30 cm for soil textural groups and sampling date (kg N· ha^{-1})

Characteristics	Spring				Autumn			
	Soil texture				Soil texture			
	Very light	Light	Medium	Heavy	Very light	Light	Medium	Heavy
Median	47,8 (25,3)	58,5 (36,6)	64,8 (44,3)	67,1 (19,5)	62,1 (41,4)	71,1 (51,3)	74,3 (54,8)	73,3 (56,9)
20% - percentile	30,3 (12,4)	37,8 (18,9)	41,6 (23,8)	42,9 (26,3)	38,2 (20,2)	43,7 (27,0)	44,7 (28,0)	44,1 (29,6)
40% - percentile	41,4 (20,2)	51,3 (30,2)	56,6 (36,9)	59,0 (41,7)	53,8 (34,0)	61,4 (42,4)	64,1 (45,3)	63,2 (47,6)
60% - percentile	55,2 (30,8)	67,0 (44,1)	73,9 (52,2)	76,8 (58,9)	71,7 (50,1)	81,9 (61,5)	86,5 (65,3)	85,6 (68,2)
80% - percentile	77,7 (49,2)	93,7 (67,7)	103 (77,5)	106 (84,3)	102 (76,1)	116 (92,2)	122 (98,2)	122 (104)
99% - percentile	187 (139)	196 (157)	202 (167)	203 (171)	235 (208)	247 (218)	248 (221)	247 (223)

Table 4. Statistical characteristics of N_{min} and $N\text{-}NO_3$ (in parenthesis) amounts in the soil profile 0-60 cm for soil textural groups and sampling date (kg N· ha^{-1})

Characteristics	Spring				Autumn			
	Soil texture				Soil texture			
	Very light	Light	Medium	Heavy	Very light	Light	Medium	Heavy
Number of samples	10924	19902	8844	8024	10913	19829	8867	8068
Median	65,3 (35,3)	81,4 (52,5)	90,9 (62,2)	94,8 (69,4)	79,3 (52,4)	89,7 (63,9)	95,4 (68,0)	96,3 (72,5)
20% - percentile	40,8 (17,0)	51,3 (26,5)	57,7 (33,1)	59,7 (36,7)	48,8 (26,2)	55,1 (33,3)	57,7 (35,5)	57,3 (37,4)
40% - percentile	56,6 (28,4)	70,7 (42,7)	79,4 (51,9)	82,3 (57,7)	69,0 (43,2)	77,5 (52,8)	82,4 (56,5)	82,7 (60,1)
60% - percentile	75,8 (43,7)	93,6 (62,9)	104 (73,9)	109 (82,0)	92,0 (63,0)	104 (76,2)	110 (81,8)	111 (87,3)
80% - percentile	107 (69,4)	131 (96,3)	145 (108)	150 (119)	132 (95,6)	148 (115)	159 (124)	157 (131)
99% - percentile	261 (196)	283 (223)	292 (234)	298 (236)	307 (253)	32 (267)	324 (271)	324 (273)

Table 5. Statistical characteristics of N_{min} and $N\text{-}NO_3$ (in parenthesis) amounts in the soil profile 0-90 cm for soil textural groups and sampling date (kg N· ha^{-1})

Numerous factors, apart from the soil texture, influenced the amount of mineral and nitrate nitrogen in soil and differences between the autumn and spring period of soil sampling. In spring, the quantity of nitrogen is affected mainly by crop grown in the previous year, while in autumn by crops grown in this particular year (Table 6). The preceding and actually grown crops have been classified by means of cluster analysis into three groups differing in the amount of N_{min} and NO_3 – N left in the soil. Cluster of crops reflects not only the type of particular crop but the agro technical measures for this crop as well, influencing the amount of mineral nitrogen in the soil. Sugar beet, potato and sometimes rape and maize are grown on farmyard manure. Leguminous crops assimilate nitrogen from the atmosphere and for intensive cereals and rape, high doses of mineral nitrogen are being commonly applied. Another factor influencing the amount of mineral and nitrate nitrogen in the spring period and the difference of these amounts between spring and autumn period is the winter precipitation. This factor will be discussed in the following part of the paper.

Cluster	Soil layer					
	0-30 cm		0-60 cm		0-90 cm	
	N_{min}	NO_3-N	N_{min}	NO_3-N	N_{min}	NO_3-N
I Intensive cereals, rape, sugar beet, maize, potatoes, leguminous crops and mixture with grasses, vegetables, grassland	Spring					
	35,7	22,7	63,8	42,2	87,1	59,5
	Autumn					
	46,3	34,7	76,7	57,4	99,4	72,7
II Extensive cereals, non-leguminous fodder crops, field grasses	Spring					
	31,3	17,7	50,0	28,0	65,7	37,3
	Autumn					
	36,2	22,4	56,2	35,6	70,6	43,6
III Set aside, bushes, forests	Spring					
	23,5	9,75	39,5	16,0	51,5	22,0
	Autumn					
	25,0	14,0	39,0	15,5	51,0	24,0

Table 6. Amounts of mineral and nitrate nitrogen in spring for clusters of preceding crops and in the autumn for crops grown in the particular year [kg N· ha^{-1}]

The largest amount of mineral nitrogen in soil has been revealed after intensive cereals (winter wheat and triticale), winter rape, row crops fertilized with manure (sugar beet, potato, vegetables and leguminous crops in a pure stand and with a mixture with grasses. The smallest quantity has been found after extensive cereals (rye, oats, cereal mixtures, non-leguminous fodder crops, grasses and on set aside land. Almost the same regularities appeared in the samples collected in autumn. The greatest differences in N_{min} and NO_3 – N quantities between autumn and spring period of soil sampling appeared in the cluster I. The quantities of nitrogen after crops of cluster III between autumn and spring periods of soil sampling in soil layers 0-60 and 0-90 cm were practically the same.

4.3 Application of N_{min} test for fertilizer recommendations

A very serious and still unsolved problem is how to include the results of N_{min} or NO_3 – N test into nitrogen fertilizer recommendations. Generally, two approaches can be considered: quantitative and qualitative. The most straightforward quantitative approach is to express the content of N_{min} in kg· ha^{-1} and to relate its amount to the amount of nitrogen from fertilizers. Based on this approach the simple model has been developed in Europe:

$N_{fer} = N_{crop} - N_{min}$ *RFV, where: N_{fer} - optimal rate of fertilizer nitrogen, N_{crop} – crop nitrogen requirement, RVF- N_{min} replacement value. As has been already pointed out Wehrman et al. (1986) assumed that RVF value is equal to 1, while in Polish investigations this value depends on crop as well as on soil and climatic conditions and changes in the range 0,6 – 1,1 (Fotyma et al., 2007). The lack of universality of RVF poses a serious drawback in quantitative application of the N_{min} soils test and since late 80ties qualitative approach gains popularity. This particular approach to soil test of mineral nitrogen has been implemented in Germany (Muller & Gorlitz 1986, 2000) and lately in Poland (Fotyma, 2010, Rutkowska & Fotyma, 2009). The first step in qualitative approach is to calibrate the soil test. Calibration is commonly understood as establishing the test critical value or several classes corresponding to the expected efficiency of a given nutrient. The classical method is to calibrate the soil test against plant indices in the series of field experiments carried on the soil differing in the test values. This is a very cumbersome and expensive procedure practically not applied any further. In Poland N_{min} soil test has been calibrated using the data from monitoring program providing representative data for different natural and agronomic conditions in the country.

The five test classes for mineral and nitrate nitrogen correspond to the percentiles of these forms in soils of different texture (Table 7). Each following interval contains 20% of ordered observations and thus five classes of N_{min} and NO_3-N were distinguished - very low, low, medium, high and very high nitrogen content. Five calibration classes have been chosen in order to comply with the numbers of availability classes for other nutrients, officially used in Poland.

Soil texture	Form of nitrogen	Class of content (percentile interval) kg $N_{min} \cdot ha^{-1}$				
		Very low 0-20%	Low 20-40%	Medium 40-60%	High 60-80%	Very high 80-100%
Very light	NO_3-N	to 17	18-28	8-11	29-44	over 70
	N_{min}	to 41	42-57	58-76	77-107	over 108
Light	NO_3-N	to 26	27-43	44-63	64-96	Over 96
	N_{min}	to 58	59-79	80-104	105-145	over 132
Medium	NO_3-N	to 33	36-52	53-74	75-108	over 109
	N_{min}	to 58	59-79	80-104	105-145	over 146
Heavy	NO_3-N	to 37	38-58	59-82	83-119	over 120
	N_{min}	to 60	61-82	83-109	110-150	over 151

Table 7. Classes of the mineral nitrogen N_{min} and nitrate nitrogen NO_3-N in spring in the soil profile 0-90 cm

Soil test for NO_3 – N seems to be more suitable for fertilizer recommendations than N_{min} test. This is because nitrates are directly available for plants. Total amount of N_{min} includes also NH_4 – N form, but NH_4^+ cation is preferred only by some specific group of plants, and besides undergoes the sorption on the soil colloids. The calibration figures are given for the soil profile 0-90 cm. For technical reasons soil sampling is sometimes limited to the depth of 0-60 cm. However data obtained from the monitoring program showed the close correlation between the amounts of mineral nitrogen in the two soil layers 0-60 and 0-90 cm.

Classes of N_{min} quantity in spring corresponds in reverse order to the classes of crop nitrogen requirements In the class of very low mineral nitrogen content, crop requirements are the highest and vice versa. In fertilizer recommendations if the content of N_{min} or NO_3–

N in the spring falls in "medium " class farmer is enhanced to stick to the standard N rate and its common splitting pattern. If the N_{min} content proved to be "very low" or "low" then the standard 1st rate of nitrogen fertilizer should be increased and applied as early as possible. The reverse is recommended when mineral nitrogen content is "high" or "very high". Usually for deep rooting and/or winter crops the content of N_{min} in the soil profile 0 – 90 cm and for shallow rooting and/or spring sown crops the content of N_{min} in the soil profile 0-60 cm is taken into consideration. Thus, this qualitative approach to mineral nitrogen soil test makes it at present rather as an auxiliary tool for nitrogen recommendations system. The calculation of standard nitrogen dose is still in foreground. This problem lies however outside the scope of this paper and is discussed elsewhere (Jadczyszyn, 2000, 2009).

4.4 Application of N_{min} soil test for environment protection

Agriculture in Poland is a main source for the leaching of nitrates to groundwater and surface waters. It is estimated that in the Baltic catchment over 60% of nitrogen in the riverine outflow originates from the diffuse sources. The diffuse outflow, as well as point sources are considered the main sources of nitrogen discharged by the Vistula and Oder rivers, the two out of seven largest rivers feeding the Baltic Sea. Fertilizer recommendations alike the problem is how to include the mineral nitrogen soils test in predicting the potential threat for ground water by excess of nitrates. It is clear that main losses of nitrate to ground water took place in winter and early spring periods due to the outflow of excess of water. Therefore, for environment protection, the content and distribution of nitrate in the soil profile in autumn are of paramount importance. On the base of data gathered from the soil monitoring program carried on in Poland since 1997, the threat from nitrate excess has been qualitatively assessed. The five classes of nitrate content in the soil profiles 0-90 cm and 60-90 cm corresponding to its pentiles distribution were distinguished (Table 8).

Soil texture	Depth cm	Class of nitrate content kg $N\text{-}NO_3$ ha^{-1}					
		Very low	Low	Medium	High	Very high	Median
Very light	0-90	≤27	27-43	44-62	63-94	≥97	52
	60 - 90	≤24	5-7	8-11	12-20	≥42	9
Light	0-90	≤233	34-53	54-76	77-115	≥116	64
	60 - 90	≤24	5-8	9-13	14-23	≥54	10
Medium	0-90	≤237	38-59	60-85	85-131	≥125	71
	60 - 90	≤25	6-9	10-15	16-27	≥60	12
Heavy	0-90	≤240	41-61	62-88	89-134	≥132	74
	60 - 90	≤26	7-10	11-17	18-30	≥62	13

Table 8. The classes of nitrate content in autumn depending on the depth in the soil profile

The low content of NO_3-N in the very light and light soils and the medium one in medium and heavy soils are proposed in Poland as the threshold ranges of soil threatening with excess of nitrogen. These limits are roughly 40 kg $N\text{-}NO_3$ ha^{-1}, and 60 kg $N\text{-}NO_3 \cdot ha^{-1}$ for the very light/ light and medium/ heavy soils, respectively. Similar limits were established in few European countries as well. In Belgium the limit is set at 90 kg $N\text{-}NO_3 \cdot ha^{-1}$ in the soil layer 0-90 cm, in Germany at 45 kg $N\text{-}NO_3 \cdot ha^{-1}$ in mineral soils and at 90 kg in organic soils. The validity of our approach was confirmed by comparing the amounts of nitrates in autumn and spring periods (Table 9). It results from the table 7 that by low content of nitrate in the very light/light soils and medium one in the medium/heavy soils the losses of nitrates from the soil profile 0-90 cm in winter period are negligible.

Class of $NO_3 - N$ in autumn	kg NO_3-N ha^{-1} in 0-90 cm	Soil texture			
		Very light	Light	Medium	Heavy
Very low	autumn	16	22	24	24
	spring	21	33	40	44
	difference	+5	+11	+16	+20
Low	autumn	34	43	45	49
	spring	31	46	54	60
	difference	-3	+3	+11	+11
Medium	autumn	52	64	68	73
	spring	36	53	65	71
	difference	-16	-11	-3	+2
High	autumn	76	92	99	107
	spring	43	62	72	81
	difference	-33	-30	-27	-26
Very high	autumn	132	156	166	172
	spring	51	72	84	92
	difference	-81	-72	-82	-79
Average for the soil	autumn	52	64	68	74
	spring	34	51	61	69
	difference	-18	-13	-7	-5

Table 9. Reserves of nitrates in autumn and following spring depending on the nitrate classes in autumn

4.5 Simulated nitrates concentrations in ground water

As has been already mentioned, for environment protection the key point is concentration of nitrate in the drinking water, which in fact originates from ground water and not the content of nitrate in the bulk of soil. The EU Nitrates Directive has stated the limit in drinking water to 50 mg· dm^3 NO_3^- or 11,3 mg· dm^3 $NO_3 - N$. However direct estimation of this concentration is laborious and difficult. The most prone to leaching processes are nitrate in the deeper soil layers being in contacts with ground water, particularly in the periods of water saturation to full soils water capacity. Soil saturation to full water capacity in the depth 0-30 cm and 30-60 cm occurs only in the early spring but the subsoil nearly through the whole vegetation period. In the paper the indirect method for estimation the safe limit of nitrate content in the bulk of soil by comparison with calculated nitrate concentration in the soil water at the depth 0 -30, 30 - 60 and 60-90 cm was proposed, using the following formulas (5) and (6):

$$\text{Concentration (mg N-NO}_3\ \text{dm}^{-3}) = \text{content (mg N-NO}_3\ \text{kg}^{-1}) * R, \quad (5)$$

$$\text{and } R=1/(W/D), \quad (6)$$

where: W – soil water content in dm^3 dm^{-3} (v/v), D- soil bulk density in kg· dm^{-3}

The soil water content W at this point depends on soil texture. Standard W values for soil saturation to full water capacity and D values for soil categories in Poland are given in table

10 (Fotyma, & Ślusarczyk, 1992). The simulated concentration of nitrates in the soil layer 0-30, 30-60 and 60-90 cm are presented in Table 11.

Soil texture	Soil bulk density G·dm-3 D	Soil water capacity v/v, dm3· dm-3 W(1)	Soil water capacity v/w, dm3· kg-1	Conversion factor **R**	Coefficient $\mathbf{W_z}$(2)
Very light	1,533	0,125	0,082	12,40	2,65
Light	1,500	0,174	0,116	8,62	1,91
Medium	1,433	0,291	0,203	4,93	1,14
Heavy	1,300	0,433	0,331	3,02	0,77

1) coefficient to recalculate the content of nitrates in bulk soil into nitrate concentration in soil solution
2) coefficient to recalculate the quantity of nitrates in the soil layer of 30 cm into concentration of nitrates in soil water

Table 10. Soil parameters for recalculation the content of nitrate in the bulk of soil to its concentration in soil solution assuming soil saturation to full water capacity

Soil texture	Soil layer 0-30 cm		Soil layer 0-30 cm		Soil layer 0-30 cm	
	autumn	spring	autumn	spring	autumn	spring
Very light	66,5	37,0	33,3	28,2	22,2	20,9
Light	59,6	36,3	29,4	28,5	21,6	27,6
Medium	37,9	26,1	19,7	19,2	11,3	18,8
Heavy	25,4	19,0	13,9	15,4	8,2	10,9
Average	48,2	30,5	23,4	22,2	13,8	16,3

Table 11. Simulated concentration of nitrates in the soil layer 0-30, 30-60 and 60-90 cm under assumption that the soil is saturated to full water capacity

The quantity of nitrates decreased from very light to heavy soils, which results in differences in their soil water capacity. Concentrations NO_3- N both in autumn and spring period diminish down the soil profile, independently of the soil texture. Except the heavy soils, the concentration of nitrates in the upper soil layer is strongly on behalf the autumn period in the intermediate soil layer still on behalf of autumn and in subsoil on behalf of spring period with except of the light soil. As a consequence of that, the downward movement of nitrates in the winter period with the water soaking into the soil profile occurs. Besides, the data in Table 13 indicate, that part of nitrates translocated from the upper soil layers is retained by the subsoil, particularly in medium and heavy soils. However, the remaining part is leached in soil profile below the rooting zone and potentially dispersed to the ground water.

In the monitoring program, soil samples to the depth 0-90 cm have been collected thus the concentration of nitrates below 90 cm can be simulated only. It has been assumed that the total quantity of lost NO_3 – N during the winter period can be found in the depth of 90-120 cm and that the soil texture on this depth is the same as in subsoil. Hence, the quantity of nitrates expressed in kg $N \cdot ha^{-1}$ removed to the soil layer 90-120 cm can be recalculate into nitrates concentration expressed in mg NO_3- N mg dm^3 in soil water using the coefficient W_z (Table 10). The value of W_z coefficient decreases from very light towards heavy soils as a result decreasing in the water capacity. For example, if the value of W_z is 2,67, it means (Table 12) that 1 kilogram NO_3 – $N \cdot ha^{-1}$ leached to the depth of 90-120 cm in very light soil increases nitrates concentration in soil water by 2,65 mg NO_3- N mg· dm^3. The value of W_z

makes possible to assess the accessible amount of nitrogen leached to the soil 90-120 cm) – 4 kg NO_3- N mg·dm^3 for very light soil, 6 kg NO_3- N·mg dm^3 for light soil, 10 kg NO_3- N mg· dm^3 for medium soil and 15 kg NO_3- N mg· dm^3 for heavy ones. Of course, ground water with an exception of shallow wells is not the drinking water. Therefore, the simulated concentrations of the nitrates give only approximate values.

The differences between the autumn and the following spring in the quantity of nitrates can be treated as nitrogen losses to ground water. Of course, it is only a simplification, because part of the nitrates might be denitrified during the autumn – winter period, nevertheless, in the paper, the terms difference and losses would be used intermittently. In autumn, the losses in the medium class over winter period from medium and heavy soils are negligible and on light soils tolerable. Only very light, coarse - textured soils reveals a dangerous for nitrate leaching to the ground water. From the data presented it can be concluded, that the downward placement of nitrates below the depth 90 cm over the winter period depends on several factors among which soil texture and reserves of nitrates in autumn are the most important. The less influencing ones are intensity of crop production and the sum of rainfall from November to March. All the factors were aggregated at the administrative levels (NUTS – 2) of voivodship (Table 12) and included into cluster analysis. The results have been presented in the following subsection.

5. Discussion

The main achievement of this work was to establish the normative for N_{min} content on the base of a soil monitoring program carried on in Poland since 1997. The normative concerning spring period of soil sampling can be used for fertilizer recommendations, and that concerning autumn period of soil sampling for environment protection. The official fertilizer recommendation system covers the whole territory of the country and is focused on individual farms. As it has been already mentioned N_{min} soil test is, at present, an auxiliary tool superimposed on the standard nitrogen doses resulting from balance approach. Due to a very large number of farms in Poland, 1,5 millions altogether including about 0,3 millions of market oriented ones performing the N_{min} soil tests for each field in each farm is out of a question. Therefore, the results of monitoring program in its spring part, which are made available for general agricultural community are treated as reference points for similar farming conditions (Jadczyszyn, 2009).

The normative concerning autumn period of soil sampling as well as a difference in mineral nitrogen content between spring and autumn periods of soil sampling are used for delineating the areas threatened by surplus of nitrogen from agriculture. Such approach is presented on the NUTS-2 scale, i.e. for administrative units the highest order, called provinces or voivodships. The whole territory of the country is divided into 16 provinces, which are further divided into 314 counties (Figure 1). This division is based rather on historical and socio-economical and not on natural grounds. However, most of the statistical Kopiński, 2006; Kuś et al., 2006). For this reason, as well as for full transparency, the problem of nitrogen surplus from agriculture is presented for administrative level NUTS – 2. Discussing this problem not only the results of the soil monitoring program for mineral data concerning agriculture are available on NUTS-2 scale only and this level is considered in most of the publications dealing with regionalization of agricultural policy (Krasowicz & nitrogen but other relevant factors have been taken into consideration. These factors are listed in Table 14 and some of them will be discussed briefly.

Voivodship	Winter rainfall (1)	Soil texture (2)	Cropping system intensity (3)	N balance difference kg N·ha^{-1} (4)	Nitrates content in autumn kg N·ha^{-1} (5)	Nitrate loss kg N·ha^{-1} (6)
Dolnośląskie DLN	147	22,1	78,0	86,3	87	14
Kujawsko-pomorskie KUJ	178	71,2	65,6	85,3	67	14
Lubelskie LUB	171	71,9	66,3	75,5	61	-1
Lubuskie LUS	191	86,9	63,3	63,0	77	32
Łódzkie LOD	220	84,3	48,0	62,9	64	18
Małopolskie MAL	201	28,2	66,1	57,6	68	2
Mazowieckie MAZ	174	88,6	50,2	54,9	51	7
Opolskie OPL	207	42,5	79,8	50,5	73	11
Podkarpackie PDK	200	23,1	72,1	50,2	65	4
Podlaskie PDL	188	83,8	39,0	48,4	45	9
Pomorskie POM	208	66,3	61,9	48,1	50	11
Śląskie SLS	257	56,1	66,0	45,3	71	7
Świętokrzyskie SWT	207	64,5	67,6	42,1	67	-3
Warmińsko-mazurskie WAM	206	57,2	60,4	40,1	44	7
Wielkopolskie WLP	186	93,1	65,8	28,7	78	27
Zachodniopo-morskie ZAP	229	78,3	63,0	22,4	65	25
Polska	187	64,7	62,8	11	**63**	12

1) sum of rainfall in winter period (January 1st – March 31st)

2) % of coarse textured soil (very light and light)

3) % of samples from fields under intensive cropping system (see table 6)

4) difference between N input and N output annually (kg N·ha^{-1}).

5) amount of autumn nitrate N in soil profile 0-90 cm in kg ha $^{-1}$

6) differences between nitrate N content in autumn (1997-2005) and spring in kg ha $^{-1}$ (1996-2006)

Table 12. The losses of nitrate in winter period and influencing factors for voivodships in Poland

Winter rainfall. Analysis of data from monitoring program reveals that the nitrate losses in the period between spring and autumn soil sampling were related to the sum of rainfall in the period November to March. In the very dry years with winter rainfall below 100 mm the losses of nitrate were negligible. In the wet years with winter rainfall over 250 mm the losses of nitrates reached over 15 kg NO_3-N·ha^{-1}.

Soil texture influenced strongly the content and losses of nitrate, and this factor was included in N_{min} normative. Particularly prone to nitrate losses are very light and light soils which share is very differentiated among the provinces. It is necessary to stress again upon the prevalence of light, sandy soils in Poland that contributes significantly to the nitrate outflow from the country.

Cropping system in Poland is partly linked to natural farming conditions, i.e. soil quality and climate. However, intensive, high demanding crops from economical reasons are often grown on light soils, and poor soil conditions are compensated by higher doses of mineral fertilizers.

Nitrogen balance is calculated in Poland according to OECD rules as a soil surface balance (OECD, 1999). N balance was positive in whole Poland. The mean N surplus was 57 kg N·ha^{-1}. The greatest surplus has been found in Kujawsko-Pomorskie and Wielkopolskie provinces. These regions are featured by a very intensive animal production and high consumption of mineral fertilizers. The smallest one has been found in Małopolskie and Podkarpackie voivodships characterized by the extensive agricultural production.

Factors listed in table 14, influencing the threat from nitrogen surplus have been included into the cluster analysis performed by Ward's method. The voivodhips were grouped in three clusters presented in table 15 and on the Figure 2.

Cluster	Voivodship	Factor 1 Winter rainfall 1)	Factor 3 Soil texture 2)	Factor 4 Cropping system intensity 3)	Factor 5 Autumn nitrate value 4)	Factor 6 Nitrogen surplus kg ha $^{-1}$ 5)	Factor 7 Nitrate loss 6)
I	KUJ, LUB, MAL, OPL, PDK, SWT	187	46,2	70,8	69,7	63,9	5,85
II	DLN, LUS, SLS, WLP, ZAP	215	78,6	64,5	72,7	39,9	22,7
III	LOD, MAZ PDL, POM, WAM,	199	76,8	51,9	50,8	50,9	10,0

Table 13. Clusters of voivodships (see Table 12)

The largest losses of nitrates were recorded in Dolnośląskie (DLN), Lubuskie (LUS), Śląskie (SLS), Wielkopolskie (WLP), Zachodniopomorskie (ZAP). Most of the area of these voivodships fall in the river Oder and/or 10 small rivers discharging directly to the Baltic Sea catchments. In the remaining voivodships grouped in clusters, I and III nitrates losses were much lower and did not exceed 10 kg $N0_3$ – N ha^{-1}. The reason for the low losses of nitrate nitrogen in voivodships in the cluster I is prevalence of good soils. In voivodships of cluster III, the low nitrate losses could be explained by low quantity of NO_3 – N in soil in autumn and the lowest share of intensive crops in crop rotation. Most of the areas of these voivodship are located in Vistula catchment. Thus, the potential losses of nitrates to the ground water are higher in North – Western part of Poland than in South – Eastern one.

Fig. 2. Regional distribution of NO_3 – N losses over winter in Poland (kg N· ha^{-1}) (see Table 12)

The most effective way for delineating the area threatened by excess of nitrate in drinking water originating from agriculture would be the monitoring program of ground water quality. Such program has been indeed launched in 1997 together with the program of soil monitoring for mineral nitrogen (Igras, 2004). However, the monitoring sites for both programs were fixed separately. Besides, the number of sites of the ground water program was much lower than for N_{min} program. Since 2007, both programs have been merged, and until now the preliminary results are available (Jadczyszyn et al., 2010). Therefore the only representative data concerning nitrate concentration in ground water are showed in the present paper. Nevertheless, these data are simulated and need validation by comparison with the "true" concentration of nitrate in the ground water from the same sampling point. For these simulated data, the cluster analyses has been performed in order to group the voivodship according to the expected concentration of nitrate in the ground water. The percent of sampling points where this concentration exceeds the limit set by Nitrate Directive (Table 14) was also taken into consideration.

The simulated concentrations of nitrates in soil water widely differ among the voivodships. On the base of Table 14 it has been concluded that in five voivodships grouped in clusters III and IV the median of nitrate concentration is very high and only about 40% of soil samples is in accordance with the Nitrate Directive. The soils of remain 11 voivodships are in the range of the safe limit below 11,3 mg NO_3 - N·dm^{-3} but still in quite numbers of samples concentration of nitrate exceed this limit. The problem of delineating areas in Poland threatened by excess on nitrate in ground water needs further investigations, particularly in smaller than NUTS-2 territorial scales.

Cluster	Voivodship in the cluster	Median value mg NO_3-N·dm^{-3} water	% samples <11,3 mg NO_3 – N dm^{-3}	Localization of voivodship in the cathment*
I	LUB, OPL, SWT	0,2	60	LUB, SWT – **V** OPL - **O**
II	DLN, MAL, MAZ, PDK, PDL, POM, SLS, WAM	8,0	55	MAL, PDK, MAZ, PDL, WAM – **V** POM – **V/R** SLS – **V/O** DLN - **O**
III	KUJ, LOD, ZAP	25,2	40	KUJ, LOD - **V/O** ZAP - **O/R**
IV	LUS, WLP	46,8	35	LUS, WLP - **O**

*V – Vistula, O – Oder, R - 10 rivers directly charged to the Baltic Sea

Table 14. Clusters of voivodships according to simulated concentration of nitrates in ground water

6. Conclusions

- In the paper the normative of mineral nitrogen content in arable soils in Poland are presented. These normative correspond to the percentile distribution of the values of NO_3-N and NH_4-N content in the country-wide monitoring programme.
- The five-classes normative (pentiles) are calculated for each of four soil textural classes separately. The method of recalculation the content of nitrate in bulk soil into its concentration in soil water (assuming soil saturation to full water capacity) is proposed.
- The method for simulating nitrate concentrations below the rooting zone was proposed.
- The regions of Poland with high risk of ground water pollution by downward placement of nitrate were designated.

7. References

Carter, J. N.; Jensen, M. E. & Bosma, S. M. (1974). Determining nitrogen fertilizers needs for sugarbeets from residual soil nitrate and mineralizable nitrogen. *Agronomy Journal,* Vol. 66, pp. 319 – 323

Enckevort van, P. L. A.; Schoot van der J. R. & ten Berge van. W. (2002). A review of potential indicators for nitrate loss from cropping and farming systems in the Netherlands. Wageningen, the Netherlands: Plant research International.

Fotyma, E. (1995). Przydatność glebowego testu N_{min} w nawożeniu zbóż ozimych. *Fragmenta Agronomica,* Vol. 3, pp. 59 – 78

Fotyma, E. (1996). Metoda N_{min} w doradztwie nawozowym. *Zeszyty Edukacyjne IMUZ,* Vol. 1, pp. 67 – 74

Fotyma, M. & Ślusarczyk, E. (1992). Optimal threshold values for agricultural soils. *Zalecenia Agrotechniczne IUNG.* Vol. 51, No. 1, pp. 41-60

Fotyma, M; Fotyma, E. & Filipiak, K. 2007. Zastosowanie modelu QUADMOD do interpretacji wyników doświadczeń z nawożeniem azotem i modelowania potrzeb nawozowych roślin. *Fragmenta Agronomica,* Vol. 1, pp. 41-48

Fotyma, M.; Kęsik, K. & Pietruch, Cz. (2010). Azot mineralny w glebach jako wskaźnik potrzeb nawozowych roślin i stanu czystości wód glebowo - gruntowych. *Nawozy i Nawożenie - Fertilizers and Fertilization*, Vol. 38, pp. 5 - 83

Igras, J. (2004). Zawartość składników mineralnych w wodach drenarskich z uzytków rolnych w Polsce. Monografie i Raporty Naukowe IUNG - PIB, Vol. 13, IUNG-PIB Puławy, ISBN 978-83-7562-2008-5

Jadczyszyn, T. (2009). The Polish fertilization recommendation system NawSald. *Nawozy i Nawożenie - Fertilizers and Fertilization*, Vol. 37, pp. 195 - 203

Jadczyszyn, T.; Kęsik, K. & Pietruch, Cz. (2010). Resortowy monitoring zawartości azotu mineralnego w glebach Polski w latach 2007-2009. *Nawozy i Nawożenie - Fertilizers and Fertilization*, Vol. 38, pp. 84 - 110

Krasowicz, S. & Kopiński, J. (2006). Wpływ warunków przyrodniczych i organizacyjno - ekonomicznych na regionalne zróżnicowanie rolnictwa w Polsce. *Raporty PIB, IUNG - PIB Puławy*, Vol. 3, pp. 81-99, ISBN 83-89576-26-0

Kuś, J., Faber, A. & Madej, A. (2006). Przewidywane kierunki zmian w produkcji roślinnej w ujęciu regionalnym. *Pamiętnik Puławski*, Vol. 124, pp. 195 - 210

Muller, S. & Gorlits, H. (1986). The relationships between soil inorganic nitrogen levels and nitrogen fertilizer requirements. *Agricultural Ecosystems and Environment*. Vol. 17, pp. 192 - 211

Nommik, H. (1996). The residual effects of nitrogen fertilizers in relation to the quantities of mineral nitrogen recovered in the soil profile. *Acta Agriculturae Scandinavica*, Vol. 16, pp. 23-35

OECD (1999). National Soil Surface Nutrient Balances: explanatory notes to interpret the data sheets. OECD Secretariat Paris.

Rutkowska, A. & Fotyma, M. (2009). Calibration of soil test for mineral nitrogen in Poland. *Communications in Soil Science and Plant Analysis*, Vol. 40, No. 1-6, pp. 987 -998

Scharpf, H. C. & Wehrman, J. (1975). Die Bedeutung des Mineralstickstoffvorrates des Bodens zu Vegetationsbeginn für die Bemessung der N - Düngung zu Winterweizen. *Landwirschaftlich Forschung* Vol. 32, pp. 110-114

Scharpf, H.,C. (1977). Der Mineralsticsktoffsgehalt des Bodens als Masstab fur den Stickstoffdungerbedarf. Disertation. Hannover

Terelak, H., Krasowicz, S. & Stuczyński, T. (2000). Środowisko glebowe Polski i racjonalne użytkowanie rolniczej przestrzeni produkcyjnej. *Pamiętnik Puławski,* Vol. 120, No. 2, pp. 455-469

Van Cleemput, O. & Hoffman, G., Vermoesen, A. (Eds). (1996). Progress in nitrogen cycling, In: *Plant and Soil Science,* Vol. 68 ,

Van Cleemput, O. ; Zapata, F. & Vanlauve, B. 2008. Guidelines on Nitrogen Management in Agricultural Systems. Training Course Series No. 29. International Atomic Energy Agency, Vienna

Van der Paauw, F. (1963). Residual effect of nitrogen fertilizer on succeeding crops in a moderate marine climate. *Plant Soil,* Vol. 19, pp. 324-331

Vandendriessche, H. (1992). Nitrogen fertilization: Nitrogen fertilization on cropland in relation to the environment. *Landbouwtijdschrift*, Vol. 45, pp. 395 - 402

Verhagen, A. & Bouma, J. (1997). Defining threshold values for residual soil N levels *Geoderma*. Vol. 85.23, pp. 199-211

17

Time-Domain Reflectometry (TDR) Technique for the Estimation of Soil Permittivity

Patrizia Savi[1], Ivan A. Maio[1] and Stefano Ferraris[2]
[1]*Dipartimento di Elettronica, Politecnico di Torino, C.so Duca degli Abruzzi 24, 10129 Torino*
[2]*Università degli Studi di Torino, Dipartimento di Economia e Ingegneria Agraria, Forestale e Ambientale (D.E.I.A.F.A.), Sezione di Idraulica Agraria, Via Leonardo da Vinci 44, 10095 Grugliasco, Torino*
Italy

1. Introduction

Time Domain Reflectometry (TDR) has been initially developed as a remote sensing electrical measurement tool and it has been widely used to detect cable faults and discontinuities of signal lines (Somlo & Hollway, 1969). Recently, TDR has been widely applied to soil science, hydrology and agronomy to estimate the electrical properties of soils (Courtney, 1998; Evett & Parkin, 2005; Nozaki & Bose, 1990; Robinson et al., 2003; Topp & Ferre', 2002).
The two major features of TDR waveforms are the travel time and the response 'late' time amplitude: they are directly related to the apparent dielectric constant K_a and to the soil electrical conductivity γ (Robinson et al., 2005). If the assumption is made that the imaginary part of the complex electrical permittivity is small compared to the real part, only the real part of permittivity changes with soil water content, and if the real part is close to the apparent dielectric permittivity K_a *(effective bulk permittivity)*, then water content can be evaluated by means of empirical formulas based on known value of K_a (Evett et al., 2005; Topp et al., 1980). Evaluation of the dielectric permittivity of a medium from its reflection response is an inverse problem in which the medium propagation behavior is used to infer its constitutive relations (Feng et al., 1999; Heimovaara, 1994; Lin, 2003; Oswald et al., 2006).
Owing to this application, over the past twenty years great efforts have been spent on the calibration of the TDR system, i.e. the relationship between apparent dielectric permittivity and soil water content (Hansson & Lundin, 2006; Roth et al., 1990; Topp et al., 1980), on the design and testing of various type of probes (Canone et al., 2009; Evett et al., 2006; Robinson et al., 2005); and on the methods to solve the inverse problem of estimating permittivity from reflection responses (Heimovaara, 1994; Oswald et al., 2006).
Regardless of the inversion method used, the model of the measurement system (relating measured waveforms to the unknown permittivity to be estimated) is the instrument enabling the permittivity estimation from the reflection response. The modeling of the TDR measurement system received, therefore, a great deal of attention in the past (Heimovaara, 1994; Lin, 2003). In spite of the advances in modeling of the TDR setup and in evaluation of permittivity from the TDR response, the intrinsic limitations introduced by the TDR

measurement setup on estimation of permittivity spectrum have rarely been addressed. If the information about electrical properties of the sample under test is lost due to the characteristics of the measurement system, it is not possible to retrieve such an information whatever the accuracy of the inversion method used.

In this study the limitations introduced by the TDR measurement setup and the effect of wave attenuation along the TDR probe are addressed. To this end, a measurement setup easy to model has been chosen. In this setup, a coaxial airline instead of the probes generally used for TDR measurements (two or three arms sensor as in Fig. 1, right) is considered. A coaxial airline can be modeled as a TEM transmission line (Pozar, 2005), whereas other types of probes generally used in TDR measurements can be only modeled by numerical techniques. Moreover, the actual signal launched by the TDR instrument and not only an ideal unit voltage step has been considered in the model as an input signal. Thanks to this accurate model and using a parametric inversion approach, it was possible to obtain a very good fitting of several reference liquids and various sand and clay samples with different water content and to determine their permittivity and conductivity. Then, with the aid of a frequency domain analysis on the same model but with some simplifications, the limitations introduced by the TDR measurement setup are studied in details. It will be shown that the probe length sets an upper limit to the frequency that can be used and that this limit is independent from the well-known impact of the TDR instrument bandwidth (Savi et al., 2010).

In more details, an equivalent circuit of the TDR measurement setup based on transmission line theory is introduced. Signals traveling along the connecting cables and probe (an open-ended ideal coaxial line) are described as Transverse Electro-Magnetic (TEM) waves and the various discontinuities (output port of the instrument, cable-probe connection) are kept into account by means of a scattering matrix description (Pozar, 2005). A parametric representation of the permittivity spectrum as a one-pole real rational model (Debye's permittivity model, (Hasted, 1973)) is assumed and its parameters are estimated by fitting the model response to the measured response. This approach has been adopted because it is less sensitive to measurements errors in the high-frequency range with respect to *direct inversion* techniques. This accurate model (in which the finite rise-time and the aberration of the real input signal, and the junction between probe and port of the TDR instrument are kept into account) and the *parametric inversion* approach lead to an accurate evaluation of permittivity.

Then, a simplified model of the TDR measurement system, where an ideal voltage step is applied to an ideal cable connected to the probe is considered. In this case, the real signal of the generator and the various discontinuities are not kept into account. The reflection coefficient at the probe input port is written as a function of the model parameters in frequency domain and decomposed into a sequence of echoes. Analysis of the second echo points out the role of the probe attenuation and a threshold on the attenuation of the working condition of the measurement setup. Two types of TDR waveforms (low-attenuation and high-attenuation) can be identified. Transition between the two types of waveforms is controlled by the wave attenuation along the probe. Starting from the ideal model of measurements setup, simple relations to evaluate the low-frequency permittivity (ϵ_s) and the high-frequency permittivity (ϵ_∞) directly from characteristic points of the waveform are introduced.

In this chapter, the following topics are presented in details: an accurate model of the TDR measurement setup based on a coaxial airline probe together with a fitting procedure to obtain the complex permittivity (see Section 2 and 3); a simplified model enabling the analysis of properties and limitations of the measurement setup (see Section 4); simple formulas to evaluate permittivity and conductivity from the TDR response (see Section 5). This analysis is applied to some reference liquids (deionized water and methanol) and various sand and clay soil with different water content (see Section 6).

2. TDR measurement system

A time-domain reflectometer (see Fig. 1) consists mainly of a very fast rise time pulse generator and an oscilloscope. The step generator produces a positive-going incident wave that is fed into the transmission system under test (probe). The step travels down the transmission line at the velocity of propagation of the line. If the load impedance is equal to the characteristic impedance of the line, no wave is reflected. If a mismatch exists at the load, part of the incident wave is reflected.
The TDR system (TDR instrument and probe) can be modeled by an ideal transmission line (Pozar, 2005) and the the reflected voltage detected from the instrument can be obtained with the help the bouncing diagram of Fig. 2. In the reflected voltage, the first falling edge represents the partial reflection coefficient at the discontinuity between the output port of the instrument and the transmission line, the stair-case is due to the multiple reflections along the line. Step amplitudes depend on permittivity, time delay between a step and the next one depends on permittivity and on the physical length of the line. Therefore, in this ideal case in which the permittivity is constant with the frequency, it is possible to find the value of the permittivity from the TDR waveform. When permittivity is a function of frequency, the TDR response is similar to an ideal case, but the problem should be faced with a frequency domain approach.

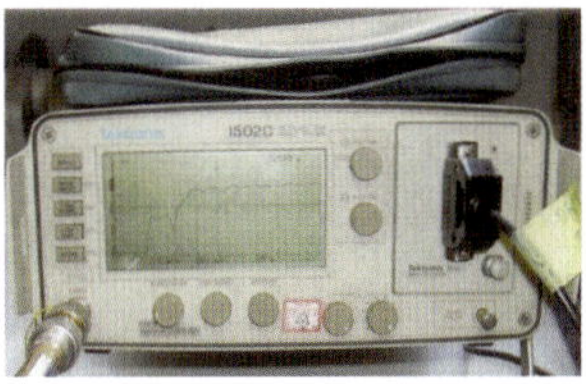

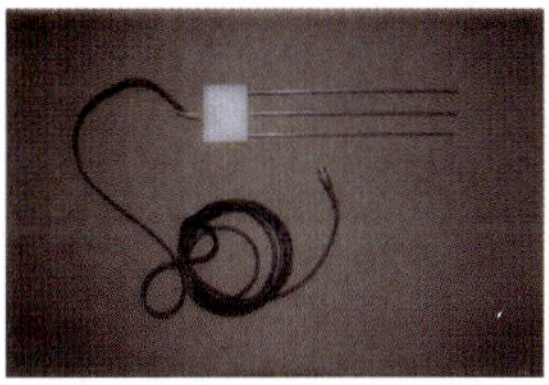

Fig. 1. Left: The frontal panel of the Tektronix 1502B reflectometer with rise time $t_r = 0.3\,\text{ns}$.
Right: A three arms probe.

The basic scheme for the analysis of the propagation along the probe in the frequency domain is shown in Fig. 3. The system's response can be calculated by means of the convolution of the input signal $v_0(t)$ and the system impulsive response $h(t)$ as:

$$r(t) = \int_{-\infty}^{\infty} v_0(t-\tau)\, h(\tau)\, d\tau \tag{1}$$

In the frequency domain the convolution can be written as:

$$R(f) = V_0(f)\, H(f) \tag{2}$$

where $R(f)$, $V_0(f)$ and $H(f)$ are the Fourier transforms of $r(t)$, $v_0(t)$ and $h(t)$, respectively. $H(f)$, and thus $h(t)$, are functions of the dielectric behavior of the soil sample and the properties of the probe, such as length and characteristic impedance. As $v_0(t)$ and $r(t)$ (and

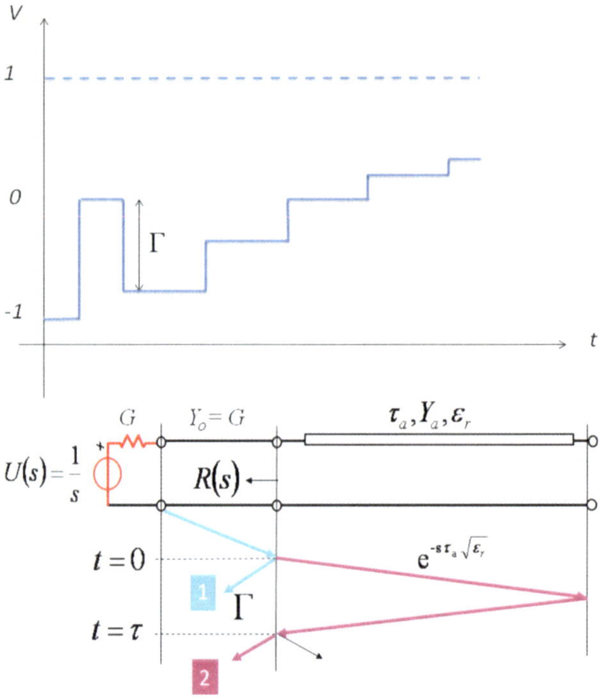

Fig. 2. TDR response and bouncing diagram of the TDR system.

therefore their Fourier transforms $R(f)$, $V_0(f)$) are known, we can calculate $H(f)$ by means of eq. (2).

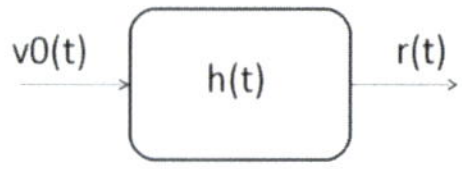

Fig. 3. Linear system analysis.

3. Accurate model of the TDR measurement setup

The TDR measurement set up has been modeled by the chain network of Fig. 4. The system has been divided into 6 parts and characterized separately. In this model, the TDR source composed by the signal generator (1) and the oscilloscope (2) is represented by a matched source with internal admittance G, driving the network with the same waveform measured in the actual TDR instrument. The finite rise time and the aberration of the real probing signal are thus kept into account. The coaxial cable (3) that connects the instrument to the probe is modeled by an ideal lossless transmission line. Part 4 is the output port of the instrument and is represented by a lumped-distributed 2-port element. This two-port element has been identified by fitting the response of the chain network to the response measured with the probe left empty (5). Part 5 and 6 are the probe itself filled with the soil sample under test and

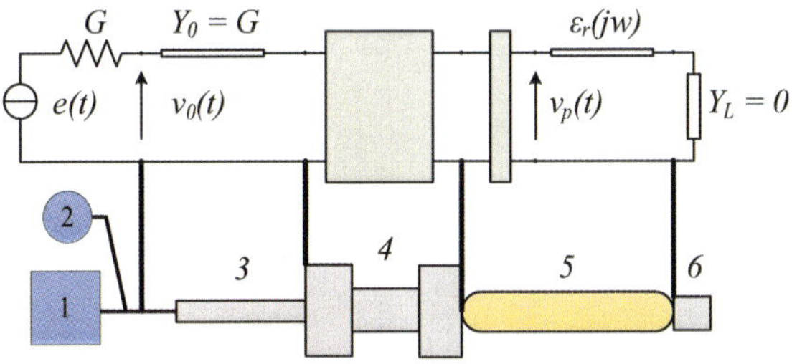

Fig. 4. Block diagram of a TDR measurement system and its chain network model.

the probe termination. Parts (1)-(3) are included in the TDR instrument, whereas parts (4)-(6) form the device whose reflection response is measured.
The unknown permittivity is described by the one-pole Debye's permittivity model modified in order to keep into account the conductivity γ of the dielectric under test:

$$\epsilon_r(j\omega) = \epsilon_\infty + (\epsilon_s - \epsilon_\infty)/(1 + \frac{j\omega}{\omega_{rel}}) + \frac{\sigma_s}{j\omega} \tag{3}$$

where ϵ_∞, ϵ_s and $\omega_{rel} = 2\pi f_{rel}$ are the Debye's parameters. The four parameters $(\epsilon_s, \epsilon_\infty, \omega_{rel}, \sigma_s)$ of eq. (3) are estimated by a parametric inversion approach, i.e. by fitting the model response to the measured one. Other models can be used, such as multi-pole equations and the Cole-Cole equations. Those methods have improved the fitting accuracy, but the accuracy gained does not worth the increase of complexity (Heimovaara, 1994; Huisman et al., 2004). As an example, the estimation of the permittivity via waveform fitting based on the transmission line model previously described is applied to two reference liquids: deionized water and methanol. Results are shown in Fig. 5 and Fig. 6, respectively. The solid line curves are the measurements obtained by an HP 54120B digitalizing oscilloscope (see subsection 6.1). The dashed line curves are obtained by fitting the response of the transmission line model described in this Section to the measured curves. The best fit of the measured responses ($T = 25°$) is obtained with $\epsilon_s = 78$, $\sigma_s = 0.3\,\mathrm{mS/m}$ for deionized water (Lide, 1992), and $\epsilon_s = 32$, $\sigma_s = 0.1\,\mathrm{mS/m}$ for methanol (Weerts et al., 2001).

4. Simplified model and probe attenuation

In order to point out the inherent properties of the TDR responses, we consider the simplified model shown in Fig. 7, where an ideal voltage step (infinite bandwidth) $u(t)$ ($U(s) = 1/s$ in the frequency domain) is applied to an ideal cable of characteristic admittance Y_o connected to the probe. The probe is an open-ended ideal coaxial line and can be modeled as a TEM transmission line of characteristic admittance and time delay of the empty airline Y_a and τ_a, respectively. In this lossless, simplified model, the measured quantity is the time evolution of the backward wave at the probe input, *i.e.*, $r(t) = L^{-1}[R(s)] = L^{-1}[S_{11}(s)(1/s)]$ where S_{11} is the reflection coefficient at the cable-probe discontinuity, L^{-1} is the inverse Laplace transform operator, and lower-case and upper-case letters are used to indicate transform pairs. The features of both the measured waveform $r(t)$ and its spectrum $R(s)$ are investigated and their relationship to the permittivity parameters (ϵ_s, ϵ_∞, ω_{rel}, σ_s) and to the probe parameters (Y_a,

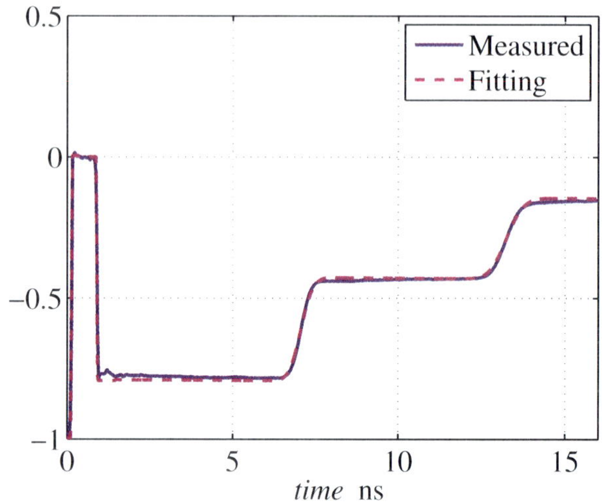

Fig. 5. TDR measurements (solid line) and fitting (dashed line) for a sample of deionized water. The measured data are obtained with an HP 54120B digitalizing oscilloscope.

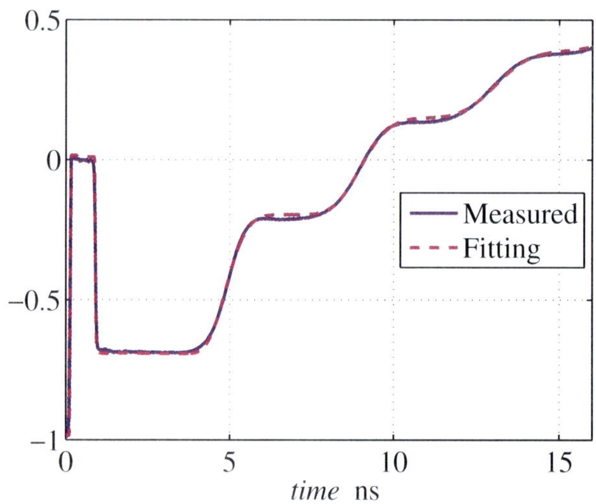

Fig. 6. TDR measurements (solid line) and fitting (dashed line) for a sample of methanol. The measured data are obtained with an HP 54120B digitalizing oscilloscope.

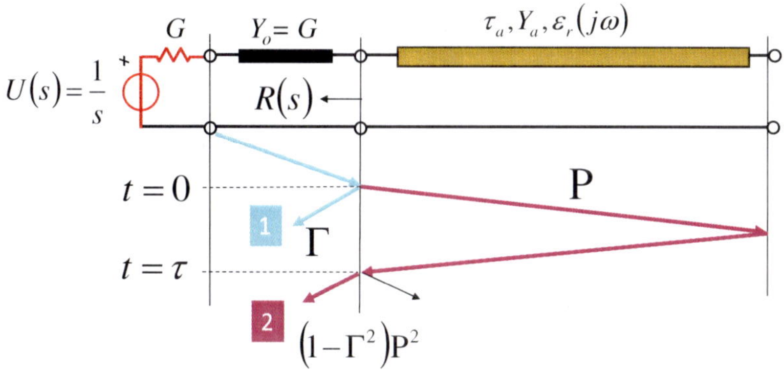

Fig. 7. Simplified transmission line model of the TDR measurement system and bouncing diagram. First contribution ($\bar{\Gamma}$: the signal reflected at the input port of the probe (partial reflection coefficient); second contribution $(1-\bar{\Gamma}^2(S))\,\bar{\mathrm{P}}^2(S)$: the signal transmitted and traveling forward and backward along the probe.

τ_a) are pointed out. To this end, the following normalized quantities are used:

$$\begin{aligned} &T = t\,\omega_{rel}, \qquad S = s/\omega_{rel}, \; \mathcal{Y} = Y_a\sqrt{\epsilon_\infty}/Y_o \\ &\mathcal{T} = \tau_a\sqrt{\epsilon_\infty}\,\omega_{rel}, \; \eta = \epsilon_s/\epsilon_\infty \end{aligned} \tag{4}$$

where T and S are the normalized time and complex frequency, respectively. The analysis is based on the decomposition of the normalized reflection coefficient $\bar{S}_{11}$ into a sequence of echoes (see the bouncing diagram in Fig. 7):

$$\begin{aligned} \bar{S}_{11}(S) = &\,\bar{\Gamma}(S) + (1-\bar{\Gamma}^2(S))\,\bar{\mathrm{P}}^2(S) \\ &(1-\bar{\Gamma}^2(S))\,(-\bar{\Gamma})\bar{\mathrm{P}}^4 + \ldots \end{aligned} \tag{5}$$

where

$$\bar{\Gamma}(S) = \frac{1-\mathcal{Y}\sqrt{\bar{\epsilon}}}{1+\mathcal{Y}\sqrt{\bar{\epsilon}}} \tag{6}$$

is the probe partial reflection coefficient (the portion of the input signal that is reflected back from the first discontinuity, see Fig. 7, contribution 1) and

$$\bar{\mathrm{P}}(S) = \exp\{-2\mathcal{T}S\sqrt{\bar{\epsilon}(S)}\} \tag{7}$$

is the propagation factor as functions of the normalized frequency and $\bar{\epsilon} = 1 + (\eta-1)/(1+S)$ and $S = j\Omega$, where Ω is the normalized angular frequency. The first term of equation (5), $\bar{H}_1(S) = \bar{\Gamma}(S)$, is the network function relating the incident wave to the wave reflected by the cable-probe discontinuity. The second term, $\bar{H}_2(S) = (1-\bar{\Gamma}^2(S))\,\bar{\mathrm{P}}^2(S)$, relates the incident wave to the wave reflected by the probe load and so on. In the time domain, these terms start at $T = 0, 2\mathcal{T}, 4\mathcal{T}, \ldots$, and give rise to the echoes forming the observed response, *i.e.*, $\bar{r}(T) = \bar{r}_1(T) + \bar{r}_2(T) + \ldots$.

4.1 Low-attenuation condition

In order to study the properties of the simplified model responses, it is convenient to focus on the second echo $\bar{r}_2(T)$, because it is the simplest term containing information on both the probe-cable discontinuity and on the propagation in the dielectric. The behavior of the second echo is decided by the network functions $\bar{H}_2(S)$ and, in particular, by the square propagation factor $\bar{P}^2(S) = exp(-2\mathcal{T}S\sqrt{\bar{\epsilon}})$ where $\bar{\epsilon} = 1 + (\eta - 1)/(1 + S)$. Figure 8a,b shows $|\bar{P}^2(j\Omega)|$ for $\mathcal{T} = 6, 15, 40$ and $\eta = 1.5, 3$. (Panel b is a close-up view of the middle lower part of panel a). For growing $\mathcal{T}$ and η values, the high-frequency magnitude of $|\bar{P}^2|$ becomes negligibly

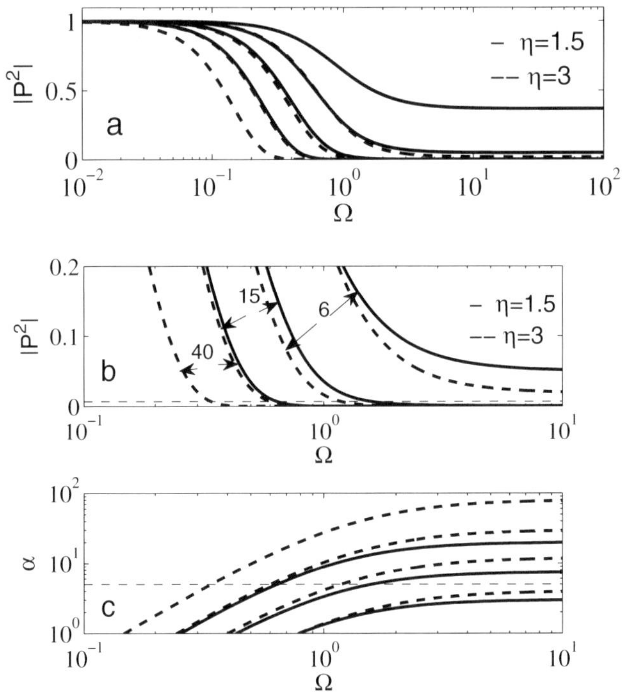

Fig. 8. Magnitude of the propagation factor (*a* and *b*) and attenuation constant (*c*) versus normalized angular frequency for $\mathcal{T} = 6, 15, 40$ and $\eta = 1.5, 3$. Panel *b* is a close-up view of panel *a*. The dashed horizontal lines of panel b and c indicate the threshold level $|\bar{P}^2| = \exp(-5)$ and $\xi = 5$, respectively.

small and, therefore, $\bar{H}_2$ becomes a bandlimited function. This bandlimiting effect is due to the imaginary part of $\bar{\epsilon}(j\Omega)$, that is responsible for the attenuation constant of $\bar{P}^2$:

$$\alpha = -\Re e\{j2\Omega\mathcal{T}\sqrt{\bar{\epsilon}(j\Omega)}\} = -2\Omega\mathcal{T}\Im m\left\{\sqrt{1 + \frac{\eta - 1}{1 + j\Omega}}\right\} \quad (8)$$

The attenuation constant is plotted vs. frequency in Fig. 8c. Every attenuation curve is composed of a quadratic low-frequency part (the constant slope part in the bi-logarithmic scales of Fig. 8c) and by a constant high-frequency part. The constant part arises from the high frequency behavior of $\Im m\{\sqrt{\bar{\epsilon}(j\Omega)}\}$, that is inversely proportional to Ω. The high-frequency asymptotic value of α is:

$$\xi = \lim_{\Omega\to\infty} \alpha(j\Omega) = \mathcal{T}(\eta - 1) \tag{9}$$

and the asymptotic value of $|\bar{P}^2|$ is $\exp(-\xi)$. For example, if $\xi = 5$, the asymptotic value of $|\bar{P}^2|$ is close to 6.7×10^{-3} and can be considered small. The relation between the $|\bar{P}^2|$ and α curves of Fig. 8 is explained by this property. Bandlimited propagation factors are associated to attenuation curves with $\xi > 5$, and the propagation factors become small where their attenuation curves overpass this threshold level (see Fig. 8a,b,c). In the following, we use the term *high-attenuation* to indicate the case $\xi > 5$. Measurement setups working in high-attenuation conditions have $\bar{H}_2$ functions with small high frequency magnitudes. In this case, the network functions of the higher order echoes have small high-frequency magnitudes as well, because they include powers of the factor $\bar{P}^2$. As a consequence, in high-attenuation conditions, the high-frequency content of the measured TDR waveforms is low and the relation with the high-frequency permittivity spectrum is expected to be weak. According to this analysis, measurement setup operating in the low-attenuation domain ($\xi < 5$) are preferable.

As an example, consider the TDR waveforms corresponding to the low and high-attenuation cases shown in Fig. 9. The parameters are $\epsilon_s = 12$, $\epsilon_\infty = 6$, $f_{rel} = 0.7\,\text{GHz}$, $Z_a = \sqrt{\mathcal{L}/\mathcal{C}} = 77\,\Omega$. In the low-attenuation condition $\tau_a = 0.34\,\text{ns}$ and $\xi = 3.7$, whereas for high-attenuation $\tau_a = 1\,\text{ns}$ and $\xi = 11$. When the wave reflected from the probe end has harmonic components with significant amplitude for frequency larger than f_{rel} the observed TDR waveforms are of low-attenuation type. This reflection behavior is illustrated in Fig. 9a, where the solid line curve is the magnitude of the network function of the first echo from the probe end for a low-attenuation case and the dotted straight line corresponds to the relaxation frequency (the bandwidth of the instrument source is larger than the relaxation frequency of the dielectric under test). Similarly, when the harmonic components of the wave reflected from the probe end have negligible amplitude for frequency larger than f_{rel}, the TDR waveform are of high-attenuation type (dashed curve of Fig. 9a). The bandwidth of the waveforms of high-attenuation type is therefore inherently limited to f_{rel}, regardless of the source bandwidth, and these waveforms are scarcely sensitive to the high-frequency parameters of the permittivity. The two types of TDR response can be easily identified because, in the low-attenuation case, the rising edge corresponding to the first echo from the probe end is asymmetric and steep, whereas, in the high-attenuation case, the edge is symmetric and gradual (see Fig. 9b, where $\rho(t) = 2v_o(t)/e_f - 1$, e_f asymptotic value of $e(t)$). Remembering that $\mathcal{T} = \tau_a\sqrt{\epsilon_\infty}\,\omega_{rel}$ and $\eta = \epsilon_s/\epsilon_\infty$, the low-attenuation condition can be expressed in physical parameters as:

$$\xi = \tau_a\omega_{rel}\frac{\epsilon_s - \epsilon_\infty}{\sqrt{\epsilon_\infty}} < 5 \tag{10}$$

or, equivalently, as:

$$2\tau_a\sqrt{\epsilon_\infty} < 10\,\epsilon_\infty/\omega_{rel}(\epsilon_s - \epsilon_\infty) \tag{11}$$

where $2\tau_a\sqrt{\epsilon_\infty}$ is the probe round trip delay. This means that $2\tau_a\sqrt{\epsilon_\infty}$, *i.e.*, the probe electrical length, must be limited in order to obtain low-attenuation operation. As a consequence of (11), two limitations on the instrument resolution and on the length of the probe should be taken

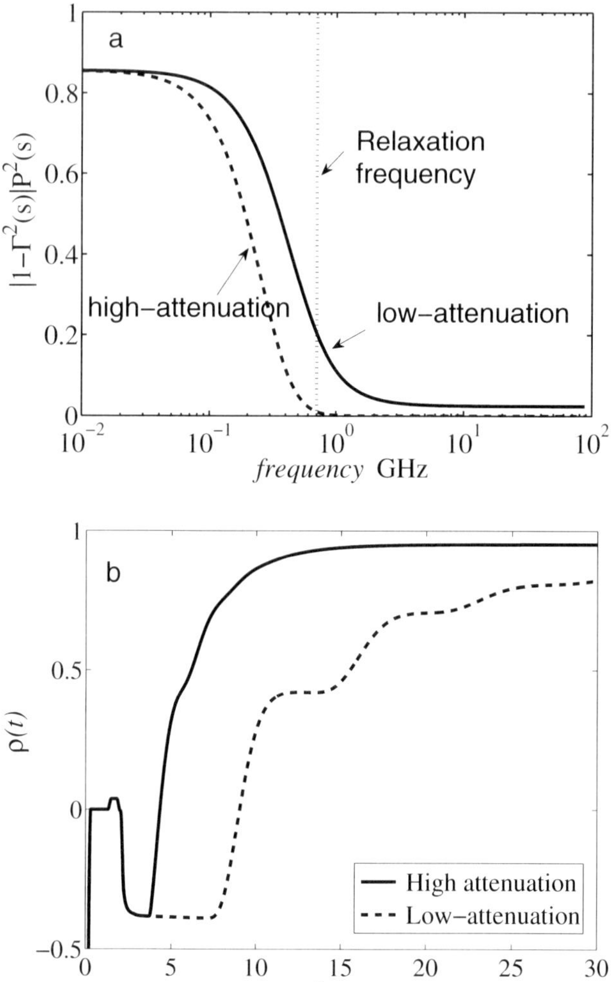

Fig. 9. Panel a : magnitude of the network function of the first echo from the probe end for a low-attenuation case (solid line) and an high-attenuation case (dashed line). Panel b: TDR waveforms corresponding to the response of panel a. Parameter values: $\epsilon_s = 12$, $\epsilon_\infty = 6$, $f_{rel} = 0.7\,\text{GHz}$, $Z_a = \sqrt{\mathcal{L}/\mathcal{C}} = 77\,\Omega$, $\tau_a = 0.34\,\text{ns}$ (low-attenuation) and $\tau_a = 1\,\text{ns}$ (high-attenuation). The responses of this figure hold for an ideal step input.

Medium	f_{rel}	ϵ_s	ϵ_∞	$10\epsilon_\infty/\omega_{rel}(\epsilon_s-\epsilon_\infty)$	ref.
deionized water	17 GHz	80	4.22	5.2ps	(Heimovaara, 1994)
wet sand	7 GHz	15	3	57ps	(Feng et al., 1999)
ethanol	0.782 GHz	25.2	4.52	445ps	(Heimovaara, 1994)
dry sand	7 GHz	3	2.8	5.25ns	(Feng et al., 1999)

Table 1. Debye's parameter values for some typical dielectrics. $2\tau_a\sqrt{\epsilon_\infty}$ is the maximum round trip delay.

into account. The instrument resolution must be significantly finer than $10\,\epsilon_\infty/\omega_{rel}(\epsilon_s-\epsilon_\infty)$. In practice, the resolution of TDR instruments is limited by the rise time of their sources that is on the order of $150 \div 200$ ps for conventional field instruments and of 30 ps for high-end instruments, and low attenuation measurements can be obtained only for dielectric with $\omega_{rel}(\epsilon_s-\epsilon_\infty)/\epsilon_\infty$ small enough. On the other hand, a too short probe ($l = \tau_a c$) cannot be considered both for practical reasons and because the transmission line model requires the cross-section to be much smaller than the length of the probe. Some examples of parameter values are listed in Tab. 1. The ethanol can be considered as a limit case. In fact, the maximum round trip delay is 445 ps and can be detected by a standard TDR instrument, and the length of the probe in this case should be 3 cm.

4.2 Parametric analysis

In this section, we analyse the variation of the parameters (ϵ_s, ϵ_∞, ω_{rel}, σ_s) of the Debye permittivity equation (3) on the TDR waveform. Results are shown in Figs.10-13. These examples are of high-attenuation type. It is worthwhile to notice that none of the parameters of the Debye permittivity affect the first falling edge of the responses. Hence, the parameter estimation is not affected by model errors influencing this part of the response (*e.g.*, the element representing the probe input section). The static permittivity, ϵ_s, dominates the propagation effects, *i.e.*, it roughly defines the probe characteristic impedance and propagation speed. In fact, Fig. 10 shows that ϵ_s modifies the position of the first rising edge (probe delay) and its amplitude (probe characteristic impedance). The parameter f_{rel} controls the slope of the rising edges 12, thereby it complicates the assessment of the probe delay and ϵ_s estimation based on probe delays (Heimovaara, 1994). Besides, when the effect of ϵ_∞ is not negligible (Fig. 11), it mixes up with the effect of f_{rel}, as they affect the same part of the responses. Finally, the electrical conductivity σ_s affects the long-time behaviors of the responses, by reducing levels and amplitudes of the reflections from the probe end (Fig. 13). This is consistent with the low-frequency lossy effect of the conductivity parameter. Unfortunately, growing values of σ_s reduce the sensitivity of responses to all parameters, thereby increasing the uncertainty on parameters estimations.

5. Simple formulas to evaluate the permittivity from the TDR response

The two major features of TDR waveforms are the travel time in the probe and the response 'late' time amplitude that are directly related to the apparent dielectric dielectric constant and to the soil electrical conductivity. Several studies in the literature have aimed at determining a good method to evaluate the travel time from the TDR curve. Starting from the ideal model of the measurements setup and the frequency domain analysis previously introduced, simple

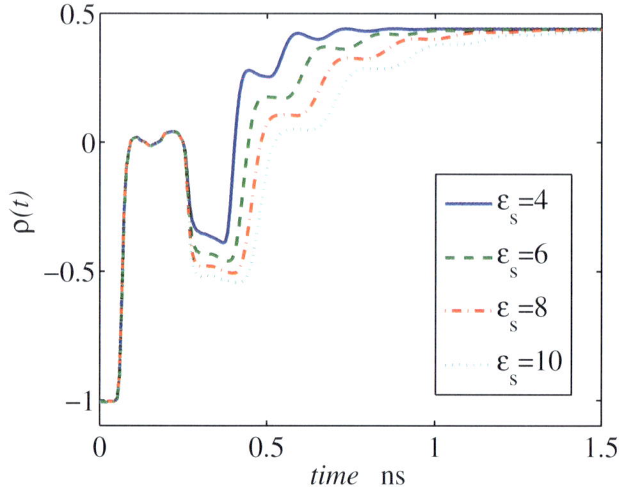

Fig. 10. Model response for different values of $\epsilon_s(4, 6, 8, 10)$. Other parameter values are $\omega_{rel} = 20 \cdot 10^9$; $\epsilon_\infty = 3$; $\sigma_s = 1 \cdot 10^{-2}\,\mathrm{S/m}$.

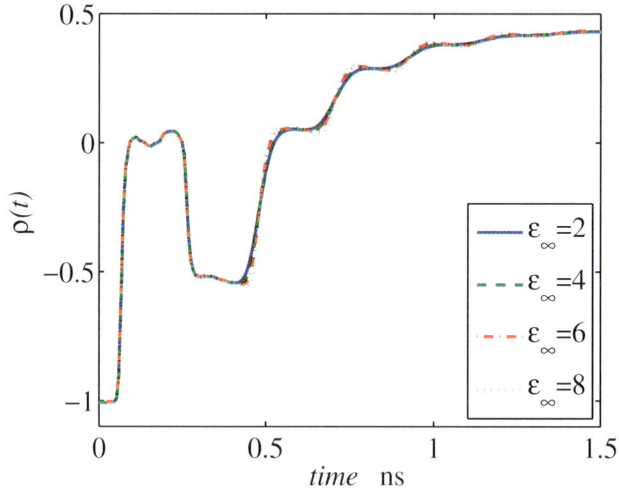

Fig. 11. Model response for different values of $\epsilon_\infty(2, 4, 6, 8)$. Other parameter values are $\omega_{rel} = 20 \cdot 10^9$; $\epsilon_s = 10$; $\sigma_s = 1 \cdot 10^{-2}\,\mathrm{S/m}$.

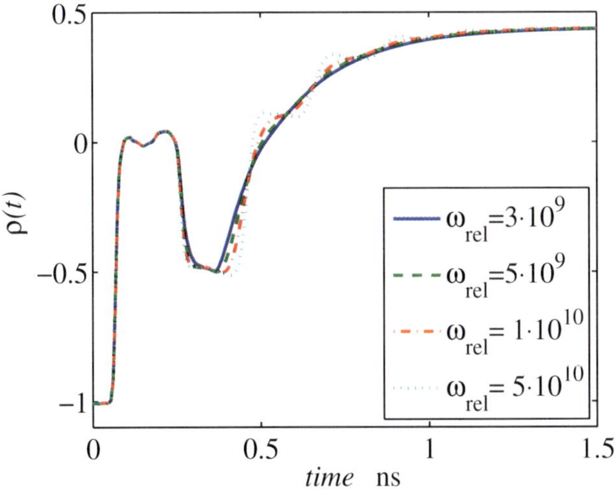

Fig. 12. Model response for different values of $\omega_{rel}(3 \cdot 10^9, 5 \cdot 10^9, 1 \cdot 10^{10}, 5 \cdot 10^{10})$. Other parameter values are $\epsilon_\infty = 8$; $\epsilon_s = 2.5$; $\sigma_s = 10^{-2}$ S/m.

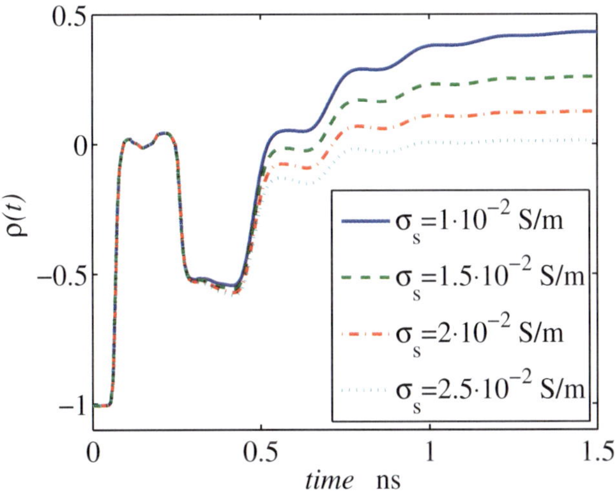

Fig. 13. Model response for different values of $\sigma_s(1 \cdot 10^{-2}, 1.5 \cdot 10^{-2}, 2 \cdot 10^{-2}, 2.5 \cdot 10^{-2})$ S/m. Other parameter values are $\omega_{rel} = 20 \cdot 10^9$; $\epsilon_s = 10$; $\epsilon_\infty = 3$.

relations to evaluate the low-frequency permittivity (ϵ_s) and the high-frequency permittivity (ϵ_∞) directly from the characteristic points of the waveform are introduced.
From equation (6), remembering that $\mathcal{Y} = Y_a\sqrt{\epsilon_\infty}/Y_o$ and that

$$\bar{\epsilon}(S) = 1 + (\eta - 1)/(1 + S) \sim \eta = \frac{\epsilon_s}{\epsilon_\infty} \tag{12}$$

the low-frequency permittivity ϵ_s can be estimated as:

$$\epsilon_s \sim \left[\frac{Y_0\,(1-h)}{Y_a\,(1+h)}\right]^2 \tag{13}$$

This formula can be used to determine the low-frequency permittivity instead of the standard method based on the delay of the inflection point of the first rising edge (t_b) that gives:

$$\epsilon_s \sim \left(\frac{t_b}{2\tau_a}\right)^2 \tag{14}$$

When a low-attenuation condition holds (small ξ values) the rising edges are asymmetric and start at $2\tau_a\sqrt{\epsilon_\infty}$. In this case, also the value of ϵ_∞ can be determined as:

$$\epsilon_\infty \sim \left(\frac{t_a}{2\tau_a}\right)^2 \tag{15}$$

where t_a is the corner of the first rising edge (see Fig. 14).

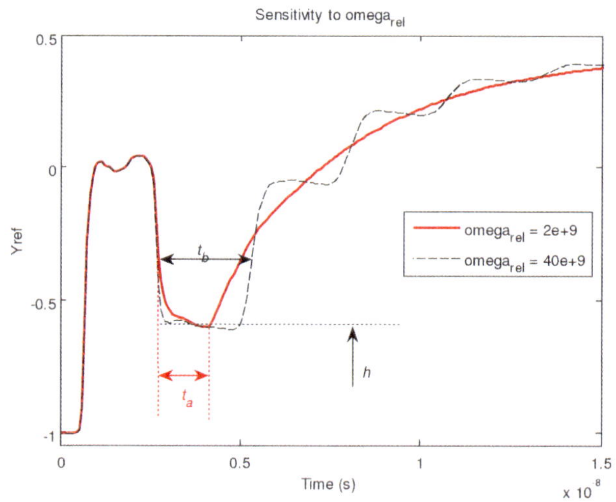

Fig. 14. Definition of the parameters h, t_a, t_b.

6. Soil measurements examples

6.1 Measurement setup description

Two different instruments were used in order to measure the time-domain response: a Tektronix 1502B reflectometer with rise time $t_r = 0.3\,\text{ns}$ (horizontal setting 0.1 m per division and a propagation velocity setting of 0.99), 13-bit data acquisition (range from 0 to 8192) (see Fig. 1); and a digitalizing oscilloscope HP54120B with rise time $t_r = 10\,\text{ps}$ (see Fig. 15 left). The samples were inserted in a coaxial probe terminated on a $50\,\Omega$ load (Maury Microwave Airline, model no. 2653S10, length $\ell = 10.5\,\text{cm}$, shield and inner conductor radii 3.5 mm and 1.5 mm, respectively, dc-resistance of inner conductor $9.4\,\text{m}\Omega/\text{m}$, see Fig. 15 right). In the case of Tektronix 1502B, the acquisition of the waveforms by the computer was done by means of a serial cable and the software WinTDR 6.1 developed by the Soil Physics Group at Utah State University, Logan Utah, USA (http://www.usu.edu/soilphysics/wintdr/). In the case of the digitalizing oscilloscope HP54120B with a GPIB communications (IEEE 488.2 standard), a data acquisition interface was developed with the help of Labview (http://www.ni.com/labview/).

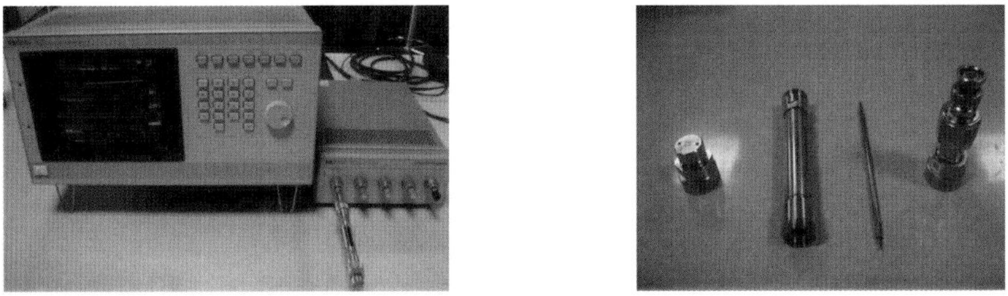

Fig. 15. The frontal panel of the HP 54120B digitalizing oscilloscope with rise time $t_r = 10\,\text{ps}$ (left). The coaxial probe (Maury Microwave Airline, model no. 2653S10, length $\ell = 10.5\,\text{cm}$ used in the measurement setup (right).

6.2 Example of parametric inversion in sand samples

In this section, the parametric inversion procedure is applied to the TDR response of an almost dry sand sample (3.4% volumetric water content) and to the response of a sand sample with 25% volumetric water content. In both cases, the TDR responses have been obtained with a Tektronix 1502B reflectometer. These waveforms are of low-attenuation waveform and, according to the analysis of the previous Section, their estimation problem is expected to be well conditioned also for the high-frequency part of the permittivity.

The measured results obtained in the first case (dry sand) are shown in Fig. 16 (solid line). The model response that best fits the measured curve is also shown (dashed line) and the estimated values of the permittivity parameters are: $\epsilon_s = 3.18$, $\epsilon_\infty = 2.69$, $f_{rel} = 413\,\text{MHz}$ and $\sigma_s = 6.36 \cdot 10^{-4}\,\text{S/m}$. These parameter values correspond to $\xi = 0.28$, confirming that the measured response is of low-attenuation type, as expected.

The TDR response of the sand sample with 25% water content is shown in Fig. 17 (solid line). The TDR response of this example as well looks like a low-attenuation waveform and its estimation problem is expected to be well conditioned for the high-frequency part of the permittivity. The model response that best fits the measurement is shown in Fig. 17

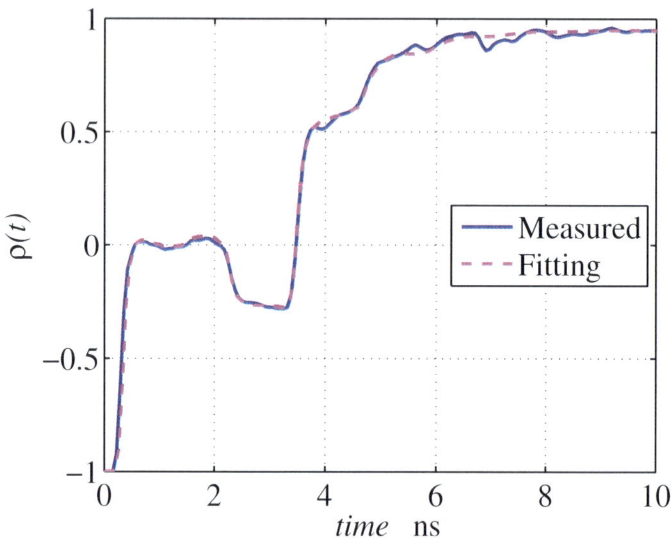

Fig. 16. Measured TDR response for a sand sample (solid line) and the best fit response of the accurate model (dashed line). Permittivity parameters: $\epsilon_s = 3.18$, $\epsilon_\infty = 2.69$, $f_{rel} = 413\,\text{MHz}$ and $\sigma_s = 6.36 \cdot 10^{-4}\,\text{S/m}$.

(dashed curve) and the estimated values of the permittivity parameters are $\epsilon_s = 12.$, $\epsilon_\infty = 8.4$, $f_{rel} = 1.38\,\text{GHz}$ and $\sigma_s = 0.007\,\text{S/m}$. These parameter values correspond to $\xi = 3.7$, confirming that the measured response is of low-attenuation type as expected.

The condition of the estimation problem is decided by the behavior of the error function around the estimated parameter values. An indicator of this behavior can be obtained by plots of the function over planes of the parameter space. The error function of the 25% water content sample over the plane $(\epsilon_\infty, f_{rel})$ is shown in Fig. 18 and looks as a well behaved convex surface, thereby suggesting the good condition of the estimation of the high-frequency parameters.

6.3 Three methods for the estimation of permittivity in sand and clay samples

In this section, three different methods for the determination of the low-frequency permittivity ϵ_s are compared. In the first method, the permittivity is calculated from the travel time of the TDR signal along the probe (as in the method described in (Topp et al., 1980)), but the starting point and ending point of the travel time is determined from the apex of the derivative of the waveform of the reflected signal (derivative method) (Robinson et al., 2005). In the second case, the permittivity is evaluated by using eq. (14-15) (low-level method), whereas in the third case the permittivity is estimated via a parametric inversion approach based on the model of the measurement setup described in Section 3 (fitting).

To this end, the TDR response of two set of sand and clay samples with different water content have been measured. After a process of weighting and drying, the water content of each sample was determined as:

$$\theta_{water} = \frac{P_w - P_d}{V_{probe}} \tag{16}$$

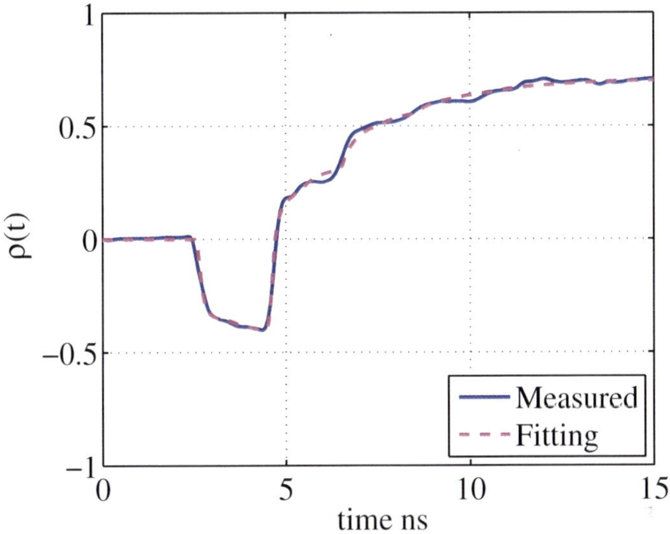

Fig. 17. Measured TDR response for a sand sample with 25% water content (solid line) and the best fit response of the accurate model (dashed line). Permittivity parameters: $\epsilon_s = 12.$, $\epsilon_\infty = 8.4$, $f_{rel} = 1.38\,\text{GHz}$ and $\sigma_s = 0.007\,\text{S/m}$.

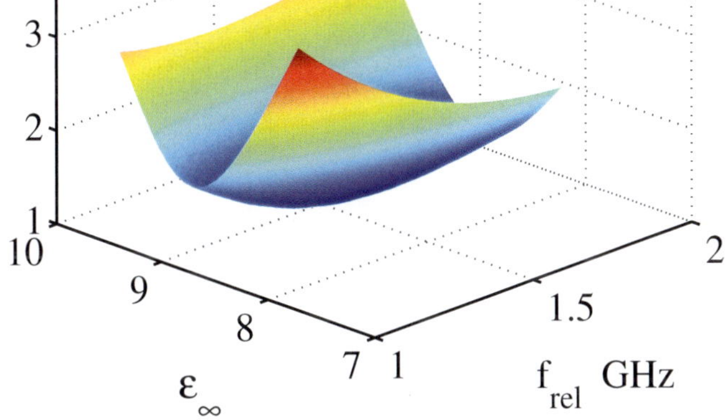

Fig. 18. Error surface in the $(\epsilon_\infty, f_{rel})$ plane for a sand sample with 25% water content.

where P_w is the wet weight, P_d is the dry weight, and V_{probe} is the volume of the probe. The results obtained with these methods are shown in Fig. 19. Sand samples correspond to a high value of ξ (high-attenuation case), whereas in case of clay samples the value of ξ is small (low-attenuation case). As it was expected, the value of ϵ_s increases with water content. The low-level method based on eq. (14-15) works as well as the more complicate derivative method and fitting method. Note that sensitivity of the low-level method increases as permittivity increases, and that sensitivity in the high-attenuation case increases with the frequency. These considerations explain the inaccuracies of the low-level method for a high water content in case of sand samples of Fig. 19, Panel a.

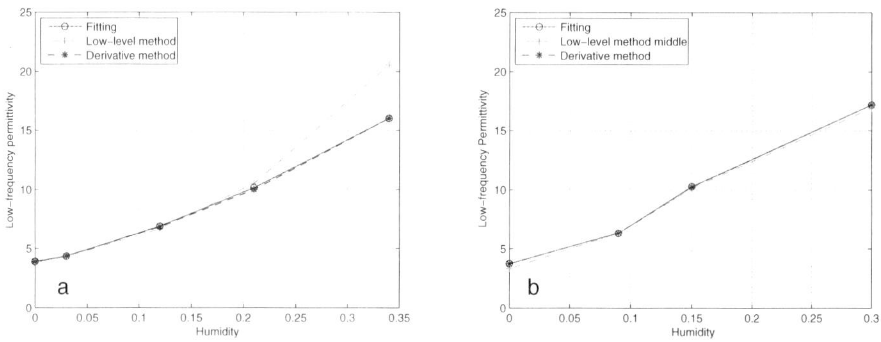

Fig. 19. Low-frequency permittivity values obtained with the fitting, low-level and derivative method for sand samples (Panel a) and clay samples (Panel b) as a function of water content.

7. Conclusion

In this analysis, an accurate model of the TDR measurement setup has been introduced. A parametric representation of unknown permittivity is introduced and its parameters are estimated by fitting the model response to the TDR measured response. This accurate model and the parametric inversion approach lead to an accurate evaluation of soil permittivity.
The values of Debye's model parameters have been related to the features of the TDR waveform. The role of probe attenuation in estimation of dielectric permittivity spectrum from TDR waveforms has been studied. The analysis is based on an simplified model of the measurement setup and on Debye's model of the unknown permittivity. Main result is that two types of TDR waveforms (low-attenuation and high-attenuation) can be identified. The shape of the edges in TDR waveform indicates wether the measurement setup works in low- or high-attenuation condition. Transition between the two types of waveforms is controlled by the wave attenuation along the probe. Low-attenuation operation can be obtained by using short probes and it yields waveforms that are more sensitive to the high-frequency behavior of the permittivity, since their harmonic content extends beyond the relaxation frequency f_{rel}. For small-ξ values, both ϵ_s and ϵ_∞ can be estimated from the TDR waveform, whereas for large-ξ values, only ϵ_s can be estimated.
Simple relations to evaluate the low-frequency permittivity directly from the TDR waveforms are introduced. These formulas are easy to use, and in good agreement with the results obtained with a derivative method based on the travel time of TDR signal along the probe

and with the results obtained using a parametric inversion approach based on an accurate model.
There are still many open problems and future works in the TDR technique applied to estimation of soil permittivity:

- Modeling of the system and in particular of the input port - probe transition, could be improved with the help of Layer Peeling techniques (Rautio, 1991; Tripathi & Jong, 1992).
- Fitting the modeled response to the measured response can be refined with numerical techniques.
- Implementation of a multi-pole Debye's equation, or another model, more suited for soil description, could be introduced.
- Several studies have been done with mixed sandy and clayey soil samples, but very few have been done with clay soils only. In this case there are some difficulties to model its behavior and a multi-pole Debye's equation and a new model suited for clay is needed.
- A new probe could be designed, in order first to obtain the majority of measurements falling in the low attenuation region (thus $\xi < 5$) and, second, to make it easier inserting soil samples into the probe. Worthless to remark that a new probe would require that the model has to be adapted to it and new parameters would rise.
- Water content is determined by means of experimental relations based on the knowledge of the low-frequency permittivity only (Topp et al., 1980). The model and its properties here introduced should be further investigated to gain a deeper in sight in the relationship between model parameters and water content.

8. Acknowledgment

This study was partially funded by the Italian Ministry of Research through project PRIN2007-SSB3WP, "Experimental measurement of atmosphere-vegetation-soil interaction processes and numerical models of its response to climate change".

9. References

Canone, D., M. Previati, S. Ferraris, M. P. & Haverkamp, R. (2009). A new coaxial time domain reflectometry probe for water content measurement in forest floor litter, *Vadose Zone J.*, 8(2): 363–372.

Courtney, C. (1998). Time domain measurements of electromagnetic properties of materials, *IEEE Transactions on Microwave Theory and Techniques*, 46(5): 517–522.

Evett, S. & Parkin, G. (2005). Advances in soil water content sensing:the continuing maturation of technology and theory, *Vadose Zone J.* 4: 986–991.

Evett, S., Tolk, J. & Howell, T. (2005). Time domain reflectometry laboratory calibration in travel time, bulk electrical conductivity, and effective frequency, *Vadose Zone J.* 4: 1020–1029.

Evett, S., Tolk, J. & Howell, T. (2006). Soil profile water content determination: Sensor accuracy, axial response, calibration, temperature dependence, and precision, *Vadose Zone J.* 5: 894–907.

Feng, W., Lin, C., Deschamps, R. & Drnevich, V. P. (1999). Theoretical model of multisection tdr measurement system, *Water Resour. Res.* 35: 2321–2331.

Hansson, K. & Lundin, L.-C. (2006). Water content reflectometer application to construction materials and its relation to time domain reflectometry, *Vadose Zone J.* 5: 459:468.

Hasted, J. (1973). *Aqueous Dielectric*, Chapman and Hall. London.

Heimovaara, T. (1994). Frequency domain analysis of time domain reflectometry waveforms 1. measurements of the complex dielectric permittivity of soil, *Water Resour. Res.* 30: 189–199.

Huisman, J., Bouten, W., Vrugt, J. & Ferré, P. (2004). Accuracy of frequency domain analysis scenarios for the determination of complex dielectric permittivity, *Water Resour. Res.* 40.

Lide, D. (1992). *Handbook of chemistry and physics*, 73rd ed. CRC Press, London.

Lin, C.-P. (2003). Analysis of nonuniform and dispersive time domain reflectometry measurement systems with application to the dielectric spectroscopy of soils, *Water Resour. Res.* 39: 6–11.

Nozaki, R. & Bose, T. (1990). Broadband complex permittivity measurements by time-domain spectroscopy, *IEEE Trans. on Instr. and Meas.* 39: 945–951.

Oswald, B., Doetsch, J. & Roth., K. (2006). A new computational techniques for processing transmission-line measurements to determine dispersive dielectric properties, *Geophysiscs* 71: k31–k35.

Pozar, D. (2005). *Microwave engineering*, 3rd ed., John Wiley Sons, Inc.,USA.

Rautio, J. (1991). A de-embedding algorithm for electromagnetics, *Int. J. of Microwave and Millimeter-wave Computed-aided Engineering* 1: 282–287.

Robinson, D., Jones, S., Wraith, J., Or, D. & Friedman, S. (2003). A review of advances in dielectric and electrical conductivity measurement in soils using time domain reflectometry, special section - advances in measurement and monitoring, *Vadose Zone J.* 2: 444–475.

Robinson, D., Schaap, M., Or, D. & Jones, S. (2005). On the effective measurement frequency of time domain reflectometry in dispersive and nonconductive dielectric materials, *Water Resour. Res.* 41: 1–9.

Roth, K., Schulin, R., hler, H. F. & Attinger, W. (1990). Calibration of time domain reflectometry for water content measurement using a composite dielectric approach, *Water Resour. Res.* 10: 2267Ű2273.

Savi, P., Maio, I. & Ferraris, S. (2010). The role of probe attenuation in the TDR characterization of dielectrics, *Electromagnetics* 30:6,: 554–564.

Somlo, P. & Hollway, D. (1969). Microwave locating reflectometer, *Electron. Lett.*, 5: 468–469.

Topp, G., Davis, J. & Annan, P. (1980). Electromagnetic determination of soil water content: Measurements in coaxial transmission lines, *Water Resour. Res.* 16: 574–582.

Topp, G. & Ferre', P.-A. (2002). *Water conten, in Method of soil analysis, Part 4, Physical methods,* SSSA Book,vol.5 ed. J.H. Dane and G.C. Topp, Soil Sci. Soc. of Am., 417-446, Madison, Wis.

Tripathi, V. & Jong, J. (1992). Time-domain characterization of interconnect discontinuities in high speed ciruits, *IEEE Transactions on Components, Hybrids and Manufacturing Technology* 15(4): 497–504.

Weerts, A., Huisman, J. & Bouten, W. (2001). Information content of time domain reflectometry waveforms, *Water Resour. Res.* 37: 1291–1299.

18

Updated Brazilian's Georeferenced Soil Database – An Improvement for International Scientific Information Exchanging

Marcelo Muniz Benedetti[1], Nilton Curi[2], Gerd Sparovek[3], Amaury de Carvalho Filho[4] and Sérgio Henrique Godinho Silva[2]
[1]*Centro de Tecnologia Canavieira,*
[2]*Universidade Federal de Lavras,*
[3]*Escola Superior de Agricultura "Luiz de Queiroz" /USP,*
[4]*Empresa Brasileira de Pesquisa Agropecuária*
Brazil

1. Introduction

The Brazilian soils represented by this database are denominated accordingly the available classification system which follows a hierarchy based on attributes and diagnostic horizons constituting the frame of the system. All the classes of soil present properties and characteristics determining a differential behavior due to the action of formation agents on the originating material. This heterogeneity has been studied for many years in order to permit a major understanding about Brazilian soils.

The classification systems allow the information organization tracking knowledge development. The classification represents the art of designing systems with some intuitive ideas about their subdivisions and priorities. In this way, it can be understood why and how soil classification systems prepared by people with varied knowledge, techniques and practical expertise differ from each other.

The soil classification is still away from the improved level of development reached by the botanical or zoological classification, but relevant progress has been achieved (Resende et al., 2007). Although there isn't a worldwide unified system available yet, attempts have been made in that direction. Among the many systems of pedological classification, the Soil Taxonomy (United States, 1975, 1999) developed in the United States must be mentioned. It constitutes the most elaborated and comprehensive system; even though, it presents some problems regarding to tropical soils (Resende et al., 2007). To allow the classification of all kinds of soil in the planet, a system with special characteristics has been continuously developed by the FAO to elaborate the map of soils of the world (FOOD AND AGRICULTURE ORGANIZATION OF THE UNITED NATIONS, 1974). In this way, there is a tendency in many countries, including Brazil, to develop their own soil classification systems keeping a coherent relation with the FAO system and the Soil Taxonomy (Resende et al., 2007).

During the most intensive period of soil surveys in Brazil, there was not a consolidated soil classification system available yet. In this way, maps and pedological surveys reports present a considerable range of differences in applied nomenclature, soil profiles classifications and legend compositions.

The Brazilian system of soil classification (Empresa Brasileira de Pesquisa Agropecuária [EMBRAPA], 1999, 2006) introduced significant modifications both in concepts and nomenclature. Because of that, an actualization and standardization is necessary to allow comparative analysis with the classifications applied in the previous surveys, which constitute the main source of data and relevant information for the system's development itself.

Updates of legends and pedological surveys have been done often with many objectives: a) just to update the legend; b) to unify legends of different maps with a specific objective, as the evaluation of agricultural suitability of lands; c) to allow fitting in other kinds of soil classifications; and d) to make easier the exchange of expertise among researchers from different countries. The update is also essential for the construction and use of a database to allow the treatment and use of information in a quick, adequate way firmed over the actualized bases.

In the last decades, the quick evolution of the methods and instrumentations used for the acquisition, storage, recovering, manipulation, analysis, access and distribution of data has facilitated the treatment of a great amount of data by the soil scientists. These techniques are indispensable for the monitoring and evaluation of the soil systems, their components and processes (Baumgardner, 1999).

The way as the data are stored in a database helps the organization, searching and updating of information. With this, the same data can be used for different applications reducing the space and effort (Assad & Sano, 1998). As examples it can be mentioned: (a) Digital Soil Map of the World (FAO, 1996); (b) SOTER - The World Soils and Terrain Database (Van Engelen, 1999); (c) CANSIS - Canadian Soil Information System (Coote & Macdonald, 1999); (d) NASIS - National Soil Information System (Soil Survey Staff, 1991); (e) Hydraulic Properties of European Soils (HYPRES) database (Nemes et al., 1999); and (f) Unsatured Soil Hydraulic Database (UNSODA), in its second version already (Nemes et al., 2001).

In this way, a database of Brazilian soils was developed starting from soil surveys done by the former "Soil's Conservation and Survey National Service" (SNLCS) of Embrapa (actual Embrapa Solos), previous institutions and the RADAM Brazil Project (Cooper et al., 2005). This basis allows just quantitative evaluation because of the lack of an updated soil's profile classification related to nomenclature changes and time based taxonomic distinction criteria.

Included in the described context, this work makes an analysis of the structure and an update of the soil's database profile classification to make available a more friendly and practical database for future research and analysis. Through a consistent and compatible identification of the database registers relatively to the soil's profiles, it is possible to update the basis comparing similar soils in distinct regions, defining central- and dispersion-values of the attributes, or generating studies focusing on the update of the classification system under analysis. Besides, this study furnishes information to help in the identification of the different classes of soil applying an approximated correspondence involving the Brazilian, American Soil Taxonomy and International Classification WRB/FAO systems.

2. Soil database

This work has used information originated from soil surveys in regional scale, covering a great portion of the Brazilian territory (Figure 1), contained in a Brazilian database soil (Cooper et al., 2005).

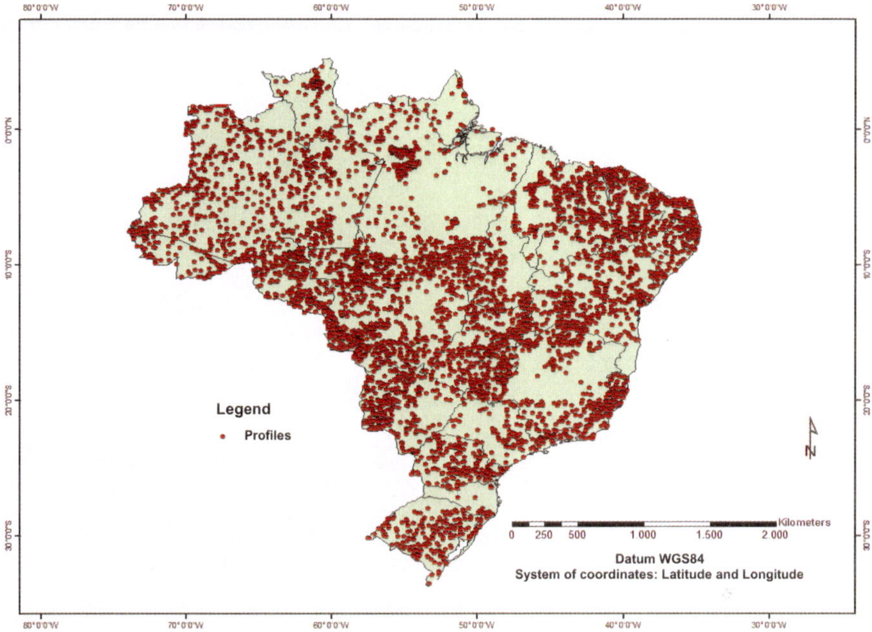

Fig. 1. Brazil map showing the political division and locations of the soil profiles used in the study

The works which originated the database came from official soil surveys chosen to assure the total coverage of the Brazilian territory. They were published between 1960 and 1986 by the former Soil's Conservation and Survey National Service (SNLCS-Embrapa), previous institutions, and the RADAM Brazil Project as described in table 1.

RADAM Volumes	Region	Year	Bulletin	Region	Year
1	São Francisco River and Aracaju	1973	34 SNLCS - BT N°60	North of Minas Gerais	1979
2	Teresina and Jaguaribe	1973	35 SNLCS - BP N°27 Vol. I	Paraná	1984
3	São Luís and Fortaleza	1973	36 SNLCS - BP N°27 Vol. II	Paraná	1984
4	Araguaia and Tocantins	1974	37 SNLCS - BP N°28	Pole Trombetas -PA	1984
5	Belém	1974	38 SNLCS - BP N°29	Pole Carajás – PA	1984
6	Macapá	1974	39 SNLCS - BP N°30	Urucará - AM	1984
7	Tapajós	1975	40 SNLCS - BP N°35 Vol. II	Maranhão	1986
8	Boa Vista / Tumucumaque	1975	41 SNLCS - BP N°36 Vol. II	Piauí	1986

RADAM Volumes	Region	Year	Bulletin	Region	Year
9	Tumucumaque /Tocantins	1975	42 SNLCS - BP N° 32	Barreirinha - AM	1984
10	Santarém	1976	43 SNLCS - BP N°18	Pole Roraima – RR	1983
11	Neblina Peak	1976	44 SNLCS - BP N°19	Tefé - AM	1983
12	Rio Branco	1976	45 SNLCS - BP N°20	Espírito Santo	1977
13	Javari/Contamana	1977	46 SNLCS - BP N°23	Ceará	1972
14	Içá – AM	1977	47 SNLCS - BP N°24	Bahia	1972
15	Juruá – AM	1977	48 SNLCS - BP N°31	Carneiro - AM	1984
16	Porto Velho	1977	49 SNPA - BT N° 12	São Paulo	1960
17	Purus – AM	1978	50 SNLCS - BP N°35 Vol. I	Maranhão and Piauí	1986
18	Manaus – AM	1978	51 SNLCS - BP N°15	Pole Pré-Amazônia Maranhense	1982
19	Guaporé	1979	52 SNLCS - BP N°16	Ariquemes - RO	1982
20	Juruema	1980	53 EPFS - BT N°13	Espírito Santo and Minas Gerais	1970
21	Fortaleza	1981	54 DPP – BT N°26 Vol. II	Pernambuco	1972
22	Tocantins	1981	55 SNLCS - BP N°26	Maranhão and Piauí	1984
23	Jaguaribe/Natal	1981	56 DPP – BT N°18	South of Mato Grosso	1971
24	Salvador	1981	57 EPFS - BT N°15	Paraíba	1972
25	Goiás	1981	58 SNLCS - BT N°45	Espírito Santo	1978
26	Cuiabá	1982	59 SNLCS - BT N°38	Left Margin of São Francisco River – BA	1976
27	Corumbá	1982	60 SNLCS - BT N°52 Vol. II	Right Margin of São Francisco River - BA	1977-1979
28	Campo Grande	1982	61 DPP – BT N°21	Rio Grande do Norte	1971
29	Brasília	1982	62 DPP – BT N°28	Ceará	1973
30	Aracaju	1983	63 SNLCS - BP N°14	Apiaú - RO	1982
31	Goiânia	1983	64 SNLCS - BP N°17	Mato Grosso	1983
32	Rio de Janeiro / Vitória / Goiânia	1983	65 SNLCS - BP N°9	Barreirinha - AM	1982
33	Porto Alegre/ Uruguaiana / Lagoa Mirim	1975	66 SNLCS - BP N°36 Vol. I	Piauí	1986

* RADAM – Exploratory Soil Surveys (Radam Brazil Project); DPP - Division of Pedologic Research; EPFS - Staff of Pedology and Soil Fertility; SNPA - National Service of Agronomic Research (Committee on Soil); SNLCS – Soil's Conservation and Survey National Service (now Embrapa Solos); BP - Research Bulletin; BT - Technical Bulletin.

Table 1. Source and year of the surveys used in the database

The database was implemented using the Microsoft Excel® format, including a set of data relative to the characteristics and analytical results of a superficial and a subsuperficial soil horizon of each profile (Table 2). The profiles are identified by codes referent to the publication source together with their original numeration, while for the horizons it was maintained the same designations used in the original works. The variables for which the data were compiled and placed in a distinct column are constituted by: code from the publication source, number of the profile, year of the survey, latitude, longitude, original soil classification, slope, drainage, symbol and depth of the horizon, moist color, macroclastic composition (pebbles and cobbles), particle size distribution (coarse sand, fine sand, silt and clay), content of oxides obtained by sulfuric acid digestion (SiO_2, Al_2O_3, Fe_2O_3), pH in H_2O and in KCl, organic C, organic matter, total N, exchangeable cations (Ca^{2+}, Mg^{2+}, Ca^{2+} + Mg^{2+}, K^+, Na^+, Al^{3+}, H^+, H^++Al^{3+}), CEC, base sum, base saturation and aluminum saturation (Cooper et al., 2005).

Starting from this initial basis, and after proceeding with several adjustments and distortion corrections, in some cases through search in the original works, it was done the actualization of the soil profiles classification accordingly to the new version of the Brazilian system of soil classification (EMBRAPA, 2006) followed by the exportation of the data to a database management system, the Microsoft Access®. In that way, it was constituted the database of Brazilian soils which presents qualitative characteristics described in a numerical format containing 10,950 horizons referring to 5,479 soil profiles and 57 columns with the above mentioned variables and the actualized classification and respective degree of reliability (Benedetti et al., 2008). Each profile has a spatial indication related to its location on the earth surface, which means, has its coordinates (latitude and longitude) in the system of coordinates DATUM: WGS 84. The access to the database with the updated classification of the soil profiles can be done in the following web address: www.esalq.usp.br/gerd.

For the update of soil profiles database classification, compatible with the new version of the Brazilian soil classification system (EMBRAPA, 2006), it were considered the characteristics of the two horizons of each profile (superficial and subsuperficial) contained in the database, considering the criteria established for the attributes and diagnostic horizons, and the concept and definition of the known classes of soil. The taxonomic fitting of the soil profiles was done up to the fourth categorical level, and for some soils it was included the distinction in fifth level referred to the occurrence of alic character (Al saturation ≥ 50% and $Al^{3+} \geq 0.5$ $cmol_c/kg$) when pertinent.

Degrees of confidence, represented by numbers from 1 to 4, were adopted to express the adequacy of the adjustment obtained in the taxonomic fitting, in one of each four categorical levels, mentioning that many profiles would need complementary data for more precisely updating the classification. The value 1 indicates taxonomic fitting with elevated degree of confidence, which reduces progressively with the increasing of the values. The value 2 expresses some doubt in relation to the classification at the considered categorical level, as a function of the unavailability of some analytical data of minor relevance, and the previous profile classification suggests the present fitting as the most probable. The value 3 follows a similar line of reasoning, but in this case the doubt is increased due to some missing data, being the present classification based exclusively on the previous denomination. The value 4 was applied for situations when it was not possible to complete the update.

Code	Variable	Description	Units
ClassfSoil	Soil Classification	Soil name according to the original document	
DecVal	Slope	Inclination of the soil surface in relation to a flat reference	%
Drain	Drainage	Velocity at which water is removed from the soil profile	
SoilDepth	Soil Depth	Lower limit of the horizon	cm
HzSimb	Soil Horizon Symbol	According to the pedological convention	
HzDepth	Horizon Depth	Depth range registering the beginning and end of the horizon	cm
Munsell Color	Munsell Color	Description according to Munsell notation	
CG	Cobble	Material> 20 mm in diameter	(%)
FG	Pebble	Material between 20-2 mm in diameter	(%)
CS	Coarse Sand	Material between 2-0.2 mm in diameter	(%)
FS	Fine Sand	Material between 0,2-0.002 mm in diameter	(%)
Sand	Sand	Material between 2-0.005 mm in diameter	(%)
Silt	Silt	Material between 0.05-0.002 mm in diameter	(%)
Clay	Total Clay	Material < 0.002 mm in diameter	(%)
SiO_2	SiO_2	Silicon oxide content	(%)
Al_2O_3	Al_2O_3	Aluminum oxide content	(%)
Fe_2O_3	Fe_2O_3	Iron oxide content	(%)
pH_H_2O	pH measured in water	pH value determined in a mixture with a specified relationship between soil and water	
pH_KCL	pH measured in KCL solution	pH value determined in a mixture with a specified relationship between soil and 1mol L^{-1} KCl	
C	Organic Carbon	Content of organic carbon in soil	(%)
MO	Organic Matter	Content of organic matter in soil	(%)
N	Total Nitrogen	Content of total nitrogen in soil	(%)
ExCa	Ca^{2+}	Content of exchangeable calcium	($cmol_c$/kg)
ExMg	Mg^{2+}	Content of exchangeable magnesium	($cmol_c$/kg)
ExK	K^+	Content of exchangeable potassium	($cmol_c$/kg)
ExNa	Na^+	Content of exchangeable sodium	($cmol_c$/kg)
ExAl	Al^{3+}	Content of exchangeable aluminum	($cmol_c$/kg)
ExH	H^+	Content of exchangeable hydrogen	($cmol_c$/kg)
Ex_H_Al	$H^+ + Al^{3+}$	Content of exchangeable hydrogen and aluminum	($cmol_c$/kg)

Code	Variable	Description	Units
CTC_pH7	Cation Exchange Capacity at pH 7.0	Sum of exchangeable Ca, Mg, K, Na, H and Al (T value)	($cmol_c$/kg)
SumB	Base sum	Sum of exchangeable Ca, Mg, K and Na (S value)	($cmol_c$/kg)
SatB	Base Saturation	Relationship between sum of bases and cation exchange capacity (V value)	%
SatAl	Aluminum Saturation	Relationship between exchangeable aluminum and sum of bases more exchangeable aluminum (m value)	%

Table 2. Code, description and units of representation of the database variables

3. Brazilian soils: A panoramic vision

The set of data, referring to physical, chemical and morphological attributes, which includes also information about location, depth and drainage, made possible the updating of the classification of almost all the soil profiles compounding the database. Figure 2 presents the representativeness of the classes of soil in elevated categorical level when compared with their total area of occupation in the Brazilian territory.

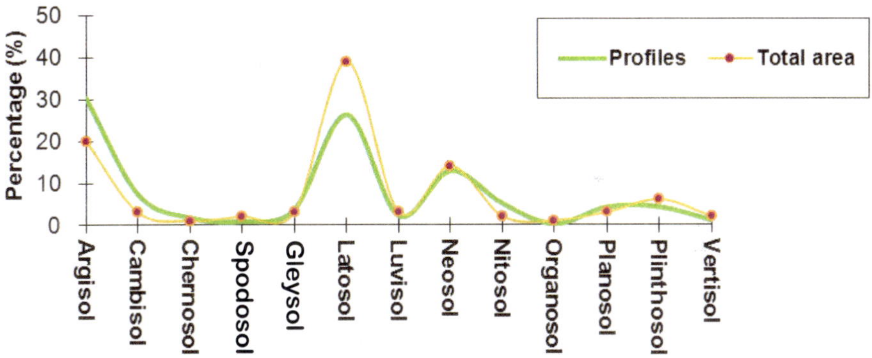

Fig. 2. Percentage of profiles in the database compared with the percentage of each class of soil in total area of Brazil

The soils were classified up to the fourth categorical level, being the alic character used when pertinent for distinction of soils in the fifth categorical level. With all classes in the most elevated categorical level (order) represented in the database, from the present actualization resulted 281 classes of soil in the subgroup level (besides three classes in suborder level referring to profiles that couldn't be classified in lower levels), for which the relation with the original classification and respective profile number are available for downloading in the following address: www.esalq.usp.br/gerd.

Through the comparison between the previous classification and the present fitting of the soil profiles in their various hierarchic levels many relevant aspects could be observed. From this evaluation emerged some characteristics expressed in the previous denomination that couldn't find a correspondence among the classes envisioned by the present SiBCS (Brazilian System of Soil Classification), both in the previous and actual versions, especially in the fourth categorical level, which resulted in loss of information. Keeping in mind to release these information to the international community and also to establish an approximated correlation between the soil classes in the Brazilian classification with the American classification "Soil Taxonomy" and with the International WRB/FAO systems, each soil class is commented individually as follows:

3.1 Argisols

With expressive participation in the database, referent to 1,660 soil profiles representing 30.29% of the total, the Argisols comprehend the majority of the soils previously classified as Podzolic soils, that present textural B horizon with low activity of clay, or with high activity of clay and low base saturation.

In this class it were also included three soil profiles previously classified as Structured Dusky Earth, that present medium texture and could not be fitted as Nitosols (class that put together almost all the former Structured Dusky Earth), and a Rubrozem profile, which by the first edition of the SiBCS would be classified as Alisol (EMBRAPA, 1999), class excluded of the present version. With the extinction of this last class and pertinent updates of the definitions of Argisols, which by now comprehend soils with elevated clay activity conjugated with dystrophic or alic character (EMBRAPA, 2006), some distortions were corrected as the case of the profile 83 in the soil survey from the "Uruguaiana/Lagoa Mirim report" (Brasil, 1975d). Originally classified as Ta Alic Gray-Brown Podzolic soil, this soil would not have taxonomic fitting accordingly to the previous edition of the system (EMBRAPA, 1999), stated that the high activity of the clay fraction would exclude it from the Argisol class, while the concentration of exchangeable Al inferior to 4 $cmol_c/kg$ made impossible its fitting as Alisol, also not fitting the requirements of any other class in the order level. The introduced modifications allowed the classification of this soil as Argisol, at the same time that the reformulation of the concepts of this class in the suborder level (2nd categorical level) leaded to its fitting as Gray-Brown soil, recently included class, and in third level as Alitic, class now incorporated to the system and that corresponds to part of the Alisols in the previous version.

The inclusion of the Gray-Brown Argisols, as mentioned above, allowed the distinction between soils which have the darkening of the superior portion of the textural B horizon as remarking characteristic, referent to the former Gray-Brown Podzolic soil. Because of the strong correspondence between the previous- and the currently-employed criteria on these soils reconnaissance, the taxonomic fitting of the soil profiles without specification of the B horizon color is based on the previous denomination. However, for some soils in this class it was not possible the fitting in lower taxonomic levels, because it is recognized only one class in the third categorical level, referring to soils with alitic character.

A good correlation was also observed between the actual classes of the second categorical level, referent to Gray-, Yellow- and Red-Argisols and the former Gray-, Yellow- and Dark Red- Podzolic soils, respectively. As a function of the current concepts, this last class presents some soils in which the color hue a little bit yellower than 2.5YR (in general 3.5YR) are now fitted as Red-Yellow Argisols. Different from the others, the former Red-Yellow Podzolic soils present relevant variation in the taxonomic fitting in the second categorical level, distincted in Yellow Argisols (with dominance of colors having hue 7.5YR or yellower) and Red-Yellow Argisols (B horizon color redder than 7.5YR and yellower than 2.5YR, except when hue is 5YR with value and chroma < 4), and also in Red Argisols (hue 2.5YR or redder). Therefore, in the absence of information about the B horizon color, it was considered for these soils a more elevated degree of uncertainty in the actualization of the classification in this categorical level relatively to the others.

In the great group level (third categorical level), the available data allow for the majority of the profiles a taxonomic fitting with increased confidence. However, for the Yellow Argisols it is notable the impossibility of distinction based on the activity of clay fraction, differently from the other classes in the same level. In this way, the profile 23 from the soil survey of the Fortaleza region (Brasil, 1981a), previously classified as Ta Alic Red-Yellow Podzolic soil, for example, fits as Yellow Argisol, due to its yellow colors, and as Dystrophic in the third level, although it presents high activity of clay. Also in the fourth categorical level some differential characteristics expressed in the former classification are not considered in the soils distinction by the new system. The profile 32 of the soil survey of Fortaleza region (Brasil, 1981a), formerly classified as plinthic abruptic Alic Red-Yellow Podzolic soil, for example, fits as abruptic Dystrophic Yellow Argisol, because it is not considered in the SiBCS the possibility of fitting soils having simultaneous occurrence of such characteristics. Besides, the occurrence of transitional character indicated by the latosolic and cambic denomination are not distinguished in some cases, as example of the profile 79 of the soil survey of Rio Branco region (Brasil, 1976c), previously classified as latosolic Alic Red-Yellow Podzolic soil, and the profile AE2 of the soil survey of the pilot area at Barreirinha-AM (EMBRAPA, 1982) denominated moderate A cambic Tb Alic Red-Yellow Podzolic soil, which following the current system fits as typic Aluminic Red-Yellow Argisol.

Establishing a high categorical level correspondence with the system proposed by the FAO, the Argisols could be approximated to the classes Acrisols, Lixisols and Alisols, while for the American classification Soil Taxonomy these soils would be approximated to the classes Ultisols and Oxisols (Kandic). These classes present as main diagnostic features the clay illuviation to the subsuperficial horizons generating a textural gradient.

3.2 Cambisols

This class covers the vast majority of soils previously identified by the same name, used in order level to designate soils having restricted pedogenetic development, characterized by the presence of incipient B horizon, excepting those of the eutrophic character, with high activity clay and chernozemic A horizon, now included in Chernosol class.

Among profiles that make up the database, 397 were classified as Cambisols, which represents 7.25% of the total. While keeping the former name in the first categorical level, even without major changes in the criteria for identification, the soils of this class have suffered a systematic structuration in all subsequent levels. The distinctions between them, represented in a low systematic way by the former nomenclature, are now distributed and hierarchically organized in the second, third and fourth categorical levels.

In the suborder level, three classes are recognized: Humic Cambisols, characterized by the presence of humic A horizon; Fluvic Cambisols, referring to the soils of alluvial origin; and Haplic Cambisols, which cover the other soils of the class, whose characteristics do not meet the requirements of the earlier ones. Of these, only the Fluvic Cambisols, class now incorporated into the classification system (EMBRAPA, 2006) and whose distinction requires a comprehensive characterization of the entire depth of the profile, or even the environment of occurrence, making its recognition difficult from the data available here, are not listed in the database. It was also observed for some profiles, a certain difficulty for the fitting as Humic, due to lack of further information about the superficial horizon.

The absence of more complete analytical data also hampered the taxonomic fitting in lower categorical levels, which increases the uncertainty about the classification. The impossibility to confirm the carbonatic nature, for example (results related to the concentration of $CaCO_3$ are not present in the database), led to the fitting as Carbonatic in the third categorical level, based on the previous name, or some characteristic that would suggest the occurrence of that character within the specified control section. Thus, the classification of the profile 244 of the soil survey of the right margin of the São Francisco River (EMBRAPA, 1977-1979) as typic Carbonatic Haplic Cambisol is based on the supposed occurrence of the carbonatic character as indicated by the previous classification (moderate A carbonatic C Tb Eutrophic Cambisol) within the first 120 cm depth.

In general, the current structuration of the classification system provided a more accurate discrimination, better representativity and a more adequate basis of comparison for Cambisols. Such result is not always observed for other classes of SiBCS. Certain distinctions afforded by the former nomenclature, however, are not represented in the current system, for example soils previously identified as Brown Cambisols, typical of the subtropical environment of southern Brazil, which currently are not being discriminated. Some characteristics are also not included in the fourth categorical level, referring to intermediate or special characteristics. Thus, for the profile 29 of the soil survey of Mato Grosso (EMBRAPA, 1983), previously classified as chernozemic A podzolic Tb Eutrophic Cambisol, the transitional characteristic expressed in the original classification was not considered, having been framed as typic Eutrophic Tb Haplic Cambisol.

The plinthic character used to distinguish soils with not enough plinthite to be characterized as plinthic horizon (EMBRAPA, 2006), is not included in classes distinction of the 4th categorical level for the Eutrophic Ta Haplic Cambisols, since the plinthic class is not considered in the fourth categorical level. Thus, soils originally classified as moderate A plinthic Ta Eutrophic Cambisols like the profile 128 of the soil survey of the State of Maranhão (EMBRAPA, 1986a), or the profile 13 of SNLCS-BP-26 (EMBRAPA, 1984), lost the reference indicative of plinthite presence (expressed both in the previous classification as in the designation of the subsurface horizon), being classified as typic in the fourth level.

The Brazilian Cambisols have close correlation with the American and International systems due mainly to their transitional feature in the formation process, characterized by a not very expressive B horizon in the soil profile. Their names are derived from the classification of the soil map of the world by FAO, where they are known as Cambisols. The American classification reports the origin of their names in the conditions of pedogenesis, calling them Inceptisols.

3.3 Chernosols

A total of 87 soil profiles, representing approximately 1.59% of the database, were classified in this class, which groups soils previously named as Brunizens, Reddish Brunizens and Rendzinas, plus some eutrophic Cambisols, with high clay activity and chernozemic A horizon. The reunion of these soils in a class on level of order reflects the importance attached to chemical and mineralogical characteristics together with the presence of chernozemic A horizon, with important implications for the sustainable use and management.

In the second categorical level, we observed a strong correspondence between Rendzinas and Reddish Brunizens with the current classes of Rendzic Chernosols and Argilluvic Chernosols, respectively, while the former Brunizens and Cambisols were classified as Haplic. However, for some profiles, the taxonomic classification was a little prejudiced because there was not enough available complementary information particularly with regard to colors of all the extension of the B horizon needed to distinguish the Ebanic Chernosol (dark color, almost black), or even because this horizon was not represented in the database. Thus, only the profile 57 of the soil survey of the south of the former State of Mato Grosso (Brasil, 1971a), formerly called Reddish Brunizem, could be classified as Ebanic Chernosol.

In the level of great group (third categorical level) the main limitation for a better taxonomical classification was related to the lack of data referent to the carbonatic character. Thus, the classification as Carbonatics was based on the previous name or even on the indication of the carbonatic character given by the subordinated designation of the subsurface horizon (EMBRAPA, 1988b).

The insufficiency of data also made it difficult to classify in the fourth categorical level (subgroup). Thus, several profiles were classified as typical as an attempt, taking into account the impossibility to confirm the occurrence of distinctive characteristics of other classes. In addition, some properties expressed in the previous name don't correspond to the classes provided by the actual system, both in its previous version (EMBRAPA, 1999) as in the current one (EMBRAPA, 2006), for not being included in the key, a class that can distinguish it, which results in loss of information. This situation is exemplified by the profile 34 of the RADAM Brazil Project (Brasil, 1981a), originally classified as abruptic solodic Reddish Brunizem, now classified as abruptic Orthic Argilluvic Chernosol. Also for profiles 48 (Brasil, 1972a) and 34 (Brasil, 1981a) of the RADAM Brazil Project, both named as vertic abruptic Reddish Brunizem, the differential characteristic due to the abrupt textural change is not addressed in the current classification.

The Chernosols occupy small areas in the Brazilian territory, but they have great expressiveness in temperate regions. They can be correlated with the Chernozems, Kastanozems, Phaeozems and Greyzems classes in the legend of the WRB / FAO. In the American system these soils correspond to Mollisols that exhibit high activity of clay.

3.4 Spodosols

Corresponding to soils that are distinguished by the presence of the spodic B horizon, this class includes the former Podzols and Hydromorphic Podzols, represented in the database by 53 soil profiles, which equals to 0.97% of the total.

In the second categorical level (suborder), these soils were classified as Humiluvic and Ferriluvic. However, the taxonomic fitting in this level was somewhat difficult due to the lack of complementary morphological information and, in some cases, the lack of data relative to the B horizon. For these profiles, the value 3 was used to represent the degree of confidence in order to provide the continuity of classification, since the distinction of classes in the subsequent categorical levels bases on equivalent concepts and denominations, especially in the great group level. In general, the classification in this hierarchical level also has a certain amount of uncertainty (degree of confidence 2), due to the lack of information about environmental conditions, once the presence of water table or saturation by water within 100 cm from the soil surface are distinctive criteria of classes. Therefore, the classification of the profiles, as Hydromorphics or Orthics, was based on the previous classification that, in general, distinguished soils subjected to hydromorphic conditions or not. Similarly, the fourth level of classification (subgroup) was also a bit impaired by the absence of complementary data of profiles that would allow a more precise taxonomic fitting . Therefore, the actualization of classification of soil profiles that belong to the class of Spodosols, in level of suborder, great group or subgroup, is about a kind of approximation, in correlation with the previous classification.

The Spodosols of American Soil Taxonomy include diagnostic characteristics similar to the spodic B horizon of the Brazilian Espodosols. The legend of the WRB / FAO retains the name formerly used by the mapping of soils in Brazil, as can be seen in the database the designation of Podzols.

3.5 Gleysols

Class of soils consisting of mineral material, whose genesis is related to strong hydromorphic conditions, expressed in the presence of gley horizon, and absence of distinctive features of Plinthosols or Planosols. This class is represented in the database by 214 soil profiles, which correspond to 3.91% of the total. They include soils formerly and generically named as Gleysols or Gley soils, which often were separated in Humic Gley, Low Humic Gley, Tiomorphic Gley and Saline Gley. It includes, in addition to those, the Gray Hydromorphic soils.

As observed for other classes, the lack of data concerning the full extension of the profiles resulted in difficulty for some taxonomic framework, inclusive in the level of order, due to their subsurface horizons contained in the database do not show enough coloration to distinguish the gley horizon . In these cases, the classification in this class was based on the previous name, in the assumption that other horizons of the profiles in question meet the requirements of this diagnostic horizon.

All classes of the second categorical level defined by the current system are contained in the database, although for some profiles the taxonomic classification at this level also provides some degree of uncertainty, especially for Salic Gleysols, since no data are presented relative to the electrical conductivity. Also for some soil profiles classified as Tiomorphic Gleysols, in which the pH of the horizons contained in the database are slightly above 3.5, therefore insufficient to characterize sulfuric horizon or sulfidric materials, requiring evaluation of other horizons for a more precise definition, the classification in the second level was established based on the former designation, with value 2 for the degree of confidence.

The distinction of these soils in the third categorical level was satisfactory, without significant limitations. Already the fitting in the fourth level, highlights the impossibility of differentiation based on some characteristics formerly included in the distinction of these soils, such as the great capacity of swelling and shrinkage of the soil material, expressed in the vertic designation, such as the profile 146 of the soil survey of the State of Maranhão (EMBRAPA, 1986a), previously classified as moderate A solodic with tiomorphism vertic Ta Eutrophic Gleysol, which was classified as solodic Orthic Tiomorphic Gleysol.

It is also important to emphasize the question of the current fitting of the former Gray Hydromorphic, whose distinction related to the presence of textural B horizon, is addressed only in the fourth categorical level, referring to the class of argisolics. However, in the event of sufficient characteristics for classification as plinthic, such as the profile 132 of the RADAM Brazil Project (Brasil, 1981b), that distinction is impossible, since these ones, as well as the leptics, take precedence in the sequence of the identification key. Moreover, it is not possible to differentiate soils with other special characteristics, such as the presence of fragipan, as occurs with the profile 37 of the soil survey of the left margin of the São Francisco River (EMBRAPA, 1976c).

The Gleysols are developed over unconsolidated materials (sediments and saprolites) and are greatly influenced by the occurrence of prolonged flooding due to the position of the water table near the surface. They are recognized in the legend of the WRB / FAO as Gleysols, except for classes of Salic Gleysols, that correlate to Solonchaks. Unlike what occurs in the two systems mentioned above, the American classification Soil Taxonomy identify the inherent properties of these soils in lower levels of classification. The Gleysols can be correlated with the Entisols (Aqu-alf-and-ent-ept), while Gleysols having saline character can best approach classes of the Aridisols and Entisols (Aqu-sulfa-hydra-salic).

3.6 Latosols

Characterized by a very advanced stage of development, as evidenced by the insignificant occurrence or even absence of weatherable primary minerals, and the presence of latosolic B horizon, the Latosols constitute the largest class of territorial expression and agricultural potential of the country, being explored with various crops, reforestation and pasture (Ker, 1997). They represent 25.79% of the database, with a total of 1,413 soil profiles.

For this class, which name was maintained in the level of order in the current system, some subdivisions were already established since the 60's, as exemplified the denominations of Red-Yellow Latosol, Dark Red Latosol and Dusky Red Latosol, to which, over time, it were added Yellow Latosol, Brown Latosol, Ferriferous Latosol and Una Latosol. These designations were, partially, maintained in the distinction of classes in the second categorical level, although the differentiation criteria have changed. In the current system, the distinction of these soils in this level is based only on the color of the B horizon, while previously competed in addition to this, other characteristics such as the iron oxide content obtained by sulfuric acid digestion, which in low contents (< 8 % of Fe_2O_3) is a sure indicator of strong cohesion ("hard-setting soils"), condition very far from the granular structure (high friability), typical of most of the Brazilian Latosols. In this context, the previous classification system was more informative. Thus, the former Dark Red-, Dusky Red-, and Ferriferous- Latosols, generally refer to the currently Red Latosols, with the distinction of the iron oxide content used in the third categorical level. The former denominations Red-Yellow-, Yellow-, and Brown- Latosols still remain, but while for this last class there is a strong correspondence between past and present distinctive criteria, for the other two classes there are not, beyond nomenclature, a direct relationship with the classes of the current system. The Yellow Latosols tend to stay in the homonymous class, while the former Red-Yellow Latosols are distributed in the classes of Yellow-, Red-, and Red-Yellow Latosols; otherwise the Una Latosols from the database were all classified as Yellow Latosols. In view of this, the lack of information about the color of the B horizon constituted one of the main difficulties for updating the classification of these soils in the second categorical level. In this case, the taxonomic fitting was oriented by the previous name, with differentiation concerning the degree of confidence, according to the following pattern: Dusky Red- and Dark Red-Latosols were classified as Reds, and Yellow- and Brown-Latosols in the homonymous classes, all with degree 2 of confidence; similarly, the Red-Yellow Latosols were fitted in the class of the same denomination, but with degree 3 of confidence, due to the increased uncertainty of this adjustment.

In the third categorical level distinction, that is based on the characteristics of the exchange complex, in association with the contents of Fe_2O_3 from the sulfuric acid digestion, the taxonomic framework was very satisfactory as well as the subsequent level (subgroup), which seems to indicate major concern in the establishment of possible distinction of these soils, perhaps because of their high potential for agricultural use, as well as their large territorial expression in the country.

The word Latosol implies soils formed under conditions of significant weathering-leaching, resulting in materials with a residual concentration of secondary minerals, including kaolinite, gibbsite, and iron- and aluminum- oxides, in different proportions. The understanding of their genesis facilitates the identification of their corresponding names in the American Classification Soil Taxonomy and legend of WRB/FAO, as Oxisols and Ferralsols, respectively.

3.7 Luvisols

Grouping of soils characterized by the presence of textural B horizon or nitic B horizon, with high clay activity and high base saturation, excluding those with chernozemic A horizon. This class is represented in the database by 118 soil profiles (2.15% of the total).

Encompasses soils originally classified as Non-Calcic Brown soils and Red-Yellow Podzolic soils, with little participation of Gray-Brown Podzolic soils and Dark Red Podzolic soils.

It also includes some profiles formerly called Reddish Brunizens, whose A horizons do not have enough base saturation (≥ 65%) to characterize the chernozemic A horizon, according to the current criteria.

Although having achieved a very satisfactory taxonomic fitting in the level of order, some loss of accuracy in the update was already observed from the second categorical level, as a function of the available data do not allow an precise evaluation of the dominance of colors of the B horizon, necessary to distinguish between Chromic- and Haplic- Luvisols. This fact resulted in a higher degree of uncertainty regarding the taxonomic fitting, also reflected in lower categorical levels. Another aspect that difficulted the classification of some soils, due to the unavailability of data for the entire extension of the profiles, is relative to the depth of the *solum* (horizons A + B), characteristic used to distinguish, in the third categorical level (great group), the class of Palic Chromic Luvisols.

For the same reason, the distinction of some soils in the fourth categorical level was also slightly affected, considering the utilized criteria to identify the classes of saprolithics and lithics. Moreover, some differential characteristics indicated in the previous denomination are not considered by the current classification, as the profile 58 of the soil survey of the State of Pernambuco (Brasil, 1972e), currently classified as planosolic Orthic Haplic Luvisol, in which the presence of solodic character and vertic characteristic expressed in the previous classification (solodic vertic planosolic Non-Calcic Brown soil), are not considered in the distinction of this type of soil in the current system.

The Luvisols can be framed in the class of Luvisols of the WRB/FAO legend, that presents the B horizon with accumulation of high activity clays. In the American classification the best correspondent classes of soils would be the Alfisols and Aridisols (Argids).

3.8 Neosols

Soils generally young, with little development and in process of formation, the Neosols are characterized by the absence of any diagnostic B horizon. With 684 soil profiles, that represent 12.48% of the database, are included in this class four major soil groups recognized by the classification scheme previously adopted in Brazil, that are the Litholic Soils, Regosols, Alluvial Soils and Quartzous Sands.

In close correspondence with the above groups four classes are recognized in the second categorical level: Litholic Neosols, Regolithic Neosols, Fluvic Neosols and Quartzarenic Neosols, all represented in the database. Because of this, as well as the close correlation of distinctive criteria adopted by the current classification system and the previous scheme, there were few profiles that presented some problems for taxonomic fitting in the suborder level.

In this sense, should be highlighted some soils previously classified as Litholic Soils, currently framed in the class of Regolithic Neosols, due to present the C horizon extending below 50 cm depth. However, some profiles of Quartzous Sands, might not even be fitted as Neosols, due to the change of the criteria adopted in distinguishing sandy soils, that from the 80's replaced the lower limit of 15% clay by the requirement of the granulometric compositions of the sandy and loamy sand textural classes. Consequently, the classification of these profiles was not updated.

The fitting of these soils in the third categorical level was also done without difficulty, in some cases guided by the previous classification, like some profiles of Hydromorphic Quartzous Sands for which the color of the subsuperficial horizon is not presented. However, the profiles AE9 and AE10 of the soil survey of the pilot area of Barreirinha-AM (EMBRAPA, 1982d), previously classified as solodic Ta Alic Alluvial Soils, that currently fit as Fluvic Neosols, could only be classified up to the second categorical level, because it is not included in the current system the distinction of soils of this class that present dystrophic character together with high activity clay. In this way, by correlation with similar soils, it is suggested the inclusion of the class of the Dystrophic Ta to the fitting of Fluvic Neosols in the third categorical level.

In relation to the fourth categorical level, in general, the distinction of the soils presented high agreement with the differentiations expressed in the previous denomination, although with certain degree of uncertainty in the taxonomic fitting, due to the absence of complete data of the profiles. For the profile P06 of the RADAM Brazil Project (Brasil, 1981c), it was not possible, however, the distinction regarding the occurrence of concretions, as indicated in the previous classification (concretionary Eutrophic Litholic Soil), characteristic that deserves to be evaluated as a differential characteristic of classes in the subgroup level.

The Quartzarenic Neosols updated in this database can be correlated with the Arenosols of the WRB/FAO legend, while in the American Soil Taxonomy this class would fit in the third categorical level (great group) as Quartzipsamments. Regolithic Neosols have been recognized as Regosols by WRB/FAO. Regarding their correlation with the American system this can be done only in the second categorical level (suborder) as Psamments. Soils that present lithic contact within 50 cm depth (Litholic Neosols), can be correlated to

Leptosols of the WRB/FAO legend. The American classification uses this property in the fourth categorical level (subgroup) and can be related to Lithic of Orthents suborder and also with Lithic of great group Psamments. The Fluvic Neosols are soils originated from alluvial sediments with close correlation with the Fluvisols of the WRB/FAO legend, while in the American Soil Taxonomy they would be fitted in the second categorical level as Fluvents.

3.9 Nitosols

In the database of soils of Brazil, this class is represented by 260 profiles, which equals to 4.75% of the total. Encompasses soils previously referred as Structured Dusky Red Earth, Structured Brown Earth, Similar Structured Dusky Red Earth and Similar Structured Brown Earth, beyond some profiles recognized as Red-Yellow Podzolic soils which present low textural gradient. More specifically for these last soils, there is a higher degree of uncertainty regarding the classification as Nitosols, because the database do not include information on morphological characteristics necessary for unambiguous identification of the nitic B horizon, distinctive of this class, whose requirements include structural development and significant presence of clay skins.

With the re-structuration promoted by the new version of SiBCS, regarding the inclusion of the class of Brown Nitosols in the second categorical level, many of the soils previously classified as Haplics, according to the first edition of the system (EMBRAPA, 1999), are now fitted in this new class.

In this work we option for the fitting as Brown Nitosols only the soils formerly called Structured Brown Earth or Similar Structured Brown Earth, provided they meet the requirement of specified color, or, in the absence of this information, assigning higher value for the degree of confidence (greater degree of uncertainty). This situation highlights the need for a more precise definition of the distinctive criteria of the class, where the comparison of related soils, members of the database under study could contribute. It should be registered also the impossibility of classification in lower categorical levels of the profile C48 of soil survey of the State of Paraná (EMBRAPA, 1984), originally identified as chernozemic A Eutrophic Structured Brown Earth, because in the current system classes that allow taxonomic fitting as Brown Nitosols with eutrophic character are not considered.

Relatively to other classes in the second categorical level, we observed a strong correspondence between the former Structured Dusky Red Earth and Red Nitosols, that beyond these soils also comprises a small proportion of Red-Yellow Podzolic soils. However, some soils previously recognized as Structured Dusky Red Earth, many of which having high iron oxide content, were classified as Haplic Nitosols, due to present 3.5YR or 4YR hues in the B horizon.

Regarding the fourth categorical level, as observed for other classes, some differences shown in the previous classification do not found correspondence which classes of the current system, such as the latosolic Eutrophic Similar Structured Dusky Red Earth (Brasil, 1982d), which fits as typic Eutrophic Haplic Nitosol. Also the profile C63 of the soil survey of the State of Paraná (EMBRAPA, 1984), previously classified as cambic Alic Similar Structured Brown Earth was fitted as typic Aluminic Brown Nitosol, for not being included possibilities of distinction relative to intermediate characteristics represented in the previous designation.

In this class are classified clayey to very clayey soils with slight increase of clay in the subsurface horizon, not establishing enough textural gradient for fitting as textural B horizon, but being possible to observe clear clay skins on soil aggregates. These observations can be found in the classes of the Nitisols, Lixisols and Alisols of the world soils legend of the FAO, while for the American classification Soil Taxonomy, it can be observed the correlation encompassing Ultisols, Oxisols (Kandic) and Alfisols.

3.10 Organosols

Due to the limited expressiveness of these soils in the country, generally restricted to small areas, as well as more difficult access, they tend to be contemplated with a few profiles sampled in more generalized pedological surveys as the ones considered here. Thus, the presence of the class of Organosols in the database was minimal, covering only 11 profiles, which equals to 0.20% of the total.

Characterized by its essentially organic constitution, the distinctive criteria of this class in order level present an adequate correlation with the set of soils previously included under the general denomination of Organic Soils. Thus, we obtained a very satisfactory taxonomic fitting in the highest categorical level, although for some profiles, considering the unavailability of complete data, the fitting in the current classification system has been performed by direct correlation with the previous classification.

For the subsequent categorical level, all profiles were classified as Haplics. Only for the profile 37 of RADAM Brazil Project (Brasil, 1981d), which presents very low values of pH in water in the subsurface horizon, there is about the possibility of occurrence of sulfuric horizon or sulfidric material (distinctive characteristics of Tiomorphic class), thus requiring more data for confirmation.

The main difficulty in upgrading the classification of these soils was observed in the third categorical level (great group), because the distinctive criteria involve determinations not performed in epochs before the publication of the first edition of the Brazilian System of Soil Classification (EMBRAPA, 1999), which makes very difficult a more accurate evaluation, even from complete morphological and analytical characterization of soils profiles. Classification at this level was therefore undertaken in an attempt form, fitting all the profiles in the class of Saprics, which groups soils constituted of organic material in an advanced stage of decomposition.

Also in the fourth categorical level, the absence of analytical data, especially concerning the electrical conductivity, reduces the confidence of taxonomic fitting. Only the profile C137 of the soil survey of the State of Paraná (EMBRAPA, 1984), referred in SiBCS (EMBRAPA, 2006) as representative of the terric Sapric Haplic Organosols, was classified with higher degree of confidence in all categorical levels.

Most organic constituents, in various stages of decomposition, in relation to the mineral ones are the basic criteria for the distinction of Organosols in the Brazilian system of classification and Histosols by the American system and FAO.

3.11 Planosols

In relation to the grouping of soils with planic B horizon, which is characterized by abrupt transition in association with sharp increase of clay in relation to the overlying horizon, the Planosols tend to occur in environments of restricted drainage, susceptible to hydromorphism at least temporarily. This class represents 4.22% of the updated of profiles of the database, which is equivalent to 231 profiles. Refer to soils formerly called Planosols and Solodized-Solonetz, now included in the same class in the order level, distinguished in the second categorical level as Haplic Planosols and Natric Planosols , respectively. A very satisfactory adjustment between the classes previously recognized and the current ones was achieved.

However, despite the strong correlation between the current criteria for distinguishing these soils and the previously criteria used, because there are not available morphological data and analytical results of the entire profiles, some of them were classified based on the original name, such as the profile 66 of the technical bulletin 16, of the former Division of Research Pedologic (Brasil, 1972e), called weak A Ta Eutrophic Planosol, for which the subsurface horizon of the database refers to the C horizon, with no indication of the occurrence of colors having low chroma. Due to this, this soil was tentatively classified as typic Eutrophic Haplic Planosol.

The non-availability of analytical data also contributed for reducing the confidence of the update of the classification in the subsequent categorical levels. The distinction of Salic Soils, at the subgroup level, for example, was not possible, due to lack of data on electrical conductivity.

Major limitations for the taxonomic fitting were observed, however, in the fourth categorical level. For example, for the profile 89 of soil survey of the State of Ceará (Brasil, 1973a), originally called weak A vertic solodic Planosol, it was not possible with the available data, the confirmation of the solodic character, although it may be present in some horizon within the first 120 cm depth. Thus, this soil was classified as vertic Eutrophic Haplic Planosol. Similarly to observed for the other classes, some differential characteristics represented in the previous name are not included in the distinction of soils by the current system. As a result, the soil represented by the profile 200 of the RADAM Brazil Project (Brasil, 1981b), formerly designated plinthic Tb endoeutrophic Planosol, was classified as typic Eutrophic Haplic Planosol, because it is not included the plinthic class.

The soils that in the first edition of the Brazilian classification system (EMBRAPA, 1999) were differentiated in the second categorical level, such as Hydromorphic Planosols, were excluded from the current version (EMBRAPA, 2006). Actually they are classified as gleissolics in the fourth categorical level, but by the ordering of the classes in the identification key, in some cases they can not be distinguished as such. The profile 81 of the RADAM Brazil Project (Brasil, 1981a), for example, even though there are enough characteristics for its fitting as gleisolic, by the current criteria it is fitted as solodic Eutrophic Haplic Planosol, due to the solodic character take precedence in the distinction of soils at this level. As for the Dystrophic Haplic Planosols, the gleisolic class take precedence in relation to plinthic and solodic in the classification key at the fourth level, not allowing the distinction in the case of simultaneous occurrence of soils with these characteristics, as occurs for the profile 97 of the RADAM Brazil Project (Brasil, 1982b), originally classified as moderate A plinthic Tb Alic Planosol, by the current system fits as gleisolic Dystrophic Haplic Planosol. Considering the above comments, and considering the strong influence of hydromorphic conditions on sustainable use and management of these soils, as well as environmental and pedogenesis aspects, it is strongly suggested the distinction of Planosols with these characteristics in higher categorical level. This has practical application and considerable geographical expression mainly in the state of Rio

Grande do Sul (Costa et al., 2009). It seems appropriate to consider the possibility of keeping the distinction in the second categorical level, as adopted by the previous version of SiBCS (EMBRAPA, 1999).

The suborder of Natric Planosols can be identified in the American classification Soil Taxonomy and WRB/FAO as Natr-(ust-ud)-alf and Solonetz, respectively. Classes that do not fall into this suborder are denominated at second categorical level as Haplics and can be correlated with the classes of Planosols (WRB/FAO), and Albaquults, Albaqualfs and Plinthaqu (alf-ept-ox-ult)s of Soil Taxonomy.

3.12 Plinthosols

In this class are included soils characterized by expressive plinthization, presenting significant occurrence of plinthite, either in its mild or hard (petroplinthite) form, resulting from changes in the formation environment. In the database of soils of Brazil this class has a participation corresponding to 5.40%, which is equal to a total of 296 soil profiles. It covers the vast majority of the former Hydromorphic Laterites, which were later recognized as Plinthosols, a designation that was kept in the current system, and, with the modifications introduced in its latest version, referring to the concept of concretionary and lithoplintic horizon, now include those soils previously listed under the generic name of Concretionary Soils, including the Lateritic Concretionary Soils.

The main difficulty of current fitting of the soils in this class precisely refers to the last soils, which, because of the lack of more precise criteria for their identification, constituted a large and highly heterogeneous group. Thus, they were classified as Plinthosols according to the reference of the occurrence of concretions (petroplinthite) expressed in its name, giving high degree of confidence only for those soils with at least one horizon with 30 cm or more thickness and volume of pebbles and cobbles equals to 50% or more, enough for characterizing the concretionary horizon. For the others, considering that often the coarse material was discarded during sampling, or it was not measured, and the available data do not cover the entire extension of the profiles, it was considered greater uncertainty in the taxonomic fitting, with a moderate degree of confidence (value 2) in case there is some indication of the occurrence of petroplinthite (given by the horizon designation, for example), or lower degree of confidence (value 3), otherwise. All these soils, also including a smaller number of profiles originally called Concretionary Plinthosols or Concretionary Laterites, were grouped in a single class at second (Petric) and third categorical level (Concretionary), being distinguished only in the suborder level, due to the occurrence of some diagnostic horizon (incipient B or cambic B, latosolic B or textural B), or other differential characteristic, as mentioned in the previous name.

Other Plinthosols of the database were classified, in the second categorical level, as Argilluvics and Haplics, with great domain of the first in relation to the second ones. The change in the latest version of the system, referring to the argilluvic character (defined as having a texture ratio superior to 1.4) instead of the presence of textural B horizon to distinguish the Argilluvics (EMBRAPA, 2006), seems to be mainly responsible for this fact. In strong contraposition to the criteria adopted for Petric Plinthosols, the distinction of both classes in the subsequent categorical level is based on the saturation of exchange complex, and subsidiarily on clay activity for soils with very high aluminum content, differentiating soils with Alitic-, Aluminic-, Dystrophic- or Eutrophic-character, all of these classes represented in the database.

Also in the fourth level it is clearly demonstrated the great difference in the criteria used in distinguishing between Argilluvic Plinthosols and Haplic Plinthosols, and the Petric Plinthosols, which constitute a very distinct group within this class.
The occurrence of appreciable amounts of plinthite in the soil profile is also identified in the class of the Plinthosols in the WRB/FAO international classification. The American system of classification adopts this character in lower categorical level (subgroup) and can be found correspondence of Plinthosols with different classes of the Soil Taxonomy, which present the plinthic character. These classes comprise the Oxisols, Ultisols, Alfisols, Entisols and Inceptisols.

3.13 Vertisols

The soils of this class, due to the dominance of very high activity clays, are characterized by a great capacity of swelling and shrinkage of soil material with wetting and drying cycles, a factor that restricts their pedogenetic development. Having little occurrence in Brazilian territory, Vertisols represent only 1.00% of the profiles in the database, comprising a total of 55 profiles.
As the identification of soils in this class includes criteria based on morphological characteristics evaluated in the field, the first categorical level was established by direct correlation with the original classification, whose name was maintained by the current system, just as the essence of the distinctive criteria, which allowed a satisfactory fit. For some profiles, however, there was some difficulty in taxonomic fitting, due to the presence, sometimes in the surface horizon, other times in subsurface horizon, of clay content below 300 g / kg, established as thresholds to distinguish the class of Vertisols (up to 20 cm depth, after mixing), but also for the vertic horizon (EMBRAPA, 2006). Once in the database are not included all the horizons of the profile, thereby preventing a more adequate evaluation, it was maintained their classification as Vertisols, assigning the value 2 for the degree of confidence in case of one of the horizons do not meet the specified requirement, and value 3 when both had amounts of clay below the minimum limit required. This situation, however, shows some discrepancy between the definitions of vertic horizon and of the class of Vertisols, as well as the need to adjust them. It may be appropriate to assess, from the characteristics of the profiles contained in the database, the adequacy of a minimum amount of clay established for distinction of these soils.
In the second categorical level, it were recognized classes of Hydromorphic Vertisols, Ebanic Vertisols and Haplic Vertisols. Some difficulty in classification was observed mainly for fitting of some soils as Hydromorphics, due to the subsurface horizons refer to depths greater than 50 cm, being necessary information on colors of the overlying horizons to confirm the classification (in this case it was assigned the value 2 to express the degree of confidence). Other profiles, on the other hand, did not present information on moist color, being classified as Haplics, with value 3 for the confidence degree.
Also in the great group level it was observed some difficulty in taxonomic fitting of the profiles, mainly due to the lack of relevant data, such as the contents of equivalent $CaCO_3$. This form, the fitting in the class of the Carbonatics was based on previous classification, corroborated by the subordinated designation of horizons of the profile (EMBRAPA, 1988b), being the other profiles classified as Orthics, since the previous name did not present indication of the occurrence of salic character.

The absence of analytical data has hampered the taxonomic fitting in the fourth categorical level, as example of the electrical conductivity, necessary for the identification of saline character.
The characteristics of vertic horizon identify typical pedological features due to shrinkage and swelling of the minerals of this class. These features are called friction surfaces ("slickensides"), and are common both in the U.S. classification for the Vertisols as well as in the international legend of the FAO for those soils also called Vertisols.

4. Conclusions

The database used in this work reveals an adequate representativeness of the soil classes distributed over the Brazilian territory.
This database can function as an advance to stimulate the exchanging of information and experience among researchers at international level.

5. References

Assad, E. & Sano, E. (1998). *Sistema de informações geográficas: Aplicações na Agricultura.* EMBRAPA, Brasília, Brasil.

Baumgardner, M. (1999). Soil databases. In: SUMNER, M.E., ed. *Handbook of soil science.* Boca Raton, CRC Press, pp. H1-H4.

Benedetti, M. ; Sparovek, G.; Cooper, M.; Curi, N. & Carvalho Filho, A. (2008). Representatividade e potencial de utilização de um banco de dados de solos do Brasil. *Revista Brasileira de Ciência do Solo.* , v.32, No.6, pp. 2591-2600.

Brasil (1972a). *Estudo expedito de solos nas partes central e sul do Estado da Bahia, para fins de classificação e correlação.* Ministério da Agricultura. Departamento Nacional de Pesquisa Agropecuária. Divisão de Pesquisa Pedológica. (DNPEA. Boletim Técnico, 24 Sudene. Série Pedologia, 17), Recife, Brasil.

Brasil (1958). *Levantamento de reconhecimento dos solos do Estado do Rio de Janeiro e Distrito Federal: contribuição à carta de solos do Brasil.* Ministério da Agricultura. Serviço Nacional de Pesquisas Agronômicas. Comissão de Solos. (Boletim, 11), Rio de Janeiro, Brasil.

Brasil (1960). *Levantamento de reconhecimento dos solos do Estado de São Paulo: contribuição à carta de solos do Brasil.* Ministério da Agricultura. Serviço Nacional de Pesquisas Agronômicas. Comissão de Solos. (Boletim, 12), Rio de Janeiro, Brasil.

Brasil (1971a). *Levantamento de reconhecimento dos solos do sul do Mato Grosso.* Ministério da Agricultura. Departamento Nacional de Pesquisa Agropecuária. Divisão de Pesquisa Pedológica. (Boletim Técnico, 18), Rio de Janeiro, Brasil.

Brasil (1973h). *Levantamento exploratório de solos do Estado do Ceará.* Ministério da Agricultura. Departamento Nacional de Pesquisa Agropecuária. Divisão de Pesquisa Pedológica. 2v. (Boletim técnico, 26; SUDENE. DRN. Série Pedologia, 14), Recife, Brasil.

Brasil (1972e). *Levantamento exploratório-reconhecimento de solos do Estado de Pernambuco.* Ministério da Agricultura. Departamento Nacional de Pesquisa Agropecuária. Divisão de Pesquisa Pedológica. 2v. (DNPEA. Boletim Técnico, 16; SUDENE. DRN. Série Pedologia, 14), Recife, Brasil.

Brasil (1971b). *Levantamento exploratório-reconhecimento de solos do Estado do Rio Grande do Norte.* Ministério da Agricultura. Departamento Nacional de Pesquisa Agropecuária. Divisão de Pesquisa Pedológica. (DNPEA. Boletim Técnico, 21; SUDENE. Série Pedologia, 9), Rio de Janeiro, Brasil.

Brasil (1970c). *Levantamento exploratório dos solos da região sob influência da Companhia Vale do Rio Doce.* Ministério da Agricultura. Escritório de Pesquisas e Experimentação. Equipe de Pedologia e Fertilidade do Solo. (Boletim Técnico, 13), Rio de Janeiro, Brasil.

Brasil (1972f). *Levantamento exploratório-reconhecimento de solos do Estado da Paraíba. II. Interpretação para uso agrícola dos solos do Estado da Paraíba.* Ministério da Agricultura. Escritório de Pesquisas e Experimentação. Equipe de Pedologia e Fertilidade do Solo. (Boletim Técnico, 15; SUDENE, Série Pedologia, 8), Rio de Janeiro, Brasil.

Brasil (1979c). *Levantamento exploratório-reconhecimento de solos do norte de Minas Gerais.* Ministério da Agricultura. EMBRAPA-SNLCS/SUDENE DRN (Boletim Técnico, 60), Recife, Brasil.

Brasil (1973-1983). *Levantamentos Exploratórios do Projeto Radam Brasil - Volumes 1 a 33.* Ministério das Minas e Energia. Departamento Nacional de Produção Mineral. 1973a, b, c, 1974a, b, c, 1975a, b, c, d, 1976a, b, c, 1977a, b, c, d, 1978a, b, 1979, 1980, 1981a, b, c, d, e, 1982a, b, c, d, 1983a, b, c; (Levantamento de Recursos Naturais), Rio de Janeiro, Brasil.

Cooper, M.; Mendes, L.; Silva W. & Sparovek, G. (2005). *A national soil profile database for Brazil available to international scientists.* Soil Science Society of America Journal, Madison, v. 69, pp. 649-652.

Coote, D. & Macdonald, K. (1999). *The Canadian soil database.* In: SUMNER, M.E., (Ed.). Handbook of soil science. Boca Raton: CRC Press, pp. H41-H51.

Costa, A.; Araújo, E.; Marques, J. & Menezes, M. (2009). *Avaliação do risco de anoxia para o cultivo de eucalipto no Rio Grande do Sul utilizando-se levantamento de solos.* Piracicaba: Scientia Forestalis, 37 pp. 367–375.

Embrapa (1988b). *Definição e notação de horizontes e camadas do solo.* Serviço Nacional de Levantamento e Conservação de Solos. 2. ed. (Embrapa-SNLCS. Documentos, 3), Rio de Janeiro, Brasil.

Embrapa (1999). *Sistema Brasileiro de Classificação de Solos.* Centro Nacional de Pesquisa de Solos. Rio de Janeiro, Brasil.

Embrapa (2006). *Sistema Brasileiro de Classificação de Solos.* Centro Nacional de Pesquisa de Solos. 2 ed, Rio de Janeiro, Brasil.

Embrapa (1984). *Levantamento de reconhecimento de média intensidade dos solos e avaliação da aptidão agrícola das terras da área do Pólo Trombetas, Pará.* Serviço Nacional de Levantamento e Conservação de Solo. (EMBRAPA-SNLCS. Boletim de Pesquisa, 28), Rio de Janeiro, Brasil.

Embrapa (1984), *Levantamento de reconhecimento dos solos e avaliação da aptidão agrícola das terras de uma área de colonização no Município de Careiro, Estado do Amazonas.* Serviço Nacional de Levantamento e Conservação do Solo. (Boletim de Pesquisa, 31), Rio de Janeiro, Brasil.

Embrapa (1983), *Levantamento de reconhecimento de media intensidade dos solos e avaliação da aptidão agrícola das terras de 21.000 hectares no município de Tefé, Amazonas.* Serviço

Nacional de Levantamento e Conservação do Solo. (Boletim de Pesquisa, 19), Rio de Janeiro, Brasil.

Embrapa (1983), *Levantamento de reconhecimento de média intensidade dos Solos e avaliação da aptidão agrícola das Terras do Pólo Roraima.* Serviço Nacional de Levantamento e Conservação do Solo. (Boletim de Pesquisa, 18), Rio de Janeiro, Brasil.

Embrapa (1984). *Levantamento de reconhecimento dos solos e avaliação da aptidão agrícola das terras de uma área de colonização do Município de Barreirinha, Estado do Amazonas.* Serviço Nacional de Levantamento e Conservação de Solo. (Boletim de Pesquisa, 32), Rio de Janeiro, Brasil.

Embrapa (1984). *Levantamento de reconhecimento dos solos e avaliação da aptidão agrícola das terras do Municipio de Urucará, Estado do Amazonas.* Serviço Nacional de Levantamento e Conservação de Solo. (Boletim de Pesquisa, 30), Rio de Janeiro, Brasil.

Embrapa (1983). *Levantamento de Reconhecimento de Solos do Mato Grosso.* Serviço Nacional de Levantamento e Conservação de Solos. (Boletim de Pesquisa, 17), Rio de Janeiro, Brasil.

Embrapa (1982). *Levantamento de reconhecimento dos solos do Município de Ariquemes, Estado de Rondônia.* Serviço Nacional de Levantamento e Conservação do Solo. (Boletim de Pesquisa, 16), Rio de Janeiro, Brasil.

Embrapa (1982), *Levantamento de reconhecimento dos solos do Pólo Pré-Amazonia-Marenhense, Estado do Maranhão.* Serviço Nacional de Levantamento e Conservação do Solo. (Boletim de Pesquisa, 15), Rio de Janeiro, Brasil.

Embrapa (1984). *Levantamento de Reconhecimento de Solos e Aptidão Agrícola das Terras da Área do Pólo Carajás Estado do Pará.* Serviço Nacional de Levantamento e Conservação de Solo. (EMBRAPA-SNLCS - Boletim de Pesquisa, 29), Rio de Janeiro, Brasil.

Embrapa (1977). *Levantamento de reconhecimento dos solos do Estado do Espírito Santo.* Serviço Nacional de Levantamento e Conservação de Solos. (Boletim de Pesquisa, 20), Rio de Janeiro, Brasil.

Embrapa (1978b). *Levantamento de reconhecimento dos solos do Estado do Espírito Santo.* Serviço Nacional de Levantamento e Conservação de Solos. (Boletim técnico, 45), Rio de Janeiro,Brasil.

Embrapa (1984). *Levantamento de reconhecimento de solos do Estado do Paraná.* Serviço Nacional de Levantamento e Conservação de Solos. (EMBRAPA-SNLCS - Boletim de Pesquisa, 27; IAPAR. Boletim Técnico, 16), Londrina, Brasil.

Embrapa (1979b). *Levantamento de reconhecimento dos solos do centro-sul do Estado do Paraná (área 9): informe preliminar.* Serviço Nacional de Levantamento e Conservação de Solos. (Boletim Técnico, 11), Curitiba, Brasil.

Embrapa (1977-1979). *Levantamento exploratório-reconhecimento de solos da margem direita do rio São Francisco, Estado da Bahia.* Serviço Nacional de Levantamento e Conservação de Solos. 2v. (Embrapa-SNLCS. Boletim Técnico, 52; SUDENE. Série Recursos de Solos, 10), Recife, Brasil.

Embrapa (1976c). *Levantamento exploratório-reconhecimento de solos da margem esquerda do rio São Francisco, Estado da Bahia.* Serviço Nacional de Levantamento e Conservação de Solos. (Embrapa-SNLCS. Boletim Técnico, 38; SUDENE. Série Recursos de Solos, 7), Recife, Brasil.

Embrapa (1982). *Levantamento e reconhecimento de baixa intesidade dos solos e avaliação da aptidão agrícola das terras do projeto de colonização Apiaú - Território Federal de Roraima*. Serviço Nacional de Levantamento e Conservação de Solo. (EMBRAPA-SNLCS. Boletim de Pesquisa, 14), Rio de Janeiro, Brasil.

Embrapa (1982). *Levantamento de reconhecimento de baixa intensidade dos solos e avaliação da aptidão agrícola das terras de área piloto do município de Barreirinha - Estado do Amazonas*. Serviço Nacional de Levantamento e Conservação de Solo. (EMBRAPA-SNLCS. Boletim de Pesquisa, 9), Rio de Janeiro, Brasil.

Embrapa (1986a). *Levantamento exploratório - reconhecimento de solos do Estado do Maranhão*. Serviço Nacional de Levantamento e Conservação de Solos. 2v. (EMBRAPA-SNLCS. Boletim de Pesquisa, 35. Brasil. SUDENE-DRN. Série Recursos de Solos, 17), Rio de Janeiro, Brasil.

Embrapa (1986b). *Levantamento exploratório - reconhecimento de solos do Estado do Piauí*. Serviço Nacional de Levantamento e Conservação de Solos. 2v. (EMBRAPA-SNLCS. Boletim de Pesquisa, 36. Brasil. SUDENE-DRN Série Recursos de Solos, 18), Rio de Janeiro, Brasil.

FAO (1996). *The digitized soil map of the world including derived soil properties*. Rome, FAO, 1 CD-ROM.

FAO (1994). *Revised legend with corrections*. FAO, Rome: 137p. (World Soil Resources Report, 60).

FAO (1974). *Soil map of the World: 1:5.000.000 legend*. Paris: Unesco, v. 1, Roma, Itália.

Ker, J. (1997). Latossolos do Brasil: uma revisão. *Geonomos*, Belo Horizonte, v. 5, nº 1, pp. 17-40.

Nemes, A.; Schaap, M.; Leij, F. & Wosten, J. (2001). *Description of the unsaturated soil hydraulic database*. UNSODA version 2.0, Journal of Hydrology. Amsterdam, n. 251, pp. 151-162.

Nemes, A.; Wosten, J.; Lilly, A. & Oude Vos Harr, J. (1999). *Evolution of different procedures to interpolate particle – size distributions to achieve compatibility within soil databases*. Geoderma. Amsterdam, n. 90, pp. 187-202.

Resende, M.; Curi, N.; Rezende, S. & Corrêa, G.(2007). *Pedologia: base para distinção de ambientes*. 5. ed. Editora UFLA, Lavras, Brasil.

Soil Survey Staff – SSS (1991). *National soil information system (NASIS): soil interpretation and information dissemination sub-system*. Draft requirements statement. Lincoln, USDA, Natural Resources Conservation Service, National Soil Survey Center. Washington, USDA.

United States (1975). *Soil Taxonomy: a basic system of soil classification for making and interpreting soil surveys*. Department of Agriculture. Soil Survey Division. Soil Conservation Service. Soil Survey Staff. Agriculture Handbook. Washington, USDA.

United States (1999). *Soil taxonomy: a basic system of soil classification for making and interpreting soil surveys*. Department of Agriculture. 2nd ed. (Agriculture Handbook, 436), Washington, USDA.

Van Engelen, V.W.P. (1999). *Soter: the world soils and terrain database*. In: SUMNER, M.E., (Ed.). Handbook of soil science, Boca Raton, CRC Press, pp. H19-H28.

19

An Application Approach to Kalman Filter and CT Scanners for Soil Science

Marcos A. M. Laia[1,2] and Paulo E. Cruvinel[1,2]
[1]*Embrapa Instrumentation*
[2]*Physics Institute of São Carlos, University of São Paulo*
Brazil

1. Introduction

Over the years, experts on soil science have brought together researchers from various fields with the aim of pooling efforts in order to characterize the properties of soils. The use of these results in agriculture through various activities can be directed towards a development whose natural resource base can be maintained in a long term.
Conservation, minimization of soil pollution together with the development of irrigation and more efficient and more cost effective drainage systems optimize the efficiency of water use and nutrients in agricultural production.
The soil can be preserved through management methods, which seek to prevent deterioration, induced erosion and deposition of sediments into rivers. Environment degradation can be caused naturally and by humans. The way in how to use the soil can cause degragation and excessive compression, salinisation and acidification.
Several conventional techniques have been used to find answers to the various physical mechanisms and also the chemical and biological processes that occur in soils. Among those techniques are included: neutron probe, gravimetry, direct transmission of rays, plotters, microscopy and mercury intrusion, however these mechanisms have limitations.
New techniques that enable an accurate prediction aim at managing the flow of contaminants through unsaturated soil zone. With the goals of acquiring non-invasive samples and reaching higher resolutions on γ and x-rays computerized tomography (CT) and on Nuclear Magnetic Ressonance (NMR), which provide cross-section images from the analysed objects. NMR, however, presents strong restrictions for its use in porous media that contain paramagnetic materials (Crestana & Nielsen, 1990) and, besides this, it is difficult or even impossible to quantify the results by correlating the NMR signal and the content of water. However, by means of CT it is possible to figure out a good correlation between the x-ray linear attenuation coefficients and the water content in soils.
The quality of an image is one of the key requirements for its analysis and it is desirable that the reconstructed object is very close to the tested sample. The use of algorithms developed for the areas of human knowledge has grown considerably and has improved image processing for visual information analysis and human interpretation or the automatic perception of machines. In human interpretation, x-ray images are used not only in medicine, but also in geology, in archaeology, and in soil science for agricultural purposes.

On the other hand, the perception of automatic machines for present automatic recognition of faces, characters, fingerprints, computer vision, control of robots for surveillance, automatic processing of satellite images for fire recognition, climate change and identification of storms and hurricanes.
Image processing aims at modeling the characteristic of the human eye, study processed images, such as Fourier transform and other separable image transforms. It also allows designing filters used to retrieve an image and use masks so that the processed image is more applicable than the original (real).
This study aims at improving the quality of tomographic images by filtering the signals before their reconstruction and reaching an image quality in the reconstructed slice obtained by the projections close to the real one.
Digital processing algorithms can be used to work with the image by using computer vision techniques such as segmentation (for feature extraction) and Hough transform (used in order to detect geometry, ie, detection of pores), which allow the count of pores in the soil. Classification algorithms can also be used to characterize the type of soil as well as its chemical components based on the values of pixels and density.
Both kinds of algorithms can be an auxiliary tool for a reliable analysis on the impacts of the use of agricultural machinery on the soil, which result in the increase of soil density due to the compression caused by the machinery axis. This increase is directly linked to the reduction of larger diameter pores that cause a decrease in water and in nutrients and also to the gas diffusion and root penetration.
Besides, this chapter aims at presenting the use of unscented Kalman filter (UKF) and the algorithm used to separate a noise from a signal. This will be done by showing that the unscented Kalman filter together with artificial neural networks is the best option for filtering. There will be an overview and a specification after each equation of the algorithm. Moreover, this study complemented previous works that aimed at creating a better algorithm by modeling and testing results with different types of Kalman filter and with different approaches in the physical model and kinds of noise (Laia et al., 2007; Laia & Cruvinel, 2008a; Laia et al., 2008b; Laia & Cruvinel, 2009; Laia & Cruvinel, 2011), which also present a comparison with an artificial neural network (ANN) solution.

2. Computerized Tomography scanner

In radiology, computed tomography (CT) consists of an image that is derived from computerized processing of data obtained from a series of x-ray angled projections which reproduce a cross-section (a "slice") of the object under study. CT, such as conventional radiology, is based on the fact that x-rays are partially absorbed by various materials. While materials like plastic and water are easily traversed by x-rays, others, like metals, are not.
This technique had already been widely applied to medical areas, however its use in soil science was introduced by Petrovic (Petrovic et al., 1982), Hainswoth and Aylmore (Hainswoth & Aylmore 1983) and Crestana (Crestana, 1985). Petrovic made it possible to use x-ray CT to measure the density of soil volumes, while Crestana demonstrated that CT can solve problems related to studies of the water physics in the soil. From these studies, a project that developed a scanner for soil science has been created (Cruvinel, 1987; Cruvinel et al., 1990; Cruvinel et al., 2009).

2.1 Computerized Tomography in soil science

The application of CT in soil science to investigate the same physical properties plays an important role in studying the transportation of water and solutes within this environment. The direct transmission of γ or x-rays provides a major contribution in order to solve the various problems in the field of soils with results on a scale of the order of millimeters, however many responses are still expected on the level of particles, macropores and micropores.

Figure 1 illustrates different CT scanners that are dedicated to soil science and were developed and installed at Embrapa Instrumentation (São Carlos, SP - Brazil), which are based on sources of γ and x-rays for the study of soils and plants and also allow the use of various sources of radiation and intensities of energy. Several studies have been developed to improve the visualization of the images acquired and the reconstruction algorithm, as well as the equipment developed (Venturini, 1995; Minatel, 1997; Granato 1998; Mascarenhas et al, 1999).

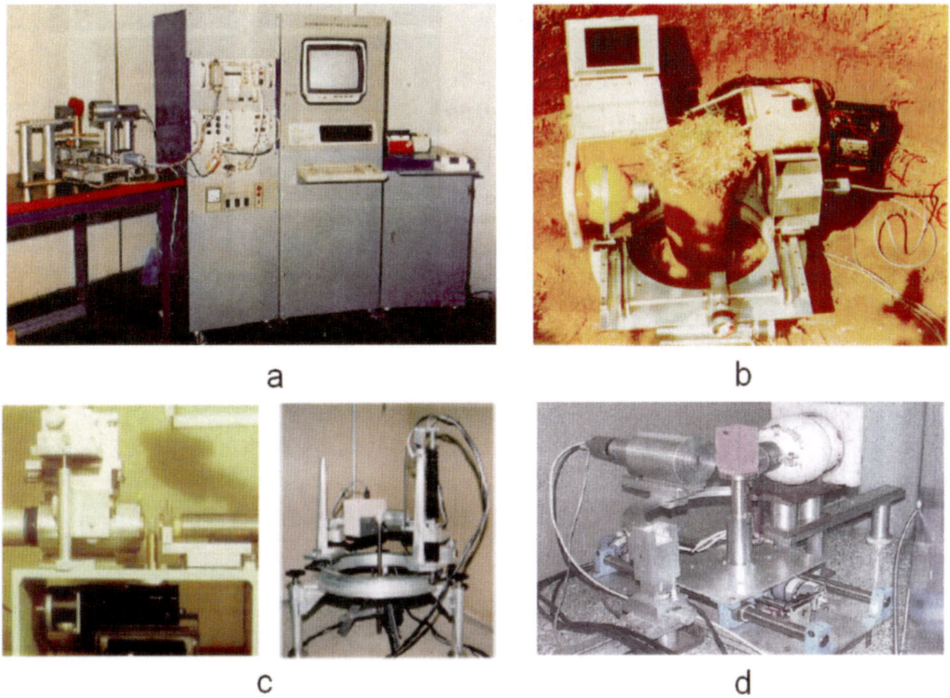

Fig. 1. Scanners developed at Embrapa Instrumentation: a. mini CT scanner developed for agricultural applications (Cruvinel, 1987); b. portable CT scanner for field use (Naime, 1994); c. tomography with micrometric precision (Macedo, 1997) and d. Compton scattering scanner (Cruvinel & Balogun, 2006)

Compared to classical methods, such as the direct transmission of γ-ray and gravimetric tests, the CT scanner presents the advantage of measuring a) heterogeneities in the soil b) the density of the soil and c) the moisture content pixel by pixel, and also gets a two-dimensional picture or three-dimensional soil samples from non-invasive and independent of the geometry and shape, even by using different energies and radioactive sources (Cruvinel et al. 1990).

Basically, a CT indicates the amount of radiation absorbed by each portion of the analyzed section, translates these variations into gray scale and also produces an image. As the x-ray absorption capacity of a material is closely related to its density, different density areas will present different intensities, which can be seen in monochromatic color and if you apply a mask of pseudo-color, it allows their distinction more clearly. Thus, each signal value corresponds to the average absorption of the tissue in the area, expressed in Hounsfield units (named after the creator of CT machines). Each projection represents an average and each set is stored in a projection matrix. Based on the intensity emitted by the x-ray source and the intensity captured by the detector at the other end of the propagation line, one can determine the attenuation weight due to the object that is located between the source and the detector. The data on the attenuation weight is crucial to the reconstruction process and enables a mapping of the linear attenuation coefficient of the object cross section.

This coefficient mapping is represented by pixels whose values are given by the so-called CT Numbers. These numbers are normalized according to the water attenuation coefficient μ_{water}. In other words, the CT numbers are defined in equation 1:

$$CT\ Number = \frac{\mu - \mu_{water}}{\mu_{water}} \times 1000 \quad (1)$$

where μ is the linear attenuation coefficient of the analyzed body.

From this number, it is possible to obtain a map of the attenuation coefficients, which allows a more detailed analysis of the body being studied. In medicine, it was agreed that the water CT number is equivalent to 0 (zero).

The integration function of the object along the ray is represented by a line integral. Each set of line integrals of parallel rays form a parallel projection (Figure 2), which can be treated having the signal theory as its base.

The main advantage of CT is to allow the study of cross sections without intrusion, that is, an unparalleled improvement in relation to techniques of soil analysis, which, in general, are invasive and can destroy important features that could be preserved. It is important to obtain a perfect image quality that prevents materials found in the soil from being erroneously interpreted. This is achieved when the material in focus is the same found by an intrusive manual search.

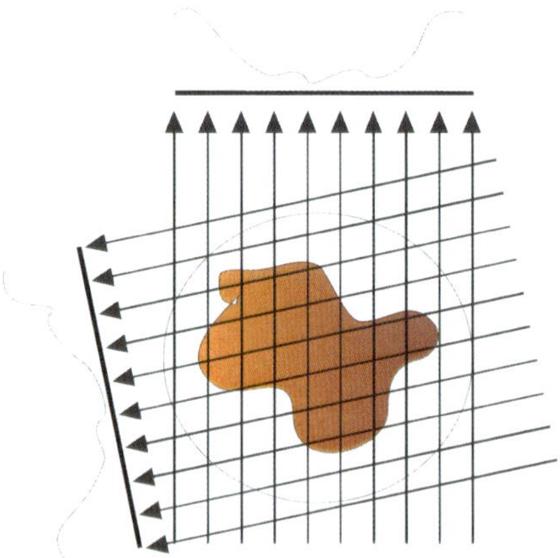

Fig. 2. Two parallel projections of an object expressed by a two-dimensional function

2.2 Noise that can occur in CT

In CT there are three main processes related with the interaction of the radiation with the matter, which are the photoelectric effect, the Compton effect and pair production effect (Cruvinel et al. 1990). Besides the issues that are related to the effects arising from the energy range used in the source, others also influence the measurement in computed tomography, like the statistics of photon counting. The probability of detecting photons in a range of exposure time t can be estimated by the distribution function of Poisson (Deremack & Crowe, 1984), (Cruvinel, 1987).

$$P_S(\lambda, t) = \frac{(\bar{\lambda})^n}{\lambda! e^n} \tag{2}$$

where λ is the number of photons and $\bar{\lambda}$ is the average number of photons emitted in the time interval t and n indicates the efficiency of the photomultiplier, as shown in the following expression:

$$\bar{\lambda} = \xi M t \tag{3}$$

where M is the average ratio of photons (photons / second) and ξ is the quantum efficiency of the photomultiplier. The uncertainty or noise is given by the standard deviation

$$\sigma = \sqrt{\bar{\lambda}} = \sqrt{\xi M t}. \tag{4}$$

Therefore, the signal to noise ratio (SNR), presented by the incident signal, is:

$$SNR = \frac{\bar{\lambda}}{\sigma} = \sqrt{\bar{\lambda}}. \tag{5}$$

In this ratio, it is estimated that, for a small number of photons, there can be considerable noise, however, as $\bar{\lambda}$ increases, the noise might be negligible. The thermionic emission of electrons in the photo multiplier cathode can cause an increase in the noise. Considering that the photo-cathode emits electrons randomly due to the cathode current, a new signal to noise ratio equation is given by equation 7:

$$\sqrt{(\xi M + M)t} \tag{6}$$

$$SNR = \frac{\sqrt{t}\xi M}{\sqrt{\xi M + M}} \tag{7}$$

In the display of a tomographic image, there is granularity, which is significant for viewing low contrast objects. During the analysis of a CT image, there is presence of granularity, which is significant for viewing low contrast objects. The term noise in tomography images refers to the variation of the attenuation coefficient at the average value and it is obtained when an image of a uniform object is used. (Hender, 1983). The image noise can be based on the calculus of the standard deviation and also on the Weiner power spectrum of the noise, which is seen as a function of spatial frequency. In other words, it allows observing the intensity and the type of noise that involves the system also influences the image obtained.

The noise in CT images involves rounding errors in the reconstruction program (algorithm noise), electronic noise and noise caused by the display system. The main source of noise in

CT images is the change in quantum (*quantus mottle*), defined as the statistical spatial and the temporal variation in the number of x-ray photons absorbed by the detector.
The noise of the algorithm depends on the pixel size of the display device and also influences the image noise, since it leads to larger pixels to reduce the image noise, however with less quality resolution. Reconstruction algorithms typically make use of anti-aliasing filters to minimize the visual effect of the noise but also present some loss in the spatial resolution.
The electronic noise can be generated by non-ideal electronic devices such as non pure resistors and capacitors, non ideal contact terminals, leakage current of transistors, Joule, and it may also be independent of the signal, due to external interference (electrical or mechanical) (Ziel, 1976).
Besides the noise, the CT images are subject to various artifacts and distortions as poly-chromatic sources (non-ideal or monochromatic), caused by the following effects: beam hardening, aliasing, different materials in the same voxel (partial volume), displacement of the sample or of the equipment (Duerinckx & Macocski, 1978; Joseph & Spital, 1978; Ibbott, 1980; Granato, 1998).
The low-pass and the median filters can be used to solve problems of signal noise, however they can cause a crucial losses of informations. Systems having different sources of noise do not present an optimal solution when using these types of filters.

3. The Kalman Filter

The Kalman Filter is a mathematical tool, created more than 30 years ago, which is widely used to solve statistical problems. It is considered a good estimator for a large class of problems and an effective and useful estimator for other classes. It has recently been used in computer graphics for applications involving the simulation of musical instruments in virtual reality (VR), for the reading of speaker's lips in video sequences, among other applications (Pereira, 2000). In 1960, Rudolf Emil Kalman published a paper describing a recursive solution to the problem of linear filtering for discrete data (Kalman, 1960).
Stanley F. Schmidt, who worked on the Apollo project at NASA, has been the first to apply it in a practical way, since he aimed at taking a spacecraft to the moon and bring it back. For this, he had to solve problems in the estimation of trajectories and control. Schmidt worked with what would be the first full implementation of the Kalman filter and it had the same integral control system of Apollo and from this experience, it has been used in most on board estimation systems and also in trajectory control of aircrafts.
The Kalman filter is considered to be an advance concerning the estimation theory. It is used, for example, to estimate a linear-quadratic Gaussian problem, which addresses the difficulty of state estimation of a dynamic linear system disturbed by white Gaussian noise, by using measurements that are linearly related to these states and also corrupted by this noise. This filter consists of a set of mathematical equations that provide a significant and efficient computational solution (recursively) to estimate the state of a process, so that the mean square error can be minimized. The filter allows that states (past, present and even future) are estimated and may do so even when the precise nature of the modeled system is unknown (Welch & Bishop, 2004).
Several models can be found, such as: increased, extended, dual estimation, joint estimation, unscented, among others.

3.1 Discrete Kalman filter

The process to be estimated solves the general problem of estimating a state x, which is a process controlled in a discrete time and generated by a linear stochastic differential equation, ie:

$$x_k = Ax_{k-1} + Bu_k + n_{k-1}, \tag{8}$$

with a measurement z, which is

$$z_k = Hx_k + v_k. \tag{9}$$

The random variables n_{k-1} and v_k represent the process noise and the measurement (respectively). They are assumed to be independent (between them) and are considered white noise with the normal probability distributions:

$$P(n) \sim N(0, Q) \tag{10}$$

$$P(v) \sim N(0, R) \tag{11}$$

In practice, the covariance matrixes of the process noise (covariance Q) and of the measurement noise (covariance R) can change at each measurement.
The discrete Kalman filter (DKF) estimates a process by using a kind of feedback, in which it estimates the process state at some time and then obtains feedback in the form of (noisy) measurement. Thus, the equations can be divided into two stages: the ones to update the time and the ones to update the measures.
The equations that update the time are responsible for projecting an *a priori* (in time) estimate of the current state and of the error covariance to obtain an a priori estimate for the next step.
The equations that update or correct measures are responsible for the feedback - for example, the incorporation of a new measure into the a priori estimate for an a posteriori estimate.
The equations that update or predict the time can also be considered prediction equations, while the equations that update the measures can be treated as correction equations. Thus, the estimation algorithm is a predictor-corrector algorithm for solving numerical problems.
Time update equations (predictor)

$$\hat{x}_k^- = A\hat{x}_{k-1} + Bu_k \tag{12}$$

$$P_k^- = AP_{k-1}A^T + Q. \tag{13}$$

Note that these equations refer to time and covariance estimation over time from step $k-1$ to step k.
Measurement update equations (corrector)

$$K_k \doteq P_k^- H^T (HP_k^- H^T + R)^{-1} \tag{14}$$

$$\hat{x}_k = \hat{x}_k^- + K_k(z_k - Hx_k^-) \tag{15}$$

$$P_k = (I - K_k H)P_k^-. \tag{16}$$

The first task during the update of the measure is to compute the Kalman gain, K_k . The next one is to measure the update process and find the z_k value and, then, generate an a posteriori state estimate by incorporating it as in equation 15. The final step is to obtain an estimated covariance error by using equation 16.

3.2 Extended Kalman filter

A solution to nonlinear systems is the extended Kalman filter (EKF) (Welch and Bishop 2004). By analyzing the prediction function of the Kalman filter, illustrated in equation 12, it is possible to observe that the filter behaves linearly. By applying a nonlinear function, one can obtain a good prediction for the next states, the extended Kalman filter. This algorithm applies the Kalman filter to nonlinear systems simply by linearizing all nonlinear models, so that the traditional equations of the filter can be applied. The nonlinear system can be rewritten as:

$$x_k = f(x_k, n_k, u_k), \tag{17}$$

$$z_k = h(x_k, v_k). \tag{18}$$

The algorithm is also adapted to solve nonlinear problems through the use of functions, as seen below:

Time update equations (predictor)

$$\hat{x}_k^- = f(\hat{x}_k, 0, u_k) \tag{19}$$

$$P_k^- = F_k P_{k-1} F_k^T + Q. \tag{20}$$

Measure update equations (corrector)

$$K_k = P_k^- H_k^T \left(H_k P_k^- H_k^T + R\right)^{-1} \tag{21}$$

$$\hat{x}_k = \hat{x}_k^- + K_k(z_k - h(x_k^-, 0)) \tag{22}$$

$$P_k = (I - K_k H_k) P_k^- \tag{23}$$

For the propagation of variances, one must know the Jacobian or the Hessian transition matrices and observe the state functions given by Equations 17 and 18, respectively.

3.3 Unscented Kalman filter

The unscented Kalman filter (UKF) is similar to the extended version (Julier & Uhlmann, 1997). The distribution of states is represented by a Gaussian random variable, however it is specified by using a minimum set of sampling points, which are chosen carefully. The sampled points capture the true mean and the covariance of a random variable and when it propagates through a truly non-linear system, it captures the mean and the covariance accurately to promote a third order estimation for any nonlinearity. Thus, this is done through the use of unscented processing.

The unscented transformation is a method used to calculate the statistics of a random variable that undergoes a nonlinear transformation (Wan & Merwe, 2000). Considering a spread of a random variable (with dimension L) through a nonlinear function $y = g(x)$, one assumes that it must present a mean x and a covariance P_x to calculate the statistics y. It should form a matrix X of $2L + 1$ sigma vectors (with corresponding weights W_i), according to the following:

$$X_i = \overline{x} \tag{24}$$

$$X_i = \overline{x} + (\sqrt{(L+\lambda)P_x})_i, \text{ for } i = 1,\dots,L \tag{25}$$

$$X_\rangle = \overline{x} - (\sqrt{(L+\lambda)P_x})_i, \text{ for } i = L+1,\dots,2L \tag{26}$$

$$W_o^{(m)} = \lambda/(L+\lambda) \tag{27}$$

$$W_o^{(c)} = \lambda/(L+\lambda) + (1-\alpha^2+\beta) \tag{28}$$

$$W_i^{(m)} = W_i^{(c)} = 1/\{2(L+\lambda)\}, \text{ for } i = 1,\dots,2L \tag{29}$$

where $\lambda = \alpha^2(L+k) - L$ is a scalar parameter. The variable α determines the sigma scattering point around the mean $\overline{x}$ and is always a minimum positive value. k is a secondary scalar parameter defined as 0 and β is used to incorporate the a priori knowledge of the distribution of x (for Gaussian distributions, $\beta = 2$ is optimal). $(\sqrt{(L+\lambda)P_x})_i$ is the i-ism line of the square root matrix. The sigma vectors are propagated through the nonlinear function:

$$y_i = g(X_i) \quad \text{for } i = 0,\dots,2L \tag{30}$$

The mean and the covariance y_i are approximated by using the mean and the covariance of the sample (Figure 3) of the posterior sigma points

$$\overline{y} \approx \textstyle\sum_{i=0}^{2L} W_i^{(m)} y_i \tag{31}$$

$$P_y \approx \textstyle\sum_{i=0}^{2L} W_i^{(c)} \{y_i - \overline{y}\}\{y_i - \overline{y}\}^T. \tag{32}$$

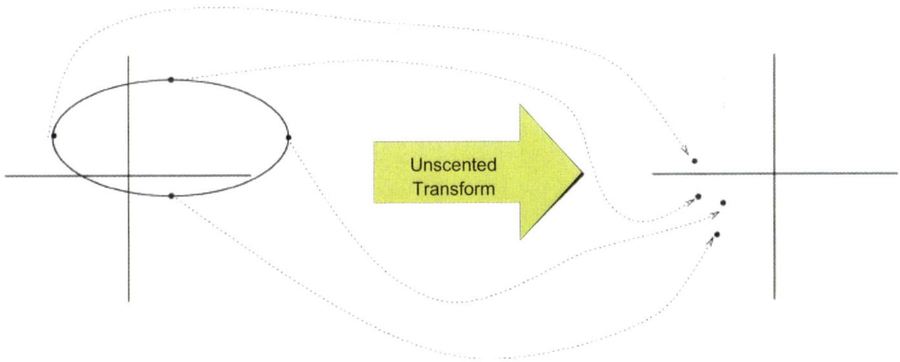

Fig. 3. The unscented transform

This method differs from the general methods of sampling (the Monte-Carlo method such as particle filters), which requires orders of magnitude with more sampling points as an attempt to define and propagate the state (possibly non-Gaussian) distributions. The unscented approaches result in more accuracy for the third order of Gaussian inputs for all nonlinearities.

The unscented Kalman filter is a direct extension of the unscented transform for the equation recursive estimation

$$\hat{x}_k = \hat{x}_k^- + K_k \cdot [y_k - \hat{y}_k^-)] \quad (33)$$

where the state of the random variable is redefined with the concatenation of the original states and the noise:

$$x_k^a = [x_k^T v_k^T n_k^T]^T. \quad (34)$$

The selection of sigma points is applied to a new random variable state to select and calculate the corresponding sigma matrix x_k^a. The unscented Kalman filter is pointed out by equations 35, 36, 37, 38, 39, 40, 41, 42, 43, 44, and 45.

The Jacobian or the Hessians matrices do not need to be calculated and the calculation total numbers are the same of the extended filters related to the nonlinear controls that require feedback from the states. In these applications, the dynamic model is a physically-based parametric model, which is assumed as known.

Unscented Kalman filter algorithm

Sigma points calculation:

$$X_k^n = [\hat{x}_{k-1}^a \quad \hat{x}_{k-1}^a \pm \sqrt{(L+\lambda)P_{k-1}^a}] \quad (35)$$

where X is the set of points with unscented transformation based on the mean and on the *a priori* covariance.

Prediction equations:

$$\hat{x}_k^- = \sum_{i=0}^{2L} W_i^{(m)} x_{i,k|k-1}^x \quad (36)$$

where $W^{(m)}$ represents the set of sigma point weights used for true mean reconstruction.

$$P_k^- = \sum_{i=0}^{2L} W_i^{(c)} [x_{i,k|k-1}^x - \hat{x}_k^-] [x_{i,k|k-1}^x - \hat{x}_k^-]^T \quad (37)$$

where $W^{(c)}$ represents the set of sigma point weights used for true covariance reconstruction.

$$X_{k|k-1}^x = F(X_{k-1}^x, X_{k-1}^v) \quad (38)$$

where F is the function of sigma propagation for state transitions.

Correction equations:

$$Y_{k|k-1} = H(X_{k-1}^x, X_{k-1}^n) \quad (39)$$

where H is the system function of sigma point generation for observation states Y.

$$\hat{y}_k^- = \sum_{i=0}^{2L} W_i^{(m)} Y_{i,k|k-1}^x \quad (40)$$

where y is the observed state estimation reconstructed for the sigma points.

$$P_{\hat{y}_k \hat{y}_k}^- = \sum_{i=0}^{2L} W_i^{(c)} [Y_{i,k|k-1} - \hat{y}_k^-] [Y_{i,k|k-1} - \hat{y}_k^-]^T \quad (41)$$

$$P_{x_k \hat{y}_k} = \sum_{i=0}^{2L} W_i^{(c)} [X_{i,k|k-1} - \hat{x}_k^-] [Y_{i,k|k-1} - \hat{y}_k^-]^T \quad (42)$$

$$K = P_{x_k y_k}(P_{\hat{y}_k \hat{y}_k})^{-1} \tag{43}$$

where K is the Kalman gain obtained through the noisy covariance

$$\hat{x}_k = \hat{x}_k^- + K(y_k - \hat{y}_k^-) \tag{44}$$

this equation represents the correction of *a priori* and

$$P_k = P_k^- - K(P_{\hat{y}_k \hat{y}_k})K^T \tag{45}$$

represents the correction of the *a priori* covariance.

The Kalman filter was originally designed to solve a problem of state estimation and has been used in many applications. This newer filter also provides a better performance. Thus, the unscented Kalman filter and the extended filter present the same order of complexity.

Due to the numerical instability of the noise and the use of the Cholesky factorization to determine the square root of the probability matrix, van der Merwe and Wan have developed the Square-Root Unscented Kalman Filter (SRUKF) (Merwe & Wan, 2001), which allows a better control of the variance matrix values and bypasses the problem of becoming a negative or indefinite matrix. As the original unscented Kalman filter, the square root filter is initialized by calculating the square root of the covariance matrix states by the Cholesky factorization:

$$S_0 = chol\{E[(x_0 - \hat{x}_0)(x_0 - \hat{x}_0)^T]\} \tag{46}$$

However, the spread factor and the Cholesky update are then done in subsequent iterations to directly form the sigma points. In the equation below, the update time of the Cholesky factor is calculated by using a QR decomposition of the matrix composed of the propagated sigma point weights and the square root of the covariance matrix of the additive noise case:

$$S_k^- = qr\{[\sqrt{W_1^{(c)}}(X_{1:2L,k|k-1}^* - \hat{x}_k^-)\ \sqrt{R^v}\ ]\} \tag{47}$$

A subsequent Cholesky update (or regression) in the equation below is needed since the weight zero is perhaps negative:

$$S_k^- = cholupdate\{S_k^-, X_{0,k}^* - \hat{x}_k^-, W_0^{(c)}\} \tag{48}$$

These two steps replace the time update. They are also used to calculate the Cholesky factorization and the error covariance of the observation:

$$S_{\bar{y}\,k} = qr\{[\sqrt{W_1^{(c)}}(Y_{1:2L,k} - \hat{y}_k^-)\ \sqrt{R^n}\ ]\} \tag{49}$$

$$S_{\bar{y}\,k}^- = cholupdate\{S_{\bar{y}\,k}^-, Y_{0,k} - \hat{y}_k^-, W_0^{(c)}\}. \tag{50}$$

Unlike the way in which the gain of Kalman filter is calculated in standard unscented, it is calculated here by using two inversions:

$$K_k(S_{\bar{y}\,k} S_{\bar{y}\,k}^T) = Px_k y_k \tag{51}$$

Since it is square and triangular, efficient replacements can be used to solve it directly, without having to invert the matrix. Finally, the Cholesky factoration updates the state covariance in the equation below and it is calculated by applying sequential Cholesky regressions:

$$S_k = cholupdate\{S_k^-, U, -1\} \tag{52}$$

The vectors are represented by the columns of the equation regression 53. This update replaces the previous equation 45:

$$U = K_k S_{\bar{y}\,k} \tag{53}$$

By knowing the process function and with a Kalman filter that supports non linear functions, it is possible to have a significant improvement in the signal. Another solution is to use a neural network to promote a better function of the mapping process by reducing the projections noise. For an estimation of the neural network weights together with the state estimates, two methods of filtering can be used: the estimation and the dual estimation. These arrangements are used to determine the filtering when the initial weights are known and the next state is obtained through a linear mapping as the previous one.

3.4 Joint estimation

Since the transfer function is not known and with the aim of increasing the filter order, a new estimation method has been used, in which it is possible to estimate new parameters from the states of the hidden chains of the Markov model. The main problem involves the identification of functions required to estimate states and parameters. The prediction equations can be described as:

$$x_k = f(x_{k-1}, n_k) \tag{54}$$

$$z_k = h(x_k, v_k) \tag{55}$$

The parameter estimation involves the determination of a nonlinear mapping

$$y_k = g(x_k, W) \tag{56}$$

where x_k is the input, W is the weight and y_k is the output. The nonlinear mapping g is parameterized by the vector W. The nonlinear mapping can be done by an artificial neural network. The learning corresponds to estimating the parameters of W. Training can be done with pairs of samples, consisting of a known input and a desired output (x_k, d_k). The machine error is defined by equation 57. The learning objective is to minimize the expected squared error.

$$e_k = d_k - g(x_k, W) \tag{57}$$

The UKF can be used to estimate the parameters by using a model for network training that writes a new state-space representation:

$$w_k = w_{k-1} + v_k \tag{58}$$

$$y_k = g(x_k, w_k) + e_k \tag{59}$$

where the parameter w_k correspond to a stationary process with an identity matrix of state transition, governed by a procedural noise v_k (the choice of variances determines the filtering performance). The output y_k corresponds to a nonlinear observation w_k. The extended Kalman filter can be applied directly as an efficient technique for the correction of second order parameters.

As the problem in focus is to work with unobserved input x_k and requires an estimation of states and parameters, one should consider a dual estimation problem, taking a dynamic and discrete-time nonlinear system into account

$$x_{k+1} = f(x_k, v_k, W) \tag{60}$$

$$z_k = h(x_k, n_k) \tag{61}$$

where states are x_k and the set of parameters W of the dynamic system should be estimated only from the noisy signal y_k.

The dynamic system can be seen as a neural network, where W is the set of weights and function f corresponds to a function of neural network that uses an input x_k. Thus, by applying these equations to the unscented Kalman filter, one can have a new function for estimating and observing new states. One approach to neural networks can be seen in (Laia & Cruvinel, 2008b; Laia & Cruvinel, 2009).

4. System modeling

The physical model of photon counting is defined by the equation:

$$I_k = I_0 e^{-\mu_k d} \tag{62}$$

where I_0 is the number of photons leaving the source, μ is the material degree of absorption, d is the distance between the source and the detector and I is the number of photons that cross the material and reach the detector. The counting of photon is affected by a Poisson-type noise. For a closer model to the physical one, each projection is analyzed individually, as if they were in time-varying positions. This classical approach enabled to develop a dynamic estimation of noise-free projections (Figure 4):

$$I_{p_k} = f(I_{p_{k-1}}, q_{k-1}) \tag{63}$$

$$I_{s_k} = h(I_{p_k}, r_k) \tag{64}$$

where I_p is a noise-free projection and I_s, a projection disturbed by noise. The variables q and r represent the white noise, ie, present distribution $q \sim N(0, Q)$ and $r \sim N(0, R)$, respectively. Function f can be used as a mapping neural network or as a transfer state matrix. Function h hides the unobserved states and can also be an array. It can also be adjusted by using the Poisson noise and the Anscombe transform. Some of the previous studies that focused on this approach and obtained good results are (Laia et al, 2007; Laia & Cruvinel, 2008b; Laia & Cruvinel, 2009).

A filtering proposal is needed for a new model (based on the physical one) to determine the process variables and how the observation is carried out. The equation of a process defines the previous state x_{k-1}, and through a transformation influenced by a function f and white noise q_k, one can reach a new state x_{k-1}. These states can be hidden from the system output. Thus, it is possible to define a new function g that transforms this variable according to what is observed.

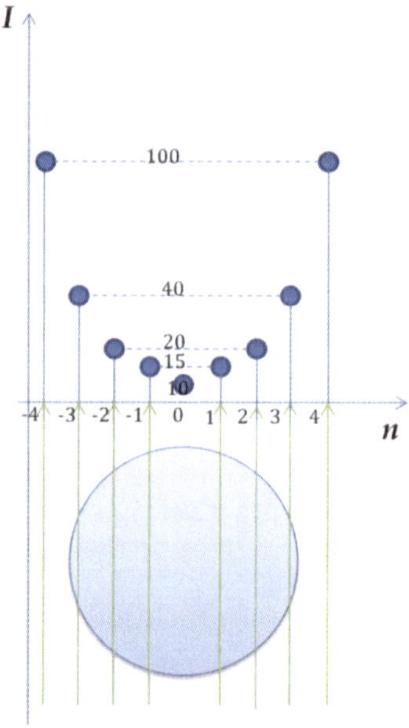

Fig. 4. The process for projections acquisition considering an object based in previous studies

The uncertainty estimation process of the current value has a confidence interval that consists of photon counting and is given by Poisson noise. This uncertainty can be filtered with an estimation of other measures of the equation that are independent on the confidence interval. Thus what ends up being filtered is the noise from the detector, ie, mechanical and electronic noise.

In order to increase confidence and get more reliable values, one can change the focus on what is being estimated.

The different distributions can be mapped with specific non-linear transformations that alter the Gaussian distributions, such as the Anscombe (Poisson) or the inverse Box-Muller (Uniform), however they are linear approximations that present cumulative errors. The Kalman filter is limited to solving stochastic problems, given by the following equations:

$$\mu_k = f(\mu_{k-1}, q_k) \tag{65}$$

$$I_k = h(I_0, \mu_k, d, r_k) \tag{66}$$

Equation 65 is the transition function that updates the measure of the attenuation coefficient or the material degree of absorption while variable q is the uncertainty of this process equation. Equation 66 is the observation function that generates the final projection from the degree of absorption and d the distance between the source and the photon detector. Variable r is the noise that is transferred to the projection. This model can be approximated by a second order filter in order to estimate the variables of the process equation and to use the Anscombe transform to the observation equation:

$$\begin{bmatrix} \mu_k \\ \Delta\mu_k \end{bmatrix} = \begin{bmatrix} 1 & 1 \\ 0 & 1 \end{bmatrix} \begin{bmatrix} \mu_{k-1} \\ \Delta\mu_{k-1} \end{bmatrix} + \begin{bmatrix} 0 \\ q_k \end{bmatrix} \quad (67)$$

$$I_k = h(I_0 e^{-\mu_k d}, r_k) \quad (68)$$

Function h is related to the Anscombe transform (Anscombe, 1948), which transforms a Poisson noise into a Gaussian noise, with variance value close to 1. As the noise variation presents a Poisson distribution, the mean and the variance of values are equivalent to the counting of photons. A new system can be defined based on an equation, in which the ray sum μk is the variable used in the process (Equation 76) and it also allows the observation of the projection, as it is presented in the array of projections:

$$\mu_k = f(\mu_{k-1}, u_k, q_k) \quad (69)$$

$$I_k = h(I_0 e^{-\mu_k d}, r_k) \quad (70)$$

Variable u_k consists of an external input that is related to the prediction of new states. In order to promote a better estimation of the states without noise, one can use neural network to determine the behavior of the equation of the process that uses the Kalman filter for a dual estimation:

$$\mu_k = f(\mu_k, w_k, q_k) \quad (71)$$

$$I_k = A(I_0, r_k) e^{-\mu_i d} \quad (72)$$

$$w_k = w_k + a_k \quad (73)$$

$$e_k = g(\mu_{k-1}, w_k) - \mu_k \quad (74)$$

where a_k is used to update the weight and e_k indicates the error between inputs and the desirable outputs.

To promote a better estimation of the transition states of the process, an artificial neural network can be used:

$$\begin{bmatrix} \mu_k \\ W_k \end{bmatrix} = \begin{bmatrix} g(\mu_{k-1}, u_k, W_{k-1}) + q_k \\ W_{k-1} + a_k \end{bmatrix} \quad (75)$$

$$I_k = h(I_0 e^{-\mu_k d}, r_k). \quad (76)$$

Now, by focusing on the observation equation, the observation noise variance can be treated with the Anscombe transform through the use combined with its inverse.

$$I_k = A^{-1}(A(I_0 e^{-\mu_k d}) + r_k) \quad (77)$$

where A represents the transformation and its inverse A^{-1}. This makes it possible to work with a Gaussian noise with distribution $r \sim N(0,1)$.

In this model, the input has been used as the current observation of the system (ie, the attenuation coefficient of the noise) so that one could take advantage of the neural network functionality which provides a nonlinear mapping. The neural network itself consists of the interaction between the value measured before and the noise presented now, in order to better predict the data.

An equation based on equation 5 can be developed in order to determine the Poisson noise given by the following expression:

$$I \pm \sqrt{I}\,. \tag{78}$$

Thus, the observation equation can be written as follows:

$$I_k = I_0 e^{-\mu_k} + \sqrt{I_0 e^{-\mu_k d}}\,\mathrm{r}_k. \tag{79}$$

where *r*, since it presents a Gaussian distribution, can assume a negative or zero value.
This alternative is allowed to deviate what is considered to be only an approximation from the use of the Anscombe transform.
Another model for the equation could be simplified as:

$$I_k = I_0 e^{-\mu_k d} + \mathrm{r}_k. \tag{80}$$

where *r* is replaced by a variance given by the signal noise ratio of

$$\frac{I_\mathrm{k}}{\sqrt{I_\mathrm{k}}}\,. \tag{81}$$

This approach allows the inclusion of other noises in the system of photon counting through the propagation of errors. In order to set the noise variance *R* being treated, one has to take the number of primary quantities into account. It is measured from the observation of the system variables *{Io, μ}*. The value of *I* depend on the relationship between these variables. In formal language, that is:

$$I = G(I_0, \mu, d). \tag{82}$$

In case the errors with the measured magnitudes I_0 , μ and d are ΔI_0 , $\Delta \mu$ and Δd, the photon counting error ΔI is given by the expression:

$$\Delta I = \left|\frac{\partial I}{\partial I_0}\right| \Delta I_0 + \left|\frac{\partial I}{\partial \mu}\right| \Delta \mu. \tag{83}$$

The values of ΔI_0 and $\Delta \mu$ are given by mean standard deviation or by their estimator as there are many or few magnitude measures I_0 and μ.
When the sample size is adequate, it can determine the statistical error of independent magnitudes and these can become the dependent magnitude variance for calculating the statistical error.

$$\sigma_I^2 = \left(\frac{\partial I}{\partial I_0}\right)^2 \sigma_{I_0}^2 + \left(\frac{\partial I}{\partial \mu}\right)^2 \sigma_\mu^2 + \left(\frac{\partial I}{\partial d\mu}\right)^2 \sigma_d^2. \tag{84}$$

Besides the uncertainty of the system variables, the detector presents a characteristic noise. This noise variation is known when the detector closes, in other words, as no photons are released to it, it still gives the score. As it is associated directly to an additive noise, you can add the noise variation into the equation as an error for variable I:

$$\sigma_I^2 = \left(\frac{\partial I}{\partial I_0}\right)^2 \sigma_{I_0}^2 + \left(\frac{\partial I}{\partial \mu}\right)^2 \sigma_\mu^2 + \left(\frac{\partial I}{\partial d}\right)^2 \sigma_\mathrm{d}^2 + \left(\frac{\partial I}{\partial I}\right)^2 \sigma_e^2. \tag{85}$$

σ_I^2 is used to define the noise variance *R* of the observation equation of the filter. Then, it is necessary to define the variance of the process. In the literature, Haykins (Haykins, 2001) has defined several ways to infer on the process variance *Q*. Since the signal is observed, the noisy signal variance can be used. As the process variance is directly linked at the μ vector, the equation can be changed for a noise μ :

$$\mu = ln\left(\frac{I_0}{I}\right)/d \tag{86}$$

From this vector, a variance Q needs to enter the filter. It can be obtained as follows:

$$Q = \sigma_\mu^2 \tag{87}$$

Another important step is the definition of the control constants: α, β and κ.
As the process variable μ magnitude differs from the observation of variable I, α = 1 has been chosen. In case this value is greater or lower than ideal, there is no possibility of filtering or causing numerical instabilities.
As suggested in the literature, for a parameter or a joint estimation filters, the variable remained β = 2, while the κ value was given 3 - the number of neurons.
With the aims of comparing the efficiency of the filter and setting up a filter to ensure a desirable image quality, both of the algorithms have been applied to various soil samples.
The first sample consists of sand grains in a Plexiglas envelope. The second and third samples are portion of natural soil. The fourth and the fifth are present in degraded soil bulks and the sixth shows a portion of naturally cemented soil. In an artificial neural network there are two neurons in the input layer, two in the intermediate and one in the output layer.
For the inputs of the filter, the same uncertainty was used, as all of the samples have been generated in the same CT scanner. The variance $\sigma_{I_0}^2$ was obtained by the maximum of vector projections. After the filtering process, it has been applied to the maximum of the projection matrix. The variance σ_μ^2 was the same used in the process variance Q. For the variance σ_d^2, the value used was 0.05, corresponding to the uncertainty of measurement in millimeters. For the errors in the photon counting detector, the value used was 100 for the variance σ_e^2.
The results obtained from this new modeling system are presented below. In Figures 5, 7, 9, 11 and 13, it is possible to visualize the comparison between the signals of a set of projections by using the samples filtered with the SRUKF: Original projections (red), Linear estimation (blue) and non linear estimation with ANN (green). In Figures 6, 8, 10, 12, 14 and 16, it is possible to visualize the reconstructed images by using the filtered back-projection algorithm.

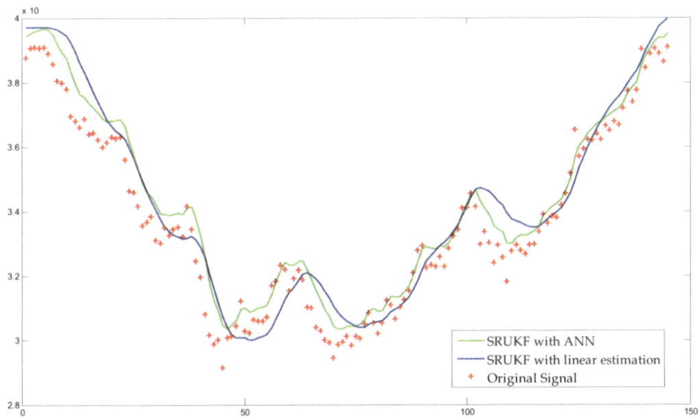

Fig. 5. Comparison between the projections of sand grains

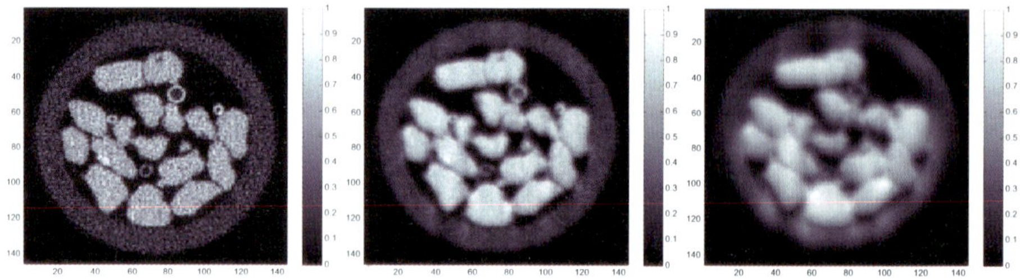

a b c

Fig. 6. Comparison between the reconstructed images of sand grains. a. Original projections, b. Projections filtered with the SRUKF and the ANN and c. Projections filtered with linear estimation

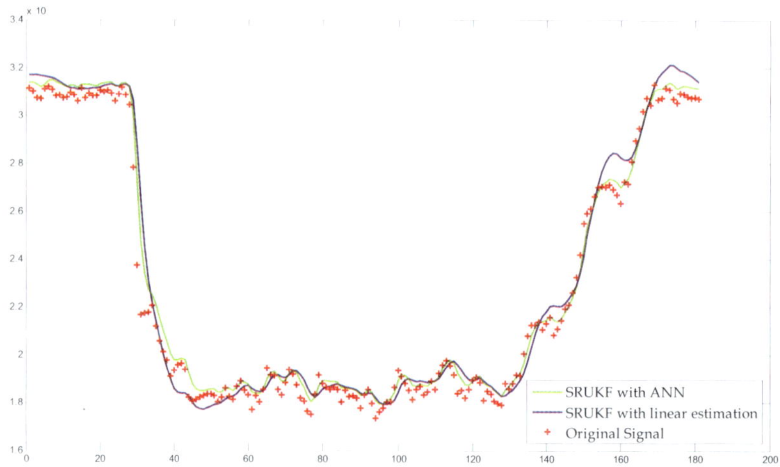

Fig. 7. Comparison between the projections of soil bulk samples

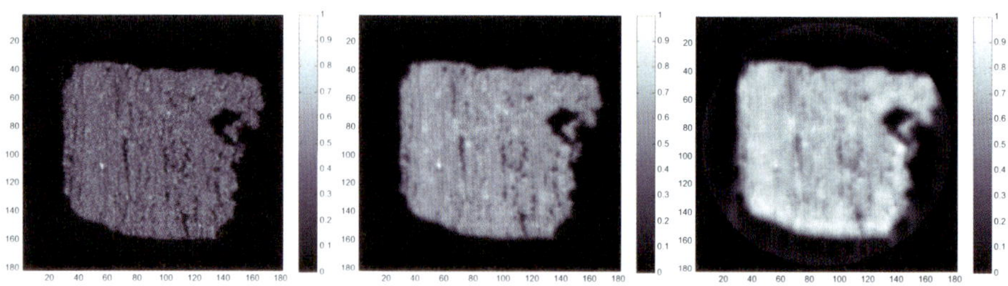

a b c

Fig. 8. Comparison between the agricultural soil reconstructed images for the visualization of bulk information. a. Original projections, b. Projections filtered with the SRUKF and the ANN and c. Projections filtered with linear estimation

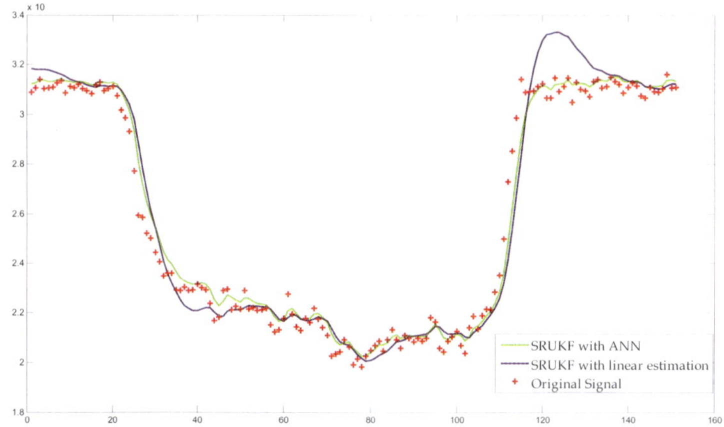

Fig. 9. Comparison between projections of the soil bulk sample

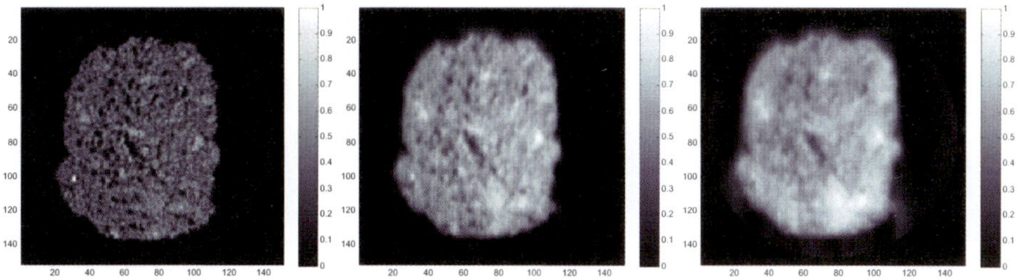

a b c

Fig. 10. Comparison between the agricultural soil reconstructed images for the visualization of bulk information. a. Original projections, b. Projected filtered projections with the SRUKF and the ANN and c. Projections filtered with linear estimation

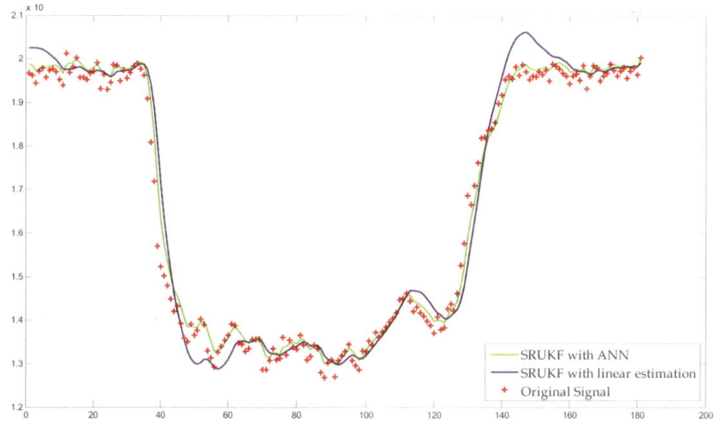

Fig. 11. Comparison between the projections of degraded soil samples

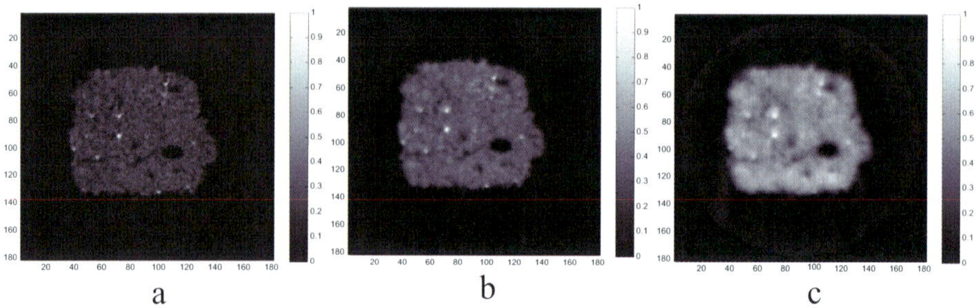

Fig. 12. Comparison between the reconstructed images of degraded soil sample. a. Original projections, b. Projections filtered with the SRUKF and the ANN and c. Projections filtered with linear estimation

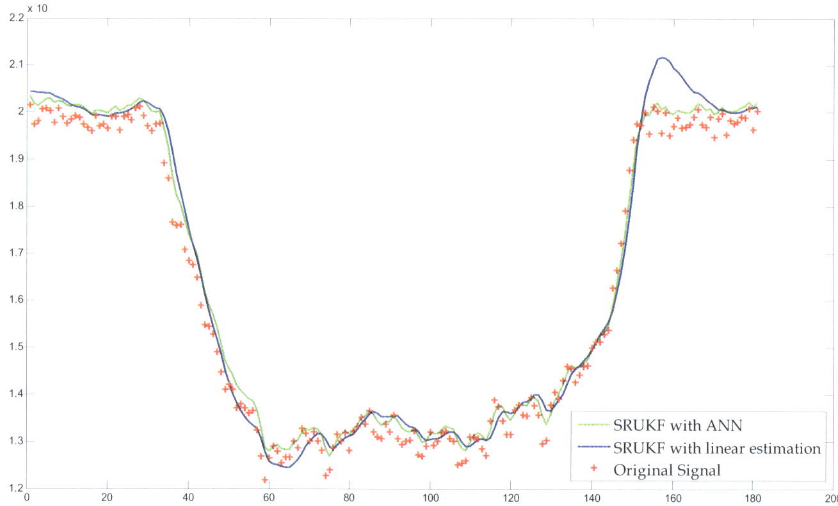

Fig. 13. Comparison between the degraded projections of a soil sample

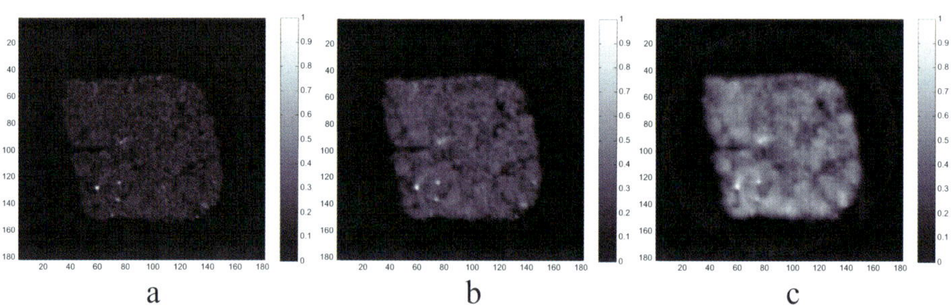

Fig. 14. Comparison between the reconstructed images: a. Original projections, b. Projections filtered with the SRUKF and the ANN and c. Projections filtered with linear estimation.

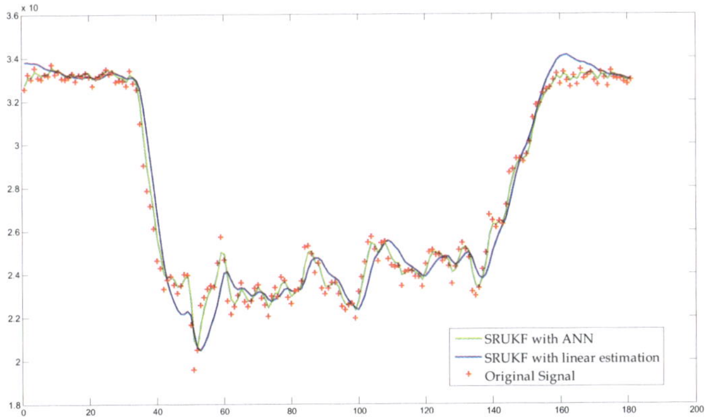

Fig. 15. Comparison between the projections of cemented soil samples

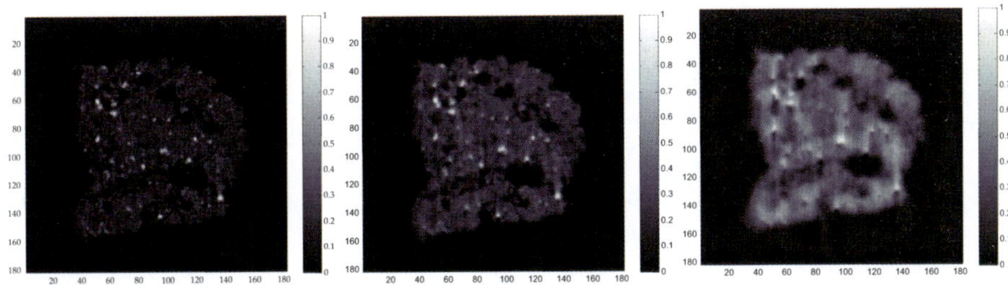

a b c

Fig. 16. Comparison between the reconstructed images of cemented soil sample: a. Original projections, b. Projections filtered with the SRUKF and the ANN and c. Projections filtered with linear estimation

In the signal comparisons it is possible to see the estimation errors. These errors in linear estimation promote false artifacts and excessive anti-aliasing. Also, the excessive filtering process with linear estimation has promoted losses in the images details. The SRKUF with the ANN eliminates any noise in a dynamic way and, thus, there is a higher predominance of low values of photon counting. Every single detail is preserved in the signal filtered with the SRUKF.

The signal filtered with the ANN algorithm presents the details better, a precise correction and the interest object (the porosity of soil bulk) is preserved while the signal filtered with the linear estimation promotes excessive anti-aliasing and detail loss. The false porosity (due to granularity of Poisson noise) has been eliminated in the images as it can be observed in the outer parts of the sample. In the image obtained by the filtering with the ANN, the contrast has been preserved as well as small-pointed (critical) elements in the image.

The low-contrast in this image is changed due to the false estimation in the projections like a high value estimated by filter. The single elements are either preserved or changed a little due to the contrast. With the linear estimation, the excessive filtering provoked a decrease in the contrast, which lead to the elimination of pores and high-contrastive elements. The outlines of the objects in the image are also lost when there is an excessive smoothing.

Another important factor is the separation between porosity and granularity of the images. The granularity can generate false micro-pores or fake elements. On the other hand, excessive smoothing can hide the pores and the major elements in the soil composition.

5. Conclusions

The Kalman filtering that uses linear estimation promotes a smoothing of values and do respect the nature of data distribution (linear attenuation coefficients), which present a uniform distribution, while the filter is limited to working with a Gaussian process.

By analyzing the results presented in Figures 4, 5, 6, 7, 8 and 9, it was possible to observe smoothed projections and estimation errors, while Figures 10, 11, 12, 13, 14 and 15 presented losses of important details, such as micro pores and other important elements of the soil, which enables to characterize it.

Nowadays, with the use of artificial neural networks, the results already show a better transition among the variables for the estimation process. Besides mapping the behavior of the sample data, the nonlinear function also makes the necessary transformation of the process uncertainties, which helps to preserve a greater number of details on an original image.

Besides, the regular presence of artifacts or distortions on the image were due to three factors: the limitation of the reconstruction algorithm because of the ramp filter, which is necessary to observe the contrast of different materials and soil porosity; the inaccurate choice of the noise variances because of the different resolutions of the samples obtained and due the equipment noise during data collection.

A better measurement of the noise variations in the detector should be made in real time. The closer the variance value is to the real value, the less it is prone to reconstruction errors or smoothing.

Tests conducted earlier have proved that the increase of neurons, layers or number of states has not affected the filter efficiency but the processing time was longer.

More recent researches involve the implementation of the filters shown in this study in an embedded system in order to ensure better results when filter variables are fed directly according to the state of the CT scanners.

6. Acknowledgment

This work was supported in part by the National Council for Research and Development (CNPq) under Grant 306988/2007-0 and Brazilian Enterprise for Agricultural Research (Embrapa) under Grant 03.10.05.011.00.01. Also we acknowledge the Coordination for the Improvement of Higher Education (CAPES, process 573963/2008-9 and 08/57870-9).

7. References

Anscombe, F. J. (1948). The Transformation of Poisson, Binomial, Negative-Binomia Data. Vol. 15, in *Biometrika*. 1948.

Crestana, S. (1985). A Tomografia Computadorizada com um novo método para estudos da física da água no solo, *São Carlos, USP, 140 p.*, 1985.

Crestana, S. & Nielsen, D. R. (1990). Investigações não-destrutivas de sistemas porosos multifásicos através de microtomografia de raios-X, gama e ressonância magnética nuclear (RMN). In: *Encontro nacional sobre escoamento em meios porosos-enemp*, 18., 1990.

Cruvinel, P. E. (1987). Minitomógrafo de raios X e raios gama computadorizado para aplicações multidisciplinares. *Campinas, UNICAMP, 329 p.*, 1987.

Cruvinel, P. E.; Cesareo, R.;Crestana, S.;Mascarenhas, S. (1990). X and γ-ray computerized minitomograph scanner for soil science, *IEEE Transactions on Instrumentation and Measurement*, V.39, N.5, p.745-750, October, 1990.

Cruvinel, P. E. & Balogun, Fatai A. (2006). Tomógrafo de espalhamento Compton para medidas agrícolas. In *Engenharia Agrícola*; vol. 26 n.1 p. 151-160, April 2006.

Cruvinel, P. E. ; Pereira , M. F. L. ; Saito, J H ; Costa, L F . Performance Improvement of Tomographic Image Reconstruction Based on DSP Processors. IEEE Transactions on Instrumentation and Measurement, v. 58, p. 3295-3304, 2009.

Deremack, E. & Crowe, D. G. (1984), Optical Radiation Detectors. John Wiley & Sons, Inc., 1984.

Duerinckx, A. J., & Macocski, A. (1978), Polycromatic Streak Artifacts in Computed Tomography Images. In *J. Comput. Assist. Tomogr.*, 1978.

Granato, L. F. (1998). Algoritmo adaptativo para a melhoria em imagens tomográficas obtidas em múltiplas energias, *São Carlos, UFSCar, 135 p.*, 1998.

Hainsworth, J. M. & Aylmore, L. A. G. (1983). The use of the computed-assisted tomography to determine spatial distribution of soil water content, *Aust. Journal Soil Res.*, N. 21, p.1435-1443,1983.

Haykin, S. (2001). *Kalman Filtering and Neural Networks*, John Wiley & Sons, Inc.,ISBN 978-0471369981,USA

Hender, W. R. (1983). The physical principals of computed tomography, USA, 1983.

Ibbott, G. S. (1980),Radiation therapy treatment planning and the distortion of CT images. Med. Phys, 1980.

Joseph, P. M. & Spital, R. D. (1978). A method for correction bone-induced artifacts of CT scanners. 1978: 100–108.

Julier, S. J. & Uhlmann, J. K.. (1997). A new extension of Kalman filter to nonlinear systems. *Symp. Aerospace/Defense Sensing, Simul. and Controls*, 1997.

Kalman, R. E. (1960). A new approach to Linear Filtering and Prediction Problems. *Transaction of the ASME - Journal of basic Engineering*, 1960.

Laia, M A. M.; Cruvinel, P. E. & Levada, A. L. M. (2007). Filtragem de projeções tomográficas da ciência do solo utilizando transformada de Anscombe e Kalman, *DINCON'07, São José do Rio Preto*, 2007.

Laia, M. A. M. & Cruvinel, P. E. (2008a), Filtragem de projeções tomográficas utilizando Kalman Discreto e Redes Neurais, *IEEE América Latina,vol. 6, ed. 1, march*, 2008.

Laia, M. A. M.; Levada, A. L. M.; Botega, L. C.; Cruvinel, P. E.; Pereira, M. F. L. & Macedo, A. (2008b). A Novel Model for Combining Projection and Image Filtering Using Kalman and Discrete Wavelet Transform in Computerized Tomography. In: *2008 11th IEEE International Conference on Computational Science and Engineering*, 2008, São Paulo. 2008. 11th IEEE International Conference on Computational Science and Engineering, 2008. v. 11. p. 219-226.

Laia, M. A. M.& Cruvinel, P. E. (2009). Applying an Improved Square Root Unscented Kalman Filtering in Tomographic Projections of Agricultural Soil Samples. In: *The 3rd Southern Conference on Computational Modeling (3MCSUL)*, 2009, Rio Grande - RS. The 3rd Southern Conference on Computational Modeling (3MCSUL),2009.v. 1.

Laia, M. A. M.& Cruvinel, P. E. (2011). Evaluation of an embedded unscented Kalman filter for soil tomography. In: *1a. Conferência de Sistemas embarcados Críticos.* (CBSEC) São Carlos - SP. 2011.

Macedo, A. (1997). Construção e uso de um tomógrafo com resolução micrométrica para aplicações em ciências do solo e do ambiente. São Carlos: USP-EESC, 1997. 129p.

Mascarenhas, N. D. A.; Santos, C. A. N. & Cruvinel, P. E. (1999). Transmission Tomography Under Poisson Noise Using the Anscombe Transformation and Wiener Filtering of Projections. *Nuclear Instruments and Methods in Physics Research Section A*, 1999.

Mcculloch, W. S. & Pitts, W. (1947), A logical calculus of the ideas immanent in Nervous Activity. Bulletin of mathematical biophysics, 1947.

Minatel, E. R. (1997). Desenvolvimento de Algoritmo para Recontrução e Visualização Tridimensional de Imagens Tomográficas com uso de Técnicas Frequenciais e Wavelets. São Carlos-SP: UFSCar, 1997.

Pereira, G. A. S. (2000).Filtro de Kalman: Teoria e Aplicações. *Centro de Pesquisa e Desenvovilmento em Engenharia Elétrica.*, 2000.

Petrovic, M.; Siebert, J. E. & Rieke, P. E. (1982). Soil bulk analysis in three dimensions by computed tomographic scanning, *Soil Sci. Soc. Am. J.*, n.46, p.445-450, 1982.

Naime, J.M. (1994). Projeto e construção de um tomógrafo portátil para estudos de ciência do solo e planta, em campo. São Carlos: USP-EESC, 1994. 87p. il..

van der Merve, R. & Wan, E. A. (2001). The square-root unscented Kalman Filter for state and parameter-estimation. *Acoustics, Speech, and Signal Processing, 2001. Proceedings. (ICASSP '01). 2001 IEEE International Conference on*, 2001: 3461-3464.

Venturini, Y. R. (1995).Análise quantitativa da qualidade de imagens digitais com o uso de espectro de Wiener. São Carlos - SP, UFSCar, 1995.

Wan, E. A. & Merwe, R. (2000). The Unscented Kalman Filter for Signal Processing. *Proceedings of Symposium 2000 on Adaptive Systems for Signal Processing*, Communication and Control, 2000, october ed.

Welch, G. & Bishop, G. (2004). An Introduction to the Kalman Filter. Chapel Hill: Departament of Computer Science, University of North Carolina, 2004.

Ziel, A. D. (1976). Noise in measurements. John Wiley & Sons, inc, 1976.

Permissions

The contributors of this book come from diverse backgrounds, making this book a truly international effort. This book will bring forth new frontiers with its revolutionizing research information and detailed analysis of the nascent developments around the world.

We would like to thank E. Burcu Özkaraova Güngör, for lending her expertise to make the book truly unique. She has played a crucial role in the development of this book. Without her invaluable contribution this book wouldn't have been possible. She has made vital efforts to compile up to date information on the varied aspects of this subject to make this book a valuable addition to the collection of many professionals and students.

This book was conceptualized with the vision of imparting up-to-date information and advanced data in this field. To ensure the same, a matchless editorial board was set up. Every individual on the board went through rigorous rounds of assessment to prove their worth. After which they invested a large part of their time researching and compiling the most relevant data for our readers. Conferences and sessions were held from time to time between the editorial board and the contributing authors to present the data in the most comprehensible form. The editorial team has worked tirelessly to provide valuable and valid information to help people across the globe.

Every chapter published in this book has been scrutinized by our experts. Their significance has been extensively debated. The topics covered herein carry significant findings which will fuel the growth of the discipline. They may even be implemented as practical applications or may be referred to as a beginning point for another development. Chapters in this book were first published by InTech; hereby published with permission under the Creative Commons Attribution License or equivalent.

The editorial board has been involved in producing this book since its inception. They have spent rigorous hours researching and exploring the diverse topics which have resulted in the successful publishing of this book. They have passed on their knowledge of decades through this book. To expedite this challenging task, the publisher supported the team at every step. A small team of assistant editors was also appointed to further simplify the editing procedure and attain best results for the readers.

Our editorial team has been hand-picked from every corner of the world. Their multi-ethnicity adds dynamic inputs to the discussions which result in innovative outcomes. These outcomes are then further discussed with the researchers and contributors who give their valuable feedback and opinion regarding the same. The feedback is then collaborated with the researches and they are edited in a comprehensive manner to aid the understanding of the subject.

Apart from the editorial board, the designing team has also invested a significant amount of their time in understanding the subject and creating the most relevant covers. They scrutinized every image to scout for the most suitable representation of the subject and create an appropriate cover for the book.

The publishing team has been involved in this book since its early stages. They were actively engaged in every process, be it collecting the data, connecting with the contributors or procuring relevant information. The team has been an ardent support to the editorial, designing and production team. Their endless efforts to recruit the best for this project, has resulted in the accomplishment of this book. They are a veteran in the field of academics and their pool of knowledge is as vast as their experience in printing. Their expertise and guidance has proved useful at every step. Their uncompromising quality standards have made this book an exceptional effort. Their encouragement from time to time has been an inspiration for everyone.

The publisher and the editorial board hope that this book will prove to be a valuable piece of knowledge for researchers, students, practitioners and scholars across the globe.

List of Contributors

E. Saljnikov and D. Cakmak
Institute of Soil Science, Serbia

J. Shamshuddin
Department of Land Management, Faculty of Agriculture, Universiti Putra Malaysia, 43400 Serdang, Selangor, Malaysia

Noordin Wan Daud
Department of Crop Science, Faculty of Agriculture, Universiti Putra Malaysia, 43400 Serdang, Selangor, Malaysia

John J. Sloan, Peter A.Y. Ampim and Raul I. Cabrera
Texas AgriLife Research, Dallas, TX, USA

Wayne A. Mackay
Mid-Florida Research & Education Center, Apopka, FL, USA

Steve W. George
Texas AgriLife Extension Service, Dallas, TX, USA

Michael A. Blazier and Lewis A. Gaston
Louisiana State University Agricultural Center, USA

Hal O. Liechty and Keith Ellum
University of Arkansas Monticello, USA

J. Castro
IFAPA Centro Camino de Purchil, Junta de Andalucía, Granada, Spain

E. Fernández
Dpto. Edafología y Química Agrícola, Facultad de Ciencias, Universidad de Granada, Granada, Spain

O.M. Nieto
IFAPA Centro Camino de Purchil, Junta de Andalucía, Granada, Spain
Dpto. Edafología y Química Agrícola, Facultad de Ciencias, Universidad de Granada, Granada, Spain

Rui Guo, Ji Zhou, WeiPing Hao, DaoZhi Gong, XiuLi Zhong and FengXue Gu
Institute of Environment and Sustainable Development in Agriculture (IEDA), Chinese Academy of Agricultural Sciences (CAAS)/Key Laboratory of Dry land Agriculture, MOA, Beijing, China

SongTao Yang
The College of Urban and Environment Sciences, Northeast Normal University, ChangChun, China

T. Watanabe and M. Nanamori
Graduate School of Agriculture, Hokkaido University, Kita-ku, Sapporo, Japan

M. S. H. Khan
Department of Soil Science, HMD Science and Technology University, Dinajipur, Bangladesh

I. M. Rao and M. Ishitani
Centro Internacional de Agricultura Tropical (CIAT), A.A.6713, Cali, Colombia

J. Wasaki
Graduate School of Biosphere Science, Hiroshima University, Higashi-Hiroshima, Japan

T. Shinano
National Agricultural Research Center for Hokkaido Region, Sapporo, Japan

H. Koyama
Faculty of Applied Biological Sciences, Gifu University, Gifu, Japan

S. Ishikawa
National Institute for Agro-Environmental Science, Tsukuba, Japan

K. Tawaraya, N. Ueki and T. Wagatsuma
Faculty of Agriculture, Yamagata University, Tsuruoka, Japan

Tomoko Nakano
Department of Economics, Chuo University, Higashi-nakano, Hachioji, Tokyo, Japan

Masato Shinoda
Arid Land Research Center, Tottori University, Hamasaka, Tottori, Japan

Magdalena Borzecka-Walker, Antoni Faber, Katarzyna Mizak, Rafal Pudelko and Alina Syp
Institute of Soil Science and Plant Cultivation-State Research Institute, Poland

X. Peng
State Key Laboratory of Soil and Sustainable Agriculture, Institute of Soil Science, Chinese Academy of Sciences, Nanjing, P.R. China
UCDSchool of Biosystems Engineering, University College Dublin, Belfield, Dublin, Ireland

S. Macdonald and N.M. Holden
UCDSchool of Biosystems Engineering, University College Dublin, Belfield, Dublin, Ireland

T. M. Murphy
Central Veterinary Research Laboratory, Young's Cross, Celbridge, Co. Kildare, Ireland

Johan van Tol, Pieter Le Roux and Malcolm Hensley
Department Soil, Crop and Climate Sciences, University of the Free State, South Africa

R. García Moreno
Departamento de la Navegación y de la Tierra, Facultad de Ciencias, Universidad de La Coruña, A Coruña, Spain

M.C. Díaz Álvarez and A. Saa Requejo
CEIGRAM (Centre for Studies and Research on Agricultural and Environmental Risk Management), School of Agricultural Engineering, Polytechnic University of Madrid, Madrid, Spain

J.L. Valencia Delfa
Escuela Universitaria de Estadistica, Universidad Complutense de Madrid, Madrid, Spain

Mareike Ließ and Bernd Huwe
University of Bayreuth, Department of Geosciences/Soil Physics, Germany

Bruno Glaser
Martin-Luther University Halle Wittenberg, Soil Biogeochemistry, Germany

Gunter Spöck
University of Klagenfurt, Austria

Ahmed Douaik
Research Unit on Environment and Conservation of Natural Resources, National Institute of Agricultural Research (INRA), Rabat, Morocco

Marc van Meirvenne
Department of Soil Management, Ghent University, Ghent, Belgium

Tibor Tóth
Department of Soil Science, Research Institute for Soil Science and Agricultural Chemistry, Budapest, Hungary
Rural, Water and Ecosystem Resources Unit, Institute for Environment and Sustainability, European Commission - Joint Research Center, Ispra, Italy

Agnieszka Rutkowska and Mariusz Fotyma
Institute of Soil Science and Plant Cultivation - National Research Institute in Puławy, Poland

Patrizia Savi and Ivan A. Maio
Dipartimento di Elettronica, Politecnico di Torino, C.so Duca degli Abruzzi 24, 10129 Torino, Italy

Stefano Ferraris
Università degli Studi di Torino, Dipartimento di Economia e Ingegneria Agraria, Forestale e Ambientale (D.E.I.A.F.A.), Sezione di Idraulica Agraria, Via Leonardo da Vinci 44, 10095 Grugliasco, Torino, Italy

Marcelo Muniz Benedetti
Centro de Tecnologia Canavieira, Brazil

Nilton Curi and Sérgio Henrique Godinho Silva
Universidade Federal de Lavras, Brazil

Gerd Sparovek
Escola Superior de Agricultura "Luiz de Queiroz" /USP, Brazil

Amaury de Carvalho Filho
Empresa Brasileira de Pesquisa Agropecuária, Brazil

Marcos A. M. Laia and Paulo E. Cruvinel
Embrapa Instrumentation, Brazil
Physics Institute of São Carlos, University of São Paulo, Brazil